THE GENUS
BETULA

To

June Ashburner

A BOTANICAL MAGAZINE MONOGRAPH

THE GENUS

BETULA

A TAXONOMIC REVISION OF BIRCHES

Kenneth Ashburner and Hugh A. McAllister

PAINTINGS BY

Josephine Hague

DRAWINGS BY

Andrew Brown

PHOTOGRAPHS BY

Peter and Maggie Williams

EDITED BY

Martyn Rix

Kew Publishing

Royal Botanic Gardens, Kew

Royal Botanic Gardens
Kew

Reprint with additions published 2016
First published in 2013 by
Royal Botanic Gardens, Kew,
Richmond, Surrey, TW9 3AB, UK
www.kew.org

Distributed on behalf of the Royal Botanic Gardens, Kew in North America by the University of Chicago Press, 1427 East 60th Street, Chicago, IL 60637, USA

ISBN 978-1-84246-141-9

British Library Cataloguing in Publishing Data
A catalogue record for this book is available from the British Library

Production editor: Ruth Linklater
Design, typesetting and page layout: Christine Beard
Maps prepared by John Stone
Publishing, Design & Photography, Royal Botanic Gardens, Kew

Front jacket illustration: *Betula insignis* subsp. *fansipanensis* painted by Josephine Hague
Back jacket illustration: *Betula luminifera* painted by Josephine Hague

Printed in the UK by Bell & Bain

The authors are donating their royalties from the sale of this book to the Stone Lane Gardens Trust, set up to safeguard Kenneth Ashburner's unique collection of birch and alder trees.

Stone Lane
garden   sculpture   nursery

National Collections of wild-origin Birch and Alder trees.
Stone Lane Gardens, Stone Farm, Chagford, Devon, TQ13 8JU
www.stonelanegardens.com

For information or to purchase all Kew titles please visit
www.kewbooks.com or email publishing@kew.org

Kew's mission is to inspire and deliver science-based plant conservation worldwide, enhancing the quality of life.

Kew receives half of its running costs from Government through the Department for Environment, Food and Rural Affairs (Defra). All other funding needed to support Kew's vital work comes from members, foundations, donors and commercial activities including book sales.

CONTENTS

LIST OF PAINTINGS

NEW TAXA AND COMBINATIONS

New taxa and combinations described in this book:

FOREWORD

Birches are the most delicate of trees and one of the great joys of the garden; from the huge early catkins of *Betula insignis* to the old gold autumn colour of *B. medwediewii*. A huge range of bark colour is an outstanding characteristic of the genus, and varies even in one species, from the purest white of *B. utilis* subsp. *jacquemontii* from Kashmir, to the almost black of some forms of subsp. *utilis* from western China.

Hergest Croft Gardens have a claim to be one of the earliest, maybe the very first, collection of Asiatic birches in Britain. From 1901 onwards, my grandfather William Hartland Banks was buying plants from Veitch's Nursery including Ernest Wilson collections; sadly birches in general are not long lived trees and the originals have died over the intervening 110 years, but many of the best were propagated and so the genes remain. For a long period of time we had one of only three specimens of *Betula potaninii* in cultivation; a dire responsibility and I am delighted to know that the species has now been reintroduced. My father Dick Banks had a special love of and interest in the genus and planted many more — there is a photograph showing him on his 80th birthday climbing a big specimen of *B. luminifera* in Sichuan to collect seed, and I have been fortunate enough to receive many plants from expeditions to China and elsewhere from 1980 onwards, which have greatly extended the collection.

However the correct identification of birches has always been a lurking problem. In most cases wild collected seed will produce the pure species, although the number of natural hybrids is quite large. Seed from cultivated plants, however, is in most cases hybridised, since birch are amongst the most promiscuous of trees; the 'lady of the forest' clearly does not put a high price on purity; the only solution for the gardener who wants the real thing is vegetative propagation.

Many of our plants have, over the years been given various different names, sometimes three or more! From amateur observation in the garden I have long wondered whether there is sufficient difference between *Betula utilis* and *B. albosinensis* to justify making them distinct species and the cline ranging from *B. pendula* in the west through to *B. szechuanica* in the east seems to include almost every intermediate which suggests that there is only one true species.

There is much more to this book than a dry taxonomic volume. It is a welcome collaboration over some 30 years between the late Kenneth Ashburner, the pre-eminent grower of birches at Stone Lane Arboretum in Devon and Hugh McAllister, the taxonomist at Ness Botanic Garden. This casts light not only on what we grow, but why we grow it and joins the lyrical and scientific accounts in a seamless whole. The illustrations by Josephine Hague are works of art, the true test of the skill of the botanical artist, while Andrew Brown's detailed line drawings of every available species described, enable the reader to recognise the details of twigs, leaves and catkins.

Taxonomists may dispute this book's classification, but as Quentin D. Wheeler wrote in 2004 in a paper for the Royal Society (Taxonomic triage and the poverty of phylogeny),

'Taxonomy is hypothesis driven science ......A species is a hypothesis'.

Ashburner and McAllister have written a coherent and consistent classification of all the species of *Betula*, which has never been done before — all of us who grow these lovely trees owe them a debt of gratitude even if some may not agree with the conclusions. Such a work is long overdue; whilst there can never be a permanent definitive account of any genus as knowledge moves on, this book will satisfy gardeners for many years, and demands to be in any tree lover's bookcase both for pleasure and instruction.

Lawrence Banks

PREFACE

This monograph was Kenneth Ashburner's idea. He had the insight to see the need for a definitive taxonomic study of the genus *Betula*. He also had the patience, as well as passion, to study over many years the characters of the trees, both during his travels and in the living collection he built up at Stone Lane Gardens. His untimely death in July 2010 means that he did not have the long awaited pleasure of seeing the results of his painstaking work published. As the surviving co-author, I acknowledge an enormous debt to Kenneth for his generous spirit and friendship throughout many years.

Kenneth first made contact with me in 1978, having seen my name on determination slips on herbarium specimens in the Natural History Museum, London. From then onwards we were in regular contact, discussing all aspects of birch, making reciprocal visits and exchanging seed and plants. His trips to Japan and Korea by the Trans-Siberian railway and to North America provided seed of many species; his Japanese contacts, particularly Prof. Hideaki Ohba and Dr Samejima, yielded species which might otherwise still not be in cultivation. Kenneth was honoured and immensely grateful to have been the recipient of a Stanley Smith Horticultural Trust grant which helped fund his 1980 seed collecting visit to Siberia and Japan.

Kenneth and I both shared the conviction that only with living trees on hand to study would any attempt at a re-classification of *Betula* make advances beyond previous work largely based on herbarium specimens. With this as our joint aim, both Kenneth and I owe much to his wife June, to whom this monograph has been dedicated. June supported Kenneth throughout his life, and without her tenacity and practical help, it cannot be said for certain that Kenneth's vision could have been fulfilled. Stone Lane Gardens' collection of birches has scientific and great horticultural worth, and also creates a perfect setting for the summer sculpture exhibition held there annually since 1992. These birches formed a large part of the base of this study.

Kenneth Ashburner was first introduced to the taxonomic confusion in birch when, working as a garden designer in London, he was sold *Betula nigra*, having ordered *B. ermanii*. However, in practice the taxonomic difficulties in *Betula* are almost confined to the white barked birches, and to a lesser extent the dwarf birches of the northern bogs and tundra which have hybridised with them. Species of other groups are mostly quite distinct from these and from one another, but most are little known except where they are native in parts of east North America, Japan, Korea, the Russian Far East, and parts of western China and the Himalaya where some are important timber trees.

Only a few of the white barked species of birch, e.g. *Betula pendula*, *B. populifolia*, *B. papyrifera*, *B. utilis*, and *B. nigra* are at all commonly grown. Very few other species, even those that are widespread in the wild, have often been introduced to cultivation outside their native areas. In the case of *B. corylifolia*, *B. globispica*, *B. potaninii*, *B. fargesii*, *B. insignis*, *B. medwediewii* and *B. raddeana* probably only two or three collections have been introduced, and for the rarer *B. chichibuensis* and *B. bomiensis* only one. The reputation of a species in cultivation is therefore based on a tiny part of the variation of that species, and it is well known in forestry how much variation there can be between individuals or populations of different provenance in hardiness, time of bud break, growth rate, and timber production characteristics (Sorensen & Miles 1982; Cannell 1987; Ødum 2003). In most species there is therefore usually a huge amount of variation in the wild from which variants can be selected suitable for cultivation in a wide variety of climatic conditions.

While this reputation for being difficult has discouraged many from attempting to identify birches, others have tackled the situation in particular geographical areas (Dugle 1966; Li & Skvortsov 1999; Palmé *et al*. 2003, 2004), or attempted an overview of the whole genus (Rehder 1940; de Jong 1993; Skvortsov 2002: Järvinen *et al*. 2004). There is therefore a wealth of published work in scientific papers and floras on which to build, and we acknowledge our debt to these works.

As has been frequently pointed out (Raven 2004; Wheeler 2004; Greene 2005), very little is known or is ever likely to be known about the majority of species which have been named. Only species of economic importance such as major crop species (Smart & Simmonds 1995; OECD 2003 for *Betula pendula*), or model species such as *Arabidopsis thaliana,* have been worked on extensively. Otherwise it is only the dominant and interesting species of rich countries which have been researched to any great extent taxonomically or ecologically. This certainly applies to birches, where little is known of many species from the mountains of China or central Asia.

In this study, we consider chromosome number to be of fundamental importance in defining the species. Where more than one chromosome number has been reported within a single species in the wild, there is clearly a taxonomic problem: either more than one breeding unit and therefore biological species, is present (as we have found in *Betula utilis* and *B. chinensis*), or interspecific hybrids between species of different ploidy level are being lumped under one of the pure species. In the case of the North American paper bark birch, *B. papyrifera* (2n=84, but 2n=56, and 70 are also reported), we suggest the explanation is hybridisation with diploid species and backcrossing, any hybrids being more or less indistinguishable from *B. papyrifera* (see pp. 73–75, 339).

Peter Crane, a former director of Kew wrote, 'If ca. 2000 plant species are described every year — and believed to be genuinely new by their authors — an important question is: how many of them will survive the detailed scrutiny of monographic revision? The starting point for such revisionary work is a list of all the species names ever published in the group of interest, and in botany we are fortunate that such a list exists and is well maintained (Nic Lughadha 2004). However, to arrive at a working list of the world's plant species, the next step and a key problem is how to detect synonymy. This is the one of the objectives of a good monograph.' (Crane 2004). The authors hope to have made an attempt at this for the genus *Betula*.

Hugh McAllister, Ness, 2013

PREFACE TO REVISED EDITION

I am most grateful for the opportunity of this reprint to correct errors in the first print run. These were largely due to my relying on chromosome counts on seedlings from cultivated trees where these appeared (but clearly were not) to be true to the parent type. Flow cytometry carried out by Drs Richard Buggs and Nian Wang of Queen Mary University suggested that all *Betula dahurica* were octoploid (2n=112), and I have confirmed this by chromosome counts. The particularly vigorous '*B. glandulosa*' (which produced tetraploid seedlings) from Goose Bay in Labrador proved to be diploid *B.* ×*minor*. Flow cytometry by Drs Buggs and Wang confirmed the dodecaploid status of several populations of *B. megrelica* in the Mount Migaria region of the North Caucasus (the previous data depending on seedlings from a single wild collected seedling grown in Moscow); and their DNA sequencing studies confirmed the reported anomalous placing of *B. michauxii* in subgenus Asperae in molecular cladograms, suggesting an interesting area for further studies to explain why.

Hugh McAllister, December 2015

ACKNOWLEDGEMENTS

A long term collaborative project such as this is not always easy within the space restrictions and job remits of the larger national botanic institutions. For this reason, I owe an enormous personal debt to the officers of the University of Liverpool, by whom I was employed for almost forty years. At Liverpool University Botanic Garden at Ness I enjoyed sufficient academic freedom to pursue my taxonomic interests. Against the recent background of tremendous growth in molecular studies as a prime source of research income for taxonomic work, several recent papers suggest the folly of carrying out molecular studies in isolation from careful morphological work and preparation of permanently archived herbarium specimens.

For allowing me this freedom to combine Kenneth Ashburner's horticultural passion for birches with my own interest in solving knotty taxonomic problems, I am immensely indebted to four individuals in particular: the late J. K. (Ken) Hulme (former Director of Ness) and Paul Cook (current Curator), neither of whom ever questioned my need for space to be given over to growing so many apparently nondescript birches; and, in the School of Biological Sciences at Liverpool University, Professor Rob Marrs and Professor Steve Edwards, both of whom trusted me to produce useful work without imposing too much administrative interference on me.

In 1975, the then Director of Ness Gardens, Ken Hulme, took me to visit the birch collection at Hergest Croft where we were warmly welcomed by Dick Banks and his wife Rosamund, shown their extensive birch collection and allowed to make numerous seed collections. Of the generosity and continued interest shown by the present Banks family, I am genuinely and deeply appreciative.

Several people have made such a key contribution to the monograph it is appropriate, after the length of time spent on components of it, to regard them as collaborators. Josephine Hague has shown infinite skill and patience in creating a faithful representation of the notoriously difficult to paint catkins, and never baulked when I turned up with yet another specimen. These paintings were expertly scanned and colour balanced for reproduction by Lukman Sinclair. Dr Andrew Brown, who drew the black and white drawings of all species, largely from classic specimens in the Kew herbarium, has demonstrated his skill in depicting the distinguishing characters of the species, but at the same time produced works of art. Dr Maggie and Peter Williams selflessly devoted much of their little leisure time to taking and documenting the majority of the photographs from the Ness and Stone Lane Gardens collections. Peter's painstaking close up shots and those depicting the fruiting catkin scales and 'seeds' are particularly useful to aid identification. I should like to take this opportunity to thank him for his inexhaustible kindness and skill with which he made this permanent record possible.

With the propensity of white barked birch species to hybridise, it was important to obtain seed from known wild sources. In this respect there are many people, both in institutions and as individual collectors, to thank for their generosity. Throughout the 1980s and1990s, expeditions from Edinburgh and Kew made valuable collections in the Sino-Himalayan region and their staff never failed to be most generous with seed and access to their living and herbarium collections.

Tony Schilling, Mark Flanagan, Susan Andrews and Melanie Thomas of Kew, and Martin Gardner, Natacha Frachon and Ron McBeath of Edinburgh deserve special mention, but others, too have contributed and helped in various ways: Dr Ian Hedge provided an interesting slide of a tree (*Betula turkestanica* = *Betula utilis* subsp. *occidentalis*) growing in the Panshir valley in north-west Afghanistan; Roy Lancaster provided early seed collections of *B. calcicola* and *B. delavayi*; the late Jim Russell of Castle Howard gave us *B. insignis* from the joint expedition with Kew staff to the Fan Jin Shan in Guizhou as well as *B. fargesii*, which he obtained through Peter Wharton in Vancouver from the 1980 American expedition to Hubei. The award of a Sargent Fellowship by the Arnold Arboretum and encouraging assistance from Michael Dosmann and other staff is most gratefully acknowledged and contributed greatly to this work including the discovery of *B.* × *dosmannii* and studies of progeny from *B. jackii*.

I am glad to be able to express my personal debt here to the late Mrs Betty 'Tinge' Horsfall who took me on a 'Siberian Safari' to Lake Baikal and Vladivostock which yielded birches of central and east Asian provenance crucial to an understanding of *Betula pendula*, as well as other collections. I am similarly indebted to The Friends of Ness Gardens who funded my participation in a trip to south-east Tibet in 1997, the trip on which *B. ashburneri* was discovered by our leader, the well-known plant collector Keith Rushforth. As a long-standing friend with whom I have many stimulating discussions, Keith has been enormously generous over the years in contributing seeds from his many other trips.

Others who have generously contributed from trips they have undertaken are: Maurice Foster, who collected and photographed in important and little visited areas in north-west Pakistan and Mongolia and contributed living *Betula skvortsovii*; Dr Makoto Amano, who collected seed of the Japanese *B. dahurica* from Nobeyama in Honshu; Professor Skvortsov, who sent seed of *B. megrelica*, *B. medwediewii*, *B. raddeana*, *B. dahurica* and *B. gmelinii*; Professor Chang of Seoul, South Korea, who contributed seed of the hexaploid *B. chinensis*; Geza Kosa of the Hungarian National Botanic Garden in Vacratot who provided seed and photographs from the wild of *B.* 'turkestanica'; and Dr Pan of the Chinese Forestry Service who contributed a very attractive provenance of *B. utilis* subsp. *albosinensis* which has been widely distributed from Stone Lane Gardens. Others who have contributed from their trips are Peter Cunnington formerly Curator of Ness Gardens, and Ian Brodie of Falsyde. Thanks are also due to Dr Dave Bufford who, together with Dr Susan Kelly, Emily Wood and Melinda Peters, sent me specimens from recent Harvard University expeditions which revealed the presence of the new species *B. skvortsovii* from the Chinese provinces of Qinghai and Sichuan. In addition to the photographers already mentioned, Isobel Raetz provided slides of *B. pendula* in the Alps, and John Grimshaw slides of *B. glandulosa* from the Altai Mountains. Tony Tollitt of the University of Liverpool provided high quality scans of slides from the pre-digital age.

It is appropriate here to express gratitude to the many anonymous collectors for botanic gardens throughout Europe, Asia and North America who, in contributing seed to their gardens' seed lists, made it available to us. The staff of the herbaria visited (Edinburgh, Harvard, Kew, Liverpool, Lund, the Natural History Museum (BM)) could not have been more helpful and often drew my attention to specimens not yet filed away in the cupboards. Dr Fred Rumsey (BM) lent me his collection from a high volcano on Flores in the Azores, a possible example of long distance dispersal from America.

The tremendous variety and, therefore, the horticultural potential within the genus *Betula* can be seen at Stone Lane Gardens, thanks to Kenneth's foresight and artistic gifts. It cannot be said too often that the best way to ensure survival in cultivation is wide distribution through introduction to commercial horticulture. For this reason, the readiness of Paul Bartlett, manager of Stone Lane

Gardens, to further this aim through developing the arboretum and commercial nursery there, has to be roundly praised. For Paul's willingness and promptness in writing the chapter on cultivars and hybrids I, personally, owe my grateful thanks. At this point, too, it is fitting to add Kenneth and June's expression of deep gratitude to various people for their genuine interest, kindness and help during the building up of the Stone Lane Gardens collection: to Lord Howick for giving Kenneth so much material from his many expeditions and for his hospitality and friendship; to Kenneth Lorentzon for all his help and knowledge; to Commander Alfred Williams and Robert Williams of Werrington Park for their hospitality and for their generosity in providing scions of their birch collections; to the late Max Walters and his wife, Lorna, for many stimulating discussions, detailed advice and generous hospitality on numerous occasions at Cambridge University Botanic Garden; and to Pru Barnes, June's niece, for accompanying Kenneth on his first seed collecting trip in Northern Spain and then to Poland and also for collecting seed in the Gaspé Peninsula, Canada and Kashmir. Her continued interest and involvement in Kenneth's legacy should ensure the preservation of Stone Lane Gardens.

At Ness Gardens also, there are many individuals to be thanked. Members of staff have been remarkably sympathetic, allowing some less than attractive but interesting species to continue to grow at the same time as clearing around promising trees to show them off to best advantage. Steve Miller and Paul Philips have been particularly helpful in removing large trees to allow the smaller birches to thrive. Other members of staff have done sterling service in watering and caring for seedlings and young trees. Sally Thompson and Tim Baxter, with their very loyal team of volunteers who, under the supervision of Wendy Atkinson, Donna Young and other colleagues at the World Museum, Liverpool, have maintained the records and managed the making of herbarium specimens as vouchers for chromosome counts.

Thanks are also due to Ted Brabin, a personal friend and chairman of the North West branch of the Rhododendron, Magnolia and Camellia group of the RHS, for not only accompanying me and driving me around New England and eastern Canada in Autumn 2010, but for his patient forbearance and good humour as I searched for answers to the dwarf birch 'problem' of *Betula glandulosa*.

Both Kenneth and I have always been immensely grateful to Dr Martyn Rix, the editor, for suggesting that the monograph be included in the *Curtis' Botanical Magazine Monograph* series. In addition, Martyn provided seed of *Betula luminifera* and *B. delavayi* in the early years of the project. He has put in a tremendous amount of work adding the introductory text on the history of the collectors of the type specimens and the authors of the names, as well as checking the nomenclature. Most importantly of all, we owe much to the staff at Kew Publishing without whose very practical support and willingness to give help and advice whenever requested, the monograph could not have been published. Gina Fullerlove, Lydia White, Ruth Linklater, Christine Beard, Lloyd Kirton, Georgina Smith and John Stone deserve special mention for their kindness and hard work through the whole publication process and for turning what could have been a rather dry botanical taxonomic work into a beautiful book which is a work of art.

I am also most grateful to Dr Mark Atkinson, author of the *Biological Flora* of *Betula pendula* and *B. pubescens*, for sending me his collection of scientific papers.

Finally, special thanks to family members: to my brother-in-law, Geoff Wainwright, for generously letting me use his London flat on repeated occasions whenever I requested it; my brother Bob for drawing the base map used for all the distribution maps; to my wife Catherine for her support, encouragement and hard work throughout the thirty years of the project to ensure that the work was completed; also Iain and Mariola, and Fiona and Alex for always providing a warm welcome, accommodation and practical support in Edinburgh and Boston.

LIST OF PHOTOGRAPHERS

Key to photographers initials in figure captions

Kenneth Ashburner (KBA)

Paul Bartlett (PB)

Nicola Browne (NB)

Carole Drake (CD)

Maurice Foster (MF)

John Grimshaw (JG)

Ian Hedge (IH)

Géza Kósa (GK)

Fiona McAllister (FMcA)

Hugh McAllister (HMcA)

Páll Petersson (PP)

Isabel Raetz (IR)

Martyn Rix (EMR)

Tony Tollitt (TT)

Maggie Williams (MW)

Peter Williams (PW)

1. INTRODUCTION

THE APPEAL OF BIRCHES

It is not surprising that birches are valued by gardeners and landscapers; their foliage is delicate, and when fresh has a translucent, shimmering quality. If the buds break particularly early (as do those of *Betula pendula* subsp. *mandshurica*) the young leaves are beautifully bright in the spring sunshine. The delicate male catkins with their muted yellow-green shades bring life and movement to the bare twigs in early spring. In autumn their leaf colours can be as striking as any maple.

But it is for their bark above all that most people are persuaded to choose a birch for their garden. It is not only visual but tactile. Birch bark can reveal all sorts of colours, not only white, but subtle gradations of pinks and yellows to copper and chocolate brown, perhaps with a misty bloom of white over the darker colour (Figs 8, 174, 192). Whatever its colour, the peeling of the thin, papery bark is a particularly fascinating phenomenon. Sheets and scrolls hang from the trees and can

Fig. 1. Snow under *Betula utilis* subsp. *jacquemontii* at Stone Lane Gardens, Chagford, Devon (PB).

Fig. 2. Birches, *Betula pubescens*, with foxgloves in Stone Lane Gardens, Chagford, Devon (PW).

be heard rattling softly in the wind. New layers of bark revealed beneath are fresh, subtly coloured, and velvety or shiny (Fig. 190). Long, horizontal lenticels streak the stems in a swirling pattern. Even when they do not peel, birch stems are distinctive, often a shining brown, or flaky, or shiny and metallic. Peeling, flaking or fascinatingly patterned barks are also found in eucalypts, planes, and in some maples, rhododendrons and cherries, but birches are the best known of all beautifully barked trees.

BIRCHES IN THE WILD

Birches need no introduction to people across the cool parts of the Northern Hemisphere, from Scandinavia through northern Russia and Siberia, northern China and across the vast tracts of Canada, where they are the background to living, both beautiful and useful. Part of folklore and mythology, since ancient times they have provided timber for building, poles for fencing, bark for roofing (Fig. 47) and for canoes, tar for glue (Pollard & Heron 1996), and wood for all kinds of utensils. Boat construction, fuel, fodder, wine, human food in times of hardship, even writing materials and medicines are among their other known uses. Kingdon-Ward records how Tibetans used hanging pieces of birch bark stiffened with bamboo as bird scarers (Cox *et al.* 2001). In the eastern Alps, the 5,000 year-old 'Iceman' Ötzi had two birch bark containers among his possessions (Pollard & Heron 1996).

Fig. 3. *Betula pubescens* on sandy bluff with *Larix sibirica* in *Pinus sylvestris* forest on shore of Lake Baikal, Buchta Peschanaya (Sandy Bay) (HMcA).

While the white barked birches are among the most easily recognised trees in northern temperate regions, the thirty or so brown barked species are little known, at least in Europe. With almost all known species now in cultivation, and therefore available for study as living trees, we believe this is a good time to prepare an account of the genus.

The white barked species are attractive, medium-sized, fast-growing trees which cast only light shade, and so are ideal for average sized gardens on a range of soils in a moist, cool-temperate climate (Fig. 182). They are possibly the most familiar and admired of trees (Figs 13, 25), conspicuous in both wild and urban settings across northern regions of Europe, Asia and North America, with their distinctive white barks and their tendency to colonise rapidly any cleared or disturbed ground.

But white barked birches are only part of the story. The other four groups do not, for the most part, have white bark and range from large subtropical trees (*Betula alnoides*) to dwarf arctic shrubs (*Betula nana*), with the majority of species falling between these extremes, offering many possibilities for landscape architects and gardeners.

In addition to being known for their bark, most birches have attractive catkins in spring and many develop good autumn colour. While most of the common white barked birches are tolerant of dry sandy soil, at least in cool climates, many others grow best in heavy, permanently moist soils. Many of the dwarf species grow in swamps in the wild, although they do not necessarily require such conditions and are suitable for rock gardens and bonsai. If the rarer species become better known and are made available commercially, several of them could be every bit as desirable as the commonly planted species (Figs 76, 78–83, 95, 122, 230, 273).

Fig. 4. Several birch species in a garden setting, Stone Lane Gardens, Chagford, Devon (PW).

Fig. 5. Birches in a wild garden setting, Stone Lane Gardens, Chagford, Devon (PW).

WHAT IS A BIRCH?

There is never any question as to whether a tree is a birch or not because the unique structure of their fruiting catkins easily distinguishes them from all other trees (Figs 11, 30, 32–34). Their nearest relatives, the alders (*Alnus* species), have cone-like, woody, fruiting catkins (Fig. 22). Fossils from ancient birches and alders with their two kinds of distinctive fruiting catkins occur fairly continuously in rocks from the present to 50 million years ago. Although no older fruit fossils than this have yet been found, fossil pollen similar to that of birch and alder is known from rocks 70 million years old. This allows us to estimate that birches and alders have been two quite separate evolutionary lines for at least 50, and possibly up to 70 million years. *Betula* is therefore one of the earlier genera of flowering trees to appear in the fossil record, occurring before the extinction of the dinosaurs 65.5 million years ago, although it is not as old as *Ginkgo* which is thought to be about 200 million years old.

In the same family as birches, the Betulaceae, but less closely related than alders, are the hornbeams (*Carpinus*), hazelnuts (*Corylus*), hop hornbeams (*Ostrya*), and the hazel-like *Ostryopsis*. Birches are easily distinguishable from all these by their fruiting catkins.

The oldest fossil birches (known from leaves, fruiting catkins, catkin scales and nutlets) are of the brown barked type, whereas fossils with the distinctive characteristics of the white barked birches (long petioles, few-veined leaves) only appear relatively recently geologically speaking, in the Miocene about 10 million years ago, and with ever increasing frequency in more recent

Fig. 6. Light and shade effects with sculpture in birch arboretum, Stone Lane Gardens, Devon, England (PW).

fossil deposits (see p. 43). The white barked birches appear to have evolved relatively recently and consequently are less differentiated from one another, and this is one reason why they are not easy to tell apart. The darker barked species, having been in existence for very much longer, are often more different from one another and therefore easier to identify. Furthermore, they do not hybridise as readily, even when brought together in cultivation. They are also mostly species of mature closed forest, rather than fast growing weedy trees like the white barked birches.

BIRCH ACROSS THE NORTHERN HEMISPHERE

Seeing trees in their native habitats cannot help but impress on the enthusiast the astonishing range among the birch species. The contrasts in physical appearance among the different species are remarkable, from stately 30 metre forest trees in some dark barked species, through tall, often slender, white barked species, to small-leaved, squat, tundra shrubs. Few other genera show similar extremes, with perhaps only willows (*Salix*), rowans (*Sorbus*) and *Rhododendron* possessing such variety. In these cold northern regions, drier climates and poorer soils usually favour conifers whereas moister climates and richer soils favour birches and other broadleaved trees (Bond 1989). While a sub-arctic pine (*Pinus sylvestris*, *P. banksiana*) usually takes about eleven years to produce its first cone and a Norway spruce (*Picea abies*) about forty, a four-year-old birch may already be producing thousands of seeds. It is a matter of no surprise that the white birches are regarded as the weed trees of the landscape in these northern latitudes.

The northern coniferous forest is the only major vegetation zone in the world today not dominated by angiosperms (flowering plants); it extends in a ring around the Northern Hemisphere, with the tundra to the north of it and broadleaved deciduous forest or grasslands to the south. Throughout the coniferous forest, especially where it is disturbed, are species of a few broad-leaved tree genera: birches (*Betula*), alders (*Alnus*), rowans (*Sorbus*), willows (*Salix*), and poplars, especially aspens (*Populus*). Where there has been wholesale destruction of the conifer forest there may be no seeds left for the conifers to regenerate, and where this has happened, as in parts of Russia and Scotland, birches will form pure stands in place of the conifers.

Coniferous forests with white barked birches are also found at progressively higher altitudes on mountains further south, as in high valleys in the Alps, the Caucasus, and even in North Africa, but especially in the great mountain ranges of western China, the Himalaya and in Japan.

South of the coniferous forest belt, or at lower altitudes on mountains, the conifers gradually give way to deciduous forest with fewer conifers and an increasing variety of broad-leaved trees. White barked birches extend into this deciduous forest zone where they grow alongside oaks (*Quercus*), beeches (*Fagus*), limes (*Tilia*), maples (*Acer*), ashes (*Fraxinus*), hornbeams (*Carpinus*), the brown barked birch species, and other genera of broad-leaved trees. Because of the drying-out of the interior of the northern continents, this broad-leaved deciduous forest zone is now much less continuous than it was in the recent geological past, and is interrupted by the steppes of Siberia, Russia and eastern Europe and the prairies of North America.

Fig. 7 (left). *Betula pendula* (right) and *B. ermanii* (left) in Stone Lane Gardens, Devon, England (PW).
Fig. 8 (right). *Betula utilis* subsp. *albosinensis* in Stone Lane Gardens, Devon, England (CD).

Fig. 9 (left). *Betula ermanii* in birch Arboretum, Stone Lane Gardens, Devon, England (PW).
Fig. 10 (right). Snow under *Betula utilis* subsp. *jacquemontii* at Stone Lane Gardens, Chagford, Devon (NB).

The white barked birches are often the first colonisers of gaps in the forest (Figs 25, 262). They are especially common on nutrient poor, light, sandy soils, so they are sporadic, confined to remnant patches of woodland, or forming crowded populations in abandoned industrial sites such as mine wastes, quarries and disused railway sidings. It is clear that white barked birches are much assisted by man's disturbance of original forest. They also occupy other ecological niches. They occur in wet habitats, as on raised bogs or the margins of shallow lakes (Fig. 234). Their shallow rooting habit enables them to survive until their trunks become too heavy for the soft, wet soil to support and they then fall into the developing peat to be preserved as 'bogwood' (Fig. 16).

North of the limit of dense coniferous or birch forest, in the transition to tundra or at high altitude above the main tree-line, white barked birches may form an open parkland forest of well spaced, slow-growing, short, stunted trees with low, thick branches (Figs 180, 203). This habit is very different from the tall, straight trunks typical of birches in dense pure stands or gaps in forests (Fig. 166).

Very different from the long-stalked, triangular-ovate leaves of the white barked birches, and less familiar, are the groups of birches with hornbeam- or cherry-like leaves. Their barks are dark, sometimes almost black, often with a shiny, glossy or even metallic sheen. They live within the species-rich broad-leaved deciduous forests and usually grow as scattered individuals, though sometimes they form pure stands. The North American cherry birch, *Betula lenta*, may invade clear felled areas, and the yellow birch, *B. alleghaniensis*, sometimes forms sub-alpine dwarf forest (Fig. 48) in the ecological role more often filled by *B. pubescens* in Europe, *B. utilis* in the Himalaya, or *B. ermanii* in Japan. Except when in catkin in spring (Fig. 30), these dark barked species might not immediately be recognised as birches and could be passed over as belonging to some other genus. They are only found in the deciduous temperate to sub-tropical forest zones in eastern North America, eastern Asia, and in a very small area of the Caucasus and north-east Turkey (Maps 1–6). These are areas to which many other formerly widespread tree genera have been confined by the climatic cooling and drying which culminated in the recent Ice Ages (Tiffney 1985; Pickering 2000; Donoghue & Smith 2004). That these birches are relicts from an earlier geological age and were formerly more widespread is also suggested by their association with other genera with a clear fossil record: *Fagus* (beech) (Denk & Grimm 2009), *Magnolia*, *Davidia*, *Nyssa*, *Platycarya*, *Pterocarya* (wingnut) and the conifers *Metasequoia*, *Taxodium* (swamp cypress), and *Sciadopitys* (umbrella pine).

A third group, the dwarf birches, are also unfamiliar. In the wild these are found in shrub tundra north of and above the tree-line, as well as in gaps in the northern coniferous forest, where they grow in association with willows and ericaceous shrubs. Some dwarf birches are particularly tolerant of water-logging and are seen bordering swamps or creeks in both the tundra and forest zones where their autumn orange and red leaf colour is striking against the dark evergreen conifers (Lev-Yadun & Holopainen 2009) (Figs 14, 15).

Fig. 11 (top right). *Betula utilis* subsp. *utilis* (*Forrest* 19505). Leaves and horizonally held young fruiting catkins (PW).

Fig. 12 (right). Dry *Betula pendula* forest on south-east shores of Lake Baikal with *Pinus sylvestris* and steppe vegetation (HMcA).

Fig. 13. *Betula pendula* forest on south east shores of Lake Baikal with blackening from forest floor fire (HMcA).

ECOLOGY

Because birches grow in so many different habitats, detailed discussion of their ecology will be dealt with under the individual species. However, there are some generalisations which can be made about the genus as a whole, compared with other trees which grow in association with birches.

Birches are often the smallest leaved broad-leaved trees in the forests in which they occur. This makes them relatively tolerant of exposure compared to species with larger leaves which are more susceptible to physical damage. Especially when held on long stalks, as in the case of the white barked birches, the small leaves are even more resistant to damage in strong winds (Vogel 2009).

Birch fruits are also smaller, lighter, and usually more widely dispersed than the wind dispersed seeds of the conifers with which they often grow and compete. All three main genera of angiosperm trees which have invaded the northern coniferous forest zone, willows (*Salix*), poplars and aspens (*Populus*), and birches, have considerably smaller seeds than the conifers, making these broad-leaved trees the primary colonisers and weed tree species of the cool temperate Northern Hemisphere.

In sub-tropical forests in the foothills of the Himalaya eastwards through China to Vietnam, *Betula alnoides* grows in forests with oaks of the genus *Lithocarpus* and the alder *Alnus nepalensis* alongside such herbaceous sub-tropical ground flora plants as begonias and gingers. At the other extreme, in the cold northern tundra, the dwarf birch, *Betula nana* and its relative *B. glandulosa* grow alongside the dwarf arctic willows and ericaceous shrubs such as bilberries (*Vaccinium* spp.) and those species of *Rhododendron* which used to be included in *Ledum* (Figs 15, 138, 239).

The same species of birch can even occur in areas of very different climate. For instance, in North America *Betula papyrifera* is found from coast to coast, in the highly oceanic climates of the Maritime Provinces of eastern Canada and New England, and western North America, but

Fig. 14. *Betula humilis* in autumn colour in clearing in *Larix sibirica* forest, at eastern end of Lake Khovsgol, NW Mongolia (MF).

Fig. 15. *Betula humilis* in autumn colour with scattered *Larix sibirica*, *Salix* spp. *Rhododendron parvifolium*, *Vaccinium myrtillus*, *V. vitis-idaea*, NW Mongolia (MF).

also occurring through the continental climates of the Prairies and Rocky Mountains. Similarly, in Eurasia *B. pubescens* grows in extreme oceanic climates in western Ireland and Scotland, but also in central Siberia on the shores of Lake Baikal. Silver birches of the *B. pendula* aggregate occur alongside *B. pubescens* in these areas, but also extend through eastern Siberia to the oceanic areas of the north Pacific region, Japan, Kamschatka, Alaska and British Columbia (Maps 8 and 10). These climate differences may affect the way trees of the same species perform in cultivation; e.g. trees of *B. pendula* of continental provenance (from subalpine Switzerland) have grown poorly in the more oceanic climate of Devon.

At high latitudes in northern Canada *Betula glandulosa* (Fig. 17), and in Scandinavia and Siberia *B. nana* (Figs 21, 265, 266; 267) and related species occur in the tundra dwarf shrub community in both wet and dry habitats, while the related *B. humilis* has a more southerly distribution in wet habitats in eastern Europe and throughout Asia (Figs 14, 284, 286). In cultivation, however, *B. humilis* grows quite well in dry situations, suggesting that it is confined to wet situations in the wild only because in drier sites it is out-competed by taller growing species (Bannister 1964).

BIRCHES AS SOIL IMPROVERS

Particularly on poor, light soils, birches have a reputation for improving the soil, raising its fertility and productive capacity (Gardiner 1968; Mitchell *et al.* 1997; 2007). This characteristic is particularly noticeable where birches colonise heathland or replace coniferous forest. It appears that birches absorb minerals from relatively deep (compared to e.g. heather, *Calluna vulgaris*) in the soil and recycle them to the surface in their leaf fall, the leaves releasing the mineral nutrients at the soil surface when they rot.

Fig. 16. *Sphagnum* mat from margin of pool growing into birch wood so that trees are destabilised by increasingly wet soil and fall into *Sphagnum* to be preserved as bogwood, Norway (HMcA).

Fig. 17. *Betula glandulosa*, Banff National Park, near Lake House (KBA).

THE GENUS BETULA

THE GENUS BETULA — DIRECTION OF EVOLUTION

Before discussing the classification of *Betula* species, we will try to deduce the direction of evolution of the various characters within the genus. Current thinking is that shared **recently evolved** characters which have probably evolved only once, (called **apomorphies**) e.g. white bark, long petioles, few leaf veins, a thin wide wing to nutlets, and the recurved lateral lobes of fruiting catkin scales, indicate close genetic relationship; whereas shared **ancestral** characters (called **pleisiomorphies)**, e.g. absence of white bark, presence of oil of wintergreen, woody fruiting catkin scales which persist attached to the axis of the catkin after the nutlets have been shed, nutlets with thicker and narrower wings, catkin scales with three similar lobes, are not a reliable indication of closeness of relationship, and that these 'primitive' characteristics may have survived in distantly-related species.

　　This approach to the study of evolutionary relationships is termed **cladistics**. Although it has many merits, its uncritical application to plants has been criticised largely because of the high frequency of parallel evolution in plants (Cronquist 1987). The taxonomist working on any particular genus must therefore make a more or less **subjective** decision about characters, based largely on knowledge of related genera.

　　The genus most closely related to birch is the alder (*Alnus* spp.) (Kikuzawa 1982; Furlow 1990; Chen *et al*. 1999). Very significantly, both *Betula* and *Alnus* species have chromosome numbers which are multiples of 14, whereas other related genera have base numbers of 8 (*Carpinus*, *Ostrya*, *Ostryopsis*) or 11 (*Corylus*). This is strong confirmation of the close relationship between *Betula* and *Alnus*, and the

Fig. 18. *Betula pendula* subsp. *szechuanica* in autumn colour in conifer forest, south-east Tibet (HMcA).

distance of their relationship from the other genera. *Betula* and *Alnus* are usually placed in the family Betulaceae, while the other four genera are often placed in the separate family Corylaceae.

Therefore, by comparison with *Alnus*, the direction of evolution of various characters within *Betula* can be determined with some degree of certainty. The characters which are found in some species of both genera are most likely to stem from common ancestry. If this were not the case it would be necessary to postulate character reversals (of course possible), or that these similar character states had evolved independently in each genus, also possible but less likely. The genus *Alnus* is considered to have been the first to split away from the common ancestor, and the first of the six genera to become a distinct evolutionary line (Chen *et al.* 1999). *Betula* is considered to be the second genus to have differentiated, with the other four genera differentiating later. This order coincides with the sequence in which the genera appear in the fossil record.

The presence or absence and the number of protective scales on the winter buds show a clear trend. The buds of some alders have no protective scales, two small unexpanded, more or less normal, leaves covering the growing point. However, other alders have one pair of bud scales protecting their winter buds. Birches usually have three pairs of bud scales while the genera of the Corylaceae have up to twelve pairs.

Alnus and *Betula* are thought to have been the first genera to differentiate, are the first to appear in the fossil record (see Phylogeography), and only these genera have evolved species with sufficient cold tolerance to colonise areas with a cold-temperate to sub-arctic climate and dominate some vegetation types in these regions.

The leaves of some sub-tropical to warm temperate alders and birches (*Alnus firma*, *A. pendula*, *Betula alnoides*, *B. cylindrostachya*, *B. luminifera*, and *B. insignis*) (Plates 10, 11, 13; Figs 110, 111, 141) are mostly large, ovate-lanceolate, with many veins and numerous fine marginal teeth, much as in species of hornbeam (*Carpinus*), hop hornbeam (*Ostrya*) and what is thought to be the hazel closest to the ancestral type, *Corylus ferox* (Forest & Bruneau 2000), and so this is thought to be the most probable ancestral state.

Similarly, all alders and some birches have indehiscent (which don't break up on ripening) fruiting catkins, narrowly winged nutlets with opaque wings more than one cell thick, and short-stalked leaves (Plate 8; Figs 32–34, 121). It is therefore thought that these states are **ancestral** in birches. The generally more shade tolerant brown barked, slower growing, birch species which are components of some sub-tropical and warm temperate forests are thought to be more similar in appearance and ecology to the ancestral birches.

Fig. 19 (left). Small leaved *Betula pubescens*, Stone Lane Gardens, Chagford, Devon (PW).

Fig. 20 (top right). Small leaves of birch shimmering in sunlight with sculpture, Stone Lane Gardens, Chagford, Devon (PW).

Fig. 21 (right). *Betula nana* with *Vaccinium uliginosum*, *V. myrtillus*, and *Empetrum hermaphroditum*, Thorskafjordur, N of Inarsstadir, west Iceland (HMcA).

Table 1

The ancestral and derived characters in *Betula* as deduced from similarities with related genera, especially *Alnus*.

Character	Pleisiomorphy (Ancestral characteristic)	Apomorphy (Derived characteristic)
Bark colour	Non-white	White
Twig	Glands non-resinous, not warty	Glands resinous, warty
Twig, presence of oil of wintergreen (methyl salicylate)	Present (subgen. *Acuminatae*; subgen. *Asperae* sect. *Lentae*; B. corylifolia, B. globispica)	Absent (subgen. *Betula*; subgen. *Asperae* sect. *Asperae*)
Leaf vein number	Intermediate (6–12)	Few (< 6) or very many (>12)
Leaf size	Large	Small
Leaf shape	Ovate-lanceolate	Triangular, obovate, or orbicular
Leaf texture	Normal	Coriaceous (leathery)
Petiole (leaf stalk)	Short, petiole to blade ratio < 1/5	Long, petiole to blade ratio >1/5
Fruiting catkin arrangement	Clustered on more or less specialised twigs with small leaves	Scattered over crown of tree
Twigs bearing female catkins	Usually dying after fruiting, no buds in axils of leaves on fruiting spur shoots	Not dying after fruiting, usually with buds in axils of leaves on fruiting spur shoots
Male and female catkins often borne on same twigs	Yes	Not always
Perianth	4 tepals in male flowers (*Acuminatae*)	Less than 4 tepals in male flowers
Fruiting catkins	Indehiscent (not breaking up on ripening), often persistent after seed dispersal	Dehiscent (breaking up on ripening) as seed is dispersed
Lateral fruiting catkin scale lobes	Forward pointing, similar in form to central (Fig. 32)	Recurved, different in shape from central (see Fig. 34 (43–48) e.g. B. pendula, most B. papyrifera, B. dahurica
Fruit wing	Narrow, opaque, 2-cells thick, formed from extension of ovary and contiguous with styles	Broad, translucent, 1-cell thick, outgrowth from ovary and not contiguous with styles
Ecology	Component of mature forest	Weedy coloniser
Shade tolerance	Somewhat shade tolerant	Light demanding
Growth rate	Slow	Fast
Male catkin	Not well protected from desiccation	Well protected from desiccation
2er bracts present in each male flower	Yes	No (only subgen. *Acuminatae*)
Number of female catkins arising from each bud	Several (most *Acuminatae*; occasionally in B. utilis, B. ashburneri; rarely in other species)	One

The most **advanced** (i.e. changed from the ancestral) state in birches is probably that seen in the white barked birches which have fruiting catkins which break up to disperse the nutlets soon after ripening, small nutlets with thin, wide translucent wings only one cell thick, triangular–deltoid leaves with long stalks (characteristics not found in any alder), very rapid growth when young, and early maturation (Figs 32–34).

In the seedling state, all species of *Betula* and *Alnus* have few-veined leaves and few, relatively coarse, marginal teeth. This characteristic of young, juvenile, plants has been retained into the adult state in the white barked and dwarf birches, and contrasts with the many-veined adult state in the other sections. This is an example of what is known as **neoteny**, a well-known evolutionary process in many animal and plant groups in which advance takes place through the modification of the juvenile (**neotenous**) state rather than the adult state, and the derived species reach reproductive maturity while retaining many of the juvenile characteristics of their ancestors.

Fig. 22. Shoot of *Alnus nepalensis* showing typical arrangement of catkins in tree alders, with group of terminal males arising from several nodes and clustered females on reduced lateral branches also arising from several nodes and, behind these, fruiting catkins developed from the previous year's female catkins (PB).

Fig. 23. Shoot of *Alnus sieboldiana* (Section *Betulaster*) with red female and yellow-green male catkins emerging from buds (PW).

The birch species closest to what is considered to be the ancestral state are, therefore, those with many-veined, ovate-lanceolate leaves (Fig. 103), clustered male and female catkins (Figs 26, 30, 111), fruiting catkins which do not disintegrate on ripening, nutlets with narrow opaque wings, and four tepals (structures homologous with petals) present in male flowers. The species which has most of these characteristics is *Betula insignis* which is a large forest tree, and, interestingly, is very similar to the 50 million year old fossil species *B. leopoldae* from the early Tertiary (Crane & Stockey 1987; Crane & Carvel 2007). *B. medwediewii* and other species of section *Lentae*, some species of subsection *Chinenses* (*B. globispica*, *B. fargesii*), *B. corylifolia*, and species of subgenus *Acuminatae*, share a good number of these ancestral features. With the exception of subgenus *Acuminatae*, all have relict distributions in areas known to have other trees which have survived, relatively unchanged, since the early Tertiary (65.5–23 Mya) (Donoghue & Smith 2004) (Maps 1–6).

Fig. 24. *Betula utilis* subsp. *utilis* (*Forrest* 19505) young shoots with two well developed leaves (those preformed in bud) with only short internodes between them, and developing long shoots with still very immature leaves (PW).

Fig. 25. Drought stressed *Betula papyrifera* with juniper confined to area of thin sandy soil in forest and showing autumn colour in August before any other species. Gatineau, Quebec (HMcA).

Fig. 26. *Betula insignis* subsp. *insignis* (*GUIZ* 82) fruiting catkins on spur shoots with no buds in leaf axils (PW).

THE GENUS BETULA — RELATIONSHIPS WITHIN THE GENUS

Over the years there have been numerous attempts to sort the genus *Betula* into groups. Not all species are easily placed, some falling between two groups, but three species groupings have remained more or less constant:

1. Section *Betula*, the white barked birches, silver birches (*B. pendula* agg), paper bark birches (*B. papyrifera*, etc.). With their white bark, long stalked triangular to ovate leaves, and usually pendent cylindrical fruiting catkins, these have always been recognised as one natural grouping of related species.
2. Section *Apterocaryon*, the dwarf birches of the northern tundra and peaty swamps. These are easily recognisable by their low stature, small erect fruiting catkins, and small few veined leaves.
3. Subgenus *Acuminatae*. These are trees of the warm temperate to subtropical forests of the Sino-Himalayan region and Japan with very elongated cylindrical drooping flowering female and fruiting catkins.

In short, there are so many similarities among the species of each of these three groups that they appear to be natural groupings of related species.

The dark barked tree species with erect fruiting catkins are much more difficult to classify into groups of related species. They have few characters in common except those originally found in all birches. These are the trees most similar to the presumed ancestral birches of the early Tertiary and so are probably relicts only distantly related to one another. The original ancestor of all birches would most likely be assigned to the subgenus *Asperae* section *Lentae*, and evolutionary lines from within this section have probably differentiated into the other groups.

The situation in birches is further complicated by polyploidy, with some species having arisen by hybridisation of two different species followed by chromosome doubling. When the ancestral species have belonged to different groups, this results in net-like relationships among the species and blurs the boundaries between the groups (Raymond *et al.* 2002). This has certainly happened in the origin of *B. murrayana*, which is derived from *B. alleghaniensis* of section *Lentae* and *B. pumila* of section *Apterocaryon* (Barnes & Dancik 1985).

The most satisfactory published subdivision of the genus is that of Skvortsov (2002), who subdivided the genus much as in Table 2. He defined four subgenera. One, the monotypic *Sinobetula* A. K. Skvortsov, based on the enigmatic Chinese *B. gynoterminalis*, is known from only a single herbarium specimen which bears some resemblance to species of subsection *Asperae*, within which we include it. A second subgenus, *Nipponobetula* A. K. Skvortsov, is also monotypic, containing only the Japanese *B. corylifolia* (Plate 1). This species is usually included in the main group of brown barked birches, but we agree with Skvortsov that it is sufficiently distinct to be put in its own subgenus. A third subgenus, *Asperae* Nakai, contains all the other non-white barked birches with erect fruiting catkins except those which belong to the well-defined northern dwarf birch group, and is split into two sections (Table 2). His fourth subgenus, *Betula*, includes all the white barked species and the *Acuminatae*, and is further split into four sections, again in a way with which we largely agree, except that we place the *Acuminatae* in a separate subgenus.

The *Acuminatae* have elongated drooping fruiting catkins, as in subgenus *Betula*, but their male flowers retain four tepals (petal equivalents) and have two stamens as in *Alnus* but not in any other birch (Hjelmqvist 1948; Abbe 1935; Furlow 1990); they contain methyl salicylate as in section *Lentae* of subgenus *Asperae* and *B. corylifolia*; and most have female catkins borne in groups. That they are not attacked by *Semudobia* midges (Ronquist & Nylin 1990) reinforces their taxonomic isolation, although this was not the conclusion drawn by these authors.

Fig. 27 (left). *Betula ashburneri*. Groups of 1–3 more or less horizontally held mature fruiting catkins clustered towards end of twig and showing somewhat reflexed thin papery brown tip to fruiting catkin scales (PW).

Fig. 28 (right). *Betula ashburneri* recently emerged male and female catkins showing alder-like arrangement with clustered catkins of both sexes towards tip of shoot (PW). (Compare Fig. 22)

Here we divide subgenus *Betula* into four sections. In section *Dahuricae*, Skvortsov included the shaggy barked species, the first time an author had grouped the North American *B. nigra* with the east Asian *B. dahurica*, a conclusion for which we provide supporting physiological data (Chapter 7); we also include *B. raddeana*. In his section *Costatae*, Skvortsov places *B. costata*, *B. ermanii*, and the species which are here treated as belonging to the *B. utilis* aggregate (*B. jacquemontii*, *B. albosinensis*), to which we add the recently discovered *B. ashburneri*. Section *Apterocaryon* Spach is the group of dwarf arctic birches usually referred to as sect. *Humiles* or sect. *Nanae*, while section *Betula* is the main natural grouping of white barked birches.

Table 2

The main subdivisions of the genus *Betula*.

Subgenus	Section	Species	Distribution	Fruiting Catkin	Nutlet wing
Nipponobetula	*Nipponobetula*	*corylifolia*	Japan	Erect	Very narrow, opaque, part of ovary and adnate to styles
Aspera	*Asperae*	**Subsection** *Asperae* *schmidtii, potaninii, calcicola, chichibuensis, bomiensis, delavayi, skvortsovii, (gynoterminalis)*	SW China (NE Asia, Japan)	Erect	Very narrow, opaque, part of ovary and adnate to styles
		Subsection *Chinenses* *chinensis* 6x, *chinensis* 8x, *globispica, fargesii*	N & Central China, Korea, Japan		
	Lentae	*lenta, alleghaniensis, grossa, insignis, murrayana, medwediewii, megrelica*	Japan, Indo-China, Central China, Caucasus, E North America	Erect	Narrow, opaque, part of ovary and adnate to styles
Acuminata	*Acuminatae*	*luminifera, cylindrostachya, alnoides, maximowicziana*	Japan, Vietnam, China, Himalaya	Pendent	Very wide, translucent
Betula	*Dahuricae*	*nigra, dahurica, raddeana*	E Asia, SE North America Caucasus	Erect to in plane of shoot to more or less pendent	Wide, translucent
	Costatae	*costata, ermanii, utilis, ashburneri*	Himalaya, E Asia	Erect or pendent	Wide, translucent
	Betula	*pendula, populifolia, fontinalis, celtiberica, (microphylla, tianshanica) cordifolia, papyrifera, (utahensis), pubescens*	Circumboreal Central Asia, E & W North America	Pendent (sometimes erect in forms of *B. pubescens*)	Wide, translucent
	Apterocaryon	*nana, michauxii, glandulosa, pumila, humilis, fruticosa, gmelinii*	Circumboreal	Erect	Narrow

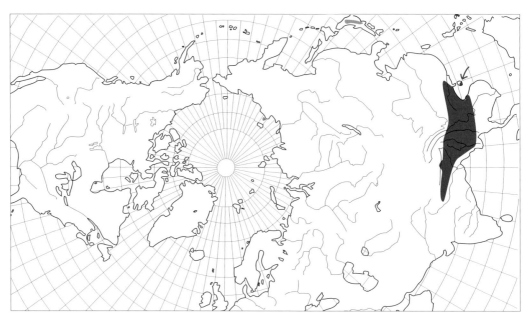

Map 1. Distribution of *Betula alnoides* ■ (subgenus *Acuminatae*) and *B. corylifolia* ■ (subgenus *Nipponobetula*). *B. corylifolia* occurs in the mountains of the northern part of the main Japanese island of Honshu. *B. alnoides* is a sub-tropical species which occurs throughout the Himalaya eastward through Burma, Laos, Thailand, Vietnam to Guangxi and Hainan provenances in southern China. Its distribution shows considerable overlap with that of its close relative *B. cylindrostachya*.

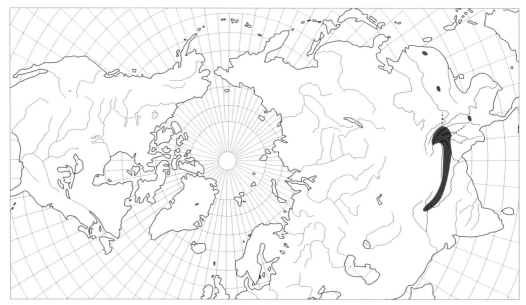

Map 2. Distribution of *Betula cylindrostachya* (subgenus *Acuminatae*) ■. Until recently *B. cylindrostachya* was much confused with *B. alnoides* (Skvortsov 1997), so the distribution mapped is likely to be an underestimate. It is often found growing in the same geographical areas as *B. alnoides*, but at higher altitudes. It occurs throughout the Himalaya, extending into Burma with outlying occurrences in northern Thailand (*Danish Expedition to Thailand* no. 1628), northern Guangxi and Fujian provinces in China (Zeng *et al.* 2008).

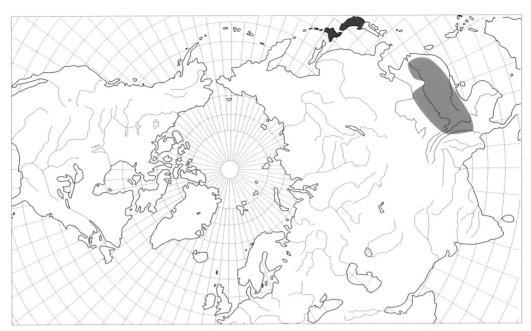

Map 3. Distributions of *Betula luminifera* ■ and *B. maximowicziana* (subgenus *Acuminatae*) ■. *B. luminifera* occurs throughout much of the warm-temperate zone in China. *B. maximowicziana* is endemic to Japan and found from central Honshu through Hokkaido to the southern Kurile Islands.

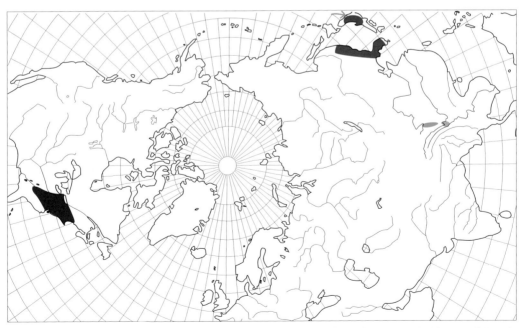

Map 4. Distributions of the diploid species of subgenus Asperae: *Betula lenta* ■, *B. schmidtii* ■, *B. chichibuensis* → (Japan), *B. potaninii* ■ and *B. calcicola* ■. *B. lenta* is clearly very closely related to *B. alleghaniensis* and has a very similar but slightly more southerly distribution. It and *B. schmidtii* are quite widespread while the other three species have much more restricted distributions.

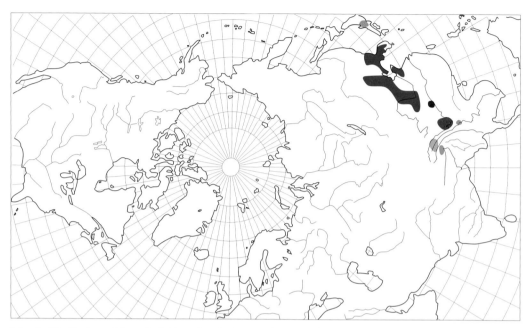

Map 5. Distributions of the polyploid species of subgenus *Asperae* section *Asperae*: *Betula bomiensis* ■, *B. skvortsovii* ■, *B. delavayi* ■, *B. chinensis* ■, *B. fargesii* ■ and *B. globispica* ■, (*B. gynoterminalis* — ploidy level unknown).

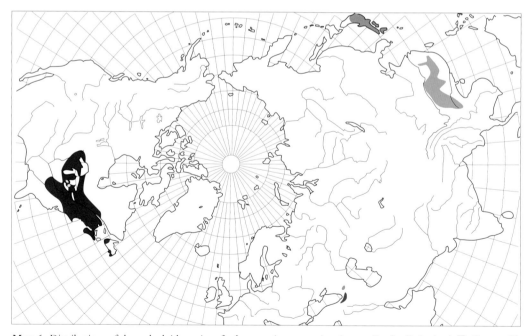

Map 6. Distributions of the polyploid species of subgenus *Asperae* section *Lentae*: *Betula alleghaniensis* ■, *B. grossa* ■, *B. insignis* ■, *B. medwediewii* ■, *B. megrelica* ■ and *B. murrayana* ■. These species are probably the most similar species to the common ancestor of all birches. Their disjunct distribution in known refugia is understood to be an indication of former more extensive distributions, except for *B. murrayana* which is a recently arisen allopolyploid known from only two localities.

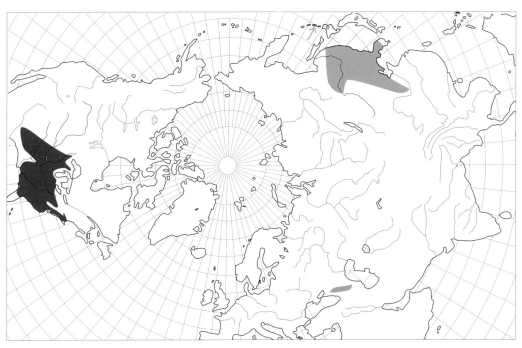

Map 7. Distributions of species of subgenus *Betula* section *Dahuricae*: *Betula nigra* ■, *B. dahurica* ■ and *B. raddeana* ■.

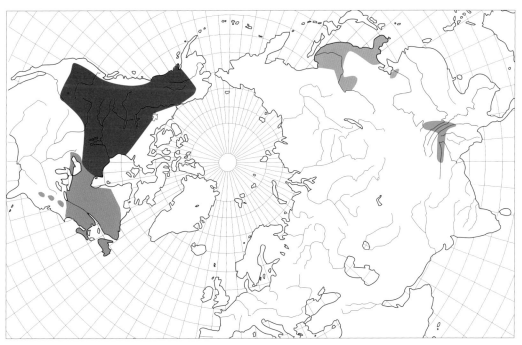

Map 8. Distributions of diploid species of subgenus *Betula* excluding the *Betula pendula* aggregate: *B. ashburneri* ■, *B. cordifolia* ■, *B. costata* ■ and *B. occidentalis* ■.

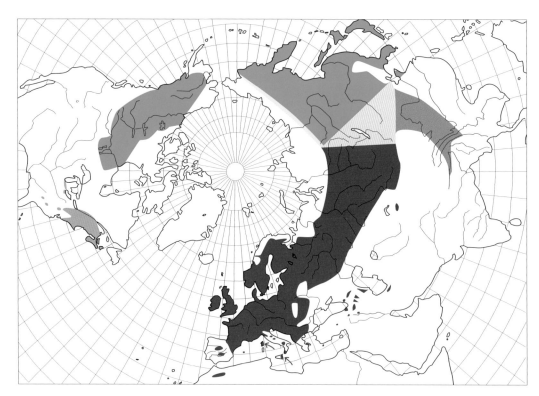

Map 9. Distribution of diploid species of the *Betula pendula* aggregate: [*B. pendula* subsp. *pendula* ■ (▧ indicates uncertain extent of distribution), subsp. *szechuanica* ■, subsp. *mandshurica* ■ (▧ indicates uncertain extent of distribution), and *B. populifolia* ■]. The geographical separation between the subspecies of *B. pendula* is approximate, and individuals and populations resembling subsp. *pendula* are often found within the region of the other subspecies and vice versa. The isolated occurrence in the western Tianshan in central Asia is based on collections made by Skvortsov (BM!). The identity of several populations in Spain and Portugal is uncertain. These are often given specific (*B. parvibracteata, B. fontqueri*) or subspecific status under *B. pendula*, but until their ploidy level has been determined, it is uncertain whether they belong here or with *B. celtiberica* (Map 10).

Map 10 (opposite, top). Distributions of tetraploid species of subgenus *Betula* sections *Betula, Costatae* and *Apterocaryon*: *Betula pumila* ■, *B. fruticosa* ◪, *B. microphylla* ■, *B. tianshanica* ■, *B. utilis* ■ (subspp. *occidentalis* ▧, *jacquemontii* ▧, *utilis* ▧, *albosinensis* ▧), *B. ermanii* ■ and *B. pubescens* ■ (▧ indicates uncertain extent of distribution). The taxonomy of this group in the Pamir to Altai region of Central Asia is very confused with isolated populations having been named as distinct species. *B. microphylla* (Altai) appears to intergrade towards the east with *B. fruticosa*, and towards the north and west with *B. pubescens*, perhaps itself being of hybrid origin between these two species. *B. microphylla* may intergrade southwards with *B. tianshanica*, and then this with *B. utilis* at the eastern end of the Himalayan chain in Afghanistan and Tajikistan, from where specimens have often been named *B. turkestanica*. Around Beijing *B. utilis* subsp. *albosinensis* may intergrade with *B. ermanii*, as *B. ermanii* appears to do with *B. pubescens* in the Lake Baikal region. *B. celtiberica* (N Spain, arrow) is probably derived from *B. pubescens* through extensive introgression from *B. pendula*.

Map 11 (opposite, bottom). Distributions of *Betula papyrifera* ■ and *B. humilis* ■. The northern and western distribution limits of *B. papyrifera* ▧ are uncertain due to confusion with *B. cordifolia* in eastern areas, and with *B. pendula* subsp. *mandshurica* in western areas and, in both areas, the probable frequent occurrence of hybrids. The central and eastern Asian distribution of *B. humilis* ▧ is uncertain due to confusion with the tetraploid *B. fruticosa*, but specimens have been seen from northern Mongolia and northern Kamtschatka.

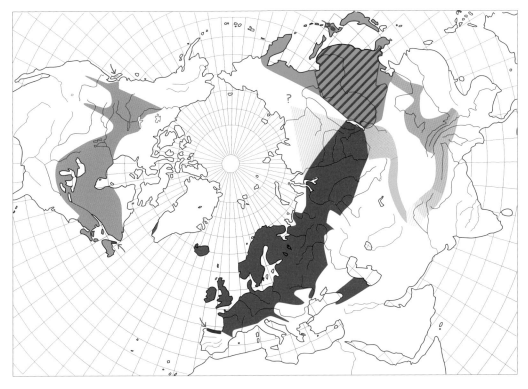

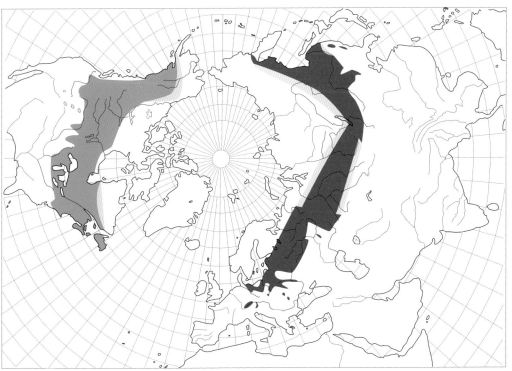

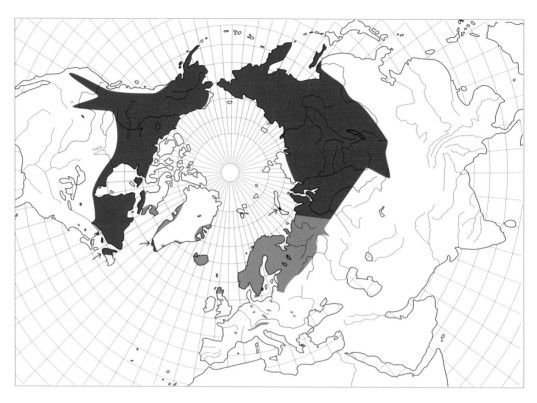

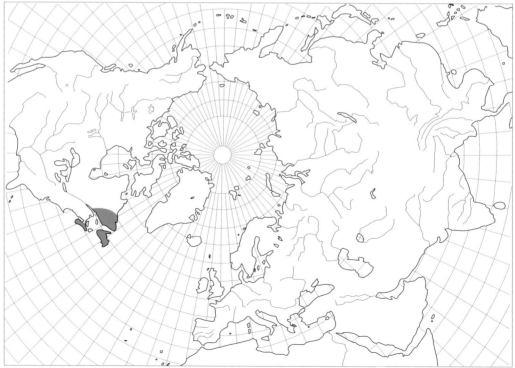

A note on the distribution maps

Maps have been drawn using a number of sources including our own herbarium studies, but most information has come from published maps in: Atkinson 1992; Bopp 1994; Browicz 1975; de Groot *et al.* 1997; Dugle 1966; Fang *et al.* 2009; Fredskild 1991; Furlow 1997; Hultén 1958, 1968, 1971; Jansson 1962; Li & Skvortsov 1999; Ohwi 1965; Peinado *et al.* 1983; Sokolov 1951; Tabata 1966; Vassiljev 1941; Zeng *et al.* 2008, reinterpreted using the taxonomy adopted in this book. We realise that dot maps in which each occurrence of a species is represented by a single dot, often the location of an herbarium specimen, on a map is the ideal. However, any species distribution mapping is totally dependent on the identification of the specimens, and in *Betula* this is often not reliable, especially towards the limits of the species' ranges and with taxonomically difficult species. Also, we have not examined specimens in the herbaria of China, Russia and continental Europe, which would be necessary to obtain more complete and accurate distribution maps.

No attempt has been made to indicate frequency of occurrence, as would be apparent from dot maps, but it can usually be assumed that species are less common towards the limits of their distribution.

Comments on problems within individual species and species groups are mentioned under each map. For a discussion of the taxonomic issues see the accounts in the relevant sections in Chapter 7.

Paler shades mean that the distribution limits in that area are uncertain (not including distribution of subspp. of *B. utilis*).

Map 12 (opposite, top). Distributions of diploid species of subgenus *Betula* section *Apterocaryon*, the dwarf arctic birches: *Betula glandulosa* ■ and *B. nana* ■. Although distributions have largely been taken from published sources, the identity of species occurring in each area under our interpretation is different from that in published maps. With the exception of Greenland and adjacent areas of Canada (Baffin Is.) where *B. nana* occurs, and Newfoundland, Labrador, Quebec, Nova Scotia, St. Pierre & Miquelon where *B. michauxii* occurs (see Map 13), all specimens in North America and eastern Siberia are believed to be *B. glandulosa* (i.e. *B. nana* subsp. *exilis* is treated as a synonym), but the position of the boundary with *B. nana* in western Siberia is uncertain.

Map 13 (opposite, bottom). Distribution of *Betula michauxii* ■ (subgenus *Betula* section *Apterocaryon*) in Newfoundland, Labrador, Quebec, Nova Scotia, St. Pierre & Miquelon. The distribution of *B. michauxii* overlaps with *B. glandulosa* in Newfoundland and Labrador, but *B. michauxii* is a species of wet bogs while *B. glandulosa* is primarily a species of drier habitats.

Fig. 29. *Betula raddeana* 'Hugh McAllister', Karachayevsk, in cultivation (PW).

REGENERATION AND LIFE CYCLE

Like young plants of most woody Angiosperms, young birches can regenerate from the base of the trunk, at least until a certain degree of maturity has been reached (Del Tredici 2001). Some shrubby species probably retain this ability throughout their life (Tabata 1966). It is this regeneration capacity which allows birches to survive frequent fires that eliminate conifers, almost all of which lack such ability to regenerate.

VEGETATIVE REGENERATION

In the wild each birch tree or shrub is usually a single individual which has arisen from a seed. No birch is rhizomatous, and only *Betula populifolia* and *B. nigra* appear to be capable of regenerating from roots (HMcA pers. obs.), and this appears to be infrequent. In other species all regrowth is from cut or otherwise damaged stumps and arises from dormant buds. This means that almost every tree or shrub seen in the wild is a separate genotype, a genetically distinct individual.

Extensive vegetative propagation in the wild has only occasionally been described in *Betula nana* and its hybrids in wet and mossy tundra where layering of horizontal shoots occurs. Patches up to five metres in diameter are reported in *B. glandulosa* and hundreds of square metres for a *B. nana* × *B. pubescens* hybrid (De Groot *et al.* 1997). In contrast, a study of the shrubby Japanese bog species *B. fruticosa* showed very little vegetative spread of individual seedling derived plants (Nagamitsu *et al.* 2004). In Greenland horizontal branches of *B. pubescens* are said to frequently become layered (Sulkinoja 1990; Hermanutz *et al.* 1989); this could well happen with other species but hardly

Fig. 30. *Betula medwediewii* twig with withering terminal male catkins and female catkins behind these on same twig with no buds in leaf axils (PW).

Fig. 31. Trunk developed from sucker shoot from root of *Betula populifolia*, Charles River, Boston, Massachussetts (FMcA).

constitutes vegetative spread, more a means of the continued existence of an existing genotype. In some shrubby species there is probably extensive and repeated regeneration of individuals from sucker shoots from dormant buds at the bases of the main stems (Rinne & Kauppi 1987). Tabata (1970), working on Japanese species, showed that seedlings of dwarf species soon developed specialised bud burls at the base of the main stems and that these buds retained the ability to sprout for many years.

With tree species it has been shown that damage to young trees often results in suckering from the base, but, after a certain age, many species lose this capability and are then destined to grow through maturity to old age and death (Del Tredici 2001). Unless regularly coppiced from a young age, a practice which can make an individual almost immortal, most tree species have a fixed lifespan which may be determined by growing conditions. The more favourable the conditions the more rapidly a tree will grow and mature and the shorter its lifespan is likely to be. *Betula pubescens* and *B. papyrifera* are quoted as losing the ability to sprout after 40–60 years but, in Alaska, Perala & Alm (1990) report that 60% of *B. papyrifera* sprouted at 140 years.

Betula pendula coppice relies on existing roots for many years, which makes it difficult to propagate by separation of suckers with their own root systems (Perala & Alm 1990), and observations suggest similar behaviour in other species.

REGENERATION FROM SEED

Although we have stressed the ecological differences between the rapid growing 'weedy' white barked birches and the slower growing species of mature forest, all birch species seem to be capable of maturing and producing seed within about ten years, even species which make large trees and can probably live to a great age, such as *Betula alleghaniensis*, *B. lenta*, *B. grossa*, *B. insignis*, and species of section *Acuminatae*. Perala & Alm (1990) state 10–15 years to first seed production for *B. pendula*, but this is clearly referring to trees in the wild in less than ideal conditions. All birches are therefore to some extent 'weedy' and thus 'r-selected'. This contrasts with 'k-selected' species of mature forest in stable conditions such as beech (*Fagus sylvatica*) and Norway spruce (*Picea abies*) which take about forty years from seed before they first flower and produce seed.

Despite what has been said above, there is no question that the 'weedy' white-barked birch species do begin to produce seed earlier and in larger quantities than the longer lived species (Table 3). Longman & Wareing (1959) and Longman (1984) showed that *Betula pendula* could be induced to flower when less than a year old by growing in continuous light, and more recent Finnish work quoted by an OECD report (OECD 2003) describes seed crops being collected eight months after seed sowing, and comments that it is easy to select for early maturation.

On the other hand, a collection of *Betula pendula* from the shores of Lake Baikal took ten years before it produced its first few fruiting catkins in cultivation at Ness, and even after nineteen years they are still fruiting very sparsely and intermittently, and *B. pubescens* from the same area has still not fruited after the same time, trees of both species having grown well to around 8 m tall. In the wild, birch trees in the lake Baikal area are reported (Molozhnikov, pers. comm.) to fruit rather erratically and often to miss some years entirely. This experience does provide a cautionary warning against extrapolating the behaviour of a species in one part of its range to its total distribution, but, of course, our experience of what happens in cultivation may differ from what happens in the wild.

The times to first fruiting of birch species in cultivation in the British Isles are given in Table 3.

Table 3

Number of years from seed germination to production of first fruit for several *Betula* species in cultivation in Britain (original provenance in brackets).

Species	Years to bearing first fruit in cultivation
B. pendula (NE France)	3
B. pendula (Lake Baikal, Siberia)	10
B. gmelinii (*apoiensis*)	3
B. pumila (Newfoundland)	3
B. pubescens (Newfoundland)	5
B. pubescens (Lake Baikal, Siberia)	>19
B. potaninii	3
B. chichibuensis	5
B. chinensis 6x	7
B. globispica	11
B. utilis (Kashmir)	>15

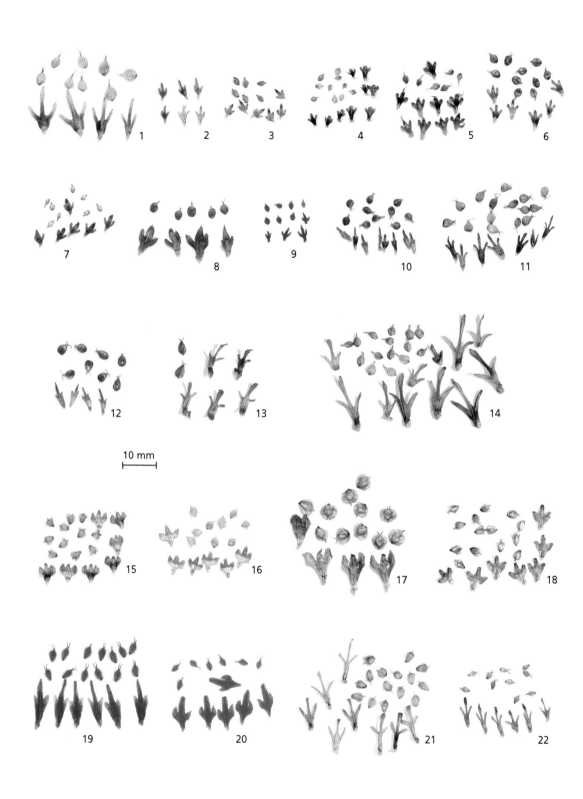

10 mm

Given that there is very little vegetative reproduction in birch tree species and that many species are relatively short lived, it is clear that regular regeneration from seed is essential to maintain the species. In all birches, fruits are produced whether viable seed has been formed or not, i.e. they are parthenocarpic.

Seed sown in cultivation usually gives a rapid and more or less complete germination of viable seed in the spring following sowing. Similarly in the wild, there is also usually a good germination of viable seeds on suitable substrates and, at least in the European white barked birches, little evidence of a soil seed bank (Ghorbani *et al.* 2003a, b). However, *Betula maximowicziana* is reported to maintain a significant seed bank for at least six years (Osumi & Sakurai 1997; Tsuda & Ide 2005). There is usually very little evidence of seedling establishment within dense birch woods, probably partly because of shade, but also possibly due to an allelopathic effect in which secretions from the existing birch trees inhibit the growth of seedlings. In open habitats, whether heathland or disused railway sidings, where birch seed is shed there is often massive seedling establishment and rapid growth.

LIFE SPAN

The commonly grown fast growing white barked birches are relatively short lived, 60–80 years usually being quoted for *Betula pendula*. However, actual lifespan depends on growing conditions, slower growth resulting in a longer lifespan (Prévosto *et al.* 1999). The slower growing species such as those in section *Lentae* are reported to be longer lived with average and maximum ages of 150 and 250 years being quoted for *B. lenta* and 150 and 300 for *B. alleghaniensis*, whereas 100 and 140 years are quoted for *B. papyrifera* (http://www.cnr.vt.edu/4H/BIGTREE/TreeAge.htm Accessed 24/8/2008).

Fig. 32 (opposite). Photographs of fruiting catkin scales and 'seeds' to show variation between subgenera, sections, species, between different trees of some species, and within individual trees (PW).

Subgenus *Nipponobetula*: 1. *Betula corylifolia* Japan, N Honshu, Mt Goyo. The 'beak' of fused styles enclosed within the wing is unique in *Betula*.

Subgenus *Asperae* section *Asperae* — seeds lacking wings, scales with lateral and median lobes of similar form: 2. *B. schmidtii* S Korea, Suanbo. Note significant difference in size and form of scales between this and following sample. **3.** *B. schmidtii* Russian Far East, Primorsk, Ussuri. **4.** *B. potaninii* (*SICH* 347) China, Sichuan, Kangding. **5.** *B. calcicola* (*CLD* 791) China, Yunnan, Lijiang. Note abnormal scales with extra lateral lobe. **6.** *B. chichibuensis* Japan, Honshu, Gunma pref, Tano-Gun, Mt Kano-san. **7.** *B. bomiensis* (*KR* 6371) SE Tibet, between Bagu and Nambu c. 120 km due W of Bomi (type locality). **8.** *B. delavayi* (*EMR* 4117) Lijiang, NW Yunnan, China. **9.** *B. skvortsovii* China, Sichuan, Dêgê Xian, W of Dêgê. **10.** *B. chinensis* 6x, S Korea, Mt Jiri. Note difference in length of lateral lobes between this and the following sample. The same variation was found between samples of the octoploid. **11.** *B. chinensis* 6x, (*HuiKim* 503), S Korea, Gangwondo, Wonju city, Mt Chiak-san. **12.** *B. chinensis* 8x, N Korea, Kumgang Mts. **13.** *B. globispica* Japan, Honshu, Tochigi Pref, Kuriyama-mura, nr San'no Rindo Kinenhi. **14.** *B. fargesii* (*SABE* 183) China, Hubei, Shennongjia Forest Distr.

Subgenus *Asperae* section *Lentae* — seeds with wings narrower than nutlets, opaque, more than one cell thick, contiguous with style bases, and scales with lateral and median lobes of similar form: 15. *B. lenta*, Canada, Ontario, Niagara region, Louth Township. **16.** *B. alleghaniensis* USA, New York, Duchess co., Millbrook. **17.** *B. alleghaniensis* Canada, Ontario, St. Williams. **18.** *B. grossa* Japan, Miyagi Pref. **19.** *B. insignis* (*GUIZ* 82) China, Guizhou, Fan Jin Shan. **20.** *B. insignis* subsp. *fansipanensis* (*KR* 2344) Vietnam, Lao Cai prov, Sapa. **21.** *B. medwediewii*, Mt Migaria region, Georgia, Caucasus. **22.** *B. megrelica* (*A. K. Skvortsov*) Mt Migaria region, Georgia, Caucasus.

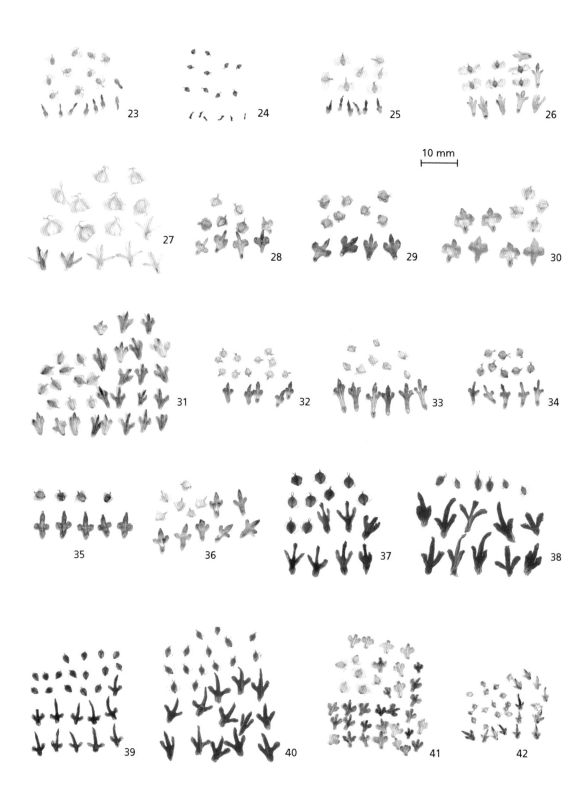

CHROMOSOME NUMBERS IN BETULA

Closely related species with the same chromosome number (ploidy level) usually interbreed freely, and species with different chromosome numbers cannot interbreed freely whether closely related or not. Therefore, in any group of closely related species within a geographical area, there is likely to be only one species with each chromosome number (ploidy level) and closely related species in any one area must have different chromosome numbers (ploidy levels), or occur in different habitats, if they are to remain distinct. These principles can help us sort out some taxonomic muddles and suggest instances where too many species may have been described.

The case of *Betula lenta* and *B. uber* growing together in Virginia is one example: both species were found to be diploid, so the rare *B. uber* would be expected to have disappeared long ago, swamped by the much commoner *B. lenta*; but *B. uber* has survived because it is merely a form of *B. lenta* probably determined by a difference in a single gene, appearing here and there in a restricted area among normal trees. It is therefore recognised as *B. lenta* f. *uber* (McAllister & Ashburner 2004) (see p. 200).

Another scenario is most strikingly demonstrated in the birches of the Caucasus and Pontic mountains of northeastern Turkey, where five ploidy levels are each represented by a single species, the three lower ploidy levels by white barked birches: diploid *Betula pendula*, tetraploid *B. pubescens* var. *litwinowii*, and hexaploid *B. raddeana*. No octoploid is known in this area, but decaploid *B. medwediewii* and dodecaploid *B. megrelica* occur in scrub around the tree-line. Several other little-

---

Fig. 33 (opposite). Photographs of fruiting catkin scales and 'seeds' to show variation between subgenera, sections, species, between different trees of some species, and within individual trees (PW).

Subgenus *Acuminatae* — seeds with translucent wings much wider than nutlets and not contiguous with styles. Scale lateral lobes much reduced except in *Betula maximowicziana*. 23. *B. luminifera* (*Rix* 4121) China, Baoxing (Moupine). **24.** *B. alnoides* (*Henry* 10,437 A) China, Hubei. **25.** *B. cylindrostachya* (*KR* 976) Bhutan, Wang Chu Valley, Chapcha view point. **26.** *Betula maximowicziana*. Japan, Hokkaido, Tashiogan, Horonobecho, Toikanbetsu, Tashio forest.

Subgenus *Betula* section *Dahuricae* — seeds with more or less translucent wings narrower to as wide as nutlets and not contiguous with styles. Scale lateral lobes often more or less square, different in shape from the central lobes except in *Betula nigra* in which lobes are similar. 27. *B. nigra* USA, Pennsylvania, Lebanon county. **28.** *B. dahurica* (Amano), Japan, Nagano Pref, Minamisaku-gun, Minamimaki-mura, Nobeyama. **29.** *B. dahurica*, Russian Far East, Primorsk, c. 30 km N of Vladivostock. **30.** *B. dahurica* 8x, S Korea, Kwangyang. **31.** *B. raddeana* Georgia, Chewi .

Subgenus *Betula* section *Costatae* — seeds with translucent wings narrower to as wide as nutlets and not contiguous with styles. Scale lateral lobes more or less similar in shape to the central lobes. **32.** *B. costata* cult. Ness 2007 KFBX, tree growing weakly with poorly developed scale base in comparison with 33. **33.** *B. costata* (*KA*) S Korea, Odae-san. **34.** *B. ermanii* (*KA*) Japan, Hokaido, Furano, Tokyo University Forest. **35.** *B. ermanii* (*KA*) Japan, N Honshu, Yamagata/Miyagi Pref, Mt Zao. **36.** *B. utilis* subsp. *utilis* (*Forrest* 19505) China, NW Yunnan, Mekong/Salween Divide, 28°N 98°50'E. **37.** *B. utilis* subsp. *jacquemontii* (*Howick & McNamara* 1846) India, Himachal Pradesh. **38.** *B. utilis* subsp. *albosinensis*. China, Shaanxi, Qinling Mt (Beijing 1986/7). The variation in the proportions of the scale lobes within a single catkin is particularly striking with shorter lateral lobes in scales from near the apex of the catkin. **39.** *B. utilis* subsp. *albosinensis* (*Purdom* 752) China, Gansu. **40.** *B. utilis* subsp. *albosinensis* China, Gansu, Kaolan, nr Lanchow 36°N 103°E. **41.** *B. utilis* subsp. *occidentalis* (as *B. turkestanica*). Tajikistan, Hissar Mts, Karakul valley. Somewhat square lateral scale lobes and wide nutlet wings might suggest introgression from *B. pendula*, perhaps via *B. pubescens*/*tianshanica*. **42.** *B. ashburneri* (*KR* 5161Q), SE Tibet, Kongbo Province, above Gyala on ridge of Namche Barwa.

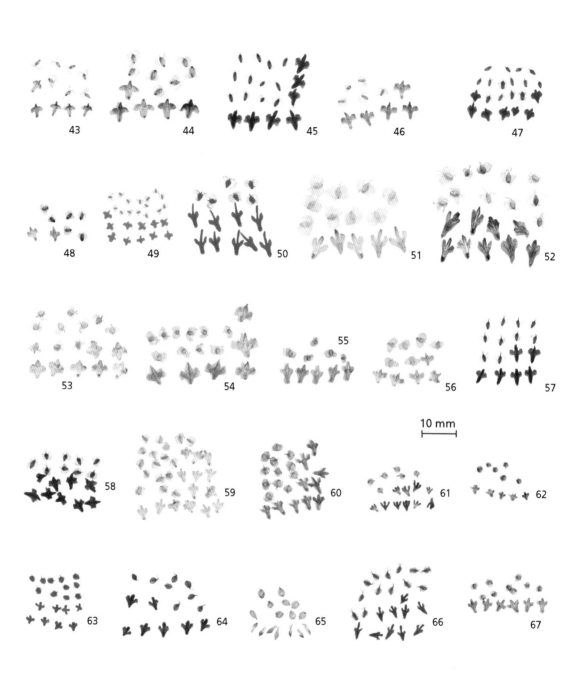

known species (*B. aischatiae*, *B. victoris* and *B. maarensis*) are described from the Caucasus by Husseinov (1971). Unless one of them is an octoploid, these three species should not coexist with the other five species without being hybridised out of existence, which suggests that they are not distinct genetically, even if they have minor morphological differences. They are probably more or less distinctive populations of the very variable *B. raddeana*.

Regular interbreeding within a population of a species maintains the similarity of the individuals and so the integrity of the species (Petit & Excoffier 2009). Being wind pollinated trees and mostly self-incompatible and therefore not self-fertile, birches can exchange pollen, and therefore genes, over considerable distances (Järvinen 2004; Järvinen *et al.* 2004). The fruits are also light and may be dispersed over large distances, so within one geographical area there are few differences between individuals or populations (Järvinen 2004).

Betula oycoviensis, *B. szaferi* and *B. pendula* (all diploid) in Poland; and *B. pubescens*, *B. carpatica*, *B. tortuosa*, *B. coriacea*, *B. callosa* and *B. obscura* (all tetraploid) in northern Europe are examples of species with the same chromosome number and overlapping distributions. They are hardly distinct, and are discussed under *B. pendula* and *B. pubescens* in Chapter 7.

Table 4 lists the chromosome numbers found in *Betula* species and some hybrids.

In *Betula* there is no known case in which a species contains both diploids and tetraploids. In other words, there is no situation where there are morphologically indistinguishable diploids (2x) and tetraploids (4x), though we come close with diploid *B. pendula* and the tetraploids *B. celtiberica* and *B.*

Fig. 34 (opposite). Photographs of fruiting catkin scales and 'seeds' to show variation between subgenera, sections, species, between different trees of some species, and within individual trees (PW).

Subgenus *Betula* section *Betula* — seeds with translucent wings wider than nutlets and not contiguous with styles. Scale lateral lobes often more or less square, different in shape from the central lobes. **43.** *Betula pendula* Italy, Sicily, Mt Etna, above Linguaglossa, 2000 m. **44.** *B. pendula* (*H. McAllister* 9/90) Listvyanka, shores of Lake Baikal, Irkutsk Oblast, Siberia, Russia. **45.** *B. pendula* subsp. *mandshurica* Japan, Hokkaido, Furano Forestry Station. **46.** *B. pendula* subsp. *mandshurica* (*KA*) Canada, Manitoba, The Pas. **47.** *B. pendula* subsp. *szechuanica* (*SICH* 346) Sichuan, Paoma Shan, a hill beyond Kangding. **48.** *Betula celtiberica* N Spain, Puerto de Pajares. **49.** *B. populifolia* USA, Rhode Island. **50.** *B. cordifolia* Canada, Gaspé Peninsula, Forillon National Park — particularly long and clearly parallel-sided central lobe to scales. **51.** *B. cordifolia* (*K. Ashburner* 9, 9 Sept. 1988) Canada, Newfoundland, Clarenville. **52.** *B. cordifolia* Canada, Newfoundland, Gros Morne Mt, screes, nr tree line. **53.** *B. papyrifera* Canada, Newfoundland, Trinity. **54.** *B. '×'utahensis* (*KA* 9) USA, Montana, N of Missoula. **55.** *B. occidentalis* (*KA*) Canada, Alberta, Jasper, Mt Edith Cavell. **56.** *B. occidentalis* (*KA*) Canada, Alberta, Trochau, Alberta, E of Calgary, Canada. **57.** *B. pubescens* UK, Scotland, Perthshire, Callander, Bracklin Falls. **58.** *B. pubescens* (*KA*) France, Vosges, Chaume de Haut Chitalet. **59.** *B. pubescens* W Iceland, Husafell. **60.** *B. pubescens* (*KA* 21) Canada, Newfoundland, Daniel's Harbour.

Subgenus *Betula* section *Apterocaryon* — seeds with more or less translucent wings narrower than nutlets and not contiguous with styles. Scale lateral lobes more or less similar in shape to the central lobes. *B. michauxii* is anomalous in lacking lateral scale lobes and having no distinct wing to the nutlet. **61.** *B. nana* (*Ian Brodie*) UK, Scotland, Wester Ross. **62.** *B. glandulosa* (*KA*) Canada, Labrador, Goose Bay. **63.** *B. glandulosa* (*HMcA*) Canada, Quebec, Gaspé Peninsula, Xalibu. **64.** *B. ×minor* (*HMcA*) Canada, Quebec, Gaspé Peninsula, Mt Xalibu. **65.** *B. michauxii* (*KA* 23) Canada, Newfoundland, Come by Chance. **66.** *B. pumila* (*KA* 19) Canada, Newfoundland, Gros Morne, W of Woody Point. **67.** *B.* cf. *pumila* (*KA* 43) Canada, Manitoba, NE of The Pas. **68.** *B. fruticosa* (*A. K. Skvortsov*) Russian Far East, Chiba, near Blagovetschensk. **69.** *B. fruticosa/microphylla* (*MF* 357) N Mongolia, 50°41'N 099°15'E. **70.** *B. fruticosa* (*tatewakiana*) Japan, Hokkaido, Sarabetsu Bog. **71.** *B. gmelinii* (*apoiensis*) Japan, Hokkaido, summit of Mt Apoi. **72.** *B. humilis* (*KA*) Poland, W of Bydgoszez, Alluvial plain of R Notec, Slesin.

tianshanica; gene flow from the diploid to the tetraploids is a more likely explanation for their similarity than autopolyploidy. In one species, *B. chinensis*, we have found both hexaploids (6x) and octoploids (8x), but the source of the extra genome present in the octoploids is not clear.

A new birch species, *Betula murrayana*, has recently arisen naturally through allopolyploidy (Barnes & Dancik 1985), and provides an example of how new species can arise (see p. 208). It is intermediate between *B. alleghaniensis* and *B. pumila*, the two species from which it is derived, but closer to *B. alleghaniensis*. This is not surprising as it has inherited 84 chromosomes from that species and only 28 from *B. pumila*. *B. murrayana* is deduced to have arisen as a backcross between an unreduced gamete (n=70) of *B. ×purpusii* (*B. alleghaniensis* × *B. pumila*) and a normal reduced gamete (n=42) from *B. alleghaniensis* resulting in this new species with 2n=112. As *B. murrayana* arose recently, its origin is clear, but it exemplifies how other polyploids probably arose in the past. Subsequent evolution and/or extinction of parents makes it difficult to work out exactly what happened, but guesses can be made based on the appearance of the trees which could be tested using molecular data.

To show how alloployploidy may (and it must be appreciated that this is purely speculative)

Table 4

The chromosome numbers of species of *Betula*.

Species	Ploidy	Chromosome No.
alnoides, ashburneri, calcicola, chichibuensis, cordifolia, corylifolia, costata, glandulosa, humilis, lenta, luminifera, michauxii, nana, nigra, occidentalis, pendula, populifolia, potaninii, schmidtii	diploid	2n=2x=28
pendula × *pubescens, nana* × *pubescens, lenta* × *pumila*	triploid	2n=3x=42 (14 + 28)
bomiensis, celtiberica, cylindrostachya, ermanii, fruticosa, gmelinii, microphylla, pubescens, pumila, skvortsovii?, tianshanica, utilis, (papyrifera × *cordifolia/populifolia/fontinalis)*	tetraploid	2n=4x=56
papyrifera × *pumila, papyrifera* × *(papyrifera* × *cordifolia/populifolia/fontinalis)*	pentaploid	2n=5x=70 (42 + 28)
alleghaniensis, chinensis 6x, delavayi, papyrifera, raddeana	hexaploid	2n=6x=84
chinensis 8x, dahurica, murrayana	octoploid	2n=8x=112
fargesii, globispica, insignis, medwediewii	decaploid	2n=10x=140
megrelica	dodecaploid	2n=12x=168
gynoterminalis	unknown	

Key
x =14, the 'base' chromosome number of the genus
n = 7, the haploid chromosome number in gametes (loosely — pollen and embryo sac)
2n = the diploid chromosome number found in all somatic (i.e. not gametic) cells of a plant

have given rise to *Betula papyrifera*, let us represent the diploid chromosome set of the *B. pendula* agg., (presumably represented by subsp. *mandshurica*) by PP. Another distinct diploid white barked species is the east North American *B. cordifolia*, whose genome may be represented by CC. The hexaploid (2n=6x=84) *B. papyrifera* is very similar in appearance and chemical composition to *B. cordifolia* (Koshy *et al.* 1972), and so probably has the genome of that species in its parentage, perhaps duplicated. The lack of platyphylloside in *B. cordifolia*, *B. populifolia* and *B. occidentalis*, and its universal presence in *B. pendula* and *B. papyrifera*, supports the suggestion that the *B. pendula* genome is present in *B. papyrifera* (Santamour & Lundgren 1996). Its genomic constitution, judging from its appearance, might therefore be represented as PPCCCC.

The appearance of the Eurasian tetraploid white barked birches, *Betula pubescens*, *B. celtiberica* and *B. ermanii*, might suggest that the *B. pendula* genome is likely to be present in all of them, but the absence of platyphylloside throws some doubt on this suggestion. However, gene silencing or elimination can take place quite rapidly after the origin of new polyploids and it is not always possible to trace particular features of parents, whether morphological or molecular, in their polyploid derivatives (Raymond *et al.* 2002; Järvinen 2004).

In some cases polyploids may have characteristics not found in any existing species of lower ploidy level, as the ancestral species of lower ploidy level may have become extinct. This would seem to be the case with *Betula medwediewii*, *B. globispica*, *B. fargesii*, *B. chinensis* and *B. insignis*.

There has been much discussion as to whether aneuploids (with chromosome numbers intermediate between precise ploidy levels) occur in *Betula* (Brown & Al-Dawoody 1977). They certainly do in other plant groups, but in *Betula* there are many reports of only exact ploidy levels in progeny of triploids, which would have been expected to yield aneuploids (Thórsson *et al.* 2001, 2007; Brown & Al-Dawoody 1977). Brown & Al-Dawoody are very honest in suggesting extensive aneuploidy, but consider it unproven due to difficulties of chromosome counting. Possible explanations for the absence of aneuploidy include the functioning in hybrids of only gametes with chromosome numbers of precise ploidy levels (Karlsdóttir *et al.* 2008), or apomixis, both unexpected situations. Therefore, although there are several reports of aneuploid plants (e.g. Dugle 1966), they have not been found in situations in which they would be expected (Anamthawat-Jonsson & Tomasson 1990; Karlsdóttir *et al.* 2008). The philosophical issues involved in chromosome counting are well discussed by Favarger (1978), great honesty in counting the chromosomes being the most important factor. The now much-used technique of flow-cytometry is very useful for detecting ploidy level, but cannot determine precise chromosome numbers.

SUMMARY OF CYTOLOGICAL METHOD

The method we have used is a modification of that described by Dyer (1963) with help from Morriset (pers. comm.). It is very simple and straightforward, and has allowed accurate counting of even large numbers of chromosomes. We are not saying there are no errors, but in most cases several root tips of each species of different provenance have been counted and, when unexpected or particularly interesting results were obtained, counts were confirmed on new samples of root tips of the same plant and on other individuals of the same population.

Pretreatment

Collect rapidly growing root tips in the afternoon and put them into vials of saturated monobromo-naphthalein solution in water overnight in the fridge at about 1°C.

Fixation

The following morning decant pretreatment solution and leave vials upside down to drain for a few minutes. Add freshly mixed 3:1 ethyl alcohol: acetic acid fixative and store for at least 12 hours in freezer or fridge ice box.

Hydrolysis

Transfer root tips to vials of 1M HCl at 60°C in a water bath for 5 mins, then transfer to 70% ethanol for storage.

Slide preparation

Under a dissecting microscope, cut off the densely cytoplasmic region of the root tip. If root tip is sufficiently large dissect out central stele column (plerome) using hypodermic needles. Place in a drop of 2:1 lactic acid:propioinic acid mountant, add coverslip, tap out to a single layer with a blunt needle while examining under dissecting microscope. Blot on absorbent paper and then squash using knuckle of middle finger to obtain adequate pressure. Examine by phase contrast microscopy using anisole (methyl phenyl ether) in place of immersion oil.

MOLECULAR PHYLOGENETICS

Unfortunately the occurrence of polyploidy and reticulate relationships, with a polyploid combining the chromosome sets of two or more species, makes the deduction of relationships from molecular data rather difficult (Mason-Gamer 2008). This problem can be got round to some extent by considering only the diploid species (Li *et al.* 2007), but even these results must be interpreted with caution and, of course, don't sort out the relationships between polyploid species.

While there is no question that molecular work has resolved many evolutionary and taxonomic issues, some published results seem so improbable that they are likely to be unreliable due to the characteristics of molecular evolution and its interpretation (Morrison 2009). There is also the probable misnaming of the material studied, the absence of voucher specimens [and taxonomists (Smith 2006)] often making it impossible to check identification of the sample (Raven 2004; Wheeler 2004; Pleijel *et al.* 2008). Several wrongly identified voucher specimens for molecular work have been found in major herbaria during the course of work for this book. Some of the relationships suggested on purely molecular or phytochemical grounds in the following papers seem so much at variance with current taxonomy based on morphology, that they cannot be summarised and the interested reader is referred to the following papers (Jiang *et al.* 2002; Järvinen 2004; Järvinen *et al.* 2004; Lahtinen *et al.* 2006; Li *et al.* 2005, 2007 and especially Schenk *et al.* 2008), but see Wang *et al.* (in press 2016).

Further work using more and reliably identified individuals of each species and complete sequencing of at least all the diploids (the DNA c-values are within the range now feasible) should resolve these problems. A balanced view of the problems involved in interpreting the results of molecular work is given by Järvinen (2004): 'A phylogenetic (gene) tree constructed from DNA sequences does not necessarily mirror the actual – – – evolutionary pathway of the species' due to 'hybridisation and introgression, recombination, lineage sorting, and gene duplication' and 'extinction of ancestral gene polymorphisms' so that 'the gene tree can be quite different from the species tree'. To this can be added the complexities introduced by extensive polyploidy.

Finally, Morrison (2009), reviewing the use of the various DNA sequence alignment algorhythms, questions the current methodology and urges caution in the interpretation of the results.

PHYLOGEOGRAPHY AND THE FOSSIL HISTORY OF BETULA

INTRODUCTION

Phylogeography is the study of the evolution of organisms and their spread to occupy their present areas of distribution. It is a combination of phylogeny (the evolution of the group) and phytogeography (how the species and species groups came to grow where they are currently found). The fossil record is very helpful where it exists, but is often fragmentary. Comparison with other genera with a similar distribution pattern can be useful in suggesting migratory pathways, although it is recognised that each species has its own individual history (Wiens & Donoghue 2004; Williams *et al.* 2007).

In this section we look at the factors which influence evolution in wind-pollinated trees such as birches, before discussing the fossil record and current distribution of birch species.

The genus *Betula* appears to have differentiated from the ancestral family line and spread rapidly throughout the northern hemisphere in warm-temperate forests. The same pattern appears to have been followed by sections *Lentae* and *Dahuricae*, and perhaps other species and groups. With climatic cooling in the late Tertiary, culminating in the Ice Ages and the drying of the continental interiors, the continuous distribution of these birches became broken up until many survived only in their present refugia. Some time in the late Tertiary, (earliest fossil evidence 10 Mya, Late Miocene), the white barked birches of section *Betula* and the dwarf arctic birches (*Apterocaryon*) evolved, perhaps from lines ancestral to the modern *B. costata*/*B. cordifolia* and *B. ashburneri*/*B. occidentalis* respectively, and spread throughout the now cold-temperate northern regions of Eurasia and North America.

MOUNTAIN BIOGEOGRAPHY

On mountain ranges very different habitats, each with its particular group of plants, are found in close proximity (Gosling & Bunting 2008). This is clearly seen on southern low latitude ranges such as the Himalaya on which the climate can change from sub-tropical to arctic over a few kilometres as altitude changes. Mountains therefore often have small populations of species dispersing into a wide range of habitats and climate zones. This will clearly be accompanied by strong selection pressure, accounting for rapid evolution and noticeable differentiation between populations. As a result, closely related populations in neighbouring valleys, or on neighbouring mountains, may look different from one another, and therefore may be given different names. Isolated populations of any plant or animal species, especially those in different habitats, tend to differentiate over time (from a few years to millions of years depending on selection pressure and generation time) into separate species or subspecies. The greatest and most rapid differentiation is found in species with the most precise pollination mechanisms (e.g. insect pollinated plants like *Rhododendron* and *Primula*) and the least efficient mechanisms of long distance pollen and seed dispersal (e.g. *Primula*) (Qian & Ricklefs 2001). *Betula*, being exclusively wind pollinated and having very efficient dispersal of the tiny, mostly winged, fruits, is probably much slower to differentiate into morphologically distinct populations and different species.

COMPARISONS WITH OTHER GENERA OF COOL TEMPERATE CLIMATES

Unlike the majority of the flowering plants which, even in the Late Cretaceous and early Tertiary (90–34 million years ago), evolved a close association with insects for pollination, and birds and mammals for seed and fruit dispersal, *Betula* (and *Alnus*) appear to have reverted to wind pollination

Geological Time Scale of sequence and dates of periods and stages mentioned in text. Dates modified from Gradstein *et al.* (2004)

The Tertiary, although not presently recognised by the International Commission on Stratigraphy, is a commonly used term for the Paleogene and Neogene periods. In this scale age is shown in Mya before present and stage names for the Paleogene, Neogene and Quaternary have been omitted.

PERIOD	EPOCH	STAGE	AGE	FOSSIL RECORD
Quaternary	Pleistocene			
			1.8	
Neogene	Pliocene			*B.* cf. *pendula, B.* cf. *nana* N Greenland *B. nigra* SE N America
			5.3	
	Miocene			*B.* cf. *utilis* China *B. cristata* (*Betula*) Iceland *B. islandica* (*Lentae*) Iceland
			23	
Paleogene	Oligocene			
			34	
	Eocene			*B. leopoldae* (sect. *Lentae*) BC, Canada
			56	
	Paleocene			*Betula*-like pollen throughout N Hemisphere
			65.5	
Cretaceous	Upper	Maastrichtian		*Betulaepollenites & Betulaceopollenites Paraalnipollenites* in Inner Mongolia & Montana
			70.6	
		Campanian		*Betulaepollenites & Betulaceopollenites* (?*Betulaceae*) in Japan
			83.5	
		Santonian		*Alnipollenites* (?*Betulaceae*) in Japan Carbonised *Normapolles* (Fagales) flowers
			85.8	
		Coniacian		
			89.3	
		Turonian		
			93.5	
		Cenomanian		*Normapolles* (?Fagales) pollen appears
			99.6	
	Lower	Albian		
			112	
		Aptian		
			125	
		Barremian		Fossil Angiosperm pollen first appears
			130	

(Dodd *et al.* 1999) and wind seed dispersal. Wind also provides the pollination and seed dispersal mechanism of by far the majority of conifers, including those of the family Pinaceae: *Pinus* (pines), *Picea* (spruces), *Abies* (firs), *Tsuga* (hemlocks) and *Larix* (larches) which were evolving and diversifying in temperate regions at the same time as *Betula* in the early Tertiary. The larches are particularly comparable as, like the birches, they are deciduous, intolerant of shade and have rapid seedling growth. Of the other broad leaved tree genera common in the conifer forest zone, *Alnus*, *Salix*, *Populus* (especially aspens), and *Sorbus* (especially rowans), birches are most similar to alders and poplars, because willows are predominantly insect pollinated and rowans are insect pollinated and have bird dispersed fruit.

Therefore, in their patterns of speciation and distribution, birches are, as might be expected, similar to alders, poplars and conifers, especially larches, with some species being common and continuously distributed over wide areas in northern forests, while there are other species, often with restricted distributions, in more southerly mountain ranges (Hultén 1971). In the widespread and common species of birch (*Betula pendula*, *B. pubescens*, *B. papyrifera*, *B. utilis*, *B. nana*, *B. glandulosa*) variation is usually very great and continuous and there are no major breaks in morphology or distribution that allow them to be divided into clearly distinct species.

Birches, often growing with willows, are a major component of the tundra scrub vegetation and are among the most successful colonising weedy species of cold temperate forests. Birches and willows are equally ready to hybridise, even quite distantly related species being able to produce at least partially fertile hybrids, and both have a wide range of ploidy levels. Such hybridisation and polyploidy are thought to have facilitated adaptation and speciation and enabled these genera to colonise the northern regions of the Northern Hemisphere in a period of rapid climate change.

The more southerly warm temperate conifer and broadleaved forests of western and eastern North America, the Caucasus and south Caspian region, the Eastern Himalaya and western China, and the Russian Far East, Korea and Japan are relicts of the formerly continuous Tertiary forests. Many of the genera and even sometimes the species they contain are very similar, because dispersal was possible between them at various times during the Tertiary. These more southerly forests are now separated by the dry, grass-dominated plains of the prairies in the middle of North America and the grasslands and deserts of Central Asia, as well as by the Pacific and Atlantic oceans. It was through this species-rich, more or less continuous Tertiary forest that the early birches spread and differentiated, with some evolving cold tolerance on mountains, or in arctic regions, as the climate cooled.

Particularly interesting is the parallel between the distribution of *Betula* section *Lentae* (probably the basal group of *Betula*) and that of the genus *Fagus* (beech). Both appear to have originated in Beringia (NE Asia & NW North America) in the Eocene, and spread westwards through Eurasia to Iceland and eastwards through North America, surviving today in the same areas: Japan, south China, south-west Asia (with spread into Europe for *Fagus*) and east North America, with extinction in Beringia, Central Asia including the Himalaya, and central and west North America (Denk & Grimm 2009). Whereas the slow maturing and dispersing, forest dominant *Fagus* has remained species poor, the basal group of *Betula* has given rise to a range of species adapted to very different habitats and ecological strategies and now placed in different subgenera.

FOSSIL HISTORY OF THE FAMILY BETULACEAE AND THE GENUS BETULA

CRETACEOUS (130–65.5 MILLION YEARS AGO)

Angiosperms (flowering plants) can first be identified for certain in the fossil record in the early Cretaceous around the Late Barremian-Early Aptian boundary, about 120 millon years ago (Mya) (Friis *et al.* 2006), although, of course, it is likely that they had been in existence for some considerable time before this. The earliest and most plentiful angiosperm fossils are of pollen which is produced in large quantities, especially by wind-pollinated genera like *Betula*. However, especially with these very early fossils, it can be difficult to assign pollen to a particular group of angiosperms, and often, unless pollen grains can be found within fossilised flowers, we have no idea from what kind of plant any particular pollen originated.

It is thought, on the basis of pollen structure, that the order Fagales (to which the family Betulaceae belongs) arose out of a group of plants which produced pollen of the *Normapolles* type. This first appeared in the Mid-Cretaceous, in the Cenomanian (c. 96 Mya) (Friis *et al.* 2003). The *Normapolles* province, in which this pollen type is common, included east North America, Europe, and west Asia, and was separated by the Turgai Strait (in what is now western Siberia) from the *Aquillapollenites* province which included central and east Asia and west North America, separated from east North America by the epicontinental sea (Fig. 35). In the late Cretaceous these two provinces (land masses) were separated by ocean with little opportunity for dispersal of plants or animals between them. However, in the early Tertiary, the mixing of the formerly distinct floras and faunas indicated that the barriers to dispersal were breaking down (Wolfe 1973).

Recently, fossil flowers containing pollen of the *Normapolles* type have been found in Portugal in Western Europe in rocks of Cretaceous (Santonian–Maastrichtian) age (c. 83.8–65.5 Mya). These flowers have been described as *Endressianthus miraensis* and *E. foveocarpus*, and the authors state that they

Fig. 35. Distribution of land and sea in the Eocene Period about 45 million years ago. Stippling indicated high ground in this geological period. Note Turgai Strait separating West Asia from central Asia. By this time the epicontinental sea has retreated from central North America so that land dispersal between east and west is possible.

show many features suggestive of a close affinity with the Betulaceae (Friis *et al.* 2003). They were borne in catkin-like structures and, unlike other flowers of the period containing *Normapolles* pollen, they are unisexual, the ovaries are bicarpellate, and the fruits are small nutlets about 0.5–1.3 mm × 0.5–0.9 mm. From their structure the flowers are presumed to have been largely wind pollinated. It is thought that the earliest angiosperms were insect pollinated (Dodd *et al.* 1999) and that wind pollination evolved from this in response to a climate with a windy, dry atmosphere with infrequent rain and in open vegetation. The *Normapolles* group is thought to have evolved from Hamamelid (witch hazel group) stock which shows, in *Corylopsis*, a transition to a catkin-like form of inflorescence.

In western Europe in the late Cretaceous it is thought that the climate was warm, and at least seasonally dry, with dinosaurs as the dominant animals living in open conifer forest. What the *Endressianthus* plants looked like is unknown as the only fossils found so far are pollen and minute flowers and fruits preserved in three dimensions because they were carbonised by fire.

From the Late Cretaceous (Santonian 85.8–83.5 Mya), pollen with *Alnus*-like characteristics (*Alnipollenites*) is found in Japan, and shortly afterwards becomes very common in Eurasia and North America (Chen *et al.* 1999). Again, the type of plant bearing this pollen is unknown, but presumably resembled *Alnus*. A pollen type somewhat intermediate between that of *Alnus* and *Betula* (*Paraalnipollenites*) is found towards the very end of the Cretaceous (Maastrichtian 70.6–65.5 Mya) in Inner Mongolia (China) and Montana, (USA); and fossil pollen resembling that of *Betula* (*Betulaepollenites* and *Betulaceopollenites*) first appears in the Late Cretaceous in Japan (Campanian 83.5–70.6 Mya) and Inner Mongolia (Maastrichtian). Thus, as far as can be judged from the pollen evidence, by the end of the Cretaceous the family Betulaceae was common throughout the main land masses of the Northern Hemisphere, North America and Eurasia, and trees, probably classifiable as belonging to the genus *Betula*, existed in eastern Asia.

Kikuzawa (1982) proposes that the family Betulaceae evolved in a Mediterranean type climate on the shores of the Tethys sea, of which the Mediterranean, Black, and Caspian seas are remnants, but which extended through the region now occupied by the Himalaya and central China.

Though many modern (*i.e.* extant) plant ***families*** are known from the Cretaceous, very few fossils of this age can be referred with any degree of certainty to modern ***genera***, even in non-angiosperms. The maples (*Acer*) are among the few modern angiosperm genera reliably recorded as macrofossils from the Cretaceous (Boulter *et al.* 1996). Fossils reliably referable to modern genera only begin to appear in significant numbers following the extinction event at the Cretaceous / Tertiary (K/T) boundary (65.5 Mya). At this time the dinosaurs, and indeed all larger land animals and many other kinds of organisms, became extinct, almost certainly due to the Earth being struck by a large meteorite and the after effects of this event (Collinson 2000; Pickering 2000; Kring 2007).

Early Tertiary (65.5–23.03 Mya): Palaeogene (Paleocene, Eocene, and Oligocene)

Although the flora was not as dramatically affected as the fauna by the K/T extinction event, the vegetation in many areas appears to have changed rapidly from the dry, warm, open forests of the Cretaceous to very dense rainforest (Spicer 1989; McElwain & Punyasena 2007). In the early Tertiary the world climate is therefore thought to have been generally warm and wet with no significant accumulations of ice on land anywhere. Throughout much of the Tertiary the temperature gradient between the equator and the poles was much less than at present (0.3°C per degree of latitude as against 1°C today), and so the vegetation zones were much broader. Also suggested is greatly decreased seasonality, so that genera and species which are today exclusively tropical or temperate could occur in the same environments (Archibald & Farrell 2003). All the fossil evidence suggests that, on the whole, in any one climatic zone, this forest was remarkably uniform with much the same genera, if not species, across Eurasia and North America (Wolfe 1973; Collinson 2000; Pickering 2000).

In the early Tertiary there were warm temperate polar forests within 20° latitude of the North Pole on the most northerly land on the shores of the Arctic Ocean in northern Greenland, Canada and Siberia. Despite three months of continuous darkness in winter, the fossil evidence suggests that these polar forests did not have particularly low winter temperatures, for example in the *Metasequoia — Glyptostrobus* swamp forests on Axel Heiberg Island in Arctic Canada which contained two species of *Betula* (Francis 1988). Fossil pollen resembling that of *Betula*, and similar to that found in the Late Cretaceous, is found in the Russian Far East, Europe and North America in the Paleocene (65.5–55.8 Mya) (Chen *et al*. 1999), so *Betula*-like trees were presumably widespread in the northern hemisphere in the Paleocene.

Throughout this early part of the Tertiary (Paleogene = Paleocene, Eocene, and Oligocene; 65.5–23.03 Mya) there are numerous fossil leaf floras known, and also fruit and seed floras like that of the Eocene London Clay. However, the identification of many of the leaves is often very uncertain (Wolfe 1973), and the London Clay Flora has its closest affinities with modern tropical floras of south-east Asia (Collinson 2000), suggesting a tropical climate in western Europe at that time. Although many leaves have been identified as belonging to genera of the Betulaceae, in the absence of reproductive structures these identifications cannot usually be regarded as reliable, even to family level, and certainly not to genus. The leaves of some birches, especially fossil species, are more or less indistinguishable from those of some alders and hornbeams (Wolfe 1973). Therefore, it is only when fruits are found that Tertiary fossils of Betulaceous leaves can be certainly identified to genus.

The discovery in Eocene (~50 Mya) deposits in British Columbia of leaves and associated reproductive structures of *Betula leopoldae* (Crane & Stockey 1987) indicates that the genus was well differentiated from *Alnus* by that time. This material is particularly well preserved and includes nutlets, which are rarely reported from the fossil record, as well as fruiting catkins and fruiting catkin scales. In overall appearance the fossilised remains of *B. leopoldae* are very similar to living *B. insignis* (Plate 10; Figs 32–34), although in the fossil species the leaf shape is very variable and the oblong, parallel-sided lateral lobes of the fruiting catkin scales more closely resemble those of the living *B. medwediewii*, *B. globispica* and *B. fargesii*. This early fossil birch from western North America fits in well with ideas of what a common ancestor of the genus might have looked like as deduced from knowledge of living species. The elongate lanceolate leaves are similar to those of some *Alnus* (alders) and many *Carpinus* (hornbeams); the fact that whole fruiting catkins and attached groups of fruiting catkin scales are found fossilised suggests that the fruiting catkins were relatively indehiscent, as in the living *B. insignis*, *B. medwediewii*, *B. globispica* and *B. fargesii*. This state is very different from that found in the white barked birches of subgenus *Betula* in which the fruiting catkins usually disintegrate rapidly on ripening, with the result that usually only individual dispersed scales are found on the forest floor in a modern northern birch wood. Also, there is evidence of the presence of only a narrow wing to the nutlets, as in the living *B. insignis*, *B. medwediewii*, *B. globispica* and *B. fargesii*, and unlike the white barked species.

To sum up, it would seem that during the early Tertiary genera and subgeneric groupings within the Betulaceae evolved in one area of the Northern Hemisphere, and fairly soon thereafter spread rapidly throughout the Arcto-Tertiary forest to acquire a circumboreal distribution. Following each successive radiation, initially of genera then of subgeneric groupings, speciation occurred with the isolation of populations by opening oceans or climatic barriers. This pattern is also found in a wide range of temperate tree genera such as *Fagus* (Denk & Grimm 2009), *Acer* (Renner *et al*. 2007), *Zelkova* (Denk & Grimm 2005), *Fraxinus* (Jeandroz *et al*. 1997), and *Tilia* (Manchester 1994).

LATER TERTIARY (23–1.8 Mya): NEOGENE (Miocene and Pliocene)

In the later Tertiary (Neogene) the fossil record is initially much as in the Paleogene. Apart from detached leaves (probably) and pollen (more certainly) assignable to *Betula*, the earliest well dated Neogene records we have been able to trace of fossil fruiting catkin scales associated with leaves are from the Miocene of Iceland (Denk *et al.* 2005; Grimsson *et al.* 2007). Subgenus *Asperae* is recorded from beds 15 Mya as leaf fossils; while in fossils from 12 Mya the associated fruiting catkin scales are often in groups, suggesting delayed disintegration of the fruiting catkins. This material (named *B. islandica*) has long narrow fruiting catkin scale lobes like those of the Eocene *B. leopoldae*, but the associated leaves are broader with cordate bases and, as is characteristic in this group, have short petioles and the secondary veins parallel to one another and much stronger than any of the tertiary veins. They also appear to have a distinct tertiary order of more or less parallel veins running between the secondary veins. This fossil *B. islandica* is very similar in most respects to the living *B. medwediewii* from the Caucasus (Fig. 30), itself probably a close relative of *B. insignis*. Also present in the 15 Mya Icelandic beds are *Glyptostrobus*, *Sciadopitys*, *Tetracentron*, *Sequoia*, *Aesculus*, *Tilia*, *Magnolia* and *Cercidiphyllum*, and in the younger 12 Mya beds *Liriodendron*, *Sassafras* and *Pterocarya*, among others (Grimsson *et al.* 2007, 2008).

In younger (~ 10 to 6 Mya) Icelandic fossil beds, *Betula* leaves, named *B. cristata*, have relatively longer petioles (petiole/leaf blade ratio 1:4 to 1:3), similar to living white barked birches. Fruiting catkin scales (*B. subnivalis*) which occur in deposits of the same age are referred to *B. cristata* in Denk *et al.* (2007) and closely resemble those of *B. cordifolia*. The fossil *B. cristata/subnivalis* therefore seems to have the characteristics of species of section *Betula*, which would indicate that the white barked birches of section *Betula* had evolved at least by the late Miocene.

Fig. 36. Habitat of *Betula nana* (behind lower part of man's legs) with Ericaceous shrubs and *Salix* species, Sermermiut, about 1 km SW of Ilulissat, W Greenland (MW).

Throughout the Miocene (23.03–5.33 Mya), there are numerous records of fossil birch leaves. Graham (1963) reports the following species from Oregon, the suggested most similar living species being given in brackets after the name of the fossil species: *B. thor* (*B. papyrifera*), *B. vera* (*B. alleghaniensis*), and *B. fairii* (*B. luminifera*). In the Late Miocene of Poland, Kolman-Adamska *et al.* (2004) record *B. salzhausensis*, which they regard as similar to modern *B. pubescens*. Olivares *et al.* (2004) report a *B. pendula*-like fossil from the Neogene of North Spain. It is difficult to interpret the relationships of Polish brown coal Miocene fossils of *B. longisquamosa*, *B. plioplatyptera*, *B. salzhausensis*, and *B. similis*.

Also from the Miocene are records from Tibet (Tao & Du 1987) of *Betula* cf. *utilis* and the somewhat similar (judging from the description) *B. mankongensis*. Unfortunately these are only leaf fossils so it is not possible to be certain about their taxonomic affinity, but it does indicate that further fossil finds in eastern Asia have the potential to yield much information.

By 3 Mya trees indistinguishable from modern *Betula nigra* were growing in southeastern USA in the area where they still grow (Stults & Axsmith 2009).

Two birch species are reported from the very species rich fossil assemblage in the Late Pliocene (~2.5 Mya) Willershausen sinkhole in northern Germany (Ferguson & Knobloch 1998), but unfortunately we have not been able to trace illustrations of the material.

Also in the Late Pliocene, the white barked *Betula* 'alba' s.l. (*B. pendula* + *B. pubescens*) and *B. nana* are known from log and leaf fossils from the far north of Greenland (Bennike & Bocher 1990) in association with *Picea*, *Larix*, *Thuja* and *Taxus* and the tundra plants *Dryas*, *Empetrum*, *Menyanthes* and *Rubus chamaemorus*. These species were therefore present in the high arctic before the onset of the Pleistocene Ice Ages about 1.8 million years ago.

Species of *Betula* sections *Lentae* and *Dahuricae* which are currently confined to East Asia, the Caucasus and eastern North America therefore had continuous distributions linking these areas, presumably occurring throughout the whole circumboreal region. On the other hand, there is no evidence that species of section *Asperae* and subgenera *Acuminatae* and *Nipponobetula* ever occurred outside eastern Asia. The species of these groups currently occur in Japan, the mountains of western China, and the Himalaya, and have no known fossil record. New finds of fossils may extend known past distributions (Grimsson *et al.* 2007, 2008).

During much of the Tertiary, the cold temperate and arctic floras we know today could not have existed in the areas they now occupy as these were too warm. It is thought that many of the species today growing in colder climate zones evolved at high altitudes on mountains further south, these being the only habitats similar to those in the present day arctic until about 3 million years ago (Abbott *et al.* 2000). As the earth cooled, these mountain-evolved, cold-tolerant organisms migrated, probably along the north-south running mountain ranges, to colonise the arctic lowlands.

Evidence for this scenario is the common occurrence of cold tolerant genera with great diversity of species on southern low latitude mountain ranges (e.g. Rocky Mountains, Appalachians, Caucasus, Himalaya, Japan, and the Heng Duan mountains of Yunnan, Sichuan and surrounding provinces in SW China), which appear to have given rise to species which have colonised the Arctic. Many of these subsequently appear to have spread to become more or less circumboreal in distribution (Hultén 1962, 1971; Hedberg 1997; Eidesen *et al.* 2007a, b; Brochmann *et al.* 2003; McAllister 2005b). There is little evidence of extensive speciation in the Arctic as there are few endemics, and most arctic species have their closest relatives in more southerly regions (Hoffmann & Röser 2009).

On the whole birches fit this picture, with species of subgenera *Nipponobetula* and *Asperae* and subgenus *Betula* sections *Dahuricae* and *Costatae* occurring only in the lower latitude mountain areas. The more northerly white barked birches of section *Betula* and dwarf birches of section *Apterocaryon* have no obvious ancestral species in these mountain ranges unless *Betula costata* and *B. cordifolia* (ancestral to section *Betula*?) and *B. occidentalis* of the Rocky Mountains and Himalayan *B. ashburneri* (ancestral to section *Apterocaryon*?) might represent modern descendents of these ancestral species.

It is interesting that in eastern North America the apparently more cold tolerant diploid *Betula cordifolia* occurs further south than the presumably closely related hexaploid *B. papyrifera* (Furlow 1997). This, combined with its similarity to the 10 Mya Icelandic fossils (mentioned above), suggests that *B. cordifolia* might be close to the ancestral line of the northern white barked birches of section *Betula*. Low latitude populations of *B. pendula* subsp. *szechuanica* of section *Betula* in the Sino-Himalayan region seem to be the result of recent southward migration rather than relics (see Chapter 7).

RECENT FOSSIL HISTORY

QUATERNARY 1.8 Mya–present

The cold and dry climates of the Ice Ages, with their rapid climatic swings to interglacials and back to glacials, eliminated large numbers of species in northern areas. Survival of many species was only possible in low latitude mountains where they could migrate short distances up or down to track climatic change (Willis & Niklas 2004), or where north-south orientated mountain ranges (as in east Asia and east and western North America) allowed easy north-south migration (Pautasso 2009).

Quaternary fossil records from glacial and interglacial deposits are usually referred to modern species and are often within their present ranges although, interestingly, tree birches have been shown to have occurred further north than they currently grow in Siberia (Blyakharchuk *et al.* 2004).

Much of the work on Quaternary fossil birch concerns the possibility or otherwise of being able to distinguish between the pollen of *Betula nana* and that of the tree birches *B. pubescens* and *B. pendula*. Clearly, if only *B. nana* is present this indicates an arctic tundra vegetation, whereas the presence of one of the tree birches indicates some sort of forest or woodland, a very important distinction when interpreting ancient environments. Attempts have been made to distinguish the pollen by size (Huntley & Birks 1983), and Freund *et al.* (2001) distinguish between *B. nana* and *B. humilis* on nutlet characteristics. Where leaf macrofossils are present the problem is easily resolved as the small, orbicular, crenate toothed leaves of *B. nana* are immediately recognisable (Figs 269, 271) and quite distinct from the larger, ovate to triangular leaves of the tree birches.

The pattern of recolonisation following the most recent deglaciation (c. 15,000 years ago to present) has been studied in two ways: the fossil, mainly pollen, record (Huntley & Birks 1983), and the molecular signature in present day populations. There is now evidence for small refugial populations surviving the last ice age well to the north of populations with a continuous fossil pollen record, and these are likely to have been involved in recolonisation following deglaciation (Stewart & Lister 2001; Fedorov & Stenseth 2002; Jaramillo-Correa *et al.* 2004; Brubaker *et al.* 2005; Provan & Bennett 2008; Stewart & Dalén 2008; Binney *et al.* 2008; Svenning *et al.* 2008; Teacher *et al.* 2009; Stewart *et al.* 2010).

In Europe the fossil record indicates the presence of birch not far from the ice fronts during glacial periods and, in interglacials, rapid recolonisation of deglaciated areas. Most of this evidence is from the pollen record and doesn't distinguish between the tundra *Betula nana* and the tree birches, but macrofossil evidence has recently confirmed the presence of tree birches in Hungary during the last Full Glacial about 18,000 years ago (Willis *et al.* 2000), probably the coldest period during the whole of the recent ice ages, and there is evidence that small populations might have existed even as far north as western Norway (Stewart & Dalén 2008). Molecular evidence (Palmé *et al.* 2003) suggests that recolonisation by *B. pendula* in Europe occurred rapidly in the northwest and southeast from populations surviving near the maximum ice front, not from further south as in many other tree species and many animals (Hewitt 2000). Interestingly, the molecular evidence also suggests that the isolated southern populations of *B. pendula* such as those in North Africa, Spain (Carrion 2002) and Mt Etna in Sicily did not contribute to the recolonisation and have remained isolated since at least the last (Eemian) interglacial, only migrating short distances up or down mountains in response to warming or cooling (Palmé *et al.* 2003). Whether these isolated southern populations should be recognised as distinct taxa (species or subspecies), or lumped within *B. pendula*, depends on their degree of distinctness — and here we have not given them taxonomic status (see p. 277). The Mt Etna population is sometimes distinguished as *B. aetnensis*, and the Iberian and North African populations as *B. fontqueri* (Lascoux *et al.* 2003). There is often more genetic diversity among such small southern populations than in the huge continuous populations to the north, which are often derived from only the most northerly "periglacial" populations (Hewitt 2000).

In *Betula* it is probably only the white barked birches of section *Betula* and dwarf arctic birches whose distributions have massively contracted and expanded with the growth and melting of the ice sheets during the last 1.8 million years. Birches are the trees which can survive closest to the ice sheets, and they were therefore in a favourable position to colonise land laid bare by the melting of the ice (Willis *et al.* 2000; Palmé *et al.* 2003; Willis & Niklas 2004). Towards the end of the most recent Ice Age there is evidence of a very rapid temperature rise of up to 10°C in ten years (Post 2003; Bard *et al.* 2010), so providing an opportunity for species with numerous, small, wind dispersed seeds and a short generation time like species of *Betula* and *Salix*, to colonise rapidly.

During interglacials mixed deciduous forest developed in warmer regions with the zone of birch in the cooler regions between this forest and the tundra, as at present. Towards the end of an interglacial, as the climate cooled, there was often another period of birch dominance over large areas before the development of new ice sheets. The evidence for this has been obtained largely from pollen studies of the main glacial and interglacial periods and the many more minor warm and cool periods within these (Willis & Niklas 2004).

REFUGIA FOR BETULA SPECIES

Japan and China: two special refugia

Having discussed what is known of the fossil history of birches together with their present day distributions, we will now look at three of the most significant refugial areas in eastern Asia and consider their significance: (a) the northern Beringian refugium; (b) Japan and adjacent northeast Asia, and (c) the southwest China-Himalayan area. The other two, eastern North America and the Caucasus, have many fewer relict species, *Betula nigra*, *B. lenta*, and *B. alleghaniensis* in east North America and *B. medwediewii*, *B. megrelica* and *B. raddeana* in the Caucasus.

a) Beringia

A general consensus seems to be developing that Beringia (east Siberia and west Alaska) was a major refugium for many northern species (Manchester & Tiffney 2001; Donoghue & Smith 2004; Brubaker *et al.* 2005; Eidesen *et al.* 2007a, b), with many species having greater diversity in this area than in any other (Alsos *et al.* 2005). The fly agaric (*Amanita muscaria*) (Figs 37 and 38), the usual mycorrhizal associate of mature birch, shows this pattern of great diversity in Beringia, decreasing eastwards and westwards (Geml *et al.* 2006, 2008).

Figs 37 and **38.** *Amanita muscaria* under *Betula pendula* growing at Ness (HMcA).

The larches have a distribution pattern somewhat similar to that of *Betula pendula* agg. In both there is great diversity in north-east Asia, with clear evidence of relatively recent rapid spread westward across Siberia into Europe and of recent dispersal into the Sino-Himalayan region (Wei & Wang 2004; Semerikov *et al.* 1999). Similar patterns have been demonstrated in several animals in which there is evidence for greatest diversity in eastern Asia, and recent, Pleistocene or even Post Glacial, spread westward throughout northern Eurasia: flying squirrel, *Pteromys volans* (Oshida *et al.* 2005); field mice, *Apodemus* spp. (Michaux *et al.* 2003); meadow mouse, *Microtus* (Repenning 1990); great spotted woodpecker, *Dendrocopos major* (Zink *et al.* 2002); willow tit, *Parus montanus* (Kvist *et al.* 2001); great tit, *Parus major* (Kvist *et al.* 2003); ants (Goropashnaya *et al.* 2004), and beetles (Painter *et al.* 2007).

b) Japan and Kurile Islands

As so often for temperate trees, Japan is a major centre of diversity with eleven birch species, five of which are endemic (*B. corylifolia*, *B. globispica*, *B. chichibuensis*, *B. maximowicxiana* and *B. grossa*). While two east Asiatic mainland species, *B. costata* and *B. chinensis*, do not occur in Japan, the other five birch species which occur on the adjacent Asiatic mainland (*B. pendula* subsp. *mandshurica*, *B. schmidtii*, *B. ermanii*, *B. dahurica* and *B. fruticosa*), also occur in Japan. This means that Japan has provided a refuge for ancient species which are now extinct on the Asiatic mainland (assuming they once grew there); at the same time north-east Asian mainland species which have survived in or reached Japan have not replaced the older types, as they may have done on the Asiatic mainland.

Four of the Japanese endemics, *Betula corylifolia*, *B. maximowicziana*, *B. chichibuensis* and *B. globispica* are not only very distinct from one another but have no very close living relatives anywhere else in the world. *B. corylifolia* is the only species in its subgenus in the classification proposed here and very distinct morphologically, but is placed in subgenus *Asperae* in molecular studies (Wang *et al.* in press 2016), a position perhaps supported by the characteristic of single nutlets per bract in the fruiting catkin of *B. corylifolia*, *B. globispica* and *B. chinensis*. *B. globispica*, *B. chichibuensis* and *B. maximowicziana*, have their closest relatives in *B. fargesii*, *B. potaninii* / *B. schmidtii* and *B. alnoides* respectively, all from south-west China except for *B. schmidtii* from north-east Asia. They are therefore probably of very ancient origin, examples of species pairs found in northeast Asia and Japan with closest relatives in southwest China and the Himalaya, a common phytogeographical disjunction/geographical discontinuity, exhibited also in *Picea* (Ran *et al.* 2006) and *Spiraea* (Zang *et al.* 2006).

The Japanese endemic *Betula grossa*, is presumed to be closely related to the east North American *B. alleghaniensis* and *B. lenta*, and so represents the well-known phytogeographical disjunction between east Asia and eastern North America (Tiffney 1985; Wen 2001).

Of these five Japanese endemics, two, *Betula chichibuensis*, and to a lesser extent *B. globispica*, exist as very small populations and are endangered to rare (Oldfield *et al.* 1998). In contrast *B. maximowicziana* is sufficiently common to be a significant timber tree, although man's intervention may have something to do with this (Osumi & Sakurai 1997; Osumi 2005).

It is interesting that two other Japanese species, once considered endangered endemics, are probably synonymous with species which occur on the Asiatic mainland, *Betula tatewakiana* (=*B. fruticosa*) and *B. apoiensis* (~ =*B. gmelinii*). In Japan these occur as very small populations, two in the case of the former species and one in the case of the latter (Nagamitsu *et al.* 2004, 2006). It seems that the rugged mountainous topography of Japan, together with its moist oceanic climate at a relatively southerly latitude, has allowed small populations of species to survive more or less where they are still found, possibly since the early Tertiary.

Japan provides a strong contrast with Europe, and especially the British Isles which is of roughly the same land area as Japan. Japan and Europe probably had much the same flora before the beginning of the Ice Ages about 2.5 Mya (Ferguson & Knobloch 1998), but the present five European species of *Betula* are widespread throughout Eurasia or, in the case of *Betula celtiberica*, probably of recent introgressed hybrid origin.

c) China and North East Asia

China, including Tibet, with its vast land mass and significant relict flora (Bartholomew *et al.* 1983), has about 25 species of birch, of which nine are endemic. Six of these endemics are closely related to one another, relatively rare, and confined to the mountains of southwest China: *Betula calcicola*, *B. potaninii*, *B. bomiensis*, *B. skvortsovii*, *B. gynoterminalis* and *B. delavayi*. Two other endemics, *B. insignis*, and *B. luminifera*, are widespread in central China, leaving only *B. fargesii* as a little known, probably rare, taxonomically isolated endemic.

The phytogeography of the Chinese birches follows the well-known pattern described above with a division between a group of thirteen species found only in the mountains of the southwest: the nine endemics mentioned above plus *Betula utilis*, *B. ashburneri*, *B. alnoides* and *B. cylindrostachya*; and an almost totally different group in the north eastern region which also occur in Korea and adjacent Far Eastern Russia and Japan: *B. ermanii*, *B. dahurica*, *B. costata*, *B. schmidtii* and *B. chinensis*. The only species common to both areas are the more 'weedy' *B. pendula* and *B. utilis*, with *B. pendula* subsp. *szechuanica* and *B. utilis* subsp. *utilis* in the Sino-Himalayan region, and *B. pendula* subsp. *mandshurica* and *B. utilis* subsp. *albosinensis* in northeast China. These are thought to be more recently evolved species which have probably extended their ranges only recently, with *B. pendula* having spread from the north and only reached the eastern part of the Himalayan region, and *B. utilis* subsp. *albosinensis* having reached only the most southerly regions of northeast Asia around Beijing (Map 10).

It is thought that a belt of dry climate from northwest to southeast China during, and to a large extent since, the Miocene separated these two regions, preventing the spread of moisture-requiring species between them, and many other species groups show similar breaks in their distribution (Chen *et al.* 1999; Ran *et al.* 2006; Zang *et al.* 2006).

THE PHYLOGEOGRAPHY OF SECTION BETULA, THE WHITE BARKED BIRCHES

There are five distinct diploid species in section *Betula* and several less distinct polyploids. The diploids are 1, the western North American river birch, *Betula occidentalis*, a multistemmed, brown barked species with small leaves and very sticky, warty twigs; 2 & 3, the white barked trees *B. cordifolia* and *B. populifolia* of eastern North America; 4, the almost circumboreal *B. pendula* and its subspecies, and 5, the eastern Himalayan *B. ashburneri* (Maps 8 and 9). Unfortunately the relationships between these species are uncertain and await molecular study.

The polyploids include the widespread Eurasian tetraploids *Betula pubescens*, *B. ermanii*, and *B. utilis*, and the hexaploid North American *B. papyrifera*, which may have evolved from the diploids which are still extant, but no obvious allopolyploid combinations can be agreed to have given rise to them. The enigmatic and disjunct Central Asian birch populations have been named as species such as *B. micropylla*, *B. tianshanica*, *B. turkestanica*, *B. pamirica*, *B. kirghisorum*, *B. korshinskyi*, *B. schugnanica*, *B. procurva*, *B. saposhnikovii*, *B. talassica* and *B. jarmolenkoana* (Eastwood *et al.* 2009). All chromosome counts from such birches have been tetraploid and these species are here proposed to be of hybrid origin involving *B. fruticosa*, *B. pubescens* and *B. utilis*, probably with introgression from *B. pendula*. These tetraploids appear to lack unique characteristics not present in the three or four proposed parent species, which makes their hybrid origin plausible. These central Asian populations are probably remnants of the forest which existed during wetter periods of the Pleistocene when the northern forests of the Altai and further west were continuous with those of the Caucasus and the Pamir-Alai mountains (Pakhomov 2006).

THE PHYLOGEOGRAPHY OF THE DWARF BIRCHES OF SECTION APTEROCARYON

Dwarf birches of section *Apterocaryon* are a major component of dwarf shrub tundra heaths, growing alongside species of the relatively few other woody genera of the northern tundra; *Salix*, *Dryas*, *Empetrum*, and several ericaceous genera. These birches are a natural group of clearly related similar shrubs which appear to be confined to northern latitudes with no obvious representatives in the low latitude mountains. In this, their distribution resembles that of two other tundra dwarf shrub taxa with which the birches often occur, the species of *Rhododendron* which used to be included in *Ledum* and the crowberry, *Empetrum nigrum* agg. (Hultén 1971).

As there was probably no extensive tundra until the late Tertiary it is probable that most of the present tundra species evolved on lower latitude mountains and spread throughout the arctic during the Late Pliocene, or perhaps as late as the Pleistocene glacial/interglacial cycles (Abbott *et al.* 2000; Hedberg 1992, 1997). However, evidence has recently been found of glaciation in Greenland as early as the Eocene (Eldrett *et al.* 2007), so there must have been some tundra habitat in northern mountains throughout much of the Tertiary.

There is one Himalayan birch which does bear some resemblance to the dwarf birches, the recently discovered *Betula ashburneri*. It is diploid and has uniform small leaves, small, more or less erect, fruiting catkins with small nutlets, a somewhat shrubby habit, and persistent stipules, characters otherwise found in subgenus *Betula* only in the dwarf birches. It thus combines several characteristics of the dwarf birches with the peeling bark of sections *Costatae* and *Betula* and leaves very similar to those of *B. utilis*, while the fruiting catkins show some similarities with those of the *Acuminatae* in their persistent scales with woody bases and very thin papery apices. It is clearly a key species to study in any evolutionary work on birches as it seems to show characteristics of several sections of the genus.

The western North American *Betula occidentalis*, which has a huge latitudinal range in the Rocky Mountains from California, New Mexico and Arizona to Alaska, is perhaps likely to be closer to the ancestral line of the dwarf arctic birches. Apart from the dwarf arctic birches themselves, *B. occidentalis* is almost the only species of subgenus *Betula* which is naturally multi-stemmed and shrubby, lacks a white bark (as does *B. ashburneri*), and has almost circular, few-veined leaves. In fact the North American dwarf birch *B. glandulosa* resembles a dwarf form of *B. occidentalis* in most characteristics, differing mainly in the less acute more rounded leaf teeth, narrower seed wing, smaller seed, and its often erect fruiting catkins. Both species have very glandular, usually very finely puberulent shoots with few long hairs, broad to almost circular leaves with few veins, and fruiting catkins of similar form. Although *B. glandulosa* is always classified in the dwarf arctic birches because of its small stature, crenate leaf toothing, leaf shape, and general appearance, it differs from most other species of the group in its almost solely terminal male catkins, non-persistent stipules, and longer catkin peduncles and therefore sometimes pendent fruiting catkins. *B. glandulosa* is therefore somewhat intermediate between the other dwarf arctic birches and *B. occidentalis*.

It is plausible to suggest that the ancestor of all the dwarf arctic birches might have evolved from an ancestor of *Betula occidentalis* in northwestern North America (part of Beringia), spread westwards and differentiated in western Eurasia, evolving the characteristic persistent stipules and bud scales and the catkin arrangement of almost all the other species of the group, and subsequently differentiated into *B. nana* and *B. humilis*. This pattern of evolution in Beringia followed by spread and differentiation can be quite complicated, as exemplified by studies on other woody arctic species such as *Vaccinium uliginosum*, a very common associate of dwarf birches (Figs 21, 266, 267) (Alsos *et al.* 2005; Eidesen *et al.* 2007a), and *Cassiope tetragona* (Eidesen *et al.* 2007b).

Betula glandulosa is probably the only diploid dwarf birch in mainland North America with the exception of *B. michauxii* of Nova Scotia, Newfoundland, Labrador and adjacent Quebec. *B. michauxii* with its densely hairy twigs and catkin arrangement is very similar to the Greenland and Eurasian *B. nana*, and presumably evolved from a transatlantic colonisation event and survival in a north-eastern refugium during one or more glaciations (Jaramillo-Correa *et al.* 2004). Given the proposed recent post-glacial colonisation of Greenland (and Iceland) from Europe by *B. nana* (Fredskild 1991; Brochmann *et al.* 2003; Abbott & Comes 2003), such a suggestion seems quite plausible.

The tetraploid species, *Betula pumila* (North America) and *B. fruticosa* (Eastern Asia) are presumed to be derived from the diploids. These tetraploids are found at lower latitudes, are taller growing, and have more elongated leaves. The nature of the dwarf birches from the western half of North America needs to be further studied as many seem to be morphologically more or less indistinguishable from diploid *B. glandulosa* but are reported to be tetraploid (Dugle 1966).

Betula fruticosa resembles *B. humilis*, having mostly ovate-elliptic more elongated and more sharply toothed leaves than the North American *B. pumila*, but is distinguishable from *B. humilis* by having larger leaves, fruiting catkins and seeds, and almost solely terminal male catkins. A rather enigmatic group of tetraploid shrub birches confined to eastern Asia may intergrade with *B. fruticosa*. These have more sharply toothed leaves, probably occur in drier habitats, and include the isolated, relict *B. apoiensis* in Japan and the more or less indistinguishable *B. gmelinii* of continental eastern Asia. Molecular studies (Nagamitsu *et al.* 2006a) suggest that *B. apoiensis* has arisen through hybridisation between *B. fruticosa* and *B. ermanii*, and this could also be the origin of the shrubs named *B. gmelinii* throughout eastern Asia. Several other shrubby species are described from eastern Asia, such as *B. middendorfii*, but we suspect that these are either synonymous with one or other of the species described above as suggested by Skvortsov (Li & Skvortsov 1999), or perhaps the result of gene flow to these from *B. glandulosa* or *B. humilis*.

2. BREEDING SYSTEMS AND HYBRIDISATION

SELF-INCOMPATIBILITY, SELF-COMPATIBILITY, AND APOMIXIS

Birches are hermaphrodite, each plant bearing separate male and female catkins, very often on the same branch (Figs 28, 59, 60, 65, 67). By far the majority of species are self-incompatible, and therefore outbreeding. Outbreeding is usually considered advantageous because it maintains genetic diversity, allowing evolutionary flexibility in the long term. However, it means that at least two individuals are required for seed production and therefore seed production is likely to be low in populations of widely scattered trees, and a single individual cannot found a new population. This is especially so in wind pollinated species such as the birches in which pollen cannot be targeted as in insect pollinated species (Culley *et al.* 2002).

Self-compatibility allows inbreeding and results in the reduction of genetic diversity, but can cause the loss of harmful alleles of genes, so-called 'purging' (Byers & Waller 1999), and so leave the survivors with greater fitness for the existing environment, although with less genetic variability to be able to cope with environmental change.

Self-incompatibility is often not total. In many outbreeding species, individuals may have some degree of self-compatibility and be able to produce a little seed by self-fertilisation (Kuser 1983). The mechanism for this is partly determined genetically, but also by environmental factors. Temperature at pollination, for instance, may have a significant effect on the degree of self-compatibility (Eriksson & Jonsson 1986; OECD 2003). However, even in species which have a degree of self-compatibility and give rise to a proportion of seedlings as a result of self-pollination and fertilisation, it has been shown that only offspring produced by outcrossing may survive to maturity (Petit & Hampe 2006). This is not surprising under the intense competition which usually occurs in the wild with perhaps only one in a million seeds growing into a mature reproductive new tree.

However, when populations are reduced to very small numbers, especially if individuals are widely scattered, self-compatibility will be selected for as only self-compatible individuals will be able to reproduce (Lande & Schemske 1985). This latter situation may explain the evolution of self-compatibility in some species of *Betula* (e.g. *B. globispica*, *B. fargesii*, *B. medwediewii*), and perhaps the fact that these highly self-compatible species are mostly rare. In contrast the more 'weedy' species are usually common, very variable, and highly self-incompatible.

In the absence of experimental evidence from controlled self-pollinations, it is often unknown to what extent the commoner species are self-compatible as it is not easy to find and study the offspring of isolated trees, either in the wild or in cultivation. Much of the evidence for self-compatibility or incompatibility is anecdotal, relying on information from progenies raised from single individuals of a species clearly isolated from any other trees of the same species. The evidence is often contradictory, with some studies suggesting total self-incompatibility while others on the same species report high self-compatibility on experimental selfing (Hagman 1971). However, as the light pollen of birch is more or less ubiquitous in the atmosphere when the male catkins are shedding pollen, it is very difficult to exclude ambient pollen and there is a very high risk of contamination in experimental pollinations (Hagman 1971; Alam & Grant 1972), so that many reports of self-compatibility may be erroneous.

SELF-INCOMPATIBLE SPECIES FOR WHICH THERE IS DATA

Betula maximowicziana, a diploid, appears to be totally self-incompatible as far as can be judged from the behaviour of two trees in cultivation in the Hergest Croft Arboretum in Herefordshire, England. Seed collected from the two trees in 1975 was highly fertile and came true. However, following the death of one of the trees, a subsequent seed collection in 1976 from the remaining tree was totally infertile. This suggests not only total self-incompatibility, but also incompatibility with any of the many other species present in the arboretum.

A large quantity of seed collected from the only tree of the tetraploid *Betula pubescens* found in Cwm Idwal in Snowdonia, North Wales, UK, was wholly infertile, indicating total self-incompatibility.

Evidence for *Betula pendula* is contradictory, with some studies reporting self-incompatibility (Hagman 1971; OECD 2003) while others report apparently high self-compatibility without comment (Wang *et al.* 1999).

Isolated plants at Ness of the diploid *Betula calcicola* and *B. potaninii*, hexaploid *B. delavayi*, and decaploid *B. insignis* produced no viable seed, whereas groups of more than one clone of these species produce good quantities of viable seed which comes true to the parent species.

SELF-COMPATIBLE SPECIES

The single tree of the Wilson introduction of the diploid *Betula corylifolia* (*W* 7651) in the Royal Botanic Gardens, Edinburgh, and also of a more recent introduction from Japan, produced some viable seed which came true but died after six years. There is the possibility that these seedlings may have suffered from inbreeding depression as has been demonstrated in species which exhibit a low degree of self-compatibility (Kuser 1983). Isolated single trees of decaploid *B. medwediewii*, *B. globispica*, *B. fargesii* and dodecaploid *B. megrelica* produce a very high proportion of viable seed which, with the exception of *B. megrelica*, give seedlings true to the parent species. *B. megrelica* produces a proportion of seedlings which appear hybrid with *B. medwediewii*.

An isolated tree of the hexaploid *Betula raddeana* at Ness yielded quantities of viable seed, some of which seem to be growing true to the parent, but with some seedlings clearly the result of hybridisation with another species.

The situation in many other species is uncertain as progeny of isolated individuals have not been available.

The fact that one individual of a species is self-incompatible or self-compatible does not mean that all individuals of that species will behave in the same way. However, it is fairly safe to say that in *Betula* it does seem that almost all the common species, and also some of the rare ones, are at least largely self-incompatible, whereas self-compatibility is only found in rare, and mostly highly polyploid species.

APOMIXIS

Although there have been reports of apomixis in *Alnus* and *Betula* (Woodworth 1929; Perala & Alm 1990; Santamour 1999), the lack of seed production in isolated trees of self-incompatible species of both alder and birch in the presence of many other species of these genera leads us to doubt such reports. Contamination by ambient pollen is perhaps a more plausible explanation than apomixis for unexpected seed set in the apparent absence of pollination, as apomixis has not otherwise been suggested in any genera in the Betulaceae.

HYBRIDS AND HYBRIDISATION IN BETULA

HYBRIDS IN CULTIVATION

Most birch seedlings grown from seed of cultivated trees of white barked birches will turn out to be hybrids. One reason for this is the tendency in many gardens to grow only a single clone of each species, and the high degree of self-incompatibility of most of the commonly grown species.

Even controlled pollinations, no matter how carefully performed and protected, usually result in some aberrant seedlings because of contamination with the local pollen rain (Hagman 1971; Alam & Grant 1972; Nokes 1979). Only very carefully controlled pollinations can be relied on, especially those carried out with grafted plants in a heated greenhouse before the flowering time of birches in the open in the same area (Brown & Al-Dawoody 1977; Nokes 1979; KBA's personal experience).

Clausen (1966, 1970, summarised in OECD 2003) made extensive crosses among twelve species of birch and recorded the percentage germination of the resultant seed. He records *Betula nigra* as crossing with ten species. In contrast, in a very thorough review of the evidence, Santamour (1999) questions the ability of *B. nigra* to hybridise at all with any other species. There are therefore very different conclusions arrived at from these two studies, but the high chance of contamination, and the fact that Clausen did not grow on seedlings to check their hybrid identity morphologically, may throw doubt on his results.

Experience in cultivation suggests that the subspecies of *Betula pendula* (Jansson 1962) and *B. utilis* interbreed freely within each species. These hybrids are discussed in Chapter 7. Several rare species have recently been introduced to cultivation; as they become more widely grown, new hybrids are likely to appear in the near future, and suspected hybrids of *B. globispica* and *B. medwediewii* are now being studied.

HYBRIDS IN THE WILD

Some birch species are known to hybridise freely in the wild, and this is certainly the case with the white barked birches (Johnsson 1945, 1949; Jansson 1962; Dugle 1966; Santamour & Lundgren 1996; Clausen 1970; OECD 2003). To some extent easy hybridisation among very similar species suggests that these species may be too narrowly defined. However, there is also no question that quite distantly related species can cross to produce fertile hybrids (Dugle 1966; Barnes *et al.* 1974). This is particularly true of polyploids. The North American paper bark birch, *Betula papyrifera* (subgenus *Betula*) can cross with the yellow birch, *B. alleghaniensis* (subgenus *Asperae*); the yellow birch and cherry birch, *B. lenta*, can cross with the dwarf *B. pumila* (subgenus *Betula* section *Apterocaryon*). It was even more surprising to find, in 2009 in the Arnold Arboretum, a very healthy hybrid between the presumably very distantly related *B. maximowicziana* (subgenus *Acuminatae*) and presumably *B. ermanii* (subgenus *Betula*) grown from seed collected in the wild in Japan. With the exception of the last example and *Betula lenta* × *B. pumila*, these hybrids have been reported from the wild, not just from plants in cultivation. That the hybrids between these distantly related species are often at least partially fertile indicates that exchange of genes between very dissimilar plants is possible in birch and could have contributed to the variability of at least some of the species in existence today, as is suggested by much recent molecular work (Palmé *et al.* 2004; Schenk *et al.* 2008).

In birches, the ability of species to interbreed and the fertility or otherwise of the hybrids is therefore unlikely to be a good indicator of the degree of relationship between species. This means that classic genome analysis, in which study of chromosome pairing in hybrids is used to deduce the parentage of polyploids, is unlikely to yield conclusive results in birch. Indeed, the validity of the concept of genome analysis has been questioned (Seberg & Petersen 1998), although, at least with ferns, the results do seem to be valid and to have advanced our understanding of their evolution (Gibby & Walker 1972).

In contrast to the ease with which the species of the sections *Betula* and *Apterocaryon* hybridise, few hybrids have been reported between species of other sections. This may be due partly to lack of observations in more remote parts of the world and the rarity of many of the species in cultivation. However, *Betula corylifolia*, *B. maximowicziana*, *B. medwediewii*, *B. grossa*, *B. chichibuensis*, *B. potaninii*, *B. calcicola*, *B. delavayi*, *B. globispica*, *B. fargesii*, *B. insignis* and *B. humilis* all seem to come true from seed even when grown within a single garden (e.g. Ness Gardens) in the presence of many other birch species (in the case of the self-incompatible species several clones were being grown close to one another). This is not to say that they cannot hybridise, just that they do not appear to do so readily even when grown within the confines of a 25 hectare garden containing many birch species.

WILD HYBRIDS BETWEEN SPECIES WITH THE SAME CHROMOSOME NUMBER

Hybridisation occurs most readily between species with the same chromosome number. Such hybrids are the most easily recognised (as long as the parents are not too similar!) as, having the same number of chromosomes from each parent, they are likely to be intermediate in most characteristics.

The following section discusses wild hybrids between birch species, beginning with species with the same chromosome number and followed by hybrids between species with different chromosome numbers.

DIPLOID × DIPLOID (2n=28 × 2n=28)

Betula pendula × B. populifolia
European *Betula pendula* is commonly cultivated in eastern North America within the distribution area of *B. populifolia* and hybrids are common where they grow in close proximity (Catling & Spicer 1988). The long tapered leaf tips, black lenticels on twigs, small tight persistent fruiting catkins, hairier adaxial surface of the fruiting catkin scales, and usually singly borne male catkins, distinguish pure *B. populifolia* from likely hybrids, but complete introgression seems to be taking place where the two species grow together.

Betula pendula × B. nana (B. ×plettkei)
In Europe the hybrid between *Betula pendula* subsp. *pendula* and *B. nana* is occasionally reported (OECD 2003; Stace 1975; Mäkelä 1998) although the two species normally occur in such different habitats that natural hybridisation is likely to be infrequent.

Betula pendula subsp. *mandshurica × B. glandulosa (B. ×dugleana)*
Although the west North American distribution of *Betula pendula* subsp. *mandshurica* appears to lie almost wholly within that of *B. glandulosa* (Furlow 1997), it was not until 1976 that Lepage (1976) described the hybrid as *B. ×dugleana*.

Hultén (1944, 1968) suggested that much of the shrub birch of Alaska, previously named *Betula occidentalis*, is in fact this hybrid, commenting that it occurs where the two proposed parental species are found. While stating that the hybrid is common, Furlow (1997), in the *Flora of North America*, maps *B. occidentalis* as occurring throughout much of Alaska. Dugle (1966) comments on Hultén's observations but, having examined some of his specimens of the suggested hybrid, believed them to be *B. occidentalis* or *B. ×eastwoodiae* (*B. occidentalis* × *B. glandulosa*). As all three species are diploid, hybridisation could produce a mixture of all possible combinations and backcrosses and the identity of the Alaskan populations clearly requires further study.

Betula populifolia × *B. cordifolia* (*B. ×caerulea*)

In eastern North America the morphologically very distinct diploid species *Betula populifolia* and *B. cordifolia* have hybridised and produced such distinctive populations of hybrid origin that these were described as two separate species, *B. caerulea* and *B. caerulea-grandis*. The origin of these 'species' has been demonstrated by their re-creation and study (Brittain & Grant 1967a, 1972; Grant & Thompson 1975; Guerrimo *et al.* 1970; DeHond & Campbell 1989). It is thought that much of the hybridisation followed felling of the forests and subsequent colonisation by both birch species which had been previously separated in different habitats: *B. cordifolia* on more fertile, moister soils and *B. populifolia* on dry gravelly and sandy soils. The name *B. ×caerulea* is now usually used for the stabilised populations of hybrid origin because it is a recognisable entity and appears to be behaving as a species in that the populations reproduce by seed.

Betula cordifolia × *B. glandulosa* (*B. ×minor*)

The tree species *Betula cordifolia* usually grows in woodland habitats, but is also often found above the tree-line in the very exposed mountain habitat of the shrubby tundra species *B. glandulosa* (Figs 280, 281). Where these species grow within pollinating distance, hybrids are so distinct and sufficiently frequent that they were initially treated as a distinct species, *B. minor* (Furlow 1997). That this was the origin of *B. minor* was deduced when seed of *B. glandulosa* collected in Newfoundland by KBA gave rise to seedlings of both that species and *B. ×minor*. The plants of *B. ×minor* were diploid and clearly intermediate in appearance between *B. glandulosa* and *B. cordifolia*, the only diploid species in the area. Seed collected from *B. ×minor* near Goose Bay in Labrador gave rise to a uniform progeny, suggesting that it was behaving as a species (cf. *B. ×caerulea*). However, Furlow (1997) suggests that herbarium specimens labelled *B. ×minor* might also include *B. glandulosa* × *B. papyrifera*.

Betula occidentalis × *B. glandulosa* (*B. ×eastwoodiae*)

Dugle (1966) reports that *Betula occidentalis* and *B. glandulosa* sometimes grow close together in parts of western North America and that intermediate individuals are hybrids. These hybrids backcross equally with both parents. However, the parents appear to remain distinct, probably because they are adapted to rather different habitats; *B. occidentalis* to tall streamside thickets and *B. glandulosa* to drier, stony tundra.

Tetraploid × Tetraploid (2n=56 × 2n=56)

Betula pubescens × *B. ermanii* (*B. ×avatschensis*)

Young trees grown from seed collected by Dr Molozhnikov from the Barguzin Ridge on the eastern shore of Lake Baikal showed a mixture of characteristics of *Betula pubescens* and *B. ermanii* var. *lanata*. It therefore seems that the Lake Baikal region probably marks a relatively sharp transition between *B. pubescens* to the west and *B. ermanii* to the east.

In cultivation in European botanic gardens, seed from trees of *Betula ermanii*, probably from isolated single trees, usually gives rise to hybrids with *B. pubescens*. Seed from a cultivated tree in

Tallinn botanic garden, Estonia, originally from Mt Chekhov in Sakhalin Island, gave rise to one tree resembling *B. ermanii* and about 19 probable hybrids with *B. pubescens*.

Betula pubescens × B. pumila

These species are not normally considered to be sympatric but, as described below, we believe *Betula pubescens* occurs in at least a few isolated localities in north-eastern North America

A seed sample from the population of shrub birches on limestone at Daniel's Harbour, Newfoundland, tentatively identified as *Betula 'borealis'* [an invalid name (Furlow 1997)], gave rise to variable progeny. All were tetraploid, the ploidy level of *B. pumila* which is common in bogs in the area, but clearly differed from *B. pumila* in leaf shape, habit and time to maturity. We regard *'borealis'* as *B. pubescens*, and those which somewhat resemble *B. pumila* in leaf shape and toothing and in their densely hairy shoots as possible hybrids between *B. pubescens* and *B. pumila*. Similar populations are reported from a restricted area in adjacent mainland Canada (Hultén 1958).

Betula ermanii × B. fruticosa (B. gmelinii and B. apoiensis)

Betula apoiensis is a small shrubby birch confined to exposed ultrabasic rocky ridges around the summit of Mt Apoi in Hokkaido, Japan. In recent papers Nagamitsu *et al.* (2006a, b) deduced, largely from molecular evidence, that *B. apoiensis* has evolved by hybridisation between *B. fruticosa* and *B. ermanii* or from *B. fruticosa* through introgression from *B. ermanii*. They have shown that *B. apoiensis* lacks

Fig. 39. *Betula gmelinii* ('*apoiensis*'), Japan, Hokkaido, Mt Apoi. Fruiting and old male catkins, and dead twig with male catkins in winter condition as result of winter dieback. White resinous glands are clearly visible (PW).

unique characteristics not found in either putative parent and its chloroplast haplotypes appear to be derived from those of *B. ermanii* and *B. fruticosa*. However, it is difficult to explain how a bog species, *B. fruticosa*, came to grow on an exposed serpentine mountain summit before becoming subject to hybridisation by *B. ermanii*.

Typical *Betula fruticosa* of eastern Asia is very different from the shrub of Mt Apoi and occurs in two sites in Hokkaido (Nagamitsu *et al.* 2004). It is typically a species of bogs with upright flexible stems and obovate, crenately toothed leaves (Figs 292, 293). In contrast, *B. apoiensis* grows in a dry exposed habitat, has very rigid, spreading branches and triangular–rhombic, sharply toothed leaves. *B. apoiensis* therefore resembles a dwarf form of *B. ermanii*, but its very white-warty twigs and leaf shape are distinctive (Figs 39, 296, 297). It seems closest to material from eastern continental Asia named *B. gmelinii* as described by Li & Skvortsov (1999). *B. gmelinii* may have a similar hybrid origin to *B. apoiensis*, but clearly the tetraploid, shrubby east Asian taxa (including specimens named *B. middendorfii* and *B. divaricata*) require further study.

HEXAPLOID × HEXAPLOID

Betula alleghaniensis × *B. papyrifera*

Wild *Betula alleghaniensis* × *B. papyrifera* hybrids were studied by Barnes *et al.* (1974) who suggest that hybridisation is likely where the species occur together, despite the fact that they are presumed to be only distantly related. Hybridisation is reported to be most frequent in disturbed habitats where large numbers of seedlings can become established, and the hybrids are distinguishable morphologically (Johnsson 1974). As the hybrids show partial fertility (Solomon & Kenlan 1982), introgression is possible. That there is no extensive introgression between the two species suggests that hybrid trees are relatively rare and no better adapted to any habitat than either parent, lacking in vigour, or have limited fertility.

WILD HYBRIDS BETWEEN SPECIES WITH DIFFERENT CHROMOSOME NUMBERS

Introduction

As well as hybridising with species of the same ploidy level, there is no question that many birch species are able to hybridise almost as freely with species of different ploidy levels (Dugle 1966; Grant & Thompson 1975). Such interploidy-level hybrids may often be infertile or of greatly reduced fertility, especially if of uneven ploidy level as in triploids (Stace 1975; Brown & Williams 1984). However, there are many reports of interploidy-level hybrids being fertile and able to backcross to the parent species (Brittain & Grant 1969, 1968b; Grant & Thompson 1975; Thórsson *et al.* 2001).

Within a diploid species, fertilisation involving an unreduced gamete results in plants with the triploid number of chromosomes (2n=42) as is recorded in *Betula pendula* (Johnsson 1944). However, when such unreduced gametes (2n=28) from a diploid fuse with a normal reduced gamete from a tetraploid species (n=28), it results in a hybrid with the normal tetraploid chromosome (2n=56) number. This is reported for *B. pendula* × *B. pubescens* by Nokes (1979). It is interesting that in this case, and in the *B. calcicola* (2n=28) × *B. delavayi* (2n=84) hybrid described below, it was an unreduced female gamete of the diploid species which was involved in the origin of the hybrid. Unreduced pollen is often produced by plants in varying proportions and has often been assumed to function in fertilisation (Bretagnolle & Thompson 1995), but unreduced female gametes may be more likely to be successful in giving rise to viable embryos.

The production and functioning of unreduced gametes is, of course, much rarer than the functioning of normal reduced gametes. Where crossable diploid (2x) and tetraploid (4x) species grow together, triploid hybrids will be much more common than tetraploid hybrids produced from unreduced gametes. This is illustrated by Nokes's work with controlled crosses between *Betula pendula* and *B. pubescens* from which 70 triploids and 4 tetraploids were produced. There was no question as to the hybrid origin of the tetraploids as they were derived from seed from a diploid and were clearly hybrid in their appearance (Nokes 1979).

The potential for gene flow from diploid to tetraploid, either directly by unreduced gametes from a diploid or through triploid intermediates producing diploid or triploid gametes, probably exists wherever related diploid and tetraploid birches grow together (Dugle 1966). Birches appear to be even more flexible in the exchange of genetic material in that it is also possible for gene exchange to occur in the other direction, from a triploid to a diploid, presumably through haploid gametes produced by the triploid hybrid (Anamthawat-Jonsson & Tomasson 1990).

DIPLOID × TETRAPLOID (2N=28 × 2N=56)

B. pendula (2n=28) × *B. pubescens* (2n=56) — *B.* ×*aurata*

B. nana (2n=28) × *B. pubescens* (2n=56) — *B.* ×*intermedia* (syn. *B.* ×*alpestris* Fr.)

B. humilis (2n=28) × *B. pubescens* (2n=56)

B. pendula subsp. *mandshurica* (syn. *B. neoalaskana*, *B. resinifera*) (2n=28) × *B. pumila* (2n=56) — *B.* ×*uliginosa*

B. glandulosa (2n=28) × *B. pumila* (2n=56) — *B.* ×*sargentii*

B. lenta (2n=28) × *B. pumila* (2n=56) — *B.* ×*jackii*

B. costata (2n=28) × *B. ermanii* (2n=56) and *B. utilis* (2n=56)

B. maximowicziana (2n=28) × *B. ermanii* (2n=56) — *B.* ×*dosmannii*

Betula pendula (2n=28) × *B. pubescens* (2n=56) — *B.* ×*aurata*

The occurrence of hybrids between *Betula pendula* and *B. pubescens* has often been suggested on morphological grounds (Walters 1975). In fact, Linnaeus (1753) did not distinguish between the two, treating them as a single species, *B. alba* L., probably because, even in those days, there had already been so much hybridisation between them that the variation seemed to be more or less continuous. Recent molecular work also suggests that extensive hybridisation occurs between them.

In any one area, plants of these species, (and surprisingly also *Betula nana*), show more similarity among their chloroplast haplotypes than is seen in plants of the same species in different areas (Palmé 2003; Palmé *et al.* 2004; Maliouchenko *et al.* 2007). That there should be more similarity among the chloroplast genotypes of the three species within an area than within individuals of a species from different areas was interpreted to indicate considerable hybridisation. This was perhaps a surprising conclusion considering the different ploidy levels and how distinct in appearance are typical *B. nana* and *B. pendula*, both from one another and from typical *B. pubescens*. A possible alternative explanation is very strong selection pressure for certain chloroplast genotypes within each area (Petit *et al.* 2004; Muir & Schlötterer 2006).

The 'normal' hybrids between diploid *Betula pendula* and tetraploid *B. pubescens* are triploid (2n=42) and have been reported to be sterile (Nokes 1979) or sometimes sterile and sometimes fertile (Brown & Williams 1984). On the other hand, the evidence from the variation seen, especially in *B. pubescens*, suggests extensive introgression (Atkinson 1992). However, in many wild populations where *B. pendula* and *B. pubescens* grow together, it is usually possible to pick out individuals which clearly belong to the diploid *B. pendula*, and also to find a wide range of intermediates intergrading without any discontinuities into 'pure' *B. pubescens*. This suggests that gene flow in many populations may be largely in the one direction, from diploid to tetraploid. Thus, it would seem that the diploid *B. pendula* can contribute genes to the gene pool of the tetraploid *B. pubescens* through the production of unreduced gametes by the diploids and diploid and triploid gametes from triploid hybrids. If this is all that is happening, the diploid species remain genetically 'pure' while the tetraploids may show various degrees of introgression (contamination) from the diploid species. In terms of appearance the population is likely to consist of the pure diploid species, primary tetraploid hybrids (which have resulted from unreduced gametes from the diploid) which are intermediate between the parents, having half of their chromosomes from each parent, and all intermediates between this 'half and half' hybrid and the 'pure' form of the tetraploid *B. pubescens* due to free interfertility between the primary hybrid and *B. pubescens*. It is interesting to note that there would be expected to be no intermediates between the pure diploid species and the 'half and half' hybrid (Williams 1981; Brown & Williams 1984), unless the triploids are also producing viable haploid gametes which can fertilise the diploids, as has been shown with *B. nana* in Iceland (Thórsson *et al.* 2001). The variation observed in mixed populations of *Betula pendula* and *B. pubescens* suggests that the explanation proposed in this paragraph is reasonable.

Fig. 40. *Betula pubescens* var. *pumila* behind figures and *B. ×intermedia* (*B. nana* × *B. pubescens*) in front of two figures in dwarf shrub heath of *Vaccinium uliginosum*, *B. nana*, and *Empetrum* species, Thorskafjordur, N of Inarsstadir, west Iceland (HMcA).

What the work of Nokes (1979) did show for the first time was that tetraploid offspring can arise directly from *Betula pendula* × *B. pubescens* crosses, at least with *B. pendula* as the female parent. Given all the interest and speculation there has been about the relationships between these two species and their hybrids (Walters 1975; Brown & Williams 1984) it is perhaps surprising that more definitive work has not been done. However, there are great difficulties in discounting contamination in controlled pollinations (Hagman 1971; Alam & Grant 1972) and obtaining reliable chromosome counts (Brown & Williams 1984). Also, such work is of a long term nature as trees can take 3–10 years to become fertile so that their certain identity can be checked and their fertility tested. In fact, the very careful, precise and detailed work of Nokes (1979) reported here is unpublished except in his PhD thesis, and we are very grateful to Alan Codling of Staffordshire University for making us aware of the work of his research student.

Although there is extensive morphological and molecular evidence for gene flow between *Betula pendula* and *B. pubescens*, both triploid and tetraploid hybrids are reported to produce very little, if any, viable pollen or seed. Many odd results have been obtained with offspring having unexpected chromosome numbers so that double reduction at meiosis or apomixis have been proposed as explanations, but rarely have these odd seedlings been grown on for morphological study, and pollen contamination or errors in chromosome counting are more probable explanations (Hagman 1971).

Betula nana (2n=28) × *B. pubescens* (2n=56) — *B.* ×*intermedia* (*B.* ×*alpestris*)

Where the diploid *Betula nana* and tetraploid *B. pubescens* grow together, *B. nana* usually remains distinct but some of the *B. pubescens* shows characteristics of *B. nana* (Elkington 1968; Kenworthy *et al.* 1972). However, in Iceland many populations of shrubby birches appear to be of hybrid origin and triploid hybrids have been shown to be relatively fertile and appear to be enabling two-way gene flow, presumably by producing functioning gametes with both haploid and diploid chromosome numbers (Anamthawat-Jonsson & Tomasson 1990; Thórsson *et al.* 2001).

Betula humilis (2n=28) × *B. pubescens* (2n=56) — ×*B. sukazewii*

This hybrid is reported from the wild (Walters 1975) and seed of *Betula humilis* collected by KBA near Slesin in Poland gave rise to many seedlings of *B. humilis* and one of this hybrid.

Betula pendula subsp. *mandshurica* (2n=28) × *B. pumila* (2n=56) — *B.* ×*uliginosa*

Dugle (1966) reports hybrids between these two species where they occur together. She notes that *Betula pendula* subsp. *mandshurica* remains distinct with no intermediates between this and the hybrids. However, all degrees of variation exist between the hybrids and *B. pumila* (as *B. glandulifera*). This is as would be expected if hybridisation involved one way gene flow from diploid to tetraploid.

Betula glandulosa (2n=28) × *B. pumila* (2n=56) — *B.* ×*sargentii*

Dugle (1966) describes the hybrids between these species as showing all intermediate conditions between the pure species and with intermediate chromosome numbers (2n=40, 42, 43, 45, 46, 48 — perhaps the most reliable report of aneuploids). She does comment that most of the variation is in the direction of the tetraploid *Betula pumila*, as would be expected if most of the gene flow were from the diploid to the tetraploid. However, the two parent species are similar morphologically, so a morphological assessment of direction of gene flow as carried out by Dugle is difficult.

Dugle (1966) reports that *Betula* ×*sargentii*, rather than either of its parents, is often the dominant birch in suitable habitats in the foothills of the Rocky Mountains and other 'fairly intermediate habitats'.

Betula lenta (2n=28) × *B. pumila* (2n=56) — *B. ×jackii*

This hybrid was reported by Woodworth (1929) to be triploid. Jack selected it from seedlings grown from a very large quantity of seed collected from a shrub of *Betula pumila* downwind from trees of *B. lenta*, the only other species in the vicinity, in 1887, and described the hybrid as intermediate between the parents in every way.

As well as a shrub of *Betula ×jackii*, the Arnold Arboretum (in 2009) also has two shrubs labelled *B. pumila*, noted as having been donated by Sargent but having subsequently been propagated by seed. From their appearance it was judged that these shrubs were *B. ×jackii*. It is therefore known that in both cases *B. pumila* was the female parent.

This hybrid is not mentioned in the *Flora of North America* (Furlow 1997) and does not seem to have been reported in the wild.

Betula costata (2n=28) × *B. ermanii* (2n=56) and *B. utilis* (2n=56)

Controlled crosses between *Betula costata* on the one hand and *B. ermanii* and *B. utilis* on the other were made by KBA at the Stone Lane Gardens around 1990. The seed produced was highly fertile and the trees raised clearly intermediate between their parents in morphology of both leaves and fruiting catkins. Interestingly, although the fruiting catkins were intermediate, most were more or less pendent as in the tetraploid parent.

Fig. 41. *Betula ×intermedia* (*B. nana* × *B. pubescens* var. *pumila* with *Vaccinium uliginosum*, *V. myrtillus*, and *Nardus stricta*, Thorskafjordur, N of Inarsstadir, west Iceland (HMcA).

B. ashburneri × *B. utilis*

In 2009, while surveying the birches in cultivation in the Younger Botanic Garden, Benmore, an annex of the Royal Botanic Garden Edinburgh, mature small trees of *Betula ashburneri* were found, grown from seed collected in Bhutan in 1984 by Ian Sinclair and David Long (*S & L* 5297). One tree of this collection more closely resembled *B. utilis* and a closer study showed that this was intermediate and likely to be this hybrid. Several seedlings of these cultivated *B. ashburneri* trees had 2n=42 and are probable hybrids with neighbouring cultivated trees of *B. utilis*.

B. maximowicziana (2n=28) × *B. ermanii* (2n=56) — *B.* ×*dosmannii* hybr. nov.

As the *Acuminatae* are only very distantly related to other birches, it was not expected that any hybrids would exist between species of the *Acuminatae* and other subgenera and none has previously been reported. It was therefore a surprise to find what was clearly a hybrid between *Betula maximowicziana* and presumably *B. ermanii* in the Arnold Arboretum in 2009. This tree bore a close resemblance to *B. maximowicziana* in its foliage but, in comparison with a nearby tree of *B. maximowicziana*, the leaves were smaller, subcordate rather than deeply cordate, and the teeth ending the main veins were shorter. In the absence of fruiting catkins it would not have been possible to state for certain that it was not an odd form of *B. maximowicziana*. However, the mostly solitary, long-elliptical fruiting catkins were green and lacked the protruding nutlet wings typical of *B. maximowicziana*, in general

Fig. 42. *Betula* ×*intermedia* (*B. nana* × *B. pubescens* var. *pumila* with *Vaccinium uliginosum*, *V. myrtillus*, and *Nardus stricta* (mat grass), Thorskafjordur, N of Inarsstadir, west Iceland (HMcA).

appearance resembling those of *B. papyrifera* or *B. utilis*. They were very different from the much longer, clustered, cylindrical catkins of *B. maximowicziana*, which had a brownish colour with nutlet wings visible among the catkin scales. As this tree was grown from seed collected in the wild in Japan, the only possible other parent, given the appearance of the fruiting catkins, was *B. ermanii*, to which the hybrid tree bore some resemblance in the general form of its leaves.

Betula ×*dosmannii* McAll. hybr. nov.

Hybrid tree intermediate between *Betula maximowicziana* and *B. ermanii*. It may be distinguished from *B. maximowicziana* by its smaller, subcordate to truncate leaves, with broader leaf teeth; fruiting catkins solitary to in groups of up to four, narrowly elliptic, with the wings of the nutlet not protruding. Type: Tree in Arnold Arboretum, Boston, Massachusetts, USA, grown from seed of *EHOK* 35: Japan, Hokkaido, Nakagawa Experimental Forest, Nakagawa town, 4–8 km NE of Saku, forest Section 112, c. 190 m, 44°45'37"N, 142°5'52"E, mixed deciduous/coniferous forest, parent rock welded tuff, growing on an exposed mountain ridge top with *Salix*, *Tilia japonica*, *Quercus mongolica* var. *grosseserrata* and *Acer mono*, with *Sasa* and *Petasites* abundant on the ground, SE exposure, fertile sandy loam, 22 Sept. 1997, *Expedition to Hokkaido 1997 EHOK* 35 (holotype LIV, leg. HMcA June 2009).

Tree with whitish bark intermediate in appearance between those of its presumed parents. *Twig* glandular, glabrous (but probably glabrescent). Leaf petiole c. 40 mm; blade ovate, acuminate, shallowly cordate to truncate, to 120 × 90 mm with 9(–11) pairs of veins, teeth terminating main veins with oblong parallel sided terminal portion about twice as long as broad, to 3 × 1.5 mm (more than twice as long as broad in *B. maximowicziana*), glandular on both surfaces, sparse long hairs on lower leaf margin and main veins on both surfaces, axillary hair tufts present, puberulent on main veins on upper (adaxial) surface. *Fruiting catkins* 1–4 in groups, green until maturity, c. 55 × 6 mm with nutlet wings hidden and lateral lobes of catkin scales spatulate and expanding distally. Figs 43, 298.

Hexaploid × Diploid

Betula alleghaniensis (2n=6x=84) × *B. lenta* (2n=2x=28)

As with hexaploid *Betula papyrifera* and diploid *B. cordifolia*, hexaploid *B. alleghaniensis* and diploid *B. lenta* differ three-fold in their chromosome numbers, but are so similar in their morphology that they are often confused. The hybrid, as might be expected, is much more similar to *B. alleghaniensis* than to *B. lenta* and not easy to distinguish from *B. alleghaniensis* (Sharik & Barnes 1971). The hybrid is reported to have limited fertility with seed giving about 10% germination despite an apparently normal meiosis. Although initially vigorous, the F1 first generation hybrids are reported to be smaller than their parents after 27 years (and therefore of no greater value for forestry than the species), and it was concluded that there would therefore be selection against the hybrids in the wild.

Betula calcicola (2n=28) × *B. delavayi* (2n=84)

This hybrid was raised from seed of *Betula calcicola* (Lancaster 1691), which produced one seedling of *B. calcicola* and one of the hybrid. The hybrid seedling grew much more rapidly than the *B. calcicola* seedling and was clearly intermediate in appearance between that species and *B. delavayi*, near which it grew in the wild on the Lijiang range in NW Yunnan in China. This seedling had a chromosome number of 2n=70, not the expected 2n=56, which would indicate that it resulted from the fusion of an unreduced female gamete (n=28) from the *B. calcicola* parent and a normal reduced gamete

Fig. 43. *Betula ×dosmannii* (*B. maximowicziana* × *B. ermanii*). Grown from seed collected in the wild in Japan: *EHOK* 35, 22 Sept. 1997, Japan, Hokkaido, Nakagawa Experimental Forest, Nakagawa town, 4 to 8 km NE of Saku (Arnold Arboretum) (HMcA).

(n=42) from the *B. delavayi* parent. It is interesting to note that, as in the case of Nokes' work with *B. pendula* and *B. pubescens*, the unreduced gamete came from the female parent. Unfortunately, the original seedling has died but a propagule survives in the garden of Maurice Foster in Sevenoaks, Kent. It is likely that hybrids between these two species may occur in the wild on the Lijiang range and perhaps also in cultivation from seed collected in the wild.

TETRAPLOID × HEXAPLOID

Betula alleghaniensis (2n=84) × *B. pumila* (2n=56) — *B. ×purpusii* (2n=70) and *B. murrayana* (2n=112)

The primary hybrid *Betula ×purpusii* is reported to be quite common where the two species occur together (Dancik & Barnes 1972, 1975) despite the distance in relationship between the parent species. It is unlikely that they share any genomes, the three in the hexaploid *B. alleghaniensis* probably being closely related to one another and derived from an early evolutionary line which has long been distinct from the two genomes present in the dwarf, shrubby *B. pumila*. Also, the two species usually occur in different habitats, *B. alleghaniensis* in forest and *B. pumila* in bogs.

The hybrid, *Betula ×purpusii*, is closer in appearance to *B. alleghaniensis* than to *B. pumila*. As *B. pumila* flowers before *B. alleghaniensis* and *B. alleghaniensis* is protogynous (in any one tree the female catkins are receptive before the pollen is shed), Dancik & Barnes suggest that *B. alleghaniensis* is likely to be the seed parent. However, a collection of *B. pumila* from St. Williams, Norfolk County, Ontario gave rise to two *B. ×purpusii* at Ness, indicating that *B. pumila* was the female parent of this particular *B. ×purpusii*.

Although there are often suggestions of introgression following hybridisation, Dancik & Barnes (1972, 1975) found no conclusive evidence in this case, believing that all the variants found could be interpreted as F₁ hybrids. This suggests that *Betula ×purpusii* is sterile. However, two seedlings from the cultivated *B. ×purpusii* (2n=70) have chromosome numbers of 2n=84 and 2n= c. 98, indicating that they are likely to have arisen from unreduced female gametes of *B. ×purpusii* fertilised by (n=14) and (n=28) male gametes from other birches (probably *B. pendula* and *B. pumila* judging by chromosome number and appearance). This parallels the proposed origin of *B. murrayana* [*B. ×purpusii* (n=70) + *B. alleghaniensis* (n=42) giving rise to *B. murrayana* with 2n=112], and suggests that the only viable gametes in birch hybrids of uneven ploidy level might be those with the unreduced chromosome number of the hybrid. This might explain why only precise ploidy levels are found in birch hybrids with there being no certain records of aneuploids, and indicates a possible route for the creation of further new allopolyploid species.

The fertile allopolyploid *Betula murrayana*, (2n=8x=112) is interpreted as a backcross to *B. alleghaniensis*, combining a full chromosome set (84) of *B. alleghaniensis* with a haploid set (28) of *B. pumila* (Barnes & Dancik 1985). It is discussed on p. 206.

HYBRIDS OF BETULA PAPYRIFERA

B. papyrifera (2n=84) × *B. cordifolia* (2n=28)
B. papyrifera (2n=84) × *B. populifolia* (2n=28)
B. papyrifera 2n=84) × *B. ×caerulea* (*B. cordifolia* × *B. populifolia*) (2n=28)
B. papyrifera (2n=84) × *B. pendula* subsp. *mandshurica* (= *B. resinifera* = *B. neoalaskana*) (2n=28)
    — *B. ×winteri*
B. papyrifera (2n=84) × *B. occidentalis* (2n=28) — *B. ×utahensis* (2n=56)

Fig. 44. Tetraploid (2n=56) hybrid birch, probably *Betula* ×*ungavensis* (*B. glandulosa* × *B. papyrifera*), from between Dawson and Whitehorse, Yukon, Canada (PW).

The majority of reported chromosome counts for *Betula papyrifera* are hexaploid (2n=84), but there are numerous records of tetraploids (2n=56) and pentaploids (2n=70) (Grant & Thompson 1975; Brittain & Grant 1967b). These can be interpreted as the morphologically more or less indistinguishable tetraploid (2n=56) hybrids with the diploids *B. cordifolia*, *B. populifolia* and *B. pendula* subsp. *mandshurica*, and backcrosses to hexaploid *B. papyrifera* resulting in hybrids with 2n=70.

Until Brittain & Grant (1965b) showed that it was a morphologically and cytologically distinct diploid species, *Betula cordifolia* had usually been treated as a variety of *B. papyrifera* or sunk within that species with no taxonomic recognition whatsoever despite the genetic, geographical distribution, and ecological differences between them. Mature trees of the two species may be difficult to distinguish by general appearance, despite the difference in chromosome number.

Hybrids between *Betula papyrifera* and the diploid species *B. populifolia*, *B. pendula* subsp. *mandshurica* and *B. occidentalis*, are also difficult to distinguish from *B. papyriifera*. In experimental crosses between *B. papyrifera* (hexaploid, 2n=84) and *B. populifolia* (diploid, 2n=28) and *B.* ×*caerulea* (*B. cordifolia* × *populifolia*, 2n=28), Alam & Grant (1972) showed that these different hybrids could not be distinguished from one another and not reliably from 'pure' *B. papyrifera*.

Dugle (1966) describes sporadic occurrences of *Betula* ×*utahensis* (*B. papyrifera* × *B. occidentalis*) in the presence of both parents. Morphologically the tetraploids are intermediate between the suggested parents with the sticky, glandular twigs, fewer leaf veins, dark bark and suckering habit of *B. occidentalis*, combined with the large size and vigour of *B. papyrifera*. A uniformly tetraploid population of this hybrid in the absence of the parent species is reported by Brittain & Grant (1966) from near Summerfield in British Columbia, Canada; and a morphologically uniform population of trees with almost black barks in the Kalispell/Flathead Lake area of Montana, USA, has been described as a distinct species, *B. montanensis* (Butler 1909). Chromosome counts from seed obtained by KBA from the dark-barked trees in this area are also tetraploid. These therefore appear to be stabilised populations of *B. papyrifera* × *B. occidentalis i.e. B. ×utahensis* behaving as a distinct species, which has arisen in a manner parallel to that of the loganberry (Crane 1940a, b).

Thus, with the possible exception of *Betula ×utahensis*, hybrids between *B. papyrifera* and these diploid species will usually be identified as *B. papyrifera*, and this is the approach usually taken, with 2n=56 and 2n=70 being treated as chromosome numbers of *B. papyrifera*, e.g. by Furlow (1997) in the *Flora of North America*. Primary tetraploid hybrids with 2n=56 appear to be fully fertile (Brittain & Grant 1966), and there is no suggestion that pentaploid backcrosses with *B. papyrifera* (2n=70) are not also fertile, although this is not known for certain.

Summary of *Betula papyrifera* and its hybrids with closely related diploid species:

A Pure *Betula papyrifera* is probably always hexaploid, with 2n=84.

B Hybridisation between *Betula papyrifera* (6x) and the diploid (2x) tree species *B. cordifolia, B. populifolia, B. pendula* subsp. *platyphylla* (= *B. resinifera*), and *B. occidentalis* results in tetraploid (4x) hybrid trees with 2n=56, which are difficult, if not impossible, to distinguish from 'pure' *B. papyrifera*. Such hybrids will usually be identified in the field as *B. papyrifera*.

C Stabilised hybrid populations with 2n=56 derived from *Betula papyrifera* × *B. occidentalis* (*B. ×utahensis*) occur in the absence of both parent species, so behaving as a species.

D Backcrosses between these tetraploid (4x) hybrids and hexaploid (6x) *Betula papyrifera* result in pentaploid (5x) trees with 2n=70 which are even more difficult to distinguish from *B. papyrifera*.

E Further hybridisation between these hybrids and their parent species would be expected to result in aneuploids with intermediate chromosome numbers, but these would be difficult to detect morphologically, and the accurate chromosome counts needed to confirm them are difficult to obtain. Again, it is likely that most of these, should they exist, would be morphologically indistinguishable from *B. papyrifera*.

F It has always been difficult to imagine how gene flow and introgression could occur between diploids and hexaploids, although on morphological grounds it seems to occur. The fertilisation of unreduced female gametes of pentaploids (n=70) by pollen from diploids (n=14) could give rise to hexaploids (2n=84) which would be likely to be interfertile with *Betula papyrifera*, providing a route for such gene flow. This would explain the *B. pendula* subsp. *mandshurica* characteristics found in the hexaploid trees found at Aklavik in northern Alaska (Brittain & Grant 1968).

Betula pumila (2n=56) × *B. papyrifera* (2n=84) — *B. ×sandbergii*, 2n=63, 56, although 2n=70 would be expected to be the most frequent number for the primary hybrid.

Dugle (1966) reports that this hybrid is fairly common where the two parent species occur together. Woodworth (1931) gives a chromosome number of 2n=63 while Dugle herself obtained one count of n=28, so clearly the *Betula 'papyrifera'* parent itself was a hybrid, presumably a tetraploid (see account above of *B. papyrifera* hybrids).

(Diploid × Tetraploid) × Hexaploid

Betula ×sargentii (2n=42) × *B. papyrifera* (2n=84) — *B. ×arbuscula* Dugle, 2n=56–70

Dugle (1966) reports this hybrid from three areas in Alberta: Pocahontas, Miette Hot Springs and Maligne Hot Springs in Jasper National Park, the large shrubs being intermediate in appearance between the parents and resembling *Betula ×sandbergii*, with chromosome numbers of 2n= c. 56, 60, 63, and 70 with associated *B. papyrifera* being reported to have 2n=70 and 84.

Betula glandulosa (2n=28) × *B. papyrifera* (2n=84) — *B. ×ungavensis* Le Page (2n=56)

Seed from the single isolated individual of *Betula glandulosa* found on the summit ridge of Mt Olivine on the Gaspé Peninsula, Quebec, Canada in 2010 gave rise to three seedlings of the hybrid with *B. papyrifera,* the only other birch species growing on this relatively low altitude mountain with somewhat toxic ultrabasic soils. This hybrid was first reported and named only in 1976 by Lepage. It is perhaps unusual to find the two parental species growing together as *B. papyrifera* is usually replaced by *B. cordifolia* at higher altitudes and latitudes (see under *B. × minor*) and so may rarely grow alongside *B. glandulosa*.

3. CULTIVATION

BUYING A BIRCH — ATTRACTIVE FEATURES

Of the forty six or so species of birch, only a few of the white barked species are at all commonly cultivated. The silver birches such as the European *Betula pendula* and eastern North American *B. populifolia* are the most common, but cultivars of the Himalayan *B. utilis* and its subsp. *jacquemontii*, the European *B. pubescens*, and the North American *B. papyrifera* and *B. nigra* are also frequently planted. The white barked Japanese *B. ermanii* and the Chinese *B. utilis* subsp. *albosinensis* are mainly encountered in arboreta and other specialised collections, along with some of the rarer species.

BARK

White barked birches are among the most commonly planted trees in cool temperate climates. The degree of whiteness varies considerably among species and between cultivars and does not develop until the stems are above a certain diameter, so that we cannot be sure how a birch will develop unless buying a named clone. Garden centres depend to a large extent on impulse buying, and therefore have an incentive to display trees of larger sizes which have at least begun to show some white bark. *Betula utilis* 'Inverleith' is particularly favoured in this situation because its bark becomes white at probably the smallest diameter of any birch. Although certainly one of the whitest-barked birches, its leaves are larger and its canopy denser and 'heavier' than that of silver birches of the *B. pendula* aggregate, which have a more delicate canopy and cast a lighter shade.

Some selections of *Betula papyrifera* are as good in their white bark, although this takes longer to develop than in most species. Selections of the north Spanish *B. celtiberica* have good white bark, are clearly more drought tolerant than more northerly provenances of other species, and are showy in spring as the catkins expand well before the leaves.

It is perhaps surprising that few cultivars of *Betula pendula* have been selected for the whiteness of their bark, although perhaps this reflects how attractive almost any *B. pendula* is and that it is cheaper to raise them from seed rather than to graft selected clones. Whiteness in the wild may not always be a guide to the degree of whiteness in cultivation but, in cultivation at Ness, a collection from Novosibirsk in Siberia has retained the striking whiteness typical of the Siberian birches and stands out in the winter sunshine (Fig. 214).

White barked birches look particularly attractive when planted in groups of twos and threes rather than being evenly spaced as if in an orchard. The trunks in each group bend away from one another and give a much more natural parkland look to the stand of trees than evenly spaced, upright, parallel trunks as in a dense forest.

The darker barked forms of *Betula utilis*, including those here called subsp. *albosinensis*, are only rarely seen outside arboreta. It is difficult to understand why these beautiful trees should be so seldom grown. Several very good clones with orange to copper coloured bark have been in cultivation since the early 1900s. These include introductions by Ernest Wilson such as his *W*. 4106, of which several clones are available (see Chapter 8). Recently several other excellent pinkish to very dark brown barked trees have been introduced from Bhutan, SE Tibet, NE Yunnan and Sichuan (Figs 174, 179). These have tremendous potential for a commercial grower prepared to give space to a demonstration tree to show their value for private gardens or amenity landscaping. While some of these might eventually be too large for a modern town garden, they are no faster growing than any white barked birch.

In addition to bark colour, some birches have particularly attractive bark forms. The white barked birches peel in different ways, the large peeling sheets often seen in *Betula ermanii* being particularly attractive, and the winter sun shining through any peeling bark catches the eye (Fig. 195). Particularly striking is the 'shaggy' bark of numerous fragments curling off the trunk and thicker branches in *B. nigra* and *B. dahurica* (Figs 158, 160). The *B. nigra* clones in cultivation in Europe such as 'Heritage' and 'Wakehurst', tend to have single straight trunks, but may be trained to have several stems.

The very rare (in both wild and cultivation) Japanese *Betula dahurica* from Nobeyama in Honshu has formed well shaped upright trees, at Dawyck in the Scottish borders, at Ness near Liverpool, and is also noted as the best *B. dahurica* in the Arnold Arboretum (Dosmann, pers. comm.) (Fig. 159). The inner surface of the bark flakes of these is an attractive red-brown rather than the duller brown common on most octoploid *B. dahurica* of continental Asian provenance.

Betula alleghaniensis has an attractive metallic sheen to its yellow-brown bark which peels off in small, thin curls (Fig. 133).

FOLIAGE

The attractive early spring foliage of *Betula pendula*, especially in the early leafing Chinese and Tibetan subspecies *szechuanica*, has already been mentioned. The most striking and unusual spring foliage is that of some *B. insignis* in which the young leaves are a deep purple (Plate 11; Figs 113, 115). The most distinctive foliage in a tree birch is found in the Japanese *B. corylifolia* (Plate 1). The underside of the leaf is whitish, but its most unusual feature is the large, broad leaf teeth which terminate the main veins, there often being few if any teeth between these as is found in all other birches. This gives a very distinctive appearance to the tree.

The densely hairy young shoots and leaves of *Betula calcicola* are also distinctive; whitish to honey-coloured hairs clothe the young shoots and leaf undersides and margins giving this species an unmistakable appearance in spring (Plate 7). The hairs persist throughout the growing season, but are less conspicuous once the dark green glossy leaves have fully expanded. The closely related *B. potaninii* and *B. bomiensis* have much less hairy shoots and neat leathery leaves which are very different from those of other birches and provide an interesting foliage texture in the garden (Figs 81, 82). The related Japanese *B. chichibuensis* is remarkably different in the garden context, the dense canopy of soft green leaves on pendent twigs giving an attractive effect (Fig. 76).

AUTUMN COLOUR

The white barked birches are not usually thought of as having good autumn colour, but their leaves often turn an attractive deep yellow which contributes to the autumn colour display in natural forests, particularly when contrasted with dark evergreens (Fig. 18) or grassy hillsides. In cultivation this feature is often less apparent, perhaps because, in the drier atmosphere of urban situations, the leaves usually soon dry and fall.

Several species of section *Lentae* are well known for their brilliant autumn colour. *Betula alleghaniensis* has golden-yellow autumn leaves and, after the maples, is one of the chief contributors to the fall colour in the forests of eastern North America. It is potentially a very large tree, but is slower growing than the white barked species and is tolerant of exposure and heavy soils. The North American *B. lenta* and the Japanese *B. grossa* and *B. globispica* also colour well, and in warmer climates, as at Kew and at Wakehurst Place, *B. insignis* develops brilliant autumn colour.

The Caucasian *Betula medwediewii* is well known for its golden yellow autumn colour, and would probably be much more widely planted were it better known, because of its slow-growing, shrubby habit (Fig. 122). In addition to its autumn colour, *B. medwediewii* is deep rooted and exposure tolerant, with among the largest and showiest male catkins in the genus, and unusual, upright, persistent fruiting catkins.

Also good in autumn colour are many of the dwarf birches of section *Apterocaryon*, and they differ from the above in having deep red and orange autumn leaves, thought to be a Tertiary relict characteristic with physiological and anti-herbivore significance (Lev-Yadun & Holopainen 2009) (Figs 270, 273). These are very rarely grown, but *Betula nana* behaves as a natural bonsai while *B. glandulosa* provides the most striking colour in gaps in the sub-arctic conifer forests in Canada, and some collections are not difficult to cultivate on any permanently moist soil. They do tend to develop rather a straggly habit unless pruned, and will be best suited to moist peaty soils in cool climates.

CATKINS

Betula papyrifera and *B. medwediewii* have particularly showy male catkins, but perhaps the most striking species in catkin is *B. insignis*, which has the longest catkins and most striking display of any species (Plate 10; Fig. 109).

The pale cream catkins of *Betula chichibuensis* are freely borne in dense clusters towards the ends of the twigs (Fig. 75), and the red stamens of *B. potaninii* (Plate 5; Figs 78, 79) need to be seen close up to be appreciated.

GENERAL CULTIVATION

The closely related species of subgenus *Betula* are all relatively similar in their requirements in cultivation. This is reinforced by the fact that *B. utilis* and *B. ermanii* are almost always grafted onto rootstocks of *B. pendula* or *B. papyrifera* so, whatever species is being grown, the root system is likely to be of one or other of these two species. All these species are shallow rooted, fast growing, exposure tolerant, light demanding and very intolerant of shade, tolerant of poor soils, do not transplant well, especially as large specimens, and tend to be relatively short lived, usually around 40 or 50 years.

Taking these points in turn, the shallow rooting habit means that they are suitable for shallow soils as long as the climate is not too dry. It also means that disturbance (e.g. digging, excavation) around the base of an established birch can do severe damage to its roots and may seriously stress the tree, cause it to become unstable, or cause injury leading to infection by parasitic fungi. The growth of other, especially shallow rooted herbaceous, plants under birch is inhibited, certainly through competition for water and nutrients, but possibly also through the secretion by the birch roots of substances which inhibit growth. The shallow rooting habit makes birches unsuitable for most street planting where tarmac and paving slabs would be lifted by surface roots. Only where there is a wide grass verge should birches be used to line streets, and even then the large surface roots can be damaged by mowers, potentially leading to fungal infection of the wounds as well as costly damage to machinery.

Their fast growth rate, tolerance of exposure and poor soil, together with their relatively light, open canopy of foliage, enables the white barked birches to be used to produce a rapid effect in any new planting, whether building development, reclamation site or private garden, yet allows for slower growing, deeper rooted species (e.g. species of *Quercus*, *Acer*, etc.) to grow among them and eventually replace them, as would happen in the wild.

It is their need for strong light which results in the mostly thin, open canopy of leaves. Shaded twigs and branches in the interior of the crown abort and are shed as their leaves cannot make more 'food' by photosynthesis than they use in respiration. This is why so many thin, dead twigs are found under birch trees. Intolerance of shade makes the birches of subgenus *Betula* sections *Betula* and *Dahuricae* unsuitable for planting in any low light situation, whether between tall buildings or in the shade of other trees.

Betula nigra is very different ecologically from the other species, coming from the warm temperate south-east USA where the summers can be very warm. It is a riverside species, but deep rooted, grows well in a wide range of conditions on deep soils, and is easy to transplant, even at large sizes. It has been shown to be more tolerant of waterlogging than *B. pendula* (Tripepi & Mitchell 1984).

CHOICE OF SPECIES BY CLIMATE

Widespread species, such as *Betula pendula*, *B. pubescens*, *B. papyrifera* and *B. utilis*, behave very differently in cultivation depending on their provenance, which may determine where they will grow well. The chief problems occur with (a) continental provenances grown in oceanic climates which break bud too early, so that the young shoots are damaged by spring frosts; and (b) southern provenances grown in more northerly latitudes which continue growing too late into the autumn, so that the soft tips of the shoots are immature and die back in winter, largely due to desiccation (OECD 2003; Barclay & Crawford 1982).

Sub-tropical and warm-temperate species

Species from warm climates such as Himalayan and Chinese *Betula alnoides*, *B. cylindrostachya*, *B. luminifera*, many *B. chinensis* and *B. nigra* collections, and Vietnamese *B. insignis*, require a long frost-free growing season with adequate summer warmth and rainfall evenly distributed throughout the growing season, but they are heat tolerant and apparently fairly drought tolerant on deep soils. The young shoots of these species are not frost hardy, so frosts in late spring after the buds have broken, or perhaps even before the buds have broken but after they have begun to open and the sap has begun to rise, can kill even quite large trees. The species of section *Acuminatae* seem to be unable to form new buds in the lower part of the trunk and regenerate (Del Tredici 2001).

However, it is the early end to the growing season, when the first frosts occur, which makes it impossible to grow these species successfully in climates with a short growing season. Winter desiccation is the main cause of dieback because the twigs and buds often look alive in autumn but are clearly dried up by the time of bud break in spring. The buds seem to be more susceptible than the twigs, as quite often a twig may be green under the bark but all the buds are dried up and dead. If such plants are brought into an unheated glasshouse in autumn the dieback does not occur, presumably because they are protected from drying winds and, being kept warmer, can still absorb water.

The six species mentioned above would therefore be expected to grow well in moist, warm-temperate climates such as south eastern USA, northern California, southern Japan, the North Island of New Zealand, and perhaps Tasmania. In the British Isles they could be expected to thrive in south-western, especially coastal, areas of Great Britain and much of Ireland, and all but *B. alnoides* are being successfully cultivated at Ness. Collections from the highest altitude of the species ranges are most likely to tolerate cooler, shorter summers.

Growing species from continental climates in temperate oceanic climates

In oceanic climates, as on the west coast of the USA and Canada, the British Isles, New Zealand and Japan, the temperature often hovers around freezing for up to four months so that species from these areas have a very long cold requirement to break bud or seed dormancy. This delays bud break until a significant risk of damaging frost is past. Depending on the species, once the cold requirement has been satisfied bud break and seed germination may occur whatever the temperature, as long as it is above freezing, or may only take place once a certain higher temperature has been reached.

In the British Isles, trees from continental climates such as Siberia, Central Asia or the prairie states of North America, can come into leaf too early. This means that most provenances of *Betula ermanii* var. *lanata*, *B. tianshanica*, *B. microphylla*, western Himalayan collections of *B. utilis* from Pakistan, and many eastern Asian continental collections of *B. ermanii*, *B. dahurica*, *B. schmidtii* and *B. pendula* subsp. *mandshurica* are difficult to grow in Britain.

A notable feature of some trees from continental climates is the death of weaker, shaded, lower buds and survival only of large buds on the thicker stronger twigs on the upper part of a tree (Fig. 255). This seems to be due to the winter desiccation of the buds on the lower part of the tree. In areas with particularly severe winters such as north Norway, the Kurile Islands north of Japan, and Alaska and north Canada, it is noticeable that the tree birches have thicker twigs and larger buds, which therefore have a lower surface to volume ratio than thinner twigged trees of the same species from climates with less winter desiccating stress.

It is usually not worthwhile trying to grow birches of continental provenance in mild oceanic climates. Most seedlings of *Betula pendula* and *B. pubescens* from around Lake Baikal do not make good trees in any garden in Britain, and the variant of *B. ermanii* with distinctive, grey, hairy buds (subsp. *lanata*), from slightly

Fig. 45. Dead *Betula utilis* subsp. *jacquemontii* in Punta Lara, Nerja, Malaga on the mediterranean coast of Spain with date palm, *Phoenix dactylifera*, and rubber plant, *Ficus elastica*, in background (HMcA).

further east, usually dies after a few years of repeated killing of its young shoots in spring. However, one collection of var. *lanata* from Olchovaja Mountain in Prymorsky Krai in the maritime Russian Far East is thriving at Ness.

Betula pendula subsp. *mandshurica* and *B. ermanii* of Japanese provenance grow well in many areas of the British Isles. Similarly, *B. papyrifera* and *B. cordifolia* from the maritime provinces of eastern Canada and New England and *B. papyrifera* from the British Columbian coastal ranges grow well in Britain, while collections from the central prairie states do not. On the whole, collections of Himalayan and Chinese provenance do reasonably well. In many cases, growing large numbers of seedlings of a difficult species allows individuals tolerant of the local climate to be selected.

Even in their native ranges, continental species may suffer from severe low temperatures (Miller-Rushing & Primack 2008), and buds opened or roots dehardened by warm early spring weather followed by a hard frost can be severely damaged. Such damage has been described in *Betula alleghaniensis* towards the southern end of its distribution (Bourque *et al.* 2005; Gu *et al.* 2008), and such events may occur more frequently within current ranges as the climate becomes warmer.

Species of high latitude provenance

Species originating from high northern latitudes can usually be grown successfully quite far south as long as adequate moisture is available in hot dry summer weather. However, deciduous plants from such origins tend to lose their leaves early, in September, presumably in response to the shortening

day length, and two months or more before the same species from lower latitude provenances. In cultivation at lower latitudes some collections of *Betula glandulosa* of northern provenance complete two growth periods in one growing season, breaking bud relatively early, developing autumn colour and dropping their leaves in June, and then leafing out again in July before becoming dormant for the winter in November. As is typical of shrubby species of such origin, they are small and slow growing and so require considerable care to keep them alive. Apart from scientific interest, the only purpose in growing them would be as natural bonsai as they have attractive neat foliage which colours very well.

Fig. 46. *Betula pubescens* largely confined to steep ground along streamlines where inaccessible to sheep and deer in Glen Nevis, Scotland. The natural treeline below the dark zone of heather (*Calluna vulgaris*) extends to higher altitudes in sheltered gullies (HMcA).

INTRODUCING NEW SPECIES TO CULTIVATION

In the early days of plant introduction, the growers were rarely the collectors and had little idea of the native environment of the species they were attempting to cultivate, and understanding of ecology was in its infancy (McLean 1997, 2004). Today our approach has been that seeds or collected young plants of new birches are usually first cultivated by the methods described above for other species, with propagation by cuttings being attempted as early as possible for distribution of the often limited number of clones raised from the wild collected seed. At least most of the young plants are overwintered outside to test for hardiness and allow for selection of individuals best adapted to the local climate, allowing the loss of individuals which break bud too early and suffer frost damage.

When they are large enough, young trees are planted in situations where they can be allowed to grow to maturity, for birches usually an open situation with little shade and the soil neither too dry nor waterlogged, but also in other conditions if sufficient plants are available. In this way it has been learnt that *Betula insignis* and *B. potaninii* are tolerant of at least winter waterlogging, while *B. corylifolia*, *B. calcicola*, and *B. delavayi* are not, and *B. chinensis*, *B. raddeana*, *B. medwediewii* and possibly *B. bomiensis* are drought tolerant while *B. globispica*, *B. costata* and young *B. chichibuensis* are not.

Another issue in the handling of new species in cultivation needs to be raised here. Garden staff are often trained for commercial practice in which suckers and branches arising from low on the trunk of young trees are pruned off. This is done to allow the maximum number of trees to be grown in a restricted area of either containers or open ground nursery, and after planting to allow access for people and mowers. However, with species which are naturally multi-stemmed or spreading in habit, this practice may result in poorly grown or unsightly specimens which are therefore not appreciated, neglected, and perhaps allowed to die. This is a further instance of how easily new introductions of potentially attractive shrubby species can be lost to cultivation. With new introductions about which little is known it would seem sensible, at least initially, to let seedlings develop naturally, allowing strong sucker shoots to grow if these develop. This way we learn what is the natural habit of the species and can decide later whether to treat it as a tree or a shrub.

PLANTING

INTRODUCTION

In general the white barked species of section *Dahuricae* (*Betula nigra*, *B. raddeana*, *B. dahurica*) and the non-white barked species tolerate transplanting much better than the commonly grown white barked species of sections *Betula* and *Costatae*.

The commonly grown white barked birches such as *Betula pendula* and *B. utilis* are well known to be difficult to transplant, so care must be taken if transplants are to survive, especially at large sizes. Following transplanting, heavy standard and semi-mature specimens of these often have thin canopies for many years and frequently fail to make satisfactory trees from the landscaping point of view. Such large transplants are only likely to be successful on very moist sites or when they have been transplanted with a large soil ball and are subsequently well watered.

It has been shown that trees which have been exposed to drought in the nursery often transplant more successfully, and that nutrient poor sites induce greater root production (Kozlowski & Pallardy 2002). For trees which have to be transplanted, it is therefore not desirable to grow them too 'soft' with shelter and lots of water and nutrients.

Use small transplants

Smaller transplants (e.g. 1 m 'feathers') usually soon outstrip larger specimens in height and are of a much better shape and vigour after two to three years, so it is better not to plant larger sizes of these species except in special circumstances. One-year-old seedlings are usually handled 'bare root' with no soil attached to their roots, making the plants light and easy to transport. However, in this condition it is very easy for the roots to dry out and die; even bruising the roots by dropping the seedlings, rather than gently laying them down, has been shown to affect adversely subsequent growth.

Containerised plants have the advantage of being transplanted without any root damage. This usually improves initial survival, but containerised plants often grow-on less successfully than well-handled bare-root or root-balled transplants. With all the root system in the relatively small volume of soil in the container, the water in this soil is soon exhausted and the plant may suffer from lack of water sooner than a carefully handled bare root or root-balled transplant with its roots spread through a larger volume of soil.

Planting at the correct depth is crucial

In many tree species planting at the correct depth is crucial to subsequent good growth, or even to survival. The roots should diverge from the base of the trunk more or less at soil surface level, as would those of a birch seedling which had grown naturally. There is often a temptation to plant more deeply to make the tree more stable, but burying the base of the trunk, even by only two to five centimetres, can lead to root death, very poor growth, and often the death of the tree due to poor aeration of the buried part of the base of the stem (root collar) (Smiley & Booth 2000). After desiccation following transplanting, deep planting is probably the next main cause of death or poor performance of transplanted trees, and is especially significant with shallow rooted species such as the white barked birches.

Timing

There has been much discussion as to the best time to transplant trees (see online review by Richardson-Calfee 2003). Autumn, when the leaves have more or less all fallen but before the soil is too cold for root growth, is often recommended for deciduous species. This allows for some root growth before winter, and is often considered the best time unless the site is subject to severe desiccation stress over winter or if the species are particularly susceptible to winter desiccation (e.g. evergreens, those with thin bark, tunnel grown plants with thin cuticle).

Transplanting in spring may therefore be better for particularly difficult sites or species, as long as the soil is kept adequately moist after transplanting. In spring new root growth will soon begin to establish the close root-soil contact required for efficient water absorption. However, it has been shown that significant new root growth only occurs several weeks after bud break as it depends on photosynthesis by the new leaves (Abod *et al.* 1991), and the new leaves of *Betula* seem much less able to control their water loss than leaves of some other genera.

It is therefore very important to follow the best practice when transplanting the white barked birches of the *Betula pendula*, *B. papyrifera* and *B. utilis* groups. Most commercial bare-root transplants lose about 95% of their root system when they are moved (Insley & Buckley 1985). They must, therefore, be well supplied with water until significant new root growth has taken place.

Choice of species by soil type

Whereas the commonly grown white barked birches (section *Betula*) are characteristic of light soils, many of the other tree species (e.g. section *Lentae*) require soils which are moist throughout the growing season and thrive on heavy clays.

Being weedy, the white barked species will grow reasonably well on almost any soil and are the most drought tolerant of all the birches. Of those commonly grown, *Betula papyrifera* and *B. pendula* are probably the most tolerant. However, even in their native habitats these species can be seen suffering from drought with their older leaves turning yellow and falling prematurely in August in dry summers (Fig. 25). This can happen while the deeper rooted species with which they grow show little sign of stress: *Quercus petraea*, *Q. robur* and *Ulmus glabra* in the case of *B. pendula*, and *Acer saccharum*, *Q. borealis*, and *Tilia americana* with *B. papyrifera*.

Judging from the few individuals in cultivation, the white barked *Betula raddeana* appears to be more drought tolerant than most other white barked species and it certainly has a much deeper rooting habit. Collections of *B. pendula* from Mount Etna, Sicily, (*B. aetnensis*, see p. 284), can be expected to be more drought tolerant than many other provenances of the species; the trees in cultivation are very deep rooting and appear to withstand dry spells well.

Birch roots need good aeration, so on very wet or very heavy soils they will be confined to the soil surface and trees may become unstable as they increase in size.

Betula nigra tolerates wet soils and waterlogging in winter, as do some species of section *Lentae*. *B. insignis* and *B. globispica*, and probably *B. lenta* and *B. alleghaniensis*, also grow remarkably well in heavy wet clays, and at least *B. insignis* appears to tolerate a water table near the soil surface in winter.

Betula delavayi, *B. calcicola* and *B. corylifolia* seem to be particularly sensitive to drought, but also need well-aerated soil. These are probably the most difficult species to cater for as they suffer from drought in dry conditions and yet can be killed by waterlogging in sites which would be ideal for species of section *Lentae*.

Although usually characteristic of bogs in the wild, *Betula pumila* and *B. michauxii* in North America and *B. humilis* and *B. fruticosa* in Eurasia, do not require very wet conditions in cultivation. In fact, experience suggests that they may be more tolerant of dry conditions even than species of section *Lentae*, and that they are restricted to wet places in the wild because, in drier places, they are shaded out by taller trees. Even so, it is better to grow these shrubby bog birches in moist conditions so that their colourful autumn leaves are retained for the maximum length of time. Under dry conditions, or in a windy, dry atmosphere, the colouring leaves soon turn brown and fall off.

Most birch species seem to be tolerant of a wide range of pH, although it is difficult to be certain about this with the rarer species of which we have little experience. *Betula calcicola* and *B. delavayi* grow on limestone in the wild, but seem to grow satisfactorily in less calcareous soils in cultivation. *B. papyrifera* is said to tolerate pH 3–4 on mine waste, so is useful for reclamation (Safford *et al.* 1990).

Root habit

The white barked birches are usually shallow rooted with few large roots diverging from the base of the trunk and running just under the soil surface for a long distance. This means that it is important in open ground nursery beds to transplant regularly to encourage the development of a more compact root system.

In contrast, species of sections *Lentae*, *Asperae* and *Dahuricae* have much more compact and deeper root systems, at least when young, and are therefore easier to transplant. The deeper root system, more like that of an *Acer*, means that the roots hold a better soil ball when transplanted, but also that these species do not grow so well on the shallow soils. The *Dahuricae*, and especially *Betula raddeana*, have roots which grow vertically downwards; when young, *B. chinensis* develops an extensive, deep root system, presumably an adaptation to the dry exposed ridges on which it grows.

Exposure

All birches, but especially the white barked species and dwarf species of section *Apterocaryon*, seem to be tolerant of exposure. In comparison with the species of other genera with which they grow, their leaves are usually small and tough in texture. Birch leaves are rarely found to be torn or dried up at the edges, common features even in the exposure-tolerant *Acer pseudoplatanus* and many other genera. Even those birches characteristic of dense forest, such as *Betula alleghaniensis*, *B. insignis*, *B. luminifera* and *B. cylindrostachya*, rarely show damage to mature leaves, even when grown in exposed positions. *B. alleghaniensis* forms tree-line scrub on Bromley Mountain, Vermont and Mt Mitchell, North Carolina (Fig. 48) (KBA pers. obs.). Birches, especially the fast growing white barked species, can therefore be used as a fast-growing windbreak in the most exposed situations.

In areas of high snowfall or where ice storms are common, tree birches are susceptible to damage. The fine twigs hold a lot of snow, so the whole tree may bend sideways under the weight so that it becomes more susceptible to subsequent snow-falls and eventually snaps or bends down almost to ground level as in a 'drunken forest'.

Pruning

Birches bleed sap severely from open wounds when the sap is rising in spring, so it is best not to prune them between winter and late spring. Late summer is the best time to prune as the tree will then not bleed, but still has time to repair the wound before the onset of winter. Sap will bleed from any cut, causing the loss to the tree of stored carbohydrate, and can result in subsequent poor growth and dieback (Brown & Kirkham 2004). This rising sap is deliberately bled and collected for making birch sap into a cordial or wine.

PESTS AND DISEASES

Introduction

Birches, especially the white barked species, are often a major component of northern forests and are known to support a wide variety of animal life and to be associated with a great number of fungi. There appears to be little information on the pests and diseases of birches other than the white barked species, so the following account largely refers to these. A good survey of the literature is provided by Atkinson (1992), from which much of the information quoted here is taken.

Many insects are known to feed solely on birches and so must have evolved with them, and the conspicuous bracket fungus *Piptoporus betulinus* is more or less restricted to birch. Birches are therefore a major food resource for wildlife, or susceptible to a wide range of pests and diseases, depending on your point of view.

Grazing

Up to a third of the land area of Iceland may once have been covered by birch (Figs 238, 241), but this is now down to about 1%, and the heather, grass and bracken moors for which Scotland is now so well known were at one time largely forest of *Betula pubescens* below 600 metres (Fig. 46). The dramatic impact of sheep is well illustrated when their density is decreased or they are removed altogether — the heaths rapidly revert to birch forest (Mitchell *et al.* 1997, 2007). Deer, especially red deer, hares and other grazing animals also eat birch foliage, but rarely have as significant an effect as sheep, and birches are less palatable than other species such as alder and mountain ash or rowan.

The presence of platyphylloside and betulin, and perhaps other defence compounds, in the bark and twigs makes these birches relatively resistant to grazing animals (Sunnerheim *et al.* 1988). Betulin, in particular, seems to deter rabbits, and the sticky resin glands on the spring shoots are also thought to be a defence against herbivores (Lapinjoki *et al.* 1991).

Defoliation by caterpillars or grubs of a number of insect species of Lepidoptera (moths), Coleoptera (beetles), Hemiptera (bugs) and Hymenoptera (saw-flies) can leave trees leafless in mid-summer and significantly affect productivity, but affected trees rarely die, usually leafing out again later in the season. Twelve insects are listed as causing such defoliation of birches in Britain and others may do so in Scandinavia (Atkinson 1992). There is very little which can be done practically to prevent such damage.

The bronze birch borer
(*Agrilus anxius* Gory: Coleoptera: Buprestidae)

Of the various insects which attack birch, the North American bronze birch borer is by far the most significant; this species is still confined to North America.

The bronze birch borer is a bark boring beetle whose larvae burrow under the bark which, when a burrow completely rings a stem, leads to the death of that stem. It is native in the forests of eastern North America west to Colorado and Utah, but is reported to cause little damage there. Its significant impact is on amenity trees in which dieback of branches is most likely to be due to attack by the borer. There are very confused reports of the susceptibility or otherwise of various species. Santamour (1999) suggests that most of the white barked species (*Betula papyrifera*, *B. pendula*, *B. pubescens*, *B. utilis*, *B. ermanii*) are susceptible although *B. papyrifera* perhaps less so, and that *B. alleghaniensis* is also susceptible. *B. nigra*, *B. dahurica*, *B. maximowicziana* and *B. occidentalis* are reported to be at least somewhat resistant. Susceptibility appears to be due to the presence of rhododendrin in the bark of susceptible species. However, experience in the Arnold Arboretum (Dosmann, pers. comm.) indicates that both *B. dahurica* and *B. maximowicziana* can be attacked. Nielsen *et al.* (2011), in a valuable long-term experiment, have confirmed that *B. nigra* is totally resistant and shown that other native N American species are relatively resistant compared to the Eurasian species tested.

Clearly every precaution should be made to ensure that the bronze birch borer is not introduced into Eurasia. The recent accidental introduction of the related Asian emerald ash borer (*Agrilus planipennis* Farimair) to North America and the consequent devastation of species of *Fraxinus* there illustrates the huge environmental and financial cost of such pest introductions (Cappaert *et al.* 2005).

Seed predation

Birch seeds are a major part of the diet of a number of birds such as mealy redpoll, lesser redpoll, and siskin and a smaller part of many others such as serin, bullfinch, goldfinch, great tit, blue tit, and house sparrow. Some small mammals also eat the fallen seeds, but such vast quantities are produced in most years that seed predation is unlikely to be significant in affecting regeneration and is generally regarded as a welcome support to wildlife.

Minute gall midges of the genus *Semudobia* live on either the fruiting catkin bracts (*S. skuhravae*) or the fruits (*S. betulae*, *S. tarda*, *S. steenisi* and *S. brevipalpis*) and differential susceptibility to these has been used in an attempt to construct a genealogy of birches (Roskam & Van Uffelen 1981). These never seem to be sufficiently frequent to have a significant effect on the seed crop.

FUNGI

Mycorrhizae

Like most plants, especially trees, birches have symbiotic root associations with fungi, the fungal mycelium behaving like an extension of the trees' root system (Perez-Moreno & Read 2001; Atkinson 1992). Northern birch forests are known for the profusion of toadstools which develop in the autumn. Most of these are the fruiting bodies of symbiotic mycorrhizal fungi, although some are saprophytic.

Successful early infection by mycorrhizal fungi is more or less essential for the survival of birch seedlings, which acquire some of their carbon nutrition from their fungal associates. Young birch trees often show colonisation by a succession of different fungi, with the familiar fly agaric, *Amanita muscaria*, being particularly associated with mature trees (Figs 37, 38). A brief review with references to much of the original research is provided by Atkinson (1992).

Disease-causing fungi

Birches are attacked by a wide range of fungi, from mildews and rusts on leaves to toadstool and bracket forming fungi which cause brown or white rots of the wood of living trees.

A conspicuous phenomenon on birches is the presence of dense, nest-like clusters of twigs which disfigure trees and are commonly referred to as 'witches brooms'. These are primarily caused by infection with the ascomycete fungus *Taphrina betulina* which induces the proliferation of buds and their growth to form the characteristic tangle of twigs (Jump & Woodward 1994). Another ascomycete, *Amisogramma virgultorum*, can also cause witches brooms. Treatment is simply to cut out the masses of tangled twigs.

Mildews are common, especially in shaded young shoots. They can cause dieback but are rarely significant. The coral spot fungi (*Nectria* spp.), which can cause such severe dieback on other tree genera, only cause minor damage to birch. Similarly, it seems that the agent of sudden oak death (*Phytophthora ramorum*), despite its a wide host range, poses little danger to birch species, *Betula pubescens* having been shown to be in the 'less susceptible' category (Appiah *et al.* 2004; Brasier *et al.* 2002).

The rust fungus *Melampsoridium betulinum*, on the other hand, can cause defoliation and significant damage to susceptible species. It produces orange masses of uredospores on the undersides of leaves in summer on *Betula pubescens* and *B. pendula*, and significant infections have been seen on *B. utilis* subsp. *occidentalis* under greenhouse conditions. However, it has not been noticed on species of subgenera other than *Betula*, which is not unexpected as rust fungi are usually very specific to particular species or narrow species groups.

The wood rotting bracket fungi *Piptoporus betulinus*, *Fomes fomentarius* and *Chondrostereum purpureum* are primarily wound parasites and so unlikely to attack young healthy specimens and are more commonly found on ageing trees. *C. purpureum*, a common species and the agent of silverleaf disease in *Prunus*, is reported to have its highest incidence on *Betula* and *Alnus* and be a major cause of dieback in North American forests (Setliff 2002). Several species of *Armillaria* (honey fungus) are also reported to attack birches.

PROPAGATION

INTRODUCTION

In the wild birches propagate almost solely by seed. Birches are therefore easy to propagate by seed and this is the normal method for most species. However, some species are also easily propagated by cuttings and selected clones of any species can be propagated by grafting onto suitable rootstocks (i.e. of closely related species of comparable growth rate).

SEED

Seed ripening, dispersal, and collection

All birches, except the sub-tropical autumn flowering *Betula alnoides*, flower in spring. Only *B. alnoides* produces its catkins in autumn and ripens and disperses its fruits between February and April (Skvortsov 1997). *B. cylindrostachya*, *B. luminifera* and *B. nigra* also ripen and disperse their seeds much earlier than all other species, in May to June. *B. raddeana* is the next species to ripen its seeds, usually in July. By the time the seeds of other species are maturing in late summer, these five species have usually long since shed all or most of their seed.

In most species the seed is dispersed in the autumn and over the winter. Only in species of the *Betula potaninii* group of section *Asperae* (*B. potaninii*, *B. chichibuensis*, *B. calcicola*, *B. delavayi*, *B. bomiensis* and probably *B. skvortsovii*) are the upright fruiting catkins persistent and dispersal commonly delayed until winter and spring, at least from evidence in cultivation. In these species, seed can usually be found in the catkins throughout the year. Their seeds have virtually no wings, and so the persistence of the fruiting catkins will extend the time during which the fruits are dispersed. The fruiting catkins of *B. globispica* and *B. fargesii* are probably the most persistent of all but, like those of *B. chinensis*, open sufficiently on drying to allow rapid dispersal of the wingless seeds in autumn. In species of section *Lentae* the fruiting catkins may often persist over winter and break up only slowly.

Even in the white barked birches with dehiscent fruiting catkins, dispersal of some seeds may occur over several months and throughout the winter. In fact *Betula populifolia* has small, tight, particularly persistent fruiting catkins which may not disperse seed until well into the following spring, quite atypical for a species of section *Betula*.

Seed longevity and storage

Although it is best to sow seed as fresh as possible, that is in the late winter or early spring after collection, the seed of all species retains its viability for at least two years in storage at room temperature, and longer if stored in a refrigerator (c. 4°C) or freezer. Several species (*Betula fargesii*, *B. potaninii*, *B. megrelica*, *B. insignis*) have retained almost full viability after ten years storage in a refrigerator, while refrigerator stored *B. globispica* seed was still fully viable after 14 years but rapidly lost viability thereafter. On the other hand, fridge stored seed of *B. calcicola* was unviable after four years. As viable seed is not always produced every year it is well worthwhile storing surplus in a refrigerator.

Seed sowing and germination

Germination in the wild normally occurs in the spring following dispersal and there are no reports of seed persisting in the soil (Moore & Wein 1977) except in *Betula maximowicziana* (Osumi & Sakurai 1997), but there is no information on the rarer species. Only seed of *B. nigra*, and presumably *B. alnoides* and *B. cylindrostachya*, normally germinates in summer immediately after dispersal, explaining why *B. nigra* is naturally restricted largely to riverbanks, while the Himalayan species are adapted to a summer monsoon climate. Seed of these species will store dry over winter and can be sown in the spring.

Before sowing it is often a good idea to do a quick viability test (easiest with a dissecting microscope), as otherwise seed may be sown too thinly or too thickly. Viable seed are firm to the touch and moist and oily when cut or pierced with a needle, while non-viable seed squash easily and can be seen to have only shrivelled, dried up contents. Seed of self-compatible species, such as *Betula medwediewii*, *B. globispica* and *B. fargesii*, usually has a very high viability. It is important not to sow too densely as overcrowding makes seedlings very susceptible to damping off which can rapidly kill every seedling.

The physiology of the germination of the commercially or ecologically important species *Betula pendula*, *B. pubescens*, *B. papyrifera*, *B. alleghaniensis* and *B. maxmowicziana* has been well researched (Black & Wareing 1954; Longman 1984; OECD 2003), and cold (temperatures just above freezing), light, and heat, both on their own and in various combinations, can stimulate germination; birch is one of the few trees in which germination is sensitive to photoperiod, long days promoting germination. However, there is usually no need to understand the detailed physiology of seed germination to be able to germinate the seed, as the same principles seem to apply to other species.

Seed can be sown in any humus-rich seed compost (John Innes seed composts with their high clay content are not very suitable as they tend to 'cap') and covered lightly with a little compost and/or grit. The best procedure is probably to sow in January or February and leave the seed pans outside so that they receive a few (4–8) weeks of cold (1–5°C) treatment, although up to ten weeks is sometimes recommended. The seed pans should be left outside until germination occurs. If no germination has taken place by about mid May, the seed pans can be brought into the greenhouse in the hope that the higher temperature might stimulate germination. These procedures usually result in a high percentage germination of viable seed. Excised embryos (of *Betula pendula* and *B. pubescens*) are reported to germinate without any pre-treatment, so dormancy is imposed by the seed coat.

Once the seeds have germinated, giving maximum light will produce the most healthy sturdy seedlings. As the seeds are small and have very little food store, the application of a mineral nutrient feed may be required, and this is usually preferable to early pricking out of the delicate small seedlings. As long as the seedlings are not too crowded, pricking out into individual pots can be delayed until the hypocotyls are firm and more or less woody and the seedlings have at least two or three post-cotyledonary leaves. They can then be handled more easily with less risk of damage.

Growing on

Once seedlings have grown to sufficient size, perhaps about 10 cm, they are best planted outside to be grown on in nursery rows in ordinary garden soil. For very good reasons, commercial practice may be different as it may be most efficient to keep the plants in containers, but even commercially it may be best to grow in nursery rows for a year or two and then containerise. Plants in nursery rows require much less attention with respect to watering, and especially pests such as vine weevil, and have more space to develop a sturdy habit.

CUTTINGS

The common white barked birches are not easy to root from cuttings, but *Betula nigra* is routinely propagated by cuttings and *B. dahurica* and *B. raddeana* can also be propagated in this way. Most of the non-white barked species are relatively easy to root from cuttings, especially the small-leaved species such as the dwarf arctic birches of section *Apterocaryon* and *B. potaninii*, *B. bomiensis*, and *B. chichibuensis*.

Taking cuttings

With all species, soft-wood cuttings taken as early as possible in spring to early summer are most successful. Cuttings from seedlings or young trees are usually by far the easiest to root, presumably because they are hormonally juvenile. As soon as the new growth is long enough, cuttings should be taken. Two to three centimetres is sufficient for the smaller leaved species (e.g. *Betula potaninii*, *B. chichibuensis*, *B. chinensis*) (Fig. 81) but a bit longer is preferable for species with larger leaves (e.g. *B. nigra*, *B. raddeana*) (Fig. 152). It is probably best to choose shoots with at least two more or less full-sized leaves on the elongated part of the young shoot, cutting just above the basal leaves which were preformed in the bud (see Chapter 5: Morphology). Once extension growth has ceased and the apparently terminal and lateral buds become evident, rooting may be less successful and it becomes difficult to induce further extension growth in the cutting. Without extension growth due to the breaking of buds, cuttings may fail to survive the following winter.

Thinner stemmed cuttings root more successfully and more quickly than those with thicker stems. This appears to be because the stem behaves as a 'carbohydrate sink', requiring 'food' for its maintenance (Howard & Ridout 1991a, b, 1992; Howard & Harrison-Murray 1995). There is no need to remove any leaves from the cuttings. A rooting mixture of equal quantities by volume of perlite and peat has proved most satisfactory. A mist or fogging unit is probably the best environment for rooting, but rooting can be achieved in any environment which is sufficiently moist to prevent wilting. Very young new shoots have little stored carbohydrate and so should be given as much light as possible (as long as wilting can be prevented) so that they can photosynthesise to produce carbohydrate to be able to grow roots (Howard & Ridout 1991a, b).

Growing on rooted cuttings

At the first sign of the development of roots, a cutting can be potted on into an open, low-nutrient compost. This maintains the well-aerated, optimum environment for root growth and water absorption, but at the same time allows roots access to some mineral nutrients to promote shoot growth. If the cuttings are potted on from the rooting medium when the roots are only 1–2 mm long there is little danger of breaking the brittle roots and the cuttings are easily handled. These newly rooted cuttings still have poor water absorbing capability and so will still require a moist atmosphere to prevent wilting. The potted on cuttings are therefore best returned to the rooting environment to grow on for a week or so until roots are well enough developed to allow removal to a drier atmosphere.

Good establishment is indicated by the presence of roots between the pot and the soil ball. Root growth is often very vigorous, and when roots reach the base of the pot and risk becoming inhibited by pot size, the potted cuttings should be potted on into larger pots of full nutrient strength compost — despite the fact that the top growth may seem rather small for the pot size. Failure to pot on at this stage reduces the frequency and vigour of bud break as root growth has a very significant effect on top growth (Young *et al.* 1997). The young plants should be grown on as rapidly as possible in full light and perhaps some heat until the leaves begin to take on autumn colouration. The plants should then be kept cool, preferably outside, and allowed to become dormant. This applies especially if no buds have broken and there has been no extension growth as such plants are particularly likely to die of carbohydrate starvation over the winter if kept warm.

In summary, the most important factors in rooting cuttings are:

- Take cuttings of new, thin shoots as early as possible in the season, if possible from 1–4 year seedlings or rapidly growing shoots on vigorous plants
- Cuttings should never be allowed to wilt

- As soon as rooting is noticed, pot on into an open, low-nutrient compost
- When roots reach base of pot, pot on into full nutrient compost and grow on as vigorously as possible in full light to promote bud break
- If buds fail to break, as soon as leaves begin to colour and fall in autumn put outside and keep as cold as possible

Grafting

We have no personal experience of grafting birches. The white barked species are known to be difficult to graft successfully, perhaps largely because the pressure of the rising sap due to root pressure tends to push the scion off the stock. The rooted stocks are therefore usually kept rather dry after grafting.

Grafting is usually carried out in February with the potted stocks with attached side-grafted scions being laid on their sides in a heated bed to stimulate union of the graft. An excellent summary of methods is provided by Lane (1993).

4. CONSERVATION

CONSERVATION PRIORITIES FOR BETULA

Of the forty to fifty species of birch, twelve are listed in the *World List of Threatened Trees* (Oldfield *et al*. 1998). Some of these are accepted here as species, others as minor variants, so it is worth looking carefully at the significance of each of these threatened taxa and whether they are worth conserving. A species which is the only one in its subgenus is clearly more worthy of conservation than a species or variety with many close relatives.

The birch considered most distinct from all others, and the only species in its own subgenus, is *Betula corylifolia*. It is, however, fairly widespread in the mountains of central Honshu in Japan.

From a scientific viewpoint, a very rare and distinct species such as *Betula chichibuensis* might be considered more worthy of conservation than the North African populations of *B. pendula* sometimes, but not in this book, distinguished as *B. fontqueri* Rothm. On the other hand the latter might prove to be more useful as one of the most drought and heat tolerant birches, and so of more value to man than the small, shrubby Japanese *B. chichibuensis*. Other rarities, e.g. *B. uber*, may differ from the normal species only by the possession of a variant of a single gene.

The twelve species and two subspecies of birch listed in the *World List of Threatened Trees* (Oldfield *et al*. 1998) are listed in Table 5 and some of them are discussed in more detail below:

DISTINCT SPECIES

Betula globispica

Betula globispica is a very distinct relict species from Honshu. Its self-compatibility should allow it to be propagated easily, although some seedlings seem much slower growing than others and may be suffering from the effects of inbreeding depression. If raising this species from seed produced as a result of self-pollination (only one tree survives at Ness) it may be advisable to raise large progenies and select only the largest and most vigorous young plants (Kuser 1983). *B. globispica* seems to require moist soils and be sensitive to drought, but is otherwise easy to cultivate and forms an attractive small tree with good autumn colour and unusual persistent fruiting catkins which remain on the tree over winter.

Betula chichibuensis

Unlike *Betula globispica*, *B. chichibuensis*, another relict species from Japan, seems to be relatively drought tolerant, at least once established. *B. chichibuensis* is self-incompatible, and therefore more vulnerable were it to be reduced to isolated single plants in the wild or to only one clone in cultivation. Its bushy habit suggests that it is likely to be a sprouter (Bond & Midgley 2001), regenerating readily from the rootstock if the aerial parts are damaged.

In a study of the rooting habit of Japanese birches, Tabata (1970) could not obtain information from the rare wild plants, and those he raised from seed failed to live beyond the seedling stage.

However, although the sprouting habit enables individuals to be very long-lived, the need for two individuals to survive close enough together to cross-pollinate one another could make the production of seed uncertain in small populations. The very low viability (less than 1%) of the wild collected seed received from Japan suggests that in some years there may be little viable seed production in the wild, and so a low chance of natural regeneration from seed. On the other hand, Tabata (1966) reported 94.3% germination, the highest he recorded for any Japanese birch.

Table 5

Species of Betula listed as rare to endangered (Oldfield *et al.* 1998; Walter & Gillett 1997) with comments. Only those species in bold type are quite distinct species with no question as to their specific status.

Species	Distribution	Comments
B. apoiensis	Mt Apoi, Hokkaido, Japan	Probable local variant of E Asian mainland *B. gmelinii*, and apparently of hybrid origin
B. browicziana	Turkey	Local variant of *B. pubescens*
B. chichibuensis	Chichibu Prov., Japan	Very distinct relict species
B. ermanii var. *saitoana*	Korea	Subspecific taxon
B. globispica	Honshu, Japan	Very distinct relict species
B. halophila	China	Probable synonym of *B. microphylla*
B. megrelica	Mt Migaria, Georgia	Distinct relict species
B. murrayana	Michigan, USA	Recently arisen allopolyploid species derivative of *B. alleghaniensis* and *B. pumila*
B. oycoviensis	Poland	Precocious form of *B. pendula* (heterozygote)
B. pendula subsp. *font-queri*	Morocco	Minor taxon
B. pendula subsp. *parvibracteata*	Spain	Minor taxon
B. recurvata	Turkey	Probably a local variant of *B. pubescens*
B. raddeana	Georgia, Russia	Distinct relict species
B. szaferi	Poland	Precocious form of *B. pendula* (homozygote)
B. uber	Virginia, USA	Small-leaved form of *B. lenta*

When several clones are grown close together in cultivation the seed viability is high. Apart from the 'weedy' white barked birches, *Betula chichibuensis* is the only other birch which has self-sown in Ness Gardens, albeit only immediately under the parent trees.

Despite the rarity of the species in the wild, the seedlings from the original wild collected seed showed considerable variability, each being recognisable, primarily by habit characteristics, at least when young in the nursery rows. Such genetic variability is not unexpected in a potentially long-lived species, even when the populations have been reduced to small numbers (Kuser 1983).

With a species such as *Betula chichibuensis* which is easily propagated by cuttings, it would be very easy for nurseries to propagate solely from a single plant and so for it to be reduced to one, self-incompatible clone in cultivation. This has happened to *Cosmos atrosanguineus* (Anon 1996), *Santolina chamaecyparissus* (McAllister 1987), and probably *Buddleja globosa*, *Malus tschonoskii*, *Cerastium tomentosum*, and many other self-incompatible species easily propagated vegetatively (De Jong 1996; Jeffrey 1982).

Betula raddeana

Distribution maps of *Betula raddeana* show numerous small populations scattered throughout the central north Caucasus in Russia, Georgia, and Dagestan. The current political instability in the northern Caucasus restricts up-to-date information on its status. It could be under increased threat due to exploitation, or less threat due to disruption to normal life. The opportunities to collect in this area are likely to be restricted for some time, so there is an obligation to ensure that the collections and genotypes already in cultivation are maintained. It is discussed in more detail on p. 232.

Betula megrelica

Betula megrelica from the eastern Caucasus, described and illustrated in *The Red Data Book of the USSR* (Iliashenko & Iliashenko 2000) is the most enigmatic of the species listed. It grows on Mount Migaria, to the north of the area in which *B. medwediewii* is found. The two species are treated as distinct by Kuzeneva (1936) in the *Flora of the USSR*, although most western publications give *B. megrelica* a minor mention under *B. medwediewii* (Rehder 1940; Bean 1970). However, the illustrations provided by Iliashenko & Iliashenko (2000) and evidence from studies on material in cultivation, particularly the difference in chromosome numbers (2n=168 in *B. megrelica* and 2n=140 in *B. medwediewii*), suggest that *B. megrelica* should be treated as a distinct taxon and one of the rarest and least known (McAllister & Ashburner 2007). It is discussed in detail on p. 192.

SUBSPECIES OF DOUBTFUL STATUS

The subspecific taxa listed above by Oldfield *et al.* (1998) and Walter & Gillett (1997) are clearly only minor variants. *Betula ermanii* var. *saitoana* of Korea, *B. pendula* subsp. *fontqueri* of North Africa and *B. pendula* subsp. *parvibracteata* of Spain are probably no more distinct than other outlying populations of species which current practice (e.g. Walters 1975) does not recognise at even subspecific level. These include *B. aetnensis* from Mt Etna in Sicily, dwarf Turkish mountain populations of *B. pendula*, and some of the many named variants of *B. pubescens* (Vaarama & Valanne 1973) or *B. raddeana* (Husseinov 1971). These mainly southern, isolated populations probably survived at least the most recent glaciation *in situ* (Palmé *et al.* 2003). *B. browicziana* and *B. recurvata* (Browicz 1972) are almost certainly no more than southern relict populations of *B. pubescens* which are difficult to distinguish from some more northerly populations. Such minor geographical variants have been named because they are in well-explored areas such as Europe and Turkey, while other, more distinct populations, and perhaps even species (such as the newly described *B. ashburneri* and *B. skvortsovii*), may exist in less well-explored areas of the world (see list below).

Betula dahurica

Wilson (quoted by Schneider 1916) remarks that he saw only a single tree of *Betula dahurica* during his visit to Japan. Ohwi (1965) mentions its presence in central and north Honshu and Hokkaido. Oldfield *et al.* (1998) list *B. dahurica* as threatened in Japan.

From the living trees and the numerous herbarium specimens seen, the Japanese populations appear to be usually distinguishable by the size of the bracts and nutlets in the fruiting catkins and the leaf marginal toothing. However, Japanese and mainland Asian populations are variable and distinguishing between them is difficult, so they are here treated as varieties, with the most distinct population, that from Nobeyama, Honshu, being treated as var. *parvifolia*. It is distinguishable by its small, double toothed leaves with 6–8 pairs of very regularly spaced veins.

In addition to the rather indistinct morphological differences between the Japanese and Asiatic mainland populations, there is a clear physiological difference in that the Japanese populations grow much better in cultivation in the British Isles, suffering much less from late spring frosts and winter dieback. This suggests genetic differentiation and possibly long isolation from mainland populations.

DOUBTFULLY DISTINCT SPECIES

Betula halophila

The status of *Betula halophila* is uncertain, although its unique habitat for a birch (marshy salt flats) suggest that it may be genetically distinct from any other birch, even should it not be easily distinguishable morphologically. Skvortsov believes it is likely to be a form of *B. microphylla* (Li & Skvortsov 1999).

Betula oycoviensis and *B. szaferi*

In the *World List of Threatened Trees* (Oldfield *et al.* 1998) *Betula oycoviensis* from south east Poland is described as a hybrid between *B. szaferi* and *B. pendula*, both *B. oycoviensis* and *B. szaferi* being listed as threatened on a world scale with *B. szaferi* described as now extinct in the wild, the only birch in this category. From this it might be considered that these species require conservation effort as a priority. However, there is almost irrefutable evidence that the Polish *B. oycoviensis* and *B. szaferi* are merely weak growing, precociously and heavily fruiting forms of *B. pendula*, their characteristics being due to the presence of a mutant gene in single (*B. oycoviensis*) or double (*B. szaferi*) dose. Szwabowicz (1972) and Pawlowska (1980c) describe how crosses between trees of *B. oycoviensis* give rise to seedlings of *B. pendula*, *B. oycoviensis* and *B. szaferi* in proportions of 1:2:1, the Mendelian ratio for a cross between two heterozygotes for a single allele with no dominance. *B. szaferi* is also recorded in the Czech Republic, Denmark, Poland, Romania, the Russian Federation, Slovakia Sweden and the Ukraine (Boratynski 1998), a distribution impossible for a species freely interfertile with *B. pendula*.

While raising a batch of seedlings from seeds of *Betula pendula* subsp. *mandshurica* of botanic garden origin, originally from Novo-Alexandrovsk on Sakhalin Island north of Japan, HMcA noticed that one of twelve seedlings grew into the equivalent of *B. oycoviensis*. In its third year it produced numerous female fruiting catkins from almost every bud, the fruiting shoots bearing small, diamond-

Table 6

Little known primarily Chinese and Korean species whose conservation status is uncertain.
Some or all of these might be under threat of extinction in the wild. Generally accepted species are in bold, the status of the others, many of which are known from single localities, has been questioned by Skvortsov in footnotes in the *Flora of China* (Li & Skvortsov 1999).

Species	Distribution	Comments
Distinct species		
B. fargesii	China (NE Sichuan & W Hubei)	Known only from restricted area on both sides of the Sichuan/Hubei border
B. calcicola	China (NW Yunnan)	Known for certain only from Yulong Shan
B. delavayi	China (NW Yunnan, W Sichuan)	Variable species well known from Yulong Shan, but probably also occurs elsewhere
B. bomiensis	China (SE Tibet)	Only know from type and three other collections
B. chinensis 6x	South Korea	Morphologically distinct 6x cytotype known from two high mountains in South Korea. Species otherwise 8x
B. ashburneri	Bhutan; China (SE Tibet; NW Yunnan, W Sichuan)	Known only from a few localities in Bhutan, SE Tibet and several collections from the Heng Heng Duan Mts, Yunnan, Sichuan — probably locally common
B. gynoterminalis	China (NW Yunnan)	Apparently distinct species known from only a single herbarium specimen

Doubtfully distinct species known from one or two localities
(*Flora of China* — Li & Skvortsov 1999)

B. trichogemma	Emei Shan, Sichuan, China	Doubtfully distinct from *B. potaninii*
B. rhombibracteata	Deqen Xian, Weixi Xian, NW Yunnan, China	Doubtfully distinct from *B. cylindrostachya*
B. jinpingensis	Jinping Xian, SE Yunnan, China	Doubtfully distinct from *B. utilis*
B. jiulungensis	Jiulong Xian , SW Sichuan	Doubtfully distinct from *B. bomiensis* or more probably *B. potaninii*

shaped leaves. It grew weakly in the following years, fruiting excessively. That a similar mutant should arise spontaneously in another subspecies of *B. pendula* is further evidence that *B. szaferi* and *B. oycoviensis* are highly likely to be merely forms of *B. pendula*. They are therefore regarded as of very minor conservation significance equivalent to cut-leaved, golden-leaved, and purple-leaved forms. *B. papyrifera* var. *pensilis* would seem to be a similar form of that species, or possibly of *B. cordifolia* (Hosie 1963)

Betula uber

Betula uber is merely a form of *B. lenta* (Ashburner & McAllister 2005). Almost all seedlings from presumed self-pollination of an isolated tree of *B. uber* and from the wild *B. uber* stand are indistinguishable from *B. lenta* (see Chapter 7 under *B. lenta*), only 1% of the seedlings from wild collected seed being reported to be *B. uber*.

OTHER ENDANGERED SPECIES

In the species-rich Sino-Himalayan region the little known species listed in Table 6 may be rare in the wild, judging from their representation in western herbaria, although they are not recorded as such in the World Lists or the *Chinese Red Data Book*.

Table 6 lists some little-known species whose status is uncertain. All we can do for them is raise awareness and ensure that those presently in cultivation are not lost.

THE ROLE OF BOTANIC GARDENS IN CONSERVATION

INBREEDING DEPRESSION

To conserve a species in cultivation, it is usually considered desirable, if not necessary, to obtain and preserve as much as possible of the genetic diversity of the species and so avoid inbreeding depression (Hedrick & Kalinowski 2000). This is particularly important with crop species, but often difficult with rare or endangered species of no immediate economic significance (Falk & Holsinger 1991; Stephens & Sutherland 1999). However, in the wild many species have suffered severe bottlenecks when populations were reduced to a few individuals (De Smet 1993), but many invasions have originated from one or a very few individuals (Meffert 1999; Taggart *et al.* 1990; Coates 1992; Jackson *et al.* 2004; Elton 1958; Merilä *et al.* 1996), so very small numbers of individuals can give rise to huge, viable and healthy populations.

In the wild, self-fertilisation of birches may hardly ever occur, as pollen from other individuals of the species will compete successfully with self pollen. In gardens there are often single individuals of species, so any seedlings will be the result of self-fertilisation or hybridisation. The fact that seedlings resulting from selfing are weak may not be evident unless outcrossed seedlings are available for comparison, as was described in *Metasequoia glyptostroboides* by Kuser (1983), although the occasional seedling from self-pollination was as vigorous as outcrossed seedlings. In the wild also, Petit & Hampe (2006) found evidence that seedlings resulting from selfing, although initially numerous, fail to compete with outbred individuals.

Rare species may occur as single individuals in the wild and be self-fertile, so selection of vigorous individuals in cultivation, as would happen naturally in the wild, may be necessary to maintain the vigour of species, especially if reintroduction to the wild is planned.

Breeding populations of *Betula calcicola*, *B. delavayi*, *B. potaninii*, *B. insignis* and *B. bomiensis*, all self-incompatible, potentially attractive species for horticulture, have been established at Ness. Their successful establishment in cultivation will probably depend on the maintenance of at least one breeding population of each species to provide viable seed for wide distribution. Lack of suitable close relatives to use as stocks for grafting, or shrubby habit, make grafting an unsuitable method of propagation for all of these but *B. insignis*, and cuttings of mature plants may not root easily.

5. MORPHOLOGY OF BETULA

HABIT

Birches vary from tall trees of 30 to 35 m in *Betula alnoides*, *B. utilis*, *B. costata*, *B. schmidtii* and *B. pendula*, to dwarf shrubs of less than 0.5 m in *B. nana* of the arctic tundra. Almost all birches grow from a single rootstock and, although horizontal branches may become layered by rooting themselves in peaty soils (De Groot *et al.* 1997), no species is rhizomatous and only two, *B. populifolia* and *B. nigra*, occasionally produce shoots from roots at some distance from the base of the trunk.

ROOTS

Birch roots vary enormously. The white barked species are nearly all shallow rooted and develop far spreading surface root systems with only a few thick roots arising from the base of the trunk. Gasson & Cutler (1990) report that many bear fibrous roots in the root plate, which means that there are roots capable of absorbing water and mineral nutrients immediately around the trunk and holding firmly to the soil in the root plate, giving stability to the tree. Species of section *Dahuricae* and subgenus *Asperae* are deeper rooted.

An Icelandic *Betula nana* × *B. pubescens* seedling 2 cm tall in cultivation had a root 35 cm long (McAllister, pers. obs.), indicating the capacity of such sub-arctic shrubs to root deeply in their first year where winter frost lift is a major cause of seedling death (Aradottir & Arnalds 2001).

TRUNKS AND TIMBER

In high forest, the tall tree species of the *Acuminatae* and *Lentae* often form large clean trunks with few branches in the lower part. The timber of these, and especially of the *Lentae*, is of high quality, that of *Betula alleghaniensis* being described as heavy, hard and moderately strong, close grained and fine textured (Hosie 1963) and is said to be one of the most valuable timbers (Van der Kelen 1993). The timber of *B. schmidtii* is said to be so heavy that it sinks, and that of *B. chinensis* is described as extremely hard and dense, close-grained, very fine textured, and one of the most valuable timber trees in north China (Li & Skvortsov 1999). This is perhaps surprising as it is also described as a shrub to five metres tall, rather small for the 'most valuable timber tree'.

Although not now much used in the British Isles, and perhaps little in North America, the timber of the white barked birches is of great importance in northern Scandinavia, especially Finland and Russia (Yaroshenko *et al.* 2001; OECD 2003). Much forestry breeding and selection work has been done on birches in Finland (Koski & Rousi 2005). They make good cylindrical trunks when grown in closed forest, which prevents the development of side branches in these very shade intolerant trees, but trees grown in the open develop very fluted trunks with deep depressions below the points of attachment of major branches. The timber of the white barked birches is said to vary from moderately hard, heavy and strong (*Betula papyrifera*), to light, soft, and not strong (*B. populifolia*), but is very susceptible to decay when wet (Hosie 1963).

The trunks of many species of section *Asperae*, especially *Betula schmidtii* and *B. chinensis*, have distinctive elongated raised lenticels running around the trunks which are very conspicuous.

BRANCHES

The branching habit of birches varies enormously, depending on both genetics and environment. Pendent branches are characteristic of most European and west Siberian *Betula pendula*. Trees possessing more extreme degrees of this trait have been selected and named as the cultivars 'Tristis' and 'Youngii'. More or less fastigiate clones are known in a number of species including *B. pendula*, *B. raddeana*, and *B. calcicola*.

BARK

The barks of many birch species have more or less peeling papery outer layers unlike any other temperate genus except a few species of *Prunus*. The colour of the bark varies from almost black in some *Betula lenta*, *B. nigra*, *B. dahurica* and *B.* × *utahensis*, through all shades of brown and copper, with or without a white bloom, to pure white; or shiny, metallic and yellowish in *B. alleghaniensis*, *B. luminifera* and *B. cylindrostachya*.

Whiteness is due to the presence of betulin, a pentacyclic triterpene compound with high thermal stability, in the cells of the bark (Hayek *et al.* 1989). Betulin renders the bark waterproof and also very resistant to decay, much more so than the wood, so that rotting logs of birch species can often be identified from the texture and colour of their bark even if most of the wood has rotted. In the past the bark was used extensively for many purposes such as canoe construction and covering wigwams in North America, and roof covering over planking and beneath turf in traditional Scandinavian house construction (Fig. 47). The bark of these species was also used as a food in times of famine (OECD 2003).

Fig. 47. Birch bark used over rafters and under turf in traditional Norwegian farmhouse of artist Nikolai Astrup, Astruptuenet, Sogn og Fjordane, Norway (HMcA).

Fig. 48 (left). Dwarf *Betula alleghaniensis* forest towards treeline, Mount Mitchell, North Carolina (KBA).

Fig. 49 (right). Kenneth Ashburner by trunk of one of the original trees of *Betula utilis* subsp. *albosinensis* grown from seed of *Wilson* 4106, Werrington Park, Cornwall, England (MF).

It has been suggested (Harvey 1923) that the white colour has evolved to reflect the far northern low angle sunlight to reduce the risk of overheating and scorching of the bark. This idea is supported by the severe scorching suffered on the south-facing side of young trunks of black barked *Betula* ×*utahensis* at Ness.

The characteristic peeling of the bark is due to the production by the more or less permanent cork cambium of alternate layers of very thick walled cork cells and thin walled cells (Evert 2006; Lendzian 2006; Furlow 1990). The way in which the barks peel is often characteristic of particular species groups or even individual species, large sheets in *Betula utilis*, *B. ermanii* and *B. costata*, and numerous small flakes in *B. nigra* and *B. dahurica* giving a very characteristic shaggy appearance (Figs 150, 159). The bark flakes of *B. dahurica* are much thicker than those of other species and tend to remain on the trunk longer, the greater thickness appearing to be due to the greater thickness of bark grown in each annual increment. In contrast to the thin bark sheets or flakes of other species, those of *B. dahurica* are white on the outside but reddish-brown or brown on the inner surface, giving a very characteristic appearance to the trunks.

Fig. 50. Tree of *Betula ashburneri* in the wild in Arunachal Pradesh (PB).

BUDS, BUD SCALES AND STIPULES

Most species have more or less ovate, acutely pointed buds. In the *Acuminatae* these are usually more elongated than in the other groups and curved at their apices. The typical members of the dwarf birches of section *Apterocaryon* have very small, more or less globular buds, usually narrower than the twig bearing them and with blunt apices (Figs. 51, 53), very different from those of any other species except perhaps *Betula calcicola*. This has globular (or rosette) buds, but they are quite large relative to the twig diameter and densely white hairy.

The bud structure in birches has not been studied by us in any detail, but is reported to be fairly uniform throughout the genus (Kikuzawa 1982). Two outer bud scales are modified stipules (see below), the associated leaf blade of which has aborted or is very reduced. There are then usually two further pairs of stipular bud scales with associated fully formed embryo laminae within the bud. These are the usually two (to occasionally four) leaves preformed in the bud the previous year and which will expand rapidly as the bud breaks in spring (Fig. 24).

In the Betulaceae the stipules are usually short-lived, small structures a few millimetres long which are present on either side of the base of the petiole as the shoots are elongating in spring, but usually soon fall. However, in some species of *Betula* they are persistent, which is a useful diagnostic characteristic in most of the dwarf species of section *Apterocaryon* (Fig. 51), the species related to *B. potaninii*, and in *B. corylifolia* and *B. ashburneri*.

TWIG AND SHOOT STRUCTURE

Birches have two kinds of shoots: (a) long shoots which increase the size of the tree and whose internodes are elongated so that the leaves are widely spaced along the shoots, and (b) short or spur shoots which are usually only a few millimetres long and bear usually two or three, or rarely to 6 leaves and, on fertile spur shoots, one or more fruiting catkins. The internodes between the leaves on these spur shoots do not usually elongate to any great extent so that the leaves often appear to be inserted at almost the same level on the stem (Fig. 54).

Although winter twigs appear to have a terminal bud, the terminal growing point of every shoot aborts at the end of a year's growth and the apparently terminal bud is in fact in the axil of the uppermost leaf. This behaviour is similar to that of some other temperate trees such as other genera of the Betulaceae except *Alnus* (Kikuzawa 1982).

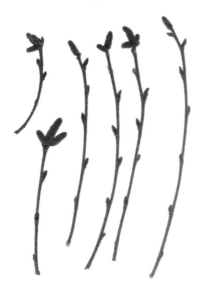

Fig. 51 (top). *Betula pumila* Canada, Newfoundland, Woody Point. Close up of short current year's shoots showing three terminal male catkins and persistent stipules. The male catkins scales are similar to the bud scales. (Compare Figs 61, 68) (PW).

Fig. 52 (above left). *Betula glandulosa* Canada, Labrador, Goose Bay with terminal male catkins and non-persistent stipules (PW).

Fig. 53 (above right). *Betula glandulosa* Canada, Quebec, Gaspé Peninsula, Xalibu, 1,000 m, exposed mountain rocks. The terminal male catkins, absence of persistent stipules, and relatively constant number of rounded teeth per leaf, largely independent of leaf size, are characteristic of the species (PW).

When the bud breaks in spring, the internode between the two pre-formed leaves hardly elongates so that the two leaves expand rapidly from the position of the bud (Clausen & Kozlowski 1965). In short shoots this is all the growth that will occur in that year and a bud usually develops in the axil of the uppermost leaf. However, in subgenus *Asperae* no bud forms on fertile short shoots and so they abort after fruiting.

Buds which are going to give rise to long shoots develop initially as described above but, after a pause during which the pre-formed leaves expand to more or less their full size, the terminal growing point begins to elongate and produce new leaves (Fig. 24). These new leaves are well spaced out as the internodes between them elongate considerably (Fig. 169). Long before growth and the production of new leaves stops, visible new buds are formed in the axils of the new leaves (Fig. 169). Only tiny, vestigial buds form in the axils of the pre-formed leaves of long shoots.

Kozlowski & Clausen (1966) showed that the pre-formed leaves at the base of a long shoot are essential for the development of that shoot, any significant damage to them often causing the death of the shoot. They suggest that the photosynthesis of the pre-formed leaves contributes primarily to the growth of the shoot, while that of the later formed leaves contributes to the development of buds in their axils.

The pre-formed leaves, whether on a short or long shoot, are usually relatively constant in form and size and are characteristic of the species. The later formed leaves which are well spaced along the long shoots often differ considerably in size and form from the pre-formed leaves, and often also

Fig. 54. Shoot structure as demonstrated by *Betula nigra*. Three long shoots on right arising from 'terminal' and subterminal buds with two mature basal leaves (which were preformed in the bud) with only short internodes between them, the less mature later formed leaves more widely spaced out along the shoots; spur shoots to left with rosettes of 2-4 leaves (which were preformed in the bud), the internodes between which have not elongated, and one of which bears a fruiting catkin (PW). (See also Fig. 152)

from one another, changing gradually towards an extreme form in the leaf at the tip of the shoot. In the key on p. 121, the leaf characteristics described are those of the pre-formed leaves, especially of vegetative short shoots.

Unfortunately the pre-formed leaves are often shed first, either under stress in a summer drought or during normal leaf fall, and may be missing in herbarium specimens of fruiting trees, so the diagnostically most useful leaves may be absent. In species of subgenus *Asperae* the short shoots bearing the female and fruiting catkins of some species often have exceptionally small leaves.

In *Betula nigra*, *B. luminifera*, *B. cylindrostachya* and *B. alnoides*, all species of low latitudes and sub-tropical to warm temperate climates, the internodes between the basal preformed leaves are much more elongated than in birches of more northerly origin. This situation also occurs on very strong shoots of almost any species. The female and fruiting catkin bearing shoots of *B. dahurica* and *B. raddeana* are particularly variable, varying, even on the same tree, from very reduced and bearing only a single small leaf to elongated and bearing up to six leaves well spaced out along the shoot. Seedlings, unless they are growing very poorly, only produce elongated shoots, and it may take several years before spur shoots are produced. This is particularly evident in the *Dahuricae*, especially in *B. dahurica* itself, in which all shoots, even in quite large (3 m) young trees, may be long shoots.

Elongated shoots are generally unbranched, except when growth is exceptionally vigorous, although *Betula dahurica* shoots show a greater tendency to branch than do those of other species.

Oil of wintergreen (methyl salicylate) is apparent in bruised or scraped fresh living twigs of species of section *Lentae*, subgenus *Acuminatae*, and *Betula corylifolia* and almost imperceptibly in *B. globispica*. The twigs of *B. lenta* used to be distilled as a source of this oil (oil of sweet birch), which was used medicinally and for flavouring (Uphof 1959).

INDUMENTUM

The young shoots with their expanding leaves may have hairs or glands which are easy to see in the living state but may be less clear on herbarium specimens, especially those which have been collected later in the season.

Hairs

Careful examination of the youngest shoots on both herbarium specimens and living plants has shown that no birch is totally glabrous. Many are, however, more or less glabrescent, with the hairs soon falling off as the shoots mature. Degree of hairiness is very variable within some species. In some populations of *Betula dahurica* hairs can be detected only on the youngest internodes of expanding shoots, while others have densely hairy young shoots with the hairs still persistent on the second year twigs. The Himalayan *B. utilis* is also very variable, with populations from the extreme ends of the distribution in the western Himalaya and N China being relatively hairless, while central Himalayan Nepalese populations may have densely hairy twigs as in *B. utilis* 'Inverleith'. Hairiness, which has often been used as a diagnostic character in keys, has therefore to be used with great caution.

Seedlings are much more hairy than more mature plants. Differential hairiness is often not evident in the seedling stage, seedlings of *Betula populifolia* and *B. pendula* being as hairy as those of *B. pubescens* and this may persist for a few years. *B. luminifera* is unusual in retaining densely hairy shoots until it is a sapling several metres tall, even though shoots on mature, catkin-bearing trees are more or less totally glabrous.

However the presence or absence, distribution and type of hairs can be of diagnostic value. The dense and very conspicuous hairiness of the stems of seedling *Betula papyrifera* seems to be absolutely constant and provides a useful character to distinguish its seedlings from those of *B. cordifolia* and

Fig. 55. *Betula papyrifera* leaf lower (abaxial) surface showing hairs. Long silky straight hairs lie on top of the midrib and similar but more wavy hairs stand out at right angles to the midrib and secondary veins, concentrated in the vein axils but, in this case, extending far up the midrib towards the next secondary vein (PW).

Fig. 56. *Betula utilis* leaf lower (abaxial) surface showing straight silky hairs similar to above and more rigid-looking axillary hairs standing out at right angles to the veins (PW).

all other species. The eastern Asian and western North American *B. pendula* subsp. *mandshurica* and hybrids usually have characteristic hair tufts in the vein axils on the leaf undersides (abaxial surface) of well-developed leaves (Figs. 55, 56). These are always absent from the European and western Asian subsp. *pendula* and nearly always from the Chinese/Tibetan subsp. *szechuanica*.

Four types of hair and one kind of gland are described in *Betula* by Hardin & Bell (1986). The largest and most conspicuous hair type, which they describe as acicular (i.e. needle-like), is unicellular and described as usually straight. In *Betula*, these are 0.5–3 mm long and can easily be seen with the naked eye or a ×10 hand lens. They may be appressed (i.e. lie along and parallel with the surface) and appear silky, or patent (i.e. stand out more or less at right angles) when they give a distinctly hairy appearance (Figs 55, 56). There seems to be no essential difference between these two types of hair except in the direction in which they lie. They are particularly conspicuous and persistent on the margins of the leaves, fruiting catkin scales, stipules and bud scales in almost all species.

Very short (0.03–0.30 mm), erect, puberulent hairs were called subulate hairs by Hardin & Bell (1986). These, when dense, form an almost velvet-like covering standing out at right angles to surfaces (Valkama *et al.* 2003) and require a magnification of at least ×20 to be detected. They are most easily seen by holding shoots or leaves up to the light and looking at the surface in silhouette.

From our observations, there appear to be hairs of all intermediate lengths between these short subulate hairs and the longer acicular ones. Intermediate lengths are particularly obvious in hairy species such as *Betula utilis*, *B. papyrifera*, *B. pumila* and *B. pubescens*. However, in other species there may be few if any hairs of intermediate length, the young shoots, petioles and leaves bearing variable amounts of silky hairs with an understorey of very short puberulent hairs or, as in *B. glandulosa*, there may be a few silky hairs on the leaf veins and a very short puberulence on the shoots with few if any longer hairs.

Hardin & Bell (1986) also describe birches as having straight or curly multicellular, uniseriate (with a single row of cells) hairs, which they describe as 'filiform', and flattened regularly twisted hairs which they term 'aduncate'. We did not detect these at all in our dissecting microscope examination of birch hairs, but we did not carry out a thorough study of the cellular structure of the hairs. On the other hand, in the hair tufts in the vein axils on the underside of the leaves of *Betula ashburneri* there are numerous flattened ribbon-like hairs to about 1 mm long and 0.036–0.05 mm wide mixed in with the needle-like hairs. This hair type has not so far been detected in any other species, but could be a non-twisted state of Hardin & Bell's aduncate hairs.

The very young shoots of a number of species (especially *Betula occidentalis*, *B. pendula* subsp. *mandshurica*, *B. ashburneri* and *B. glandulosa*) are very resinous and covered by a glutinous film which makes the detection of any hairs, especially silky ones lying along the shoots, very difficult. This sticky material can result in herbarium specimens appearing hairy due to fibres from the drying paper sticking to the specimen.

Glands

Unstalked peltate resin glands are said to be characteristic of all species of birch (Hardin & Bell 1986), but we have been unable to detect any in most of the *Betula potaninii* species group of the *Asperae*. These glands may be small and reddish and produce little if any resin as in *B. schmidtii*, or larger and produce large quantities of sticky resin as in many species of subgenus *Betula*.

The sticky secretion is usually initially clear and translucent and forms a 'blob' enclosing the gland. The resin soon solidifies or dries, initially to clear and reddish in colour, but usually eventually forming whitish 'warts'. The margins of each gland turn white first, the whiteness gradually spreading to the whole structure after one to two years depending on the species. Initially the glands are very sticky, often making the whole surface of the young leaves and twigs sticky to touch. The stickiness may function as a deterrent to aphid attack (Raatikainenen *et al.* 1992). It appears that the dried resin turns white due to the development of cracks and is probably due to the presence of air in the cracks. Poking a reddish dry gland with a needle induces cracks and the gland becomes whitish.

Fig. 57. *Betula utilis*. Current year shoot on left with small round and longitudinally elongated lenticels. Older stem on right with splitting outer bark on young branch and lenticels transversely elongated (PW).

LEAVES

Leaf shape and venation

Leaf shape in birches varies from ovate-lanceolate with a short petiole (Plate 3; Figs 114, 138), a very common leaf shape in temperate deciduous trees, to ovate, to rhombic to triangular-deltoid with a long petiole in the white barked birches (Fig. 58), to almost circular in *Betula nana* (Fig. 269). The leaf shape and length of leaf blade to petiole ratio provide useful differential characters for separating some major sections of the genus (see Key, Chapter 7). Other leaf characteristics are useful for separating smaller groups of related species, and even individual species. Most species of subgenus *Asperae* have leaves which appear pleated, especially when young, due to the veins being impressed on the upper surface and proud on the undersurface (Fig. 81). This contrasts with the usually much flatter leaves of species of subgenus *Betula* in which the veins are often proud on the upper surface. The very tough, leathery leaves with deeply impressed veins of *B. potaninii*, *B. calcicola*, *B. bomiensis*, *B. gynoterminalis* and *B. delavayi* are unique in the genus and often remain un-decomposed on the ground under the plants throughout the winter.

The leaves of *Betula insignis* and *B. cylindrostachya* are remarkably similar, although they are presumably only distantly related. Of course their catkins are very different, but non-fruiting specimens were previously very difficult to distinguish. However, the leaves of *B. cylindrostachya* have the secondary veins running straight into the major teeth and significantly thicker than the third order venation (craspedodomous), while in *B. insignis* one or two pairs of the lowermost (proximal) secondary veins are barely distinguishable from the tertiary venation towards the leaf margin and curl back into the leaf at the tooth base and send a branch into the tooth (semi-craspedromous) (Fig. 106). The leaves of these two species may be compared on the drawings in Chapter 7.

Fig. 58. Leaves of *Betula pendula* subsp. *pendula* from Mt Etna, Sicily showing typical long petioles, triangular-rhombic leaf shape and double toothing (PW).

Fig. 59 (left). *Betula luminifera*, (*Rix* 4121), Pendent long thin male and solitary female catkins, several leaves with dead, brown tips due to frost as leaves were emerging from buds. 26 April 2006 (PW).

Fig. 60 (right). *Betula cylindrostachya. KR* 976 Bhutan, Wang Chu Valley, Chapcha view point. Pendent expanded long thin male and female catkins shortly after emergence, reddish male catkins and females often in pairs. 26 April 2006 (PW).

CATKINS

Arrangement

Birches are hermaphrodite with male and female catkins borne on the same tree. The male catkins in most species are borne directly on the twigs at the ends of the long shoots, although sometimes also in the axils of leaves behind these, in which case they appear clustered towards the ends of the shoots (Fig. 114). The male catkins become visible in May to July as they begin to develop, and are usually fully formed by late summer. Female catkins develop somewhat later than the males, but within developing buds which will produce short lateral spur shoots with usually two (0–6) leaves and one or more (in most species one) female catkins at its tip in the following spring (Plates 4, 9, 10; Figs 30, 67, 162).

The arrangement of the catkins is characteristic in some groups of birch. In many species of subgenus *Asperae* (particularly *Betula insignis*, *B. alleghaniensis* and *B. medwediewii*, but also in the *B. potaninii* group) most of the female catkins are concentrated on certain shoots grown the previous year which have male catkins at their tips with all the buds behind these bearing female catkins (Fig. 30). The leaves on these fertile shoots are very much smaller than those on vegetative spur shoots and do not usually bear buds in their axils so that, following fruiting, the whole shoot dies. Particularly in *B. insignis*, these catkin bearing shoots are often rather weak shoots within the canopy (Fig. 108), and twigs bearing fruiting catkins are found on the ground under the trees during autumn and winter. In *B. medwediewii* these specialised catkin bearing shoots are often strong shoots sticking out from the canopy of the shrub, so that the large, showy, male catkins (Fig. 121) are conspicuous in spring, and the large, persistent, fruiting catkins are very obvious throughout the winter on the leafless twigs.

In the white barked birches of section *Betula* in contrast, the fruiting catkins are scattered around the canopy.

In the dwarf birches of section *Apterocaryon* (*Betula nana*, *B. michauxii*, *B. humilis*, *B. pumila*, but not *B. glandulosa*) the typical arrangement of the male and female catkins is very different, with the female catkins developing in buds towards the apex of long shoots grown the previous year, and the bud-like male catkins mostly borne laterally below these (Figs 65, 271, 282, 285, 288, 289), not terminally as in almost all other species.

Male catkins

The male catkins of tree birches usually elongate considerably when they expand in spring, while those of dwarf birches of subgenus *Apterocaryon* and *B. calcicola* remain short. The longest and most conspicuous male catkins are produced by *Betula insignis* and may reach 16 cm long.

Fig. 61 (top left). *Betula maximowicziana AGS/J* 402. Japan. Male catkin in winter state, 9th Feb. 2006 (PW).

Fig. 62 (bottom left). *Betula utilis*, (*CLD* 406) with knobbly male catkins (PW).

Fig. 63 (right). Knobbly male catkins and emerging female catkins of *Betula insignis* subsp. *fansipanensis*, (*KR* 2344) (PW).

The winter catkins of all subspecies of *Betula utilis*, occasionally *B. ermanii*, and *B. insignis* subsp. *fansipanensis*, and to a lesser extent in some trees of *B. grossa*, have embossed scales giving a knobbly appearance and feel to the tightly closed catkins (Figs 62, 63). This is much less evident on herbarium specimens, and less conspicuous once the catkins have begun to swell and elongate, but it is only between April and August when this character cannot easily be observed on living trees. It is a very useful character for distinguishing *B. utilis* in the living state from all other species and hybrids with which it might be confused, especially *B. papyrifera* and *B. pubescens*.

In some species, notably *Betula corylifolia*, but to some extent *B. delavayi*, *B. calcicola*, *B. potaninii* and *B. chichibuensis*, the male catkin scales have recurved, or at least free, apices, and the margins are not tightly appressed to one another. This gives a very distinct chaffy, dry and papery texture and 'shaggy' appearance to the male catkins in winter (Plate 2), very different from the smooth, shiny appearance of those of other species (Plate 16; Figs 61, 68). As with other male catkin characteristics, this is much more evident in living trees than on the herbarium sheet on which the dried catkin scales of all species shrivel and separate to some extent.

Female and fruiting catkins

The female catkins of birches are enclosed in the buds over winter, only emerging at bud break in the spring. They appear to be relatively similar in most species, being usually small, erect and often with reddish styles (Figs 64, 67, 243). Only in subgenus *Acuminatae* e.g. *Betula luminifera* (Figs 59, 60),

Fig. 64 (above). *Betula pubescens* var. *pumila*, (KA 21) terminal male catkins above lateral females, the normal arrangement in most *Betula* species. Canada, Newfoundland, Daniel's Harbour (PW).

Fig. 65 (right). *Betula pumila* with catkins showing arrangement on previous year's growth increment with males below females which, unlike the males, are borne on leafy shoots, the arrangement found only in *B. pumila*, *B. nana*, and *B. michauxii* (PW).

Fig. 66 (left). *Betula cordifolia*, Gros Morne, Newfoundland, Canada. Young female catkins pendent on emergence from bud in spring (PW).

Fig. 67 (right). *Betula papyrifera*, Belle Vue, Newfoundland, Canada. Young female catkins erect on emergence from bud in spring (PW).

B. corylifolia, and, surprisingly, *B. cordifolia* (Fig. 66), are they pendent from bud break, although in *B. corylifolia* they become erect later. The very young female catkins of *B. potaninii* and *B. bomiensis* (and *B. calcicola*?) point rigidly downwards as they emerge from the buds (Figs 79, 85), but they soon turn upwards and remain rigidly erect till they disintegrate (Figs 83, 86).

In all species of subgenus *Asperae* the mature fruiting catkins are erect, as are those of *Betula corylifolia*, *B. costata*, usually *B. ermanii* and *B. nigra*, and species of sect. *Apterocaryon* (except often in *B. glandulosa*). The fruiting catkins of species of section *Betula* and *B. utilis* become pendent shortly after flowering (Fig. 186), while the attitude of those of *B. ashburneri*, and species of section *Dahuricae* are very variable, from erect, to horizontal and held more or less in the plane of the leaves, to more or less pendent (Figs 69, 167).

There are usually three florets, and therefore three nutlets, borne in the axil of each scale of the female catkin. Only in *Betula corylifolia*, *B. globispica*, *B. chinensis* and *B. michauxii*, is this number normally reduced to one (Fig. 96). Exceptional scales have been found towards the base of catkins of *B. delavayi* with four and five nutlets per scale.

Fruiting catkin scales

The scales of the fruiting catkins are somewhat woody or more or less chaffy. Each has a central and two lateral lobes (Figs 32–34). In by far the majority of species the three lobes are similar in form and directed forwards, usually more or less oblong and parallel-sided. Only in subgenus *Acuminatae* (*Betula alnoides*, *B. cylindrostachya*, *B. luminifera*, *B. maximowicziana*), the Vietnamese *B. insignis* subsp. *fansipanensis*, and the east North American dwarf bog species *B. michauxii*, are the lateral lobes very much reduced or almost absent.

In the dwarf birches of section *Apterocaryon*, and in *Betula nigra*, *B. lenta* and *B. alleghaniensis*, and more rarely in other species, the lobes are usually sub-equal, whereas in most other species the middle lobe is longer. Only in *B. pendula* agg., *B. populifolia*, *B. celtiberica*, *B. papyrifera*, and *B. dahurica* do the lateral lobes usually differ markedly in shape from the middle lobe, being somewhat recurved and often more or less square. *B. pendula* has a particularly distinctive scale shape with a short triangular middle lobe and much longer and broader, usually recurved, lateral lobes. *B. papyrifera* has particularly variable fruiting catkin scales, the lateral lobes sometimes pointing forward and similar to the central, and sometimes recurved and very different in shape from the central one (Brittain & Grant 1965a; 1966; 1967b; 1968a, b)

SEEDS (FRUITS)

The birch seed consists of a flattened nutlet with two stigmas at the apex, and a papery wing on each side. What are usually called the seeds are technically fruits because they are developed from whole ovaries rather than from ovules, indicated by the presence of the stigmas. It is probable that they are 'false' fruits developed from inferior ovaries (Abbe 1938, 1974), but the flowering and fruiting structures are so reduced that precisely what they have evolved from is difficult to determine. Again, technically, there are two fruits (carpels) within each nutlet, but these are fused and usually only one produces an ovule and seed; occasionally both must develop as two seedlings are sometimes seen to emerge for a single seed. For simplicity we will refer to the whole winged structure as a "seed", and the central part containing the embryo as the "nutlet", whose dimensions are given separately from those of the wing.

The size and shape of the nutlets and wings, and especially the thickness and width of the wings, provide useful characters for distinguishing the major groups of species, and occasionally even for differentiating between related species. Skvortsov (2002) considered the differences in wing structure to be very important, regarding the wing as an extension of the ovary/fruit wall in the thick winged species in which the wing joins the styles (subgenera *Nipponobetula*, *Asperae*), but as a new outgrowth from the fruit in those groups in which the wing is thin and not joined to the style (subgenera *Betula*, *Acuminatae*). Hjelmqvist (1948) considers the wing homologous with the perianth.

Fig. 68 (left). Male catkins of form typical of most *Betula* species with closely appessed overlapping scales very different from bud scales. *B. dahurica* var. *parvifolia* (PW). (Compare Fig. 51).

Fig. 69 (right). Fruiting catkins of *B. dahurica* var. *parvifolia* lying more or less in plane of shoot and neither clearly erect or pendent (PW).

VARIATION IN SIZE OF VEGETATIVE CHARACTERS

Within a single tree, and even more so within a species, the variation in size of vegetative parts makes quoting lower limits to size ranges almost meaningless. Keys published in floras are usually constructed from measurements made on herbarium specimens, so the sizes given may bear little relationship to the maximum sizes of various organs, especially leaves, reached by plants in cultivation. Where possible keys should usually use qualitative rather than quantitative characters, ideally of reproductive organs. In the key provided here, average and maximum sizes measured are sometimes given without ranges, but it should be remembered that the sizes seen in herbarium specimens and weak shoots on cultivated plants may be very much smaller.

When the leaf measurements quoted for *Betula calcicola* by Li & Skvortsov (1999) in the *Flora of China* (2.0–3.2 × 1.5–2.5 cm) are compared with the measurements made on shrubs in cultivation, the leaves on the cultivated plants were found to be much larger (to 5.7 cm long and to 4.0 cm wide), although the maximum number of vein pairs is the same, suggesting that vein number is a genetically determined and taxonomically useful character. Herbarium specimens of this species are mostly from exposed limestone mountain cliffs, so it is not surprising that leaf measurements are much smaller on wild collected specimens than on shrubs grown in a sheltered situation in a garden. The maximum cultivated leaf size was a regular size for the shrubs, not an exceptional measurement made on only a few leaves. Similar tables could be constructed for many other species, highlighting one problem in using keys and descriptions in floras to identify cultivated plants.

The habit of a species can also vary enormously over its range. In Turkey, at the southern limit of its distribution, Browicz (1975) quotes records of *Betula pendula* growing as a 1–1.5 m shrub at high elevations, and a herbarium specimen [*A. J. Huxley* 20, (W!) from near Tatvan, Turkey, 7–9,000 ft. 1975] has small leaves similar in shape to those of *B. nana*, although there is no doubt that the specimen is *B. pendula*. Another problem in such situations is that the plants may be stressed and lack reproductive organs.

Fig. 70. Kenneth Ashburner by 3 m tall *Betula calcicola*, Howick Arboretum, Northumberland, UK (MF).

6. IDENTIFICATION AND NAMING OF BIRCHES

Correct naming is clearly necessary for scientific work, but it is also important in forestry and horticulture as different species may have different cultural requirements, and consistency of naming is important to customers. Classifications on which taxonomy and naming are based must, above all, be usable without too much difficulty (Davis & Heywood 1963). We hope that the classification of birches set out in this book will prove easy to use, so that correct naming of birch species will become more common.

The commonly cultivated white barked birches are difficult to name because in the past the species have been too narrowly defined; many white barked birches grown from seed from non native cultivated plants are hybrids, and this has added to the difficulties. Problems with the brown barked species are mainly due to their rarity in cultivation and therefore lack of familiarity with the species. We have been very fortunate that, over the last thirty years, we have been in a position to build up a living collection of almost all known species of *Betula*, growing them from wild-collected seed in at least two very different locations: Ness Gardens, Wirral, Cheshire near Liverpool in north-west England, and Stone Lane Gardens, Chagford, near Exeter, Devon, in south-west England. The growing conditions near Liverpool, low rainfall, and long, five to seven month, frost-free growing season, contrast with the more humid climate of Chagford on Dartmoor in Devon with only a four month frost-free growing season. These two gardens have enabled us to get to know the species well. We have also studied birches in cultivation at The Royal Botanic Gardens, Kew near London and its annex at Wakehurst Place in Sussex; The Royal Botanic Garden, Edinburgh and its annexes at Dawyck (near Peebles in the Scottish Borders) and Benmore (Dunoon, Argyll), and at the Arnold Arboretum in Boston, USA.

THE NAMING OF BIRCHES

White barked birches have always been difficult to name. Linnaeus put the two common European species in a single species *Betula alba* L. and his description clearly covers what are now regarded as the two distinct species *B. pendula* and *B. pubescens* (Linnaeus 1753; Tuley 1973). As Linnaeus's name cannot be assigned with certainty to either species, its use would cause confusion, and the later names *B. pendula* Roth (1788) and *B. pubescens* Ehrh. (Ehrhart 1791) are now used. In fact, *B. alba* L. has been used for both species (Tuley 1973; Govaerts 1996).

When the European silver birches were first named in the late 1700s, naturalists had not yet collected the related birches of Japan (*Betula platyphylla* and *B. mandshurica*), China and Tibet (*B. szechuanica*) or western North America (*B. neoalaskana*/ *B resinifera*). Once these silver birches were brought back and cultivated in Europe, not only did they look different from their European relatives and from one another, but also it was taken for granted that trees from such distant places were likely to be different species. At that time few if any specimens were available from the intervening geographical regions such as Siberia, Central Asia, and inland northern China, so the trees from different areas were named as distinct species.

However, we now know that silver birches are more or less continuously distributed across northern Europe and Asia in a continuous belt from Britain in the west to Japan in the east, and there seems to be no sudden change in appearance, just a gradual change from one to another. Similar situations exist in several other plant and animal species, with single species being distributed across the whole of Eurasia. Plants which live in association with birches across Eurasia include the following woody species: green alder *Alnus viridis*, juniper *Juniperus communis*; bog myrtle *Myrica gale*; wild cherry *Prunus avium*; bird cherry *P. padus*; red currant *Ribes rubrum*, gooseberry *Ribes uva-crispa*, and tufted hair grass *Deschampsia cespitosa* (Kawano 1966, Hultén 1971); and the animals moose *Alces alces* (Hundertmark 2002); great tit *Parus major* (Kvist *et al.* 2003; Päckert *et al.* 2005); harvest mouse *Micromys minutus* (Yasuda *et al.* 2005); and small tortoiseshell butterfly *Aglais urticae* (Vandewoestijne *et al.* 2004). Moreover, some of these, such as the moose, also extend across the Bering Strait into North America. It is therefore not unreasonable to suggest that the silver birch should be regarded as the same species from western Europe to Japan, and even into western North America.

A similar situation occurs with the even more variable Sino-Himalayan *Betula utilis* group of birches with a more or less continuous variation from north-east Afghanistan (*B. utilis* subsp. *occidentalis*) through subspp. *jacquemontii* and *utilis* to subsp. *albosinensis* in northern China.

Our solution to these problems has been to use species names only for taxa which are quite distinct in the wild, using the categories of subspecies or variety for some geographical variants which were formerly treated as distinct species. Thus *Betula pendula* is taken to include all the diploid white barked birches except *B. populifolia*, *B. cordifolia*, and the more distantly related *B. costata* and *B. nigra*. We hope that this will provide a more workable system that emphasises relationships and similarities.

Apart from the *Betula pendula* and *B. utilis* groups, we have left the taxonomy of most of the other species untouched, except to distinguish the Japanese, Honshu population of *B. dahurica* and to describe *B. ashburneri* and *B. skvortsovii* as new species.

We are aware that we have had to commit ourselves to some conclusions which further study may prove wrong. This is especially so with those recently-described Chinese species of which we have seen no specimens and have treated as synonyms. It may seem somewhat arrogant for us to treat many of these as synonyms, while at the same time to describe two new species and to accept as a species *Betula bomiensis* which was reduced to subspecific status in the *Flora of China* by Li & Skvortsov (1999). However, the Chinese taxa which we have reduced to synonymy are based on one or only a few herbarium specimens (few localities are quoted for them in the *Flora of China*), and have mostly been illustrated so that their overall appearance can be judged. In most cases their status as species was also doubted by Skvortsov. One of the new species we have described, *B. ashburneri*, is in cultivation, has been shown to be significantly different morphologically from any other species, and has a different ploidy level from its closest relatives.

Peter Crane (2004), formerly Director of Kew, stated: '2,000 plant species every year are *believed* (our italics) by their authors to be genuinely new to science', leading him on to advise that 'to arrive at a working list of the world's plant species, the next step and a key problem is how to detect synonymy. This is one of the objectives of a good monograph.' We have tried to fulfil this aim, defining distinct species and providing keys for others to identify them, while reducing other species names to synonymy. 948 species and subspecies names have been applied to taxa in *Betula* while we accept 46 species with only five subdivided into one to four subspecies.

Since Winkler's monograph in 1904 (Tuley 1973) there have been a few attempts to sort out the naming and classification of the white barked birches (Kuzeneva 1936; Johnsson 1945; Rehder 1940; Ashburner & Walters 1989), but no full monograph of the genus. We have an unprecedented range of living specimens in botanic institutions and private gardens to study (McAllister 1999). Living trees may show important characters which are hard to detect on herbarium specimens and knowledge of chromosome numbers is also valuable as species with different chromosome numbers will not normally interbreed.

7. TAXONOMIC TREATMENT

KEY TO THE SPECIES OF BETULA

Notes on the choice of characters used in the key

Where possible we have tried to use less variable, easily assessed characters, especially for the major subdivisions of the key. Where possible these characters are the ones used to define the subgeneric taxonomic groupings so that related species are keyed out together. Such characters are the nature of the seed wing: whether a narrow rim < 0.3 mm wide, a wider (to equal to nutlet width) more or less opaque wing more than one cell thick and adnate to the styles, or a very wide thin translucent wing one cell thick and not connecting with the styles. Similarly the shape of the lobes of the fruiting catkin bracts is usually easily assessed: whether the three lobes are similar or the laterals are of a different shape from the central lobe, and the relative lengths of the central and lateral lobes. However, even the nutlet and fruiting catkin scale size and characteristics can be extremely variable within a single fruiting catkin, especially towards the base and apex of the catkin, and with degree of maturity. Characteristics given are therefore those of scales and seeds from the middle of a well-developed catkin.

The attitude of the fruiting catkins, whether erect or pendent, is very easy to see in a living tree but can be difficult to assess in a herbarium specimen. *Betula dahurica*, and *B. raddeana* have fruiting catkins which are often horizontal and may vary from erect to pendent on the same tree. The same variability often occurs in northern forms of *B. pubescens* influenced by introgression from *B. nana*, and occasionally in a few other species.

Time of flowering is very important in distinguishing between *Betula alnoides* and *B. cylindrostachya*, and the early (summer) time of ripening of the fruiting catkins is important in identifying these species and *B. luminifera*, *B. nigra* and, to a lesser extent, *B. raddeana*.

Although size characteristics are often very variable, and it is possible to come across individuals with structures smaller or larger than the ranges quoted in keys or descriptions, it is still useful to give sizes of parts to give an idea of the appearance and scale of the particular part.

While most mature trees bearing fruiting catkins are likely to be identifiable unless they are hybrids, individual shoots or herbarium specimens may be unidentifiable to species even when of known wild source, especially within section *Betula*.

Notes

Many cultivated white barked birches of unknown wild origin are hybrids and are likely to be unidentifiable to species. Most involve 2 or more of the following species: *Betula pubescens*, *B. pendula* agg. (including *B. platyphylla* and *B. szechuanica*), *B. ermanii*, *B. utilis* agg. (including *B. jacquemontii* and *B. albosinensis*) and *B. papyrifera*.

When a species might fall within either option of a dichotomy in the key it is keyed out under both.

Leaf characters quoted are always those of the leaves on vegetative spur shoots or the basal pair of leaves on long shoots, i.e. those leaves which were pre-formed in the bud. **It is very important to remember this when using the key** as leaves on the elongated portion of long shoots (or on non-fruiting young trees, sucker or epicormic shoots) may be very different, often much more elongated with fewer pairs of veins and more sharply toothed or lobed — although unfortunately these may be the only leaves left later in the season. In species of the *Lentae* and some *Asperae* leaves on fruiting, catkin-bearing spur shoots may be very small, so characters of non-fruiting spur-shoot leaves and basal long shoot leaves are given.

Leaf vein number must always be considered at least plus or minus 1 because of the difficulty in deciding whether to count small apical veins. This is particularly problematic in species with leaves with few pairs of veins.

Silky hairs are those 0.5–2(3) mm long lying parallel to the surface. **Puberulent** hairs are **very short (< 0.05 mm), straight** hairs sticking out like a velvet pile at right angles to a surface, mostly shoot, petiole and veins. In some species (e.g. *Betula papyrifera*, *B. pumila*, *B. pubescens*, *B. utilis*) there are hairs of all lengths between these very short puberulent hairs and hairs to 2 mm long standing out at right angles to the surface on young shoots, veins on leaf undersides, and especially in the axils between the midrib and secondary veins.

Symbols used in Key

< Less than / up to
> More than
± More or less
c. about

1. Fruiting catkins rigidly erect; seed wing narrower than nutlet, opaque or translucent; lateral lobes of fruiting catkin scales ± similar to, and no wider than, central lobe; petiole short with blade:petiole ratio nearly always > 5:1 (mostly > 7:1) (except in *B. ermanii*, *B. humilis*, *B. nigra*); bark rarely white (except in *B. nigra*, *B. costata*, *B. ermanii*, *B. fruticosa*). **2**
1a. Fruiting catkins pendent; nutlet wing at least half as wide as to much wider than nutlet, thin, translucent, one cell thick, never contiguous with styles; catkin scale lateral lobes of ± different shape from (except *B. ermanii*, *B. pubescens*, *B. ashburneri*) and mostly wider than central lobe; petiole long with blade:petiole ratio < 7:1 (mostly < 5:1); bark white (most species of section *Betula*, *Dahuricae*) or not (*Acuminatae*, *B. occidentalis*, some *B. utilis*) **7**

2. Bark shaggy with numerous small curls, creamy-white (young branches) to very dark (old trunks); leaves rhomboid, conspicuously double toothed; fruiting catkins ripening and dispersing nutlets in May/July. . . . . 24. **B. nigra**
   [and sometimes other species of *Dahuricae* — see (38)]
2a. Bark never shaggy; fruiting catkins ripening in late summer to autumn **3**

3. Rare small tree or large shrub [< 4 (–10) m] from eastern Himalayan region (Bhutan) to Yunnan and Sichuan; stipules ± persistent; leaves with 6–9(–10) pairs veins, leaf blade < 5 cm; fruiting catkins erect to pendent, in clusters of 1–3; nutlet < 2 mm long . . . . .28. **B. ashburneri**
3a. Not as above, without above combination of characters. **4**

4. Bark white **5**
4a. Bark not white (whitish in *B. fruticosa*) **6**

5. Leaves lanceolate, usually > twice as long as wide, with 9–16 pairs veins, teeth terminating main veins often < 2 × as long as upper adjacent tooth, leaf base rounded to truncate; blade: petiole ratio > 4:1; fruiting catkins erect on short peduncles . . . . 27. **B. costata**

5a. Leaves triangular-ovate, usually < twice as long as wide, with 7–12 (–15) pairs of lateral veins, sharply toothed with teeth terminating main veins often > 2 × as long as upper adjacent tooth, leaf base often ± cordate; blade:petiole ratio < 4:1; fruiting catkins erect to pendent. . . . . 30. **B. ermanii**

6. Shrubs, mostly < 2 (– 4) m; twigs densely puberulent, variably long-hairy; leaves small (< 5 cm) with crenate or acute teeth and < 6 (7) pairs of veins; stipules ± persistent (except *B. glandulosa B. pubescens*); buds ovoid, ± blunt; fruiting catkin scale lobes ± equal in length, < or = twice as long as broad (or lateral lobes absent in *B. michauxii*); wing < or = nutlet and not contiguous with styles (section *Apterocaryon*) **8**

6a. Mostly trees (or multi-stemmed shrubs < 3 m with small leathery leaves and/or > 6 pairs veins and/or wingless seeds) with mostly larger leaves with acute teeth and > 6 pairs of veins; stipules persistent only in shrubby species and *B. corylifolia*; buds mostly acute; fruiting catkin scales and seeds various; seeds almost unwinged (c. 0.3 mm) to wing about half as wide as nutlet, opaque, > one cell thick (subgen. *Asperae* and *B. corylifolia*) **16**

7. Bark never truly white; mature leaves with long-acuminate teeth and > 9 pairs of veins; female catkins pendent from emergence from bud; fruiting catkins long-cylindrical (> 3 × as long as broad — often much more), pendent, 35–110 mm long; wing of seed much wider than nutlet; living scraped twig smelling of oil of wintergreen (*Acuminatae*). **34**

7a. Bark usually white (but not *B. occidentalis*, *B. ashburneri*, some *B. utilis* and *B. papyrifera*); teeth of mature leaves not long-acuminate; female catkin initially erect (except *B. cordifolia*), becoming pendent (except *B. ermanii*, *B. nigra*, *B. ashburneri*, some *B. pubescens*); fruiting catkin usually < 45 (< 50) mm long; wing of seed variable; living scraped twig never smelling of oil of wintergreen (sections *Betula*, *Dahuricae*) **37**

8. Leaf at least as broad as long (except some *B. glandulosa*), margin crenate (teeth rounded); nutlet < 2 mm long **9**

8a. Leaf usually distinctly longer than broad; margin usually serrate (teeth ± triangular), rarely crenate; nutlet > 2 mm long (except *B. humilis*) **11**

9. Fruiting catkin scales without distinct lateral lobes, one seed per scale; seed with rugose, spongy textured wing not well differentiated from nutlet; male catkins never terminal; shoot hairy; stipules persistent; leaf about as broad as long (E Canada) . . . . 42. **B. michauxii**

9a. Fruiting catkin scales with distinct lateral lobes, 3 fruits per scale; clear distinction between wing and nutlet **10**

10. Shrub to 3 m (but often much smaller); petiole conspicuous to 5 mm; leaves 5–30 mm, obovate to orbicular, truncate to cuneate at base; twigs sparsely hairy but densely puberulent and glandular; stipules not persistent; male catkins solely terminal (S Greenland, N America, central and E Asia). . . . . 41. **B. glandulosa**

10a Shrub to 1 m; petiole inconspicuous to 3 mm; leaves to 20 mm, often much less, orbicular to reniform, often broader than long and ± cordate at base; twigs densely hairy with few glands; stipules persistent; male catkins lateral below female on previous year's shoot, scales brown, rounded, largely papery and similar to bud scales (Europe, NW Asia, Greenland) . . . . 40. **B. nana**

11. Stipules not persistent, buds ± acute; preformed leaves on spur and basal on long shoots larger than leaves along long shoots; nutlet > 2 mm . . . . .35. **B. pubescens** var. **pumila**

11a. Stipules persistent or not; buds often rather blunt; preformed leaves on spur and basal on long shoots often smaller than leaves along long shoots **12**

12. Shoots and twigs thin and delicate, 1–1.5 (2) mm in diam., densely glandular, glands becoming white in second year; leaves < 25 (40) mm, coarsely and sharply toothed; stipules persistent; fruiting catkins slender, c. 4 (6) mm wide with scales to c. 3 mm; nutlet c. 1.5 mm long; with female catkins crowded on short laterals towards shoot apices with males crowded on laterals below, rarely terminal. . . . . 43. **B. humilis**

12a. Shoots and twigs stouter, variably glandular; fruiting catkins wider, > 4 mm, with scales > 3 mm; nutlet > 1.5 mm long. **13**

13. Leaves elliptic-rhomboidal to ovate-orbicular, margins serrate with acute teeth; shoots and twigs very glandular warty with conspicuous white warts on second year twigs; buds acute; fruiting catkins robust, > 5 mm across with scales 4–5 mm. . . . . .46. **B. gmelinii/apoiensis**

13a. Leaves ovate to obovate to ± circular; shoots variably warty; buds often blunt; fruiting catkins less robust **14**

14. Leaves triangular ovate to ± circular; stipules not persistent; leaf teeth acute; fruiting catkins erect to pendulous; male catkins terminal . . . . p. 63. **B. ×minor**

14a. Stipules ± persistent; fruiting catkins erect. **15**

15. Twig variably glandular and hairy, leaves elliptical-ovate to broadly elliptic, ± acute, teeth ± acute; male catkins primarily terminal, perhaps with some laterals distal to these (NE Asia) . . . . 45. **B. fruticosa**

15a. Twig variably glandular and hairy, often densely hairy in EN America; leaves obovate to almost reniform, broadly acute to rounded at apex, sometimes eglandular; teeth crenate; female catkins crowded on short laterals towards shoot apices with males crowded on laterals below, although also with some terminal on shorter shoots (N America) . . . . 44. **B. pumila**

16. Leaves whitish beneath, coarsely toothed, some of the large teeth in which the main 2er veins end being untoothed and giving a distinctive appearance to the leaf; seeds glabrous, one per fruiting catkin scale; conspicuous style enclosed within wing . . . . 1. **B. corylifolia**

16a. Leaves green beneath, more finely toothed with subsidiary teeth between those terminating the main veins; 3 seeds per fruiting catkin scale (except *B. globispica*, *B. chinensis* where 1), nutlets usually hairy, style ± absent (subgen. *Asperae*). **17**

17. Seed almost unwinged with only a barely visible narrow rim < 0.3 mm wide; lateral bracts of fruiting catkin scales with ± parallel-sided lobes; scraped fresh twigs not smelling of oil of wintergreen (except *B. globispica* in which it is only detectable with difficulty)(sect. *Asperae*) **18**

17a. Seed wing distinct, 0.5 mm, to c. half as wide as nutlet, opaque, > one cell thick; fruiting catkin bracts variable; scraped living twigs smelling of oil of wintergreen (sect. *Lentae*). **28**

18. Mature dry fruiting catkin ± globular when dry and 'open' due to scales spreading to release seeds and give 'spikey' appearance long before catkin disintegrates, scale central lobe > (2) 3 × as long as broad; one seed per catkin scale (except *B. fargesii*) (subsect. *Chinenses*) **19**

18a. Mature dry fruiting catkin longer than broad, never opening conspicuously and often retaining seeds until catkin disintegrates; fruiting catkin scale central lobe < 3 × as long as broad; three seeds per catkin scale (subsect. *Asperae*) **22**

19. Mature fruiting catkin < 25 mm long; fruiting catkin scale central lobe > 3 × length of lateral lobes; leaves triangular-rhombic, with 7–9 pairs of veins (*chinensis* agg.) **20**

19a. Mature fruiting catkin 25–35 mm long, fruiting catkin scale central lobe 2–3 × length of lateral lobes; leaves elliptic to ovate-triangular, with 8–10 pairs veins **21**

20. Bud blunt; leaf tooth mucro about half tooth length making leaf appear sharply toothed; fruiting catkin scales silky hairy abaxially . . . . 10. **B. 6x chinensis**

20a. Bud acute; leaf tooth mucro < half (about a third) tooth length; fruiting catkin scales glabrous abaxially . . . . 10. **B. 8x chinensis**

21. Leaves ovate-triangular; fruiting catkin scales very persistent for over one year, glabrous on lower surface; seeds mostly c. 4 × 3 mm, one per catkin scale (Japan) . . . . 11. **B. globispica**

21a. Leaves elliptic to broadly elliptic; fruiting catkin scales much less persistent (< 1 year), silky hairy on lower surface; seeds mostly about 3–3.5 × 2 mm, three per catkin scale (China, Sichuan & Hubei) . . . . 12. **B. fargesii**

22. Leaves with up to 16 pairs of veins, which are deeply indented only in some *B. delavayi*; leaves with conspicuous glands beneath (× 10) **23**

22a. Leaves with (5–) 10–22 pairs of deeply indented (except *B. chichibuensis*) veins; fruiting catkins scales < 5 mm; plant wholly eglandular **25**

23. Potentially large tree; leaves thin and delicate in texture, flat, ovate to lanceolate, with small brownish glands on young shoot, petiole and leaf underside, 9–10 pairs veins; twigs finely puberulent; fruiting catkin scales glabrous, not woody and persistent . . . . .2. **B. schmidtii**

23a. Small tree or shrub; leaves thin or leathery, lanceolate-elliptic to ovate to orbicular, deep green, glossy, veins ± impressed above, proud below; leaf veins run to sinuses then bend upwards to enter tooth; fruiting catkin scales ciliate, woody and persistent **24**

24. Leaves to 70 mm but often smaller, lanceolate-elliptic to almost orbicular, 2er veins usually at least somewhat indented above, 8–12 (16) pairs; fruiting catkin > 7 mm wide, scales to 8.5 × 6 mm, nutlet 2–3 mm long . . . . .7. **B. delavayi**

24a. Leaves small, to 35 × 23 mm, oblong-elliptic, flat, veins (5) 8–9 pairs; fruiting catkin much narrower, to 7 mm wide; scales to 5 × 4 mm; nutlet c. 2 mm . . . . 9. **B. skvortsovii**

25. Shrub to small tree; mature leaves and shoots hairy; leaves mid-green not leathery, triangular-ovate, with (10) 17–18 pairs of veins, mostly < twice as long as broad; female catkins erect from emergence from bud; fruiting catkins > 6 mm wide . . . . 3. **B. chichibuensis**

25a. Tree or shrub; leaves dark green, thick and leathery with (8) 10–22 pairs of deeply adaxially impressed veins; female catkins pointing downwards on emergence from bud but soon becoming erect . . .**26**

26. Shrub or small tree; leaves > twice as long as broad, with (9) 17–22 pairs of veins, mature leaves and shoots almost glabrous; fruiting catkins to about 6 mm across when mature . . . . .4. **B. potaninii**

26a. Tree or shrub; leaves < twice as long as broad, with (8) 10–16 pairs veins, mature leaves and shoots hairy; fruiting catkins > 6 mm across when mature **27**

27. Leaves ± orbicular, with (8) 15–16 pairs veins; young shoots very densely light golden brown hairy; buds spherical, blunt . . . . 5. **B. calcicola**

27a. Leaves mostly broadly-elliptic to ovate, with 10–12 pairs of veins; young shoot less densely hairy; bud ovate, acute . . . . 6. **B. bomiensis**

28. Shrubs with stiff, thick (2–3 mm) twigs and broadly elliptic to orbicular leaves **29**
28a. Trees or shrubs with thinner twigs and lanceolate to triangular-ovate leaves **30**

29. 1 year old twigs shiny yellow green, scarcely hairy except in youngest parts and around buds; leaves often ± orbicular . . . . 15. **B. medwediewii**
29a. 1 year old twigs not shiny, greyish, hairy; leaves broadly elliptic . . . . 16. **B. megrelica**

30. Fruiting catkin scale lateral lobes < a quarter length of terminal lobe . . . . 13. **B. insignis**
30a. Fruiting catkin scale lateral lobes > half length of terminal lobe **31**

31. Fruiting catkin scale surfaces usually glabrous, although margin ciliate, leaves and twigs soon glabrous; leaf margin finely toothed with usually 6 or more teeth per cm of mid-leaf margin; bark of trunk and thicker branches not yellowish, not exfoliating . . . . 17. **B. lenta**
31a. Fruiting catkin scales ciliate on lower surface, leaves persistently hairy beneath; leaf margins coarsely toothed with usually up to 5 teeth per cm of mid-leaf margin **32**

32. Leaf base usually ± cordate, spur shoot leaf underside and petiole with only silky hairs, puberulent hairs only on shoots and petiole bases on long shoots; bark not yellowish and exfoliating. . . . . 14. **B. grossa**
32a. Leaf base usually truncate to broadly cuneate, leaf underside and petiole with silky and short puberulent hairs **33**

33. Leaves with (9)12–18 pairs of veins, base usually truncate, rarely cuneate, apex acuminate; bark of trunk and thicker branches usually yellowish, exfoliating. . . . . 18. **B. alleghaniensis**
33a Leaves with 7–11 pairs of veins, base broadly cuneate, apex acute or only slightly acuminate; bark dark reddish, not exfoliating . . . . .19. **B. murrayana**

34. Leaves broadly ovate, deeply cordate at base, petiole > 25 mm; fresh winter male catkins > 3.5 mm across; lateral lobes of fruiting catkins bracts about half length of middle lobe; fruiting catkins ripening and disentegrating in autumn . . . . 23. **B. maximowicziana**
34a. Leaves ovate-lanceolate, not deeply cordate, petiole < 25 mm; fresh winter male catkins < 3.5 mm across; lateral lobes of fruiting catkins bracts very small (< a quarter central); fruiting catkins ripening and disentegrating in spring or early summer. **35**

35. Fruiting catkins mostly solitary, lateral lobes of bracts very small or absent . . . 22. **B. luminifera**
35a. Fruiting catkins mostly borne in groups of 2–5, lateral lobes of bracts more prominent **36**

36. Young shoots conspicuously glandular and often only thinly hairy; leaves lanceolate, base usually cuneate; toothing sparse with (0)1–2(3) teeth between major teeth at 2er vein ends, teeth forward pointing to incurved; flowering (Sept. to) Oct. to Jan.; fruiting catkin up to 6 mm broad; nutlet about 1.0 × 1.0 mm. . . . . .20. **B. alnoides**
36a. Young shoots eglandular or with a few glands (rarely many — in Fujian), often densely hairy; leaves ovate to oblong-ovate, base rounded or slightly cordate; toothing serrate with up to 7 teeth between major teeth at 2er vein ends, teeth pointing outwards to almost at right angles to the leaf edge; flowering March to May; fruiting catkin usually > 6 mm broad; nutlet longer than broad, about 1.5–3.2 × 1.0–1.5 m . . . . 21. **B. cylindrostachya**

37. Bark shaggy with numerous small curls (except *B. raddeana*); fruiting catkin attitude variable (erect to horizontal to often pendent when dry); fruiting catkin often ripening before those of other species (not *B. dahurica*), unripe fruiting catkins often 'spiky' with patent to reflexed central scale lobes so that wings of seeds are visible within catkin (not *B. dahurica*); leaves ± rhomboid (*Dahuricae*) **38**

37a. Bark never shaggy with numerous small curls; fruiting catkins usually obviously pendent (± erect in *B. ermani* and *B. ashburneri*); central lobes of scales of unripe fruting catkins appressed, forward pointing, not 'spiky'; leaves triangular to long-ovate (section *Betula*) **42**

38. Leaf rhomboid, long shoot leaves very distinctly coarsely double toothed, with 5–12 pairs of veins; fruiting catkin ripening very early (May/June) (Eastern N America) . . . . .24. **B. nigra**
38a. Leaf rhomboid-ovate, not or less distinctly double toothed, with 5–8 pairs of veins; fruiting catkin ripening July (*B. raddeana*) or autumn (*B. dahurica*) (Caucasus, E Asia) **39**

39. Unripe fruiting catkins 'spiky' with patent to reflexed central scale lobes so that wing of seeds are visible within catkin, ripening summer (July), scales eglandular; leaves with c. 5 pairs veins . . . . . .
. . . . .26. **B. raddeana**
39a. Unripe fruiting catkins only very rarely 'spiky', ripening in autumn, catkin scales glandular; leaves with 6–8 pairs veins (*B. dahurica* agg.) **40**

40. Male catkin scales appearing circular with margins mostly fully exposed; leaves small (< 65 × 50 mm), double toothed with fine toothing (mostly > 5 teeth per cm of mid-leaf margin); fruiting catkin scales 5–6 mm (Honshu, Japan). . . . . .25. **dahurica** var. **parvifolia**
40a. Male catkin scales overlapping apically; leaves often larger, single or double toothed with coarser toothing (mostly < or = 5 teeth per cm of mid-leaf margin). **41**

41. Fruiting catkins scales 5–6 mm (Hokkaido, Japan; Kurile Islands) . . . . . .25. **dahurica** var. **okuboi**
41a. Fruiting catkins scales (5.5) 6.5–7 mm (mainland Asia). . . . . .25. **dahurica** var. **dahurica**

42. Lateral lobes of fruiting catkin scales recurved, or at least spreading, usually longer and larger than and very different in shape from the middle lobe; twigs ± glandular, some very, mostly glabrous or glabrescent; leaves triangular-deltoid to rhombic to ovate, broadly cuneate to truncate, rarely cordate, at base **43**
42a. Lateral lobes of fruiting catkin scales not recurved but forward pointing, shorter than the middle lobe and similar or different in shape; at least young shoots and leaves conspicuously hairy (except *B. occidentalis*) and persistently minutely puberulent, glandular or not; leaves mostly ovate to long-ovate, cuneate to cordate at base. **49**

43. Young twigs and leaves very soon glabrescent; leaves usually without axillary hairs on underside; nutlets about twice as long as broad **44**
43a. Young leaves pubescent, in some cases glabrescent; axillary hairs persistent on leaf underside . . .**46**

44. Leaves usually with only a single order of toothing, often large (tree with leaves >70 × 60 mm likely to belong here), usually truncate or cuneate at base; trunk remaining white to base without development of thick dark corky bark (SW China, Tibet) . . 31c. **B. pendula** subsp. **szechuanica**
44a. Leaves usually double toothed, never large **45**

45. Large shrub or small tree; leaves long-acuminate; fruiting catkins small (< 20 × 6 mm), tight and tardily dehiscent, trunks white to base (Eastern N America) . . . . 32. **B. populifolia**
45a. Tree; leaves acuminate; fruiting catkins larger, soon dehiscent, trunk soon developing thick dark corky bark at base (Europe, W and central Asia) . . . . 31a. **B. pendula** subsp. **pendula**

46. Leaves cuneate to shallowly cordate, ± persistently silky hairy on veins beneath with persistent puberulent hairs on leaf lamina and twig, vein pairs 5–9. . . . .**47**
46a. Shoots and leaves usually soon ± glabrous except beneath in vein axils, vein number various . . .**48**

47.   Young twigs usually persistently puberulent; leaf vein pairs 6–9, seedling stems very densely hairy; nutlets < twice as long as broad, hairy in upper half (N America) . . . . .39. **B. papyrifera**

47a. Leaf vein pairs 5–7 (8), seedling stems not persistently densely hairy; nutlets mostly about twice as long as broad, with only a few hairs (E Asia, Western N America to E Ontario) . . . . .31b. **B. pendula** subsp. **mandshurica**

48.   Leaves ± truncate to cordate at base, with 8–10 pairs of veins; nutlets about twice as long as broad (Eastern N America) . . . . 33a. **B. ×caerulea**

48a. Leaves cuneate to almost truncate at base with 5–6 pairs of veins; nutlets < twice as long as broad (Cordillera Cantabrica, N Spain). . . . . .34. **B. celtiberica**

49.   Rare small tree or large shrub (< 3–10 m) from Bhutan east in Himalayan region to Yunnan and Sichuan; stipules persistent; shoot densely resinous glandular; leaves with 6–9 (–10) pairs veins, leaf blade < 5 cm; fruiting catkins erect to pendent, in clusters of 1–3; nutlet < 2 mm long . . . . 28. **B. ashburneri**

49a. Not as above; without above combination of characters. **50**

50.   Leaves triangular-ovate, usually less than twice as long as wide, with 7–12 (–15) pairs of lateral veins, sharply toothed with tooth ending main vein > twice length of neighbouring tooth, or with less sharp toothing but bud densely grey-hairy; fruiting catkins often ± erect (NE Asia, Korea, Japan) . . . . .30. **B. ermanii**

50a. Fruiting catkins pendent (except often *B. pubescens* var. *pumila* and *B. microphylla*); without above combination of characters. **51**

51.   Male catkins with embossed scales, giving knobbly feel to overwintering catkins; styles short c. 0.5 mm; wing width about equal to nutlet width; leaf blade: petiole ratio often > than 5:1; lower (proximal) secondary veins parallel, often lacking strong tertiary branches (Sino-Himalaya to N Vietnam & Beijing) (*B. utilis* agg.). **52**

51a. Male catkin scales not embossed, giving ± smooth surface to overwintering catkins; styles > 0.75 mm; wing wider than nutlet; leaf blade:petiole ratio usually < 5:1; lower (proximal) secondary veins usually more divergent, often with strong tertiary branches (Eurasia excluding Sino-Himalayan region; N America) **56**

52.   Bark white; leaves with 7–10 (13) pairs of lateral veins, slightly (rarely very) pubescent beneath (W Himalaya) **53**

52a. Bark pale orange to light or dark chocolate brown with a variable amount of white bloom, rarely almost white. Leaves ± softly downy beneath especially near midrib, with 10–14 pairs of lateral veins (Central Himalaya eastward and northward to Beijing) **55**

53.   Bark pure white; shoots densely hairy, almost eglandular; leaves ovate to broadly elliptic with < 13 pairs of veins, densely hairy on veins beneath . . . . 29b. **B. utilis 'Inverleith'**

53a. Bark variable in colour; twigs at least slightly glandular; leaves ovate-triangular **54**

54.   Buds encrusted with white resin; twigs densely warty; leaves with many resin glands, in extreme west (Tajikistan, Afghanistan, Pakistan) male catkins less knobbly, leaves smaller, more triangular and deeply toothed than in other *B. utilis* . . . . .29d. **B. utilis** subsp. **occidentalis**

54a. Buds ± resinous; twigs sparsely warty; leaves large, flat, with few resin glands . . . . 29b. **B. utilis** subsp. **jacquemontii**

55. Leaves ± glossy above (adaxially), firm and leathery (coriaceous) in texture with veins ± impressed, ovate; bark white to dark brown. . . . . .29a. **B. utilis** subsp. **utilis**

55a. Leaves matt above, ovate to lanceolate, less leathery, veins not indented; bark copper coloured, with or without white bloom. . . . . 29c. **B. utilis** subsp. **albosinensis**

56. Leaves ovate to rhomboid, veins 4–5 (8) pairs, larger blades usually < 5 (–7) cm; seed wing to 1.5 times width of nutlet, rarely protruding beyond styles (Eurasia) (*B. pubescens* group & *B. occidentalis*) . . **57**

56a. Leaves ovate to triangular, veins 6–9 (14) pairs; larger blades usually > 5 cm; seed wing > 1.5 times width of nutlet, often (*B. papyrifera*) protruding beyond styles. **63**

57. Twigs sticky; trunks several, never white barked; young shoots and leaves ± glabrous although sometimes puberulent; leaves broadly ovate to rhombic to almost circular with 2–5 pairs of veins . . . . 33. **B. occidentalis**

57a. Twigs not glutinous, trunks usually whitish-barked **58**

58. Small tree; leaves ± triangular to rhombic, truncate at base, twig tomentose and ciliate; veins 4–5 (6) pairs (Central Asia, very rare in cultivation). **59**

58a. Small to large tree; leaves ovate (in north) to ± triangular (mainly in south), veins 5–8 pairs (*B. pubescens* agg.) **60**

59. Bark yellow-white or light yellow-brown; leaves broadly ovate-rhombic, base broadly cuneate; fruiting catkin usually pendent . . . . .37. **B. tianshanica**

59a. Bark greyish-white; leaves rhombic to ovate-rhombic, base cuneate; fruiting catkin usually ± erect . . . . 36. **B. microphylla**

60. Leaves ovate to triangular to rhombic to elliptical, large, 30–50 (98) mm; apex acute or acuminate, apex angle generally 20°–40° **61**

60a. Leaves broadly ovate, small, leaf single toothed, mostly < 30 mm, apex short-acute, apex-angle generally 40°–50° **62**

61. Leaves ovate-rhombic to elliptical, often glabrescent even in vein axils although shoots and petioles usually remain puberulent, ± double toothed (Caucasus, Turkey) . . 35. **pubescens** var. **litwinowii**

61a. Leaves ovate to triangular-ovate, single or occasionally double toothed, variably persistently hairy, shoots often glabrescent; leaf base cuneate to truncate to somewhat cordate (Europe & W Asia except in north). . . . . 35. **pubescens** var. **pubescens**

62. Tree to small shrub with stiff, often densely puberulent and hairy twigs; leaves broadly ovate, leaf base cordate to rounded, crenations mainly shallow . . . . 35. **pubescens** var. **pumila**

62a. Small tree to shrub, stems fluted; shoots and leaves with few if any long hairs when mature; twigs thin, often pendent, glandular when young, long shoots persistently puberulent; buds sticky. . . . . 35. **pubescens** var. **fragrans** (north west UK, especially Scottish Highlands)

63. Fruiting catkin bract scale central lobe on average smaller and shorter than spreading to recurved lateral lobes . . . . 33a. **B. ×caerulea**

63a. Fruiting catkin bract scale central lobe longer and usually larger than the erect lateral lobes **64**

64. Fruiting catkin bract scale central lobe oblong, ± parallel sided; spur and basal long shoot leaves usually cordate at base; lenticels on young twigs large (smallest c. 1 mm); female catkins pendent from emergence from bud . . . . .38. **B. cordifolia**

64a. Fruiting catkin bract scale central lobe triangular to rhomboid; leaves only occasionally cordate at base; lenticels very variable in size, smallest about 0.25 mm; female catkins erect on emergence from bud, soon becoming pendent . . . . .39. **B. papyrifera**

BETULA Linnaeus, *Species Plantarum* 2: 982 (1753).

Trees or shrubs, deciduous (except perhaps *Betula alnoides* and perhaps sometimes *B. nigra*). *Buds* ovoid, rarely globular. *Leaves* ovate-lanceolate to rhombic to triangular to more or less circular, sometimes with two orders of toothing, bearing simple hairs and usually (except *B. potaninii* species group) stalkless, sometimes caducous, often resin-secreting glands. *Male inflorescence* elongate, cylindric, exposed over winter, expanding in spring to pendulous (except in dwarf species of section *Apterocaryon*) catkin of bracts subtending usually 2 bracteoles and 3 flowers; perianth 0–4-lobed; stamens 2; anthers 2-loculed. *Female inflorescence* 1–5 in terminal racemes on short shoots, emerging from bud in spring to form usually erect, small, more or less cylindrical catkin of numerous usually 3-lobed bracts, each subtending usually 3 (rarely 1) flower. *Fruiting catkin* erect to pendent, of leathery, usually 3-lobed, bracts subtending (1–)3 usually winged nutlets, breaking up or not on ripening and drying. *Nutlet* compressed, elliptic to almost circular, with 2 persistent styles.

Subgenus NIPPONOBETULA section NIPPONOBETULA

Small trees; *leaves* glaucous on abaxial surface, eglandular, coarsely toothed with many teeth entire and not further toothed; *male catkin* scales with reflexed apices; *female catkins* initially pendent with single fruit per bract; *nutlet* with distinct style, style and stigmas adnate to wing. One species in Japan.

1. BETULA CORYLIFOLIA

HISTORY

Betula corylifolia was first collected by Tschonoski Sungawa (1841–1925), a Japanese collector who was Maximowicz's assistant during his stay in Japan from 1860 to 1864. After Maximowicz returned to St Petersburg, Tschonoski continued to send specimens, and is commemorated in numerous familiar species, such as *Carpinus tschonoskii*, *Acer tschonoskii*, and *Malus tschonoskii* (Ohwi 1965).

    Betula corylifolia was introduced to cultivation by E. H. Wilson (*W.* 7651) in 1914 from Honshu, prov. Shimotsuke, at Yumoto, and has been represented in Britain since then by a single tree in the Royal Botanic Garden, Edinburgh. Even in the presence of a wide range of other species this tree has produced only seedlings of true *Betula corylifolia*, suggesting that it cannot hybridise with other birches.

ECOLOGY AND OCCURRENCE IN THE WILD

Betula corylifolia has a very restricted range in the wild, occurring only in central Japan. In subalpine forests of central Honshu it occurs with *B. ermanii* in areas dominated by *Abies mariesii* and *A. veitchii*, with *Tsuga diversifolia* regenerating in openings in the forest (Yamamoto 2000).

    In the Nikko region in Honshu north of Tokyo, we (KBA) saw trees of *Betula corylifolia* scattered here and there on the edge of woods, although they were never conspicuous because of their dark and only moderately shining bark.

TAXONOMY, RECOGNITION AND RELATIONSHIPS

Betula corylifolia is a very distinct birch, probably not closely related to any other living species. Skvortsov (2002) created the subgenus *Nipponobetula* for it and we follow him in this. It differs from all other species of the genus in its leaves, in that the major teeth in which the secondary veins end are broad and often lack subsidiary teeth between them. The whitish leaf underside, and matt, bluish-green upper leaf surface are also distinctive and give the foliage an appearance quite different

PLATE 1. *Betula corylifolia*
Painted by Josephine Hague

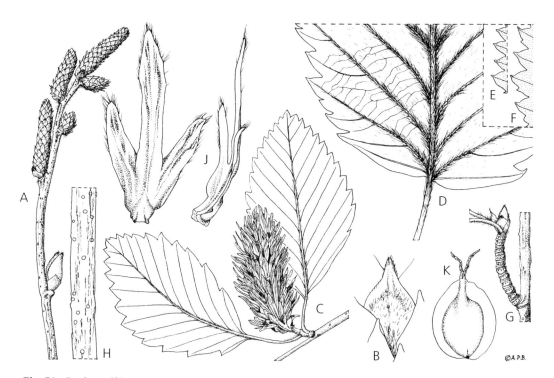

Fig. 71. *Betula coryilfolia.* **A** twig with bud and immature male catkins × 1; **B** distal portion of male catkin scale × 10; **C** fruiting shoot with mature fruiting catkin × 1; **D** underside of leaf base × 2; **E–F** marginal serrations on veins 5–8 (from base)× 1; **G** 12 year old lateral shoot (common on specimen from ~ 2000 m) × 1; **H** bark of twig × 2; **J** female catkin scale, back and side views × 5. **K** seed × 5. **A, B** from *Wilson* 7651 (Hondo, 16 Oct. 1914); **C, J, K** from *Wilson* 7502 (Hondo, 13 Sept. 1914); **D, E, H** from *Togashi* 10670 (Mt Kitadake, Yamanashi Prefecture, 6 Aug. 1967); **F, G** from *Furuse* 19972 (Tokugoo-Tooge, Hondo, 1900–2150 m, 21 Aug. 1948). DRAWN BY ANDREW BROWN.

from that of any other birch. The male catkins are also quite unlike those of other species, most resembling those of *B. calcicola* or *B. chichibuensis* in their thin, hardly fleshy scales with elongated triangular apical portions. These are recurved and so give the male catkins a very different appearance from the smooth glossiness of those of most birches. The fruiting catkins and their scales are large and only one nutlet is borne in the axil of each scale. The glabrous nutlets with a distinct beak and narrow wing also distinguish this species from all others in the genus (see Figs 32–34).

CULTIVATION

Seed collected from the tree in the Royal Botanic Garden, Edinburgh, probably in 1981, germinated to give six seedlings, which grew to over a metre in three years; five of them died, largely of drought, at Ness, and one at Stone Farm grew to four metres before dying in 2006. There is now another, younger tree in Edinburgh, probably grown from seed obtained from a Japanese nursery (as *B. maximowicziana*) of unknown wild source (Kerby, pers. comm. 1982) and several plants in Edinburgh and at Benmore from a recent, 2005, expedition to Nigata (*Brownless, P.; McNamara, B.; Bolton, T.; Jamieson, R. & Tsukie, S.*, no. 174). A recent introduction of seed from Mt Goyo, Iwate Prefecture, gave rise to two seedlings which were planted in a more favourable moist site at

Ness and both grew well to six metres and fruited, one producing viable seed by self-pollination. In spring 2006 many branches on both trees failed to break bud and the trees have since died, perhaps from waterlogging.

As a young tree, *Betula corylifolia* has a neat pyramidal habit and makes an attractive specimen with its sea-green leaves with whitish undersides. The large, erect fruiting catkins, which remain on the branches for most of the winter, are an unusual and interesting feature. It is relatively slow growing, and like many birches, perhaps especially those from mountainous areas, it appears to be very intolerant of shade and seems to grow best in moist, humus-rich soils in a relatively humid climate, in a sunny situation.

Soft cuttings, taken in early summer, root readily under mist. This is welcome, for it means that the better-growing seedlings may be selected and propagated.

PLATE 2. *Betula corylifolia* (catkin)
Painted by Josephine Hague

Betula corylifolia Regel & Maxim., *Bull. Soc. Imp. Naturalistes Moscou* 38 (2): 417 (1865). Type: In den höhern Gebirgen der insel Nippon, *Tschonoski*: (holotype LE).

ILLUSTRATIONS. *Woody Plants of Japan*: 119 (1985).

DESCRIPTION. *Trees* to 17 m. *Bark* dark reddish-brown. *Twigs* to 4.25 mm in diam., brown, with variably persistent silky hairs and patent hairs of variable length, scented of oil of wintergreen when skinned. *Buds* ovate-elliptical to oblong-ovate on 2nd year twigs, about 6–11 × 3–4 mm. *Young shoots* covered with white silky hairs; petioles and veins on leaf underside silky hairy, glabrescent. *Stipules* ovate-triangular, about 10 × 5 mm, somewhat persistent (but less so than in the dwarf birches of section *Apterocaryon*). *Leaves* with petiole 7–15 mm; blade, broadly ovate, 40–80 × 25–50 mm, giving an approximate blade:petiole ratio of 5:1 to 6:1; coarsely toothed with 0 or 1–3 subsidiary teeth between the major teeth terminating the main veins; bluish-green, whitish beneath, with 8–14 pairs of veins. *Male catkins* up to 2 terminal, with up to 2 laterals, to 35 × 6.5 mm (not including protruding recurved scale tips in width), expanding to 60 × 9 mm; scales peltate but attached in lower third, with greenish-brown fleshy (living) base gradually tapering into triangular brown (dead) recurved apex, about 3 × 1.75 mm, diamond-shaped, brown. Margins ciliate, and both surfaces surfaces silky hairy. *Fruiting catkins* erect, to 3–6 × 2 cm, cylindrical. *Scales* to 9–13 × 6 –8 mm, ciliate on margins and at base of abaxial surface. Central lobe to 6–8(–9) mm, oblong; lateral lobes 3–4(–5) mm, oblong. *Seeds* one per scale, glabrous, to 4–5 × 5 mm, nutlet 2.5–3 mm across, wing to 0.5 mm, about ¼ nutlet width, opaque, more than one cell thick, beak style c. 1 mm, stigmas c. 2 mm.

DISTRIBUTION. Japan, Honshu (Kinki district and eastwards in the mountains).

HABITAT. Subalpine zone of high mountains, mixed with other species (Tabata 1966).

FLOWERING AND FRUITING TIME. April, May; fruiting in October.

CHROMOSOME NUMBER. Diploid: 2n=28.

Subgenus ASPERAE

Trees or shrubs; short shoots bearing female and fruiting catkins with smaller and narrower leaves than vegetative short shoots and lacking axillary buds; *leaves* on long shoots usually smaller and differing in shape from those on vegetative short shoots, petioles relatively short; fruiting catkins erect and tardily dehiscent with lateral bract lobes of similar form and width to central lobe; fruits with opaque wing two cells thick which is adnate to styles.

It is likely that the common ancestor of all species in subgenus *Betula* would have been classified within this subgenus and so, technically, if we applied cladistic logic (Cronquist 1987), should be classified within the *Asperae* as otherwise the *Asperae* are paraphyletic, having given rise to the subgenus *Betula* which are not classified within it.

Here we divide subgenus *Asperae* into two sections as the species seem to fairly naturally fall into two groups, section *Asperae* with very narrowly winged nutlets restricted to China, Japan and adjacent continental NE Asia, and section *Lentae* with more broadly winged nutlets with a disjunct relict distribution in Japan, China, Caucasus, and eastern North America.

Subgenus ASPERAE section ASPERAE

Trees or shrubs with *nutlets* lacking a conspicuous wing, having only a narrow (< 0.3 mm) opaque rim.

Subgenus ASPERAE section ASPERAE subsection ASPERAE

Fruiting catkins hardly open on drying, often retaining a high proportion of the fruits until the catkin disintegrates; catkins scale lobes not obviously oblong or parallel sided.

2. BETULA SCHMIDTII

HISTORY

Betula schmidtii was described by Edward Regel in 1865 in the addendum to his monograph of *Betula*. He based the name on specimens collected by Schmidt & Maximowicz in south-eastern Manchuria, now part of China. It was introduced to cultivation by E. H. Wilson from the Nikko region of Honshu in 1914.

ECOLOGY AND STATUS IN THE WILD

Betula schmidtii is widespread in north-eastern Asia, growing in north-eastern China, Korea, the northern half of Honshu and the Ussuri region in the Russian Far East. In the wild, this species is said to make a tree of up to 35 m, with a girth of 9 m (Bean 1970; Li & Skvortsov 1999), and so it must be the tallest growing birch, although in the mountains of Korea it forms gnarled, multi-stemmed trees (Fig. 73).

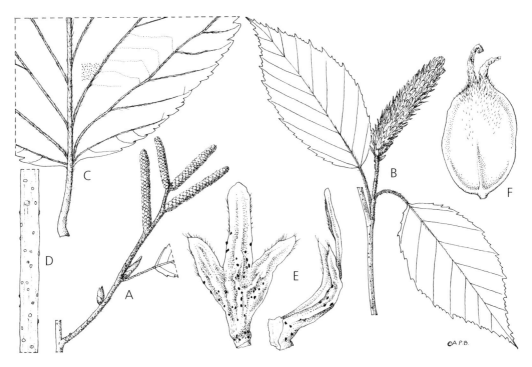

Fig. 72. *Betula schmidtii.* **A** twig with buds and immature male catkins × 1; **B** fruiting shoot with mature fruiting catkin × 1; **C** underside of leaf base (indication of gland distribution right of midrib) × 2; **D** bark of twig × 2; **E** female catkin scale, back and side views × 10; **F** seed × 15. **A, B, D–F** from *Furuse* 45797 (Japan, Kawaji-Fujiwara-choo, Shiwoya-gun, 10 Sept. 1967); **C** from *Furuse* 43524 (Japan, Kawaji-Fujiwara-choo, Shiwoya-gun, 10 Aug. 1967).
DRAWN BY ANDREW BROWN.

Fig. 73. Base of trunk of *Betula schmidtii*, Tuta-San, South Korea (KBA).

On the western side of the Korean peninsula where the coast is very indented, the rocks are igneous with scattered boulders of granite and gneiss — with the gneiss strikingly reflective in the moonlight. These rocks extend eastwards into the peninsula and the floor of the mixed forest and the rivers running through it are strewn with boulders — moss-covered under the trees. We (KBA) saw *Betula schmidtii* in this habitat in central Korea in the Odae-san National Park, growing as scattered many-stemmed trees in the mixed forest.

TAXONOMY, RECOGNITION AND RELATIONSHIPS

Betula schmidtii can be distinguished from all other species by the combination of lanceolate leaves with usually 9–10 pairs of veins and small, 20–30 mm long, erect fruiting catkins with unwinged seeds. It has thick branches and blackish bark.

The foliage is an unmistakeable shade of green, caused by very small bluish-white glands on the shoots and leaves. On young trees and seedlings, the shoots are covered by a brownish down.

The timber of *Betula schmidtii* is very hard and so dense that it is said to sink in water (Bean 1970). Under good conditions it develops yellow autumn colour, and attracts attention as a totally black barked tree.

CULTIVATION

Betula schmidtii is rare in cultivation. Introductions from the Ussuri region and some from Korea suffer from winter dieback at Ness and so do not make well-shaped trees, but they might grow better in more continental climates. There are good, large (Wilson?) specimens in the Arnold Arboretum.

Seed collected from a tree in the Odae-san National Park in Korea in 1980 (see above) germinated well and seedlings were grown in a plastic tunnel. In all innocence they were then lined out in the open ground at Stone Lane Gardens in March; all were dead by May, having suffered from repeated air frosts.

Plants of Japanese origin, particularly from more oceanic areas, should be better adapted to the British climate, and Japan, perhaps, is the origin of the well-established trees in the British Isles, such as Wilson's collection, reported in Bean (1970) to be growing well at Kew, and other trees in the Royal Horticultural Society's Garden, Wisley and in the Royal Botanic Garden, Edinburgh. At present, several young trees are to be found in the *Betula* collection at Wakehurst.

Betula schmidtii Regel, *Bull. Soc. Imp. Naturalistes Moscou* 38 (2): 412 (1865). Type: in der südöstlichen Mandschurei von *Schmidt* und *C. Maximowicz* gesammelt (holotype LE).

ILLUSTRATION. *Woody Plants of Japan*: 116 (1985). Fig. 73.

DESCRIPTION. *Tree* to 35 m. *Bark* dark grey, peeling, with conspicuous raised lenticels on younger trees, becoming plated when older. *Twig* to 1.5 mm in diam., light brown, finely brown-tomentose. *Bud* elliptic-ovate to 6 × 2.5 mm. scales ciliate at base and more or less so dorsally; more than 3 times as long as broad. *Shoot* coarsely ciliate hairy, glabrescent, stipule margins initially ciliate, soon glabrous, abaxal surface silky hairy. *Leaf* petiole 5–10 mm; blade ovate-elliptic, 40–80 × 25–45 mm with 7–10 pairs of veins; silky hairy on vein beneath; small reddish-brown glands on leaf abaxial surface and petiole underside, marked hyaline zone around veins on young leaves, 3rd order veins not regular. *Male catkin* 1–2 terminal with up to 2 laterals, to 27 mm opening to 60 mm; scales peltate, shield-shaped, about 1.5 × 1 mm, light brown, upper margins ciliate. *Fruiting catkin* erect; on peduncle of 3–6 mm; about 20–30 × 8–9 mm; scales (4–)5–6 mm, ciliate hairy, mid-lobe about 3 times as long as broad. Seed c. 2 × 1.5 mm with very narrow rim.

DISTRIBUTION. E Russia (Primorye), Japan, Honshu (central and northern district), Korea, China (E Jilin, NE Liaoning).

HABITAT. Rocky places in the mountains, in mixed forest (Tabata 1966).

FLOWERING AND FRUITING TIME. April, May; fruiting October onwards.

CHROMOSOME NUMBER. Diploid: 2n=28.

3. BETULA CHICHIBUENSIS

HISTORY, ECOLOGY AND STATUS IN THE WILD

Betula chichibuensis is very rare in the wild and confined to the Chichibu area in the mountains of central Honshu on Mt Kamo-san, near Tano-Gun, in Gunma prefecture, where it forms a multi-stemmed shrub or small tree to 10 m high. It was described as recently as 1956 by Hiroshi Hara (1911–1986).

TAXONOMY, RECOGNITION AND RELATIONSHIPS

This species is very distinct in its soft ovate leaves with up to 18 pairs of impressed lateral veins, clusters of numerous male catkins from several buds towards the ends of the twigs, and short, upright fruiting catkins with wingless seeds. Most similar is probably *Betula schmidtii*, but this has flatter leaves with fewer than 11 pairs of veins, male catkins only at the ends of the twigs, is much less hairy, and forms an upright tree. Other related Chinese species have much more leathery leaves: *B. potaninii* has smaller, darker green leaves and smaller fruiting catkins; *B. calcicola* broader leaves and much hairier shoots; and *B. delavayi* much darker green leaves and larger female catkins and seeds.

PLATE 3. *Betula chichibuensis* (leaves)
Painted by Josephine Hague

CULTIVATION

On a visit to Japan in 1980 KBA requested seed of *Betula chichibuensis*, but it was only in 1986 that Mr Tetsuo Satomi was able to visit the trees when ripe seed was available and send it to Britain through Professor H. Ohba of Tokyo University Museum. Earlier visits to the site had been unsuccessful for various reasons, in one case because a hurricane had so damaged the trees that no seed had been set. The seed was very variable in size and of low percentage viability (c. 1%), only the larger seeds being viable.

Some of the seeds were sown, and others retained in case of failure. Eight germinated, forming densely branched plants about 30 cm tall in their first year. They were wintered outside so that they would receive sufficient cold to induce bud break — many temperate woody species fail to break bud in spring if they have not been chilled (vernalised) for long enough. At the first sign of bud break in spring the young plants were brought indoors to promote early soft growth which makes the best cuttings (Howard & Ridout 1991a) and gives the rooted cuttings the longest growing period. When the shoots reached 2–3 cm, they were cut and inserted in the mist unit; almost 100% rooted within two to three weeks.

The young plants, both seedlings and vegetative propagules, grew well producing a fairly dense, compact root system which included 'droppers' — vigorous roots which penetrate deeply into the soil. This behaviour is very different from that of the common white birch seedlings which quickly form a very superficial, far-spreading root system. Young bushes of *Betula chichibuensis* up to two

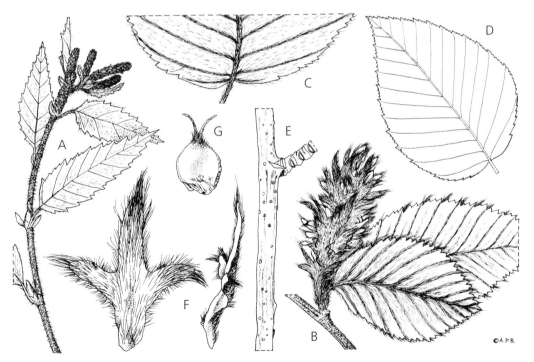

Fig. 74. *Betula chichibuensis.* **A** twig with buds and immature male catkins × 1; **B** fruiting shoot with mature fruiting catkin × 2; **C** underside of leaf base × 2; **D** outline of leaf in **C** × 1; **E** bark of twig × 2; **F** female catkin scale, back and side views × 7; **G** seed × 7. **A** from fresh material collected from Stone Lane Gardens (cult. *K. Ashburner*, 19 Aug. 2010); **B, E–G** from *Furuse* 20578 (Mt Bukoo, Hondo, 2 June 1949); **C, D** from *Furuse* 43269 (Mt Futago, Chichibu-gun, Saitama, Hondo, 6 July 1965). DRAWN BY ANDREW BROWN.

metres tall and of similar width transplant well with a well-developed root ball and grow on with little check as long as they do not suffer from drought in their first year following transplanting. Recent transplants seem to be sensitive to drought as several have died in dry situations, but well-established plants appear to be remarkably drought tolerant, probably because of their deep rooting. The species also appears to be fairly tolerant of wet soils.

Although young immature plants as much as a metre high are densely bushy and appear to be relatively shade tolerant, mature trees are very intolerant of shade. Shaded branches die and dry up over winter and become brittle and easily broken off.

Fig. 75. *Betula chichibuensis* with male catkins clustered towards ends of twigs (PW).

The canopy created by mature trees is dense due to the closely-spaced leaves, and casts a deep shade, and there are few if any shoots within the actual canopy.

Eight clones from the original seed are now in cultivation, and have been distributed by rooted cuttings taken from the original seedlings, and viable seed has also been distributed. The original seedlings are growing well at Ness and have formed small trees of up to five metres. Their foliage is remarkably handsome and so are the large bunches of creamy yellow male catkins and small red female catkins with showy tufts of violet styles in spring. They have flowered and fruited freely from five years of age. Where more than one clone is grown, a high proportion of the seed produced is fertile and self-sown seedlings are sometimes frequent underneath the trees.

Despite the rarity of this species, seedlings from the original wild-collected seed showed considerable variability, each being recognisable, primarily by habit characteristics, in the nursery rows. Two clones had a spreading habit, producing shrubs which were much wider than tall with drooping terminal branchlets. Others were much more upright, forming shuttlecock-shaped plants with much less drooping terminal twigs. Such genetic variability is not unexpected in a potentially long-lived species, even when the populations have been reduced to small numbers (Kuser 1983).

No hybrids have yet been found, but would be most likely with *Betula schmidtii*, *B. potaninii* and *B. calcicola*, the most closely related diploid species.

Betula chichibuensis H. Hara, *J. Jap. Bot.* 31: 122 (1956). Type: Honshu, prov. Musashi: circa Mae-shiraiwa (c. 1550 m), in montibus Mitsumine, Chichibu, *H. Hara*, 8 Sept. 1955, fl. juv. masc. (holotype TI).

ILLUSTRATION. *Woody Plants of Japan*: 116 (1985).

DESCRIPTION. *Tree* or *shrub* to 8–10 m (Satake *et al.* 1989) although in cultivation so far much smaller and forming a spreading small tree to about 5 m. *Twigs* light brown, nearly hairless, to 3 mm in diam., slightly rough with pale brown lenticels. *Buds* light orange-brown, to 6 × 2.5 mm, wider than the shoot especially towards apex. *Shoots* pale brown and smooth, very hairy with silky hairs; eglandular; *stipules* persistent. *Leaves* ovate, acute, rounded or slightly cordate at base, mid-green and hairy so that the leaves feel soft with long hairs along midrib and lateral veins, with (14–)17–18 pairs of veins (only to 10 or more on long-shoot leaves), indented above, but leaf largely flat; with 1–3 teeth between teeth ending secondary veins; 30–75 × 15–45 mm; petiole:blade ratio 1:7 to 1:10. *Male catkins* to 3 terminal and 2 lateral, sometimes also with similar lateral clusters on short peduncles, giving large clusters of catkins towards the ends of the twigs; individual catkins to 20 × 4 mm in winter, expanding to 60–80 × 7 mm in spring; scales attached at their bases; stamens cream coloured. *Female catkins* erect, red, with showy violet styles, to 13–20 × 7 mm on 2-leaved short shoots. *Fruiting catkin* erect, 16–26 × 8–10 mm; peduncle 3–7 mm. *Scales* 5–7 mm with terminal lobe oblong, 2.25–3(–4) mm and lateral lobes oblong, 1.5–2 mm, densely silky hairy below (on abaxial

PLATE 4. *Betula chichibuensis* (inflorescences)
Painted by Josephine Hague

Fig. 76. *Betula chichibuensis* with male and female catkins showing pleated young leaves with many veins (PW).

surface), glabrous above (on adaxial surface). *Seed* 2.5–3 × 2–2.5 mm, shiny light chestnut brown, with very narrow wing (c. 0.15 mm wide), glabrous except at apex where it tapers into styles; styles about 2.2 mm, conspicuously hairy at bases.

DISTRIBUTION. Japan, Honshu, Gunma Prefecture.

HABITAT. Limestone outcrops, 1550 m.

CHROMOSOME NUMBER. Diploid: 2n=28.

4. BETULA POTANINII

HISTORY

The shrubby *Betula potaninii* was described by Alexander Batalin from specimens collected near Kanding by Grigori Potanin (1835–1920), on his expedition to Mongolia and China in 1884–1886. It has been in cultivation for about 100 years from E. H. Wilson's introduction from Sichuan (see below), but until recently only three plants had been recorded, one at Mount Usher, Eire, and a second at Hergest Croft, Herefordshire, obtained from Veitch's sale and received on 8th October 1920. Both proved very difficult to propagate despite many attempts at taking cuttings and grafting, but Lawrence Banks has told us that a few cuttings were successfully rooted and one was planted out at Hergest Croft in 1992 (cat. 4453, 4454): the original plant has since died. The third record is from the rock garden of Mr F. J. Hanbury at Brackhurst, East Grinstead, Susex; there is a specimen from it in the Kew herbarium dated 29th June 1935, deposited by A. B. Jackson.

ECOLOGY AND STATUS IN THE WILD

Wilson's notes on the wild habitat of *Betula potaninii* are quoted by Schneider (1916): 'moist rocks and cliffs of the forest-covered higher mountains of western Szech'uan (Sichuan) this interesting Birch is common. Usually all its slender whip-like branches hang down over cliffs and boulders, but often it is a shrub from 1 to 3 m high with decumbent and prostrate branches'. The material from which Wilson introduced seed presumably grew on Wa-shan as Bean (1970) states that it was introduced by Wilson in 1909, and the only possible Wilson collection from which this could be derived is '*W*. 1140 from Wa-shan, Western Sichuan, 2100–2800 m. June and October 1908. Type of *B. wilsonii*: shrub 1.5–3 m tall hanging down over cliffs: flowers and old fruits'. The only earlier Wilson collection is from the same place in 1902 in July and is annotated as being in young fruit (Schneider 1916). The other Wilson collections (*W*. 4490; 4209, 4299) were made in 1910 and are also described as prostrate and hanging down over cliffs.

The recent reintroduction from western Sichuan (*SICH* 347) was from what appeared to be a single shrub growing to one metre tall in scrub (Howick, pers. comm.) from Paoma Shan, a hill behind Kangding, collected in 1988. The associated species were roses, an apple (*Malus* sp.), *Rhododendron decorum* Franch. and *Betula pendula* subsp. *szechuanica*, growing among young larch trees, suggesting quite a dry habitat. It is a plant grown from this reintroduction which is represented in the painting.

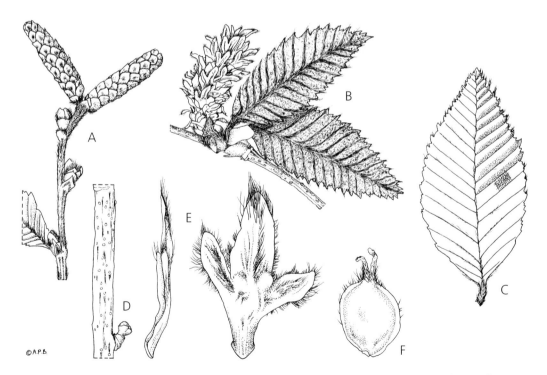

Fig. 77. *Betula potaninii.* **A** twig with emergent buds and immature male catkins × 2; **B** fruiting shoot with mature fruiting catkin, showing dense indumentum on underside of leaves × 2; **C** upper (adaxial) surface of leaf, with indication of fine vein network and deep corrugations (as shown upper right) × 2; **D** bark of twig × 2; **E** female catkin scale, side and back views × 10; **F** seed × 10. **A** from *E. H. Wilson* 4299 (W Szechuan, Oct.? 1910); **B, D–F** from *E. H. Wilson* 4490 (Mt Wu, 9,000 ft, 1910?); **C** from *E. H. Wilson* 1140 (Wushan, 1908, Type). DRAWN BY ANDREW BROWN.

PLATE 5. *Betula potaninii* (inflorescences)
Painted by Josephine Hague

The photograph (Fig. 80) was taken by Martyn Rix at 2840 to 3200 m in the valley which rises in the Zheduo Shan, south-west of Danba, north of Kangding. Here *Betula potaninii* was growing in several places on north-facing cliffs with *Tsuga dumosa*, *Pleione* and *Epimedium* species, forming much-branched, spreading shrubby bushes hanging from narrow ledges. Both *B. utilis* and *B. pendula* subsp. *szechuanica* grow nearby on open, rocky slopes.

TAXONOMY, RECOGNITION AND RELATIONSHIPS

The small, leathery, deeply veined leaves, silky-hairy beneath, are characteristic of this species and are similar to those of *Betula calcicola* from Yunnan and *B. bomiensis* from Gansu and Tibet. These two species share many other features with *B. potaninii*. They form a group of three closely related species and appear to be of very restricted distribution in the wild.

The shrubby *Betula trichogemma* described from Emei Shan in Sichuan in 1985 seems to be no more than a minor variant of *B. potaninii*, differing primarily in its hairier buds. Its distinctness from *B. jiulungensis* and *B. calcicola* is questioned by Skvortsov (Li & Skvortsov 1999). *B. jiulungensis*, described in 1979 from Jiulong Shan in SW Sichuan, differs from *B. potaninii* mainly in that it is described as a tree up to 12 m tall; it may be merely a tree form of *B. potaninii* or perhaps a distinct species.

CULTIVATION

Young plants of *SICH* 347 from near Kanding have proved easy to grow at Ness, with no dieback and no problems with early autumn or late spring frosts. This species is relatively slow growing but vigorous, as long as it is grown in reasonably high light without too much shade from surrounding

Fig. 78 (left). *Betula potaninii* (*SICH* 347) male catkins with red stamens (PW).

Fig. 79 (right). *Betula potaninii* (*SICH* 347) close up of male catkins with red stamens and initially deflexed emerging red female catkin (PW).

PLATE 6. *Betula potaninii* (leaves)
Painted by Josephine Hague

Fig. 80. *Betula potaninii*, South West of Danba, N of Kanding, Sichuan, China (MR).

trees. Although said to form long whippy growths and hanging pendent from cliffs in the wild, in cultivation it makes a spreading shrub with indications that it might become a small tree. As in most other species of this group, leading shoots soon bend over and grow horizontally, further growth in height being made by the growth of a lateral shoot. One to four or five main stems may develop. After three years, flowering begins and the terminal shoots bear male catkins, restricting lateral growth. After fifteen years some plants have reached five metres.

This species appears to be self–incompatible as the isolated trees of the earlier Wilson introductions apparently did not produce any viable seed; but when plants of different clones of the recent reintroduction are grown together, seed with a high percentage viability is produced and has come true. Cuttings of seedlings of this new introduction rooted easily and plants of several clones have been widely distributed. Whether cuttings of older plants will root as readily as those from the two year old seedlings is uncertain. Hybridisation in cultivation would be most likely with *Betula calcicola*, or perhaps *B. chichibuensis* or *B. schmidtii*, which are the most closely related diploid species. These species have not been grown close together and no hybrids have been detected among the seedlings raised. It can be expected that any hybrids would be easily recognised as the parents are quite distinct.

The history of the first introduction to cultivation of this species demonstrates how quickly and rapidly a newly introduced species can become reduced to a very few isolated individuals, which may then be difficult to propagate when the species is self-incompatible. Although there was wide knowledge of the existence of these two plants and many attempts at grafting and rooting cuttings, no-one seems to have attempted to obtain seed by cross pollination between them. If we wish such species to be maintained in cultivation it is important that breeding populations of more than one clone should be established in at least one garden. This has been done at Ness with the recent 1988 introduction.

Fig. 81. *Betula potaninii*, *SICH* 347, elongating long shoots with about two expanded leaves. Stage at which elongated portion could be taken as softwood cutting (PW).

Fig. 82. *Betula potaninii* leaves with deeply impressed veins and fruiting catkins (PW).

Fig. 83. *Betula potaninii*, *SICH* 347, fruiting catkins (PW).

Betula potaninii Batalin, *Trudy Imp. S.-Peterburgsk. Bot. Sada* 13: 101 (1893). Type: China borealis, prov. Szetschuan septentrionale, ad fluv. Honton, infra pagum San-Shei, 13 Aug. 1885, *Potanin* (holotype LE).

B. wilsonii Bean, *Bull. Misc. Inform., Kew* 1914: 30 (1914); *Trees & Shrubs Brit. Isl.* 1, 264 (1914). Type: western Sichuan, Wa-shan, 2100–2800 m, June and October 1908, *Wilson* 1140 (holotype K).

B. potaninii var. *trichogemma* L. C. Hu ex P. C. Li, *Acta Phytotax. Sin.* 17 (1): 91 (1979). Type: Sichuan, Omi-shan, Shi-shang-chih, Oct. 20 1938, *Liou Tchen-ngo* 10530 (holotype PE).

B. trichogemma (L. C. Hu ex P. C. Li) T. Hong in W. C. Cheng, *Silva Sin.* 2: 2133 (1985).

B. jiulungensis Hu ex P. C. Li, *Acta Phytotax. Sin.* 17 (1): 90 (1979). Type: Szechuan, Jiulung Hsien, Swe-ta-ba, in forest, 2400 m, Oct. 9 1960, *Ying Tsun-shen* 4779 (holotype PE).

DESCRIPTION. *Spreading shrub* to 5 m. *Twig* light brown, up to 2 mm in diam., finely hairy with very short velvety (puberulent) hairs and a few silky hairs even on one-year-old twigs. *Bud* often with two subtending persistent stipules, red-brown, ovate, blunt, about 5.5 × 3.5 mm, scales often retaining silky hairs. *Young shoot* densely silky hairy, totally eglandular. *Leaves* with petiole 3–5 mm; blade ovate-lanceolate, mostly more than twice as long as broad 20–51 × 11–24 mm, leathery, densely silky hairy beneath when young, sharply toothed, with 9–21 pairs of veins deeply impressed above and proud below. *Male catkins* 1–2 terminal with up to two laterals but often with others on short lateral peduncles on more proximal nodes, dark brown, to 35 × 3 mm, expanding to 65(–80) × 5 mm in spring. *Scales* 1–1.5 × 1–1.5 mm, peltate, densely silky hairy on margins and exposed part. *Female catkins* to 15 × 5 mm, occasionally with terminal male sections, initially deflexed but soon erect. *Fruiting catkin* erect 12–20 × 5–7 mm with peduncle about 2 mm; scales slightly hairy on lower (abaxial) surface, 3.75–6 mm with oblong terminal lobe 1.5–4 and spatulate laterals about 2 mm, tip of central lobe thin and triangular, (a character otherwise only found in related species in *B. calcicola*). *Seed* about 2–2.25 × 2 mm with styles about 1 mm; wing a very narrow rim.

DISTRIBUTION. China, in SE Gansu, Shaanxi, N & W Sichuan.

HABITAT. Shaded cliffs and scrub.

CHROMOSOME NUMBER. 2n=28

5. BETULA CALCICOLA

HISTORY

Although George Forrest collected *Betula calcicola* in Yunnan in the early part of the 20th century, it was only in 1986 that it was first introduced into cultivation by Roy Lancaster and it was collected again by the CLD (Chungtien, Lijiang & Dali) expedition in 1990. It was described in 1915 by Sir William Wright Smith as a variety of *B. delavayi*, and only raised to specific status by P. C. Li in 1979.

ECOLOGY AND STATUS IN THE WILD

Betula calcicola is found primarily on the Lijiang Range in northwest Yunnan. In British herbaria we have seen specimens only from Lijiang, but in *Flora of China* it is also recorded from southwest Sichuan. Specimens from a wider range of localities are likely to be available in Chinese herbaria but, without seeing them, we cannot say for certain that *B. calcicola* occurs outside the Lijiang area.

It grows on limestone rocks and cliffs, often in mossy crevices with *Pleione* sp., *Primula forrestii* and various small shrubs.

TAXONOMY, RECOGNITION AND RELATIONSHIPS

Betula calcicola is fairly easily distinguished by the extreme hairiness of its young shoots and almost circular leaves; the young shoots and leaves are most attractive with their dense covering of white to golden brown hairs (Plate 7); they are deep green, rather leathery in texture, with deeply impressed veins. Many leaves persist on the twigs throughout the winter and they decay slowly so that entire

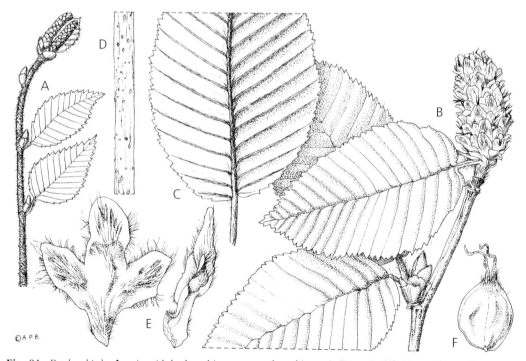

Fig. 84. *Betula calcicola.* **A** twig with buds and immature male catkins × **1**; **B** terminal fruiting catkin (sole material available), subtending leaf and lateral shoot × 2; **C** underside of leaf × 2; **D** bark of twig × 2; **E** female catkin scale, back and side views × 8; **F** seed × 8. All from *H. McAllister* (Ness BG, 30 Aug. 2010). DRAWN BY ANDREW BROWN.

Plate 7. *Betula calcicola*
Painted by Josephine Hague

leaves are always present beneath the plants. The fruiting catkins also persist over the winter and the nutlets have only a rudimentary wing which is more than one cell thick.

This species could only be confused with *Betula bomiensis*, which has much less densely hairy young shoots and smaller leaves. *B. delavayi* also has much less densely hairy shoots and leaves, larger fruiting catkins and seeds, and usually larger, more elongated, flatter leaves.

Also similar to *Betula calcicola* is *B. potaninii*, but that has much smaller, narrower leaves with more numerous veins, less hair, and much smaller fruiting catkins and occurs only in Sichuan, Gansu and Shaanxi. *B. potaninii* and *B. calcicola* are the only species with male catkin scales attached at their bases (not centrally) and thin triangular tips to the fruiting catkin scales; in other birch species (perhaps with the exception of *B. ashburneri* and species of section *Acuminatae*) the tip is thicker and the same texture as the rest of the scale,

Study of the plants in cultivation at Ness Gardens has shown *Betula calcicola* to be diploid with a chromosome number of 2n=28, while *B. delavayi* is hexaploid with a chromosome number of 2n=84, and the more recently distinguished *B. bomiensis* is tetraploid with 2n=56.

The plant illustrated (Plate 7) was grown from seed of *CLD* 791 from Lijiang. Despite the difference in chromosome number, *Betula calcicola* and *B. delavayi* probably can hybridise in the wild where they grow in close proximity in the Lijiang area. The wild collected seed of *Lancaster* 1691 from *B. calcicola*, gave rise to a diploid plant of *B. calcicola*, as well as to a pentaploid hybrid with *B. delavayi* (see chapter 2, Hybridisation). It is likely that these hybrids will be largely sterile, so their occurrence is perhaps unlikely to lead to introgression between the species. However, fertilisation of an unreduced gamete of the hybrid with n=70 by a normal reduced gamete from *B. calcicola* (n=14) could result in a plant with 2n=84 which might be interfertile with *B. delavayi*. This would be a mechanism of gene flow from *B. calcicola* to *B. delavayi*.

CULTIVATION

In cultivation seedlings of *Betula calcicola* have grown to 4 m in ten years. There is a lot of variation between individuals in habit. Some are fastigiate with strong leading shoots, making attractive, narrow, upright small trees in cultivation. Others are more spreading, with the leading long shoots arching outwards as is usual in the related *B. potaninii*, *B. bomiensis* and *B. chichibuensis*.

In cultivation this species requires light, well-aerated but moist, preferably humus-rich, soil and full sun. It does not tolerate drought or waterlogging. Although the buds do not break particularly early in spring, the young shoots are tender and are often damaged by late frosts. Many of the male catkins and the terminal part of others fail to expand in spring. This would appear to be the result of desiccation during the winter. Presumably the rather loosely packed, matt scales do not provide as much protection as the closely overlapping, densely packed, glossy (presumably thick cuticled) scales of most other species, especially the white barked species. This winter desiccation of male catkins reduces pollen production and seed set.

Betula calcicola (W. W. Sm.) P. C. Li, *Fl. Reipubl. Popul. Sin.* 21: 137 (1979).
B. delavayi Franchet var. *calcicola* W. W. Sm., *Notes Roy. Bot. Gard., Edinburgh* 8: 333 (1915). Type: China, NW Yunnan, in the crevices of limestone cliffs on the eastern flank of the Likiang range. lat. 27°20'N, 11,000–12,000 ft, June 1910, *G. Forrest* 5835 (holotype E).
Betula forrestii var. *calcicola* (W. W. Sm.) Hand.-Mazz., *Symb. Sin.* 7: 20 (1929).

DESCRIPTION. *Shrub* to small *tree*, somewhat fastigiate to spreading, to 4 m; *twig* to 2.5 mm in diam., the densely hairy epidermis is persistent for at least two years, splitting in thicker twigs to reveal lenticels c. 0.5 mm in diam. in a brown bark. *Bud* densely grey-white silky hairy, ovate, c. 9 × 4 mm. *Young shoot* densely silky hairy with honey-coloured hairs. *Leaf* with petiole 1–12 mm; blade 20–57 × 15–40 mm; veins densely hairy; with (9–)13–15(–16) pairs of veins. *Male catkins* 1–3 terminal with 1–2 lateral, to 35 × 5 mm; scales c. 4 × 2.5 mm, triangular, attached at base, not peltate, margins free, densely hairy, stamens yellow. *Female catkins* to 16 × 2 mm, pink, densely silky hairy, scales reflexed. *Fruiting catkin* cylindrical, to 20 × 12 mm, elliptic-cylindrical; scale c. 5 × 4 mm with ciliate margins; terminal lobe to 1.6 mm, triangular oblong, with a thin, triangular tip; lateral lobes to 1.6 mm, spatulate. *Seed* 2–2.75 × 2–2.75 mm, white hairy in upper half, especially around style base; styles 1–1.75 mm; wing about 0.1 mm, opaque, more than one cell thick.

DISTRIBUTION. Lijiang Range, NW Yunnan; SW Sichuan (*fide Flora of China*).

HABITAT. Thickets on limestone cliffs.

CHROMOSOME NUMBER. Diploid: 2n=28.

6. BETULA BOMIENSIS

HISTORY AND NOMENCLATURE

Betula bomiensis was described in 1983 in *Flora Xizangica* (Tibet) by P. C. Li, but unfortunately the type specimen is a mixture of two species; it consists of a fruiting twig of the new species, *B. bomiensis*, with small leaves with closely spaced, deeply impressed veins, and a vegetative twig of *B. utilis* or *B. ashburneri* with large, broadly ovate, flat leaves. After examining the type specimen and the published illustration in *Flora Xizangica* (1983), Rushforth (2003) chose the fruiting specimen as the lectotype. In the *Flora of China*, *B. bomiensis* is treated as a synonym of *B. delavayi* var. *microstachya* P. C. Li, described from Sichuan.

Fig. 85 (left). *Betula bomiensis*. Deflexed, recently emerged, female catkins, and young leaves showing apical browning due to frost damage (PW).

Fig. 86 (right). *Betula bomiensis*. Female catkins initially deflexed but becoming erect (PW).

ECOLOGY AND STATUS IN THE WILD

Betula bomiensis was described from a specimen collected in southeast Tibet near Bomi (Pomi) in 1979, and living material is in cultivation from material collected by Keith Rushforth in 1999 between Bagu and Nambu (29°59'43"N 94°38'47"E) about 120 km due west of Bomi. Apart from the type collection and the material in cultivation, the only other two specimens we have seen are *Rock* 14758 (E!) from southwest Gansu, Lower Tebbu country, upper Mayaku, 6 Sept. 1926; and *Purdom* 812 (E!) W Gansu, S of Minchow, 1 July 1911; these specimens are distinct from *B. delavayi* and *B. skvortsovii* and identified as *B. bomiensis*, primarily on the basis of the absence of any glands.

RECOGNITION AND RELATIONSHIPS

Betula bomiensis is closest to *B. potaninii*, which differs in its more elongated leaves with more numerous veins, and to *B. skvortsovii* which differs in its persistently glandular leaves which usually have a pair of weak basal secondary veins. Whereas *B. potaninii* is diploid, *B. bomiensis* is tetraploid, and a seedling of what is probably *B. skvortsovii* is also tetraploid.

CULTIVATION

At Ness *Betula bomiensis* has formed attractive small trees around 3 m tall, with arching branches bearing small, dark green leaves with deeply impressed veins. Unfortunately, most of the plants distributed to other gardens have died from what is presumed to be the effects of late spring frost.

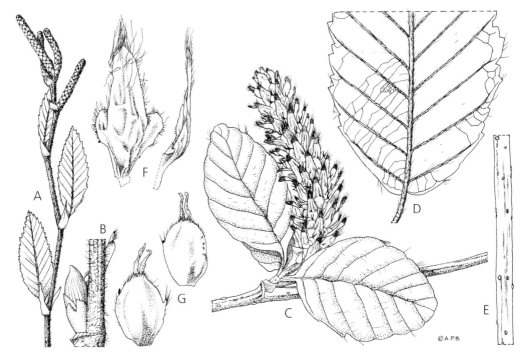

Fig. 87. *Betula bomiensis.* **A** twig with buds and immature male catkins × 1; **B** obscured bud from **A** after removal of one stipule and petiole × 4; **C** fruiting shoot with mature fruiting catkin × 2; **D** underside of leaf base × 2; **E** bark of twig × 2; **F** female catkin scale, back and side views × 8; **G** seed × 10. All from *H. McAllister* (Ness BG, 30 Aug. 2010). DRAWN BY ANDREW BROWN.

Cuttings taken from seedlings root very easily and viable seed which comes true is produced where the different clones are grown together.

Betula bomiensis P. C. Li, *Flora Xizangica* 1: 484 (1983). Type: Xizang, Pomi, alt. 3300 m, edge of forest on valley, August 1979, *Huang Yung-fu* 776 (PE) (lectotype the fruiting twig only, not the vegetative one, which is *B. utilis*).

DESCRIPTION. *Small tree* to 8 m with arching branches. *Twigs* more or less glabrous; young shoots initially densely softly silky hairy, to 2.5 mm in diam., smooth, brown, eglandular, with small (0.5–1.0 mm) lenticels, white, roundish to elliptic, longitudinally elongated. *Buds* ovoid, acute, about 6 × 3 mm. Petiole 3–6 mm. *Leaf blade* from very small to 38 × 30 mm (rarely to 40 × 34 mm), with (5–)9–13(–15) pairs of veins; veins deeply indented; secondary veins running to the sinus between the teeth and then bending up into the tooth; 1 (rarely 2) teeth between the teeth ending the secondary veins; secondary veins close 2 mm apart (–3 mm towards the apex); weak basal veins usually absent; margin recurved. *Male catkins* 1, or 2 or 3 in a cluster, to 4.7 cm; bracts densely yellow tomentose. *Female catkins* initially deflexed, clearly downward pointing, soon becoming erect (Figs 85, 86). *Fruiting catkin* up to 18(–32) × 9 mm with peduncle about 3.5–8 mm and bracts usually covered with silky hairs. *Scales* silky hairy on dorsal (abaxial) surface, with ciliate margins, and apical hairs longer and twisted together to make point to 1.5 mm long. *Seeds* c. 2 × 1–1.5 mm, hairy towards apex; style c. 1 mm.

DISTRIBUTION. Southeast Xizang.

HABITAT. Margins of broad-leaved forests; c. 2,400 m.

FLOWERING AND FRUITING. Flowering June; fruiting October.

CHROMOSOME NUMBER. Tetraploid: 2n=56.

7. BETULA DELAVAYI

HISTORY

Betula delavayi was described by Franchet in 1899, from specimens collected by Père Delavay above Mo so yn (Lankong), in Yunnan. Jean Marie Delavay (1834–1895) was one of the most prolific collectors in Yunnan in the 19th century, collecting over 200,000 specimens, of which at least 1500 were new species. From 1882 to 1888 he was based in Tapintze, near present-day Heqing, south of Lijiang in northwest Yunnan, but collected widely in Yunnan and to a lesser extent elsewhere in China. His collections were sent to the Muséum National d'Histoire Naturelle in Paris, where they were studied and the new species described by Adrian Franchet in *Plantae Delavayanae* (1889–90). Delavay returned to Yunnan in 1893 and continued to collect, but died there in 1895.

In 1915 William Wright Smith described specimens collected by George Forrest near Lijiang as var. *forrestii*, but they do not differ significantly from Delavay's collections.

ECOLOGY AND STATUS IN THE WILD

Most collections identifiable as *Betula delavayi*, and all the plants known in cultivation, come from the limestone mountains of the Lijiang range in northwest Yunnan. However, several specimens from the Zongdian area, also in NW Yunnan, (*Forrest* 22036, *Forrest* 30831) certainly belong to this species, and others from SW Sichuan (*Yü* 14722, *Yü* 14404, *Rock* 24049 and *Rock* 24516 from near Muli, *Handel-Mazzetti* 472 and *Handel-Mazzetti* 2168) also seem to be *B. delavayi*.

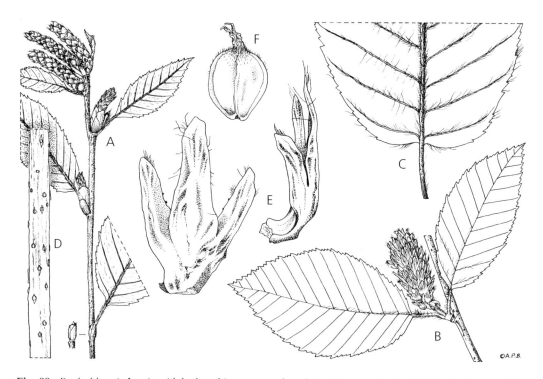

Fig. 88. *Betula delavayi*. **A** twig with buds and immature male catkins (sole vegetative bud hidden by lowest petiole is displayed separately) × **1**; **B** fruiting shoot with mature fruiting catkin × 1; **C** underside of leaf base × 2; **D** bark of twig × 2; **E** female catkin scale, back and side views (resin glands present)× 7; **F** seed × 7. All from *H. McAllister* (Ness BG, 30 Aug. 2010). DRAWN BY ANDREW BROWN.

In the *Flora of China*, *Betula delavayi* is also cited for W Hubei, E Xizang (Tibet), and annotated '?Gansu, ?Qinghai', but confusion with other species, perhaps especially *B. bomiensis*, and simple misidentifications of such a poorly understood species, makes its complete distribution difficult to determine from the literature.

In the wild *Betula delavayi* has been described as a tree of up to 12 m and *B. calcicola* as a shrub to about 2 m (R. Lancaster, pers. comm., C. Grey-Wilson 1988). The *Flora of China* gives the height of *B. delavayi* as up to 8 m, but Lancaster described climbing a tree of about 40 ft (c. 12 m) and George Forrest similarly annotated some of his specimens as being up to 40 ft in height.

TAXONOMY, RECOGNITION AND RELATIONSHIPS

Betula delavayi can be distinguished from its close relatives by its ovate-elliptic, firm, leathery leaves with mostly 10–14 pairs of veins, and especially by the numerous small reddish glands on the underside of the leaf. Most similar to *B. delavayi* are *B. potaninii*, *B. calcicola* and *B. bomiensis*, but the leaves of these three species are totally eglandular, and, although also leathery, have veins more indented than in most specimens of *B. delavayi*, giving the leaves a much more corrugated appearance. The leaves of *B. potaninii* have more numerous veins (9–21 pairs). *B. calcicola*, which for many years was treated as a variety of either *B. delavayi* or *B. forrestii*, can be distinguished by its much hairier shoots and usually much broader, more leathery leaves, and *B. bomiensis* can be distinguished by its smaller leaves.

The two other species in subsection *Asperae*, *Betula gynoterminalis* and *B. skvortsovii* are both little known and not and very rare, respectively, in cultivation.

As mentioned in Chapter 2, a hybrid between the diploid *Betula calcicola* and hexaploid *B. delavayi* has been raised from seed collected in the Lijiang mountains where the two species grow in close proximity. Wherever the distributions of these species overlap, these hybrids may be present and difficult to distinguish from the parents. However, fertilisation of an unreduced n=70 gamete in a hybrid by a *B. calcicola* (n=14) gamete could result in a hexaploid, which would likely be interfertile with *B. delavayi*, perhaps resulting in introgression.

Specimens labelled *Betula delavayi* in herbaria are very variable, but our counts of hexaploids from the Lijiang Range in NW Yunnan and tetraploids from SE Tibet (*B. bomiensis*) indicate that at least these two species might be present among specimens identified as *B. delavayi*. Indeed, it is possible to sort herbarium sheets into two groups: the majority, with larger and often flatter leaves, larger fruiting catkins and seeds, glandular young shoots and leaf lower surface are *B. delavayi*; a few smaller-leaved specimens with fewer veins, smaller fruiting catkins, and eglandular young shoots and leaf lower surface were identified as *B. bomiensis*.

From the description in the *Flora of China*, *Betula delavayi* var. *polyneura* Hu closely resembles both *B. delavayi* and *B. calcicola*, but has more numerous veins. Without seeing specimens it is impossible to draw firm conclusions as to its relationships and it may represent a distinct taxon. Similarly, *B. delavayi* var. *microstachya* P. C. Li, from western Sichuan deserves further study.

In his original descriptions of *Betula delavayi* and *B. fargesii*, Franchet (1899) made clear the distinction between them; he described *B. delavayi* from NW Yunnan as having cylindric-ovate fruiting catkins and the wings of the nutlets as very narrow (*angustissimae*); in contrast, he says of *B. fargesii* collected in Sichuan, that the fruiting catkins are ovate and the nutlet wings are a quarter the width of the nutlet. This agrees with the specimens from the Lijiang area in Yunnan being *B. delavayi* and those on the Sichuan-Hubei border about 800 km away being *B. fargesii*, and this is the interpretation accepted in the *Flora of China* (Li & Skvortsov 1999), in contrast to the interpretation of Skvortsov (1998).

CULTIVATION

Betula delavayi is undoubtedly handsome in foliage, slightly reminiscent of *Cotoneaster salicifolius* in its arching stems and acute, rather leathery leaves. So far it has been little planted but experience suggests that it requires soil which is well aerated yet never becomes too dry, and is most likely to thrive in maritime areas on humus–rich soils. In this it appears to be similar to *B. calcicola* and will succeed in conditions similar to those required by rhododendrons, although judging by its native habitat, it may survive on soils with higher pH.

Cuttings of young plants root relatively easily and viable seed has been produced in cultivation where two clones have been grown together.

Betula delavayi Franch., *J. Bot. (Morot)* 13: 205 (1899). Type: China occidentalis, Yunnan in silvis ad Koutoui supra Mo so yn, alt. 2800 m (holotype P, isotype HUH).
B. delavayi var. *forrestii* W. W. Sm., *Notes Roy. Bot. Gard. Edinburgh* 8: 322 (1915). Type: Open situations in pine forests on the eastern flank of the Lichiang Range, Lat. 27°10'N, May 1910, *G. Forrest* 5835 (holotype E, isotype K).
Betula forrestii (W. W. Sm.) Hand.-Mazz., *Symb. Sin*. 7: 20 (1929).
Betula delavayi var. *polyneura* Hu ex P. C. Li, *Acta Phytotax. Sin*. 17 (1): 90 (1979). Type: Yunnan, Likiang Snow range, 2600 m, in mixed forest, 18 August 1942, *Feng Kuo-mai* 9061 (PE).

Fig. 89. *Betula delavayi*. Fruiting catkins and foliage (HMcA).

Betula delavayi var. *microstachya* P. C. Li, *Acta Phytotax. Sin.* 17 (1): 90 (1979). Type: western Sichuan: Tehkeh Hsien, Damakou, *Tsui Yu-wen* 5957 (PE).

DESCRIPTION. *Shrub* or *small tree* 4–12 m. Young shoots densely silky-hairy with reddish glands; stipules hardly persistent. *Twigs* brown; second-year twig still pale brown with white lenticels. Buds narrowly ovoid, to 7 × 3 mm. *Leaves* with petiole to 8 mm, very dark green and moderately shiny above, yellow-green and dull or sometimes quite shiny below, elliptic-ovate, to 70(–90) × 40(–46) mm; length usually about twice the width but sometimes broader, rounded at base and acute at apex, convex and somewhat revolute at the margins, leathery with 8–12(–16) pairs of veins inserted about 5–6 mm apart, usually somewhat indented above; entering base of tooth but not running to sinus, (1–)2–3(–5) teeth between main secondary vein ends; weak basal veins usually present; sparsely hairy above, persistently silky-hairy along the midrib and veins below; reddish glands on lower surface when young; blade:petiole ratio c. 1:9. *Male catkins* to 3 terminal with up to 2 lateral, to 30 mm; scales peltate, overlapping towards the apex. *Fruiting catkins* to 32 × 15 mm, elliptic-oblong; peduncle < 5 mm; scales about 8.5 × 6 mm, margin ciliate, adaxial surface puberulent, abaxial glabrous; terminal lobe c. 4.5 mm, rhombic-oblong; lateral lobes c. 3 mm, oblong. *Seed* c. 3 × 2.5 mm; wing a narrow rim, c. 0.2 mm, ⅛ of nutlet width, opaque, more than one cell thick; styles 1 mm.

DISTRIBUTION. NW Yunnan, Lijiang, Zhongdian: SW Sichuan, Muli: also doubtfully recorded from W Hubei, E Xizang (Tibet), ?Gansu, ?Qinghai in *Flora of China* (Li & Skvortsov 1999).

HABITAT. Broad-leaved forests and thickets, usually on limestone.

CHROMOSOME NUMBER. Hexaploid: 2n=84.

8. BETULA GYNOTERMINALIS

HISTORY

Unfortunately *Betula gynoterminalis* is known from only a single specimen in the herbarium in Kunming, collected near Gongshan Xian in NW Yunnan in 1956. It is reported to grow in mixed, broadleaved forest, at c. 2,600 m, at Drungzu Nuzu Zizhixian near Gongshan, east of Lijiang in northwest Yunnan.

TAXONOMY, RECOGNITION AND RELATIONSHIPS

From the description and accompanying illustration, *Betula gynoterminalis* seems to be unique in the combination of large leaves, as large as those of *B. alnoides* or *B. insignis*, very hairy on the underside, with 16–18 pairs of raised veins beneath. The nutlets have a very narrow wing as in all species of section *Asperae*, but the fruiting catkins are probably pendulous and in a terminal raceme, a state not normally found in any birch, although occasionally as an aberrant state. It is therefore a very distinct species with a unique combination of characters. Skvortsov (2002) placed this species in a monotypic subgenus of its own, as he does with *B. corylifolia*, lumping all other birch species into two other subgenera. He therefore considered *B. gynoterminalis* a very distinct and taxonomically isolated species with no close relatives.

Fortunately an image of the type specimen is available on the internet (http://kun.kib.ac.cn/specimenimages/type/443635.jpg), and a drawing prepared from this is provided in the type description (Hsu & Wang 1983). The specimen is a second year twig with three lateral spur shoots each bearing one or two, very broadly elliptic, clearly rather leathery, ribbed leaves, perhaps most resembling those of *Betula calcicola* but larger. The four terminal fruiting catkins are more or less stalk-less and borne very close together at the end of the twig. In no other species of birch are female catkins normally borne terminally in this way, but as only one specimen of this species is known, this may be an aberrant state for this species too. Plants of *B. bomiensis* bearing two terminal fruiting catkins in a very similar manner, of *B. medwediewii* bearing three such catkins, and of *B. insignis* bearing a cluster of five have been seen in cultivation (Fig. 116).

A nutlet and fruiting catkin scale are depicted on the illustration of the type specimen. These clearly resemble those of *Betula delavayi* in size and shape and, when taken together with the leathery leaves with indented veins and dense indumentum, suggest that *B. gynoterminalis* is a member of section *Asperae* subsection *Asperae*, probably most closely related to *B. delavayi* and *B. calcicola*.

As far as is known, this species is not in cultivation.

Betula gynoterminalis Y. C. Hsu & C. J. Wang, *Acta Bot. Yunnan.* 5 (4): 381–382 (1983). Type: Yunnan, Gongshan Xian, alt. 2600 m, 26 Sept. 1956, *P. Y. Mao 521* (KUN).

DESCRIPTION. *Tree* to 7 m. *Leaves* with petiole c. 5 mm, densely hairy; blade broadly ovate or oblong, 12–13 × 7–8 cm, probably leathery, with glabrous veins indented above, densely fulvous hairy beneath, with 16–19 pairs of prominently raised veins. *Fruiting catkins* 4 in a terminal group, apparently pendulous, cylindric, 5–7 × c. 1 cm; bracts c. 8 × 4 mm ciliate at apex, lateral lobes about half as long as central lobe. *Seed* c. 2 × 1.5 mm with narrow rim.

DISTRIBUTION. China, NW Yunnan, Gongshan, Drungzu Nuzu Zizhixian.

HABITAT. Broad-leaved forest, c. 2600 m

CHROMOSOME NUMBER. Not known.

9. BETULA SKVORTSOVII

HISTORY, ECOLOGY AND STATUS IN THE WILD

This species is described here for the first time, and is named in memory of the eminent Russian botanist and expert on birch, Alexey K. Skvortsov (1920–2008). Skvortsov was born in Smolensk oblast, where his father was a country doctor who later became a psychotherapist in Moscow, as well as being an amateur field botanist. Alexey studied medicine, specialising in histology, and served as a physician in the Second World War. His doctorate, completed in 1948 was on the evolution of the vascular system of the spleen. Later that year, the Laboratory of Genetics was liquidated, and in 1951 Skvortsov was sent to the Denezhkin Kamen Preserve in the Urals to conduct a floristic survey. In 1952 he was offered a position to help set up a new Botanic Garden at Moscow University, where he was instrumental in establishing the herbarium as a world-class institute, and in improving the living collections in the garden. His study of *Salix* in the USSR was the subject of a second doctorate in 1966 and published in 1968 (Skvortsov 1968); it included the evolution and phylogeny of the genus, as well as biogeographic and paleogeographic studies in Europe and northern Asia. He travelled throughout the USSR, collecting thousands of herbarium specimens and studying many genera in the field; after *Salix*, his main interests were in *Betula* and the Onagraceae (http://salicicola. com/translations/Skvortsov.html). He was a kind, thoughtful and generous man who provided us with wild collected seed of several species.

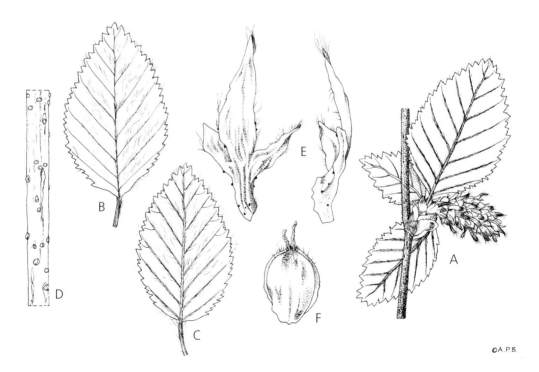

Fig. 90. *Betula skvortsovii* (no male material available for observation). **A** fruiting shoot with mature fruiting catkin × 2; **B** leaf upper surface × 2; **C** leaf lower surface × 2; **D** bark of twig × 2; **E** female catkin scale, back and side views × 10; **F** seed × 10. All from *Boufford et al.* 36417 (Sichuan, Dêgê Xian, 31°41'10"N 98°35'39"E, 14 Aug. 2006). DRAWN BY ANDREW BROWN.

Fig. 91. Seedling of *Betula skvortsovii*, *MF* 5566 (PW).

Betula skvortsovii has been recognised from among a number of herbarium specimens collected from shrubs growing in dry scrub in Qinghai and Sichuan by two expeditions from the USA in 1996 and 2006, and initially tentatively identified as *B. chinensis* or *B. bomiensis*. They had distinctive small, flattish leaves with few veins (Fig. 91), differing in shape from those of *B. chinensis* and unlike those of *B. bomiensis* being glandular.

TAXONOMY, RECOGNITION AND RELATIONSHIPS

The small, erect, persistent, fruiting catkins containing nutlets with only a narrow rim clearly place this species in the *Betula potaninii* group. It is distinguished from other species of the group by its flattish leaves with the veins on the upper surface not appearing much impressed and by its rather oblong leaf shape in contrast to the lanceolate to circular leaves of most of the other species. The presence of conspicuous reddish glands on the shoots and leaf undersides distinguishes it from all other species of the group except *B. delavayi*. In habitat, and to some extent in appearance, *B. skvortsovii* has some similarities with *B. chinensis*, but that species has ovate leaves broadest in the lower half and a longer petiole, the petiole:leaf blade ratio being about 1:4 rather than 1:7 as in *B. skvortsovii*. More recent collections, including of seed, suggest that on strongly growing specimens the leaves may be more deeply ribbed with more numerous veins, and the species may be a glandular northern vicariant of *B. bomiensis*. Another distinguishing feature appears to be the usual presence of a pair of weak secondary veins at the base of the leaf in *B. skvortsovii*, these being absent in *B. bomiensis*.

CULTIVATION

This species is extremely rare in cultivation. It probably has no particular horticultural merits beyond being an interesting, drought tolerant, neat-leaved shrub which may have attractive catkins in spring.

Betula skvortsovii McAll. & Ashburner **sp. nov.** affinis *B. potaninii* sed foliis oblongis, planis, venis superis non impressis; petiolis brevibus, differt. Typus: China, Qinghai, Nangqên Xian, Larong Gou on east side of the Zi Qu, N of Jiangxi forest Station and SE of Mozhong. Gorge with *Picea likiangensis* under *Picea* and *Betula albosinensis*, 3450 m, 32°9'N 97°3'E, 29 Aug. 1996, multi-stemmed shrub to 2.5 m, *Ho, Bartholomew, Watson, Gilbert* 2629 (holotypus LIV; isotypus HUH).

DESCRIPTION. *Shrub* with several stems to 2.5 m. *Twigs* thinly covered with patent hairs; stipules persistent. *Young shoots* initially hairy with silky and patent hairs on veins, covered with reddish glands. *Petiole* 3–5 mm, with petiole:blade ratio c. 1:7. *Leaves* from very small to 35 × 23 mm,

Plants of China

Betulaceae
Betula delavayi Franchet

Sichuan Province, Dêgê Xian: W of Road (highway 317)
from Dêgê to Jinsha Jiang (upper Chang Jiang) just W of Gongya
Xiang along Serqu Ge (serqu River). 31°41'10", 98°35'39"E;
3100 m. Dry slope above river with spinescent shrubs (ca. 2 m tall)
of Rosa, Caragana and Berberis. On steep, dry slope above river.
Shrub 1.5-2 m tall; fruit green.

D. E. Boufford, B. Bartholomew, S. L. Kelley, R. H. Ree, H. Sun,
J. P. Yue, L. L. Yue, D. C. Zhang & W. D. Zhu
36417   14 August 2006

Harvard University Herbaria

Fig. 92. *Betula skvortsovii* China, Sichuan, Dêgê Xian: W of road (highway 317) from Dêgê to Jinsha Jiang (upper Chang Jiang) just W of Gongya Xiang along Serqu Ge (Serqu River). 31°41'10"N 98°35'39"E, 3100 m. Dry slope above river with spinescent shrubs (c. 2 m tall) of *Rosa, Caragana* and *Berberis*. On steep dry slope above river. Shrub 1.5–2 m tall: fruit green. *D. E. Boufford, S. L. Kelly, R. H. Ree, H. Sun, J. P. Yue, L. L.Yue, D. C. Zhang & W. D. Zhu 36417*, 18 August 2006, HUH (PW).

breadth more than half length, oblong-elliptic, often flattish, with reddish glands on veins and lamina beneath, with 6–8 pairs of veins and nearly always with weak basal vein pair; veins beneath silky hairy; 2–3 teeth between secondary vein ends; secondary veins 2.5–3 mm apart (–3.5 mm towards apex). *Male catkins* c. 15 × 3.5 mm in dormant state; scales triangular, peltate. *Fruiting catkins* borne singly or in twos at twig ends or on spur shoots, to 14 × 7 mm, not opening on ripening so that fruits are retained within the persistent catkin for a considerable time; scale c. 5 × 4 mm with mid-lobe c. 3 mm and laterals c. 2.5 mm, parallel sided to more or less spatulate, glabrous except for ciliate margins, with lobe apices having particularly long cilia which are often twisted together. *Seed* c. 2 × 2 mm. *Style* c. 1 mm, base hairy; wing a narrow rim.

DISTRIBUTION. Qinghai, NW Sichuan.

HABITAT. Dry slopes and thickets in association with species of spiny shrubs of *Rosa*, *Caragana*, *Berberis*, *Zanthoxylum*, *Artemisia*, *Quercus*, and *Salix*.

CHROMOSOME NUMBER. Probably tetraploid: 2n=56 (counted from seedling which is probably of this species).

OTHER SPECIMENS SEEN. Qinghai: Yushu Xian, just E of Jiangxi forest Station on E side of Zi Qu, SE of Mozhong, 32°5'N 97°1'E, 2540 m, 27 Aug. 1996, *Picea* forest margin, tree to 1.5 m, *Ho, Bartholomew, Watson & Gilbert* 2448 (E). Nangqên Xian, Bêca xian, along the Ba Qu towards the Xizang from Bêca Forest Station SE of Bêcacu, 31°50'N 96°33'E, 3450 m, *Picea* forest, multi-stemmed shrub to 2 m, *Ho, Bartholomew, Watson & Gilbert* 2968 (E). Sichuan: Xinlong Xian: road (highway 217) from Xinlong

to Ganzi (Garze) along small tributary of Yalong Jiang, S of Junba Xiang (Orxoi), 31°11'41"N 100°17'51"E, 3141 m, dry thickets of *Salix, Rosa, Zanthoxylum* and *Artemisia* in narrow ravine with small stream just upstream from Yalong Jiang, on slope, shrub 2–5 m tall, herbarium specimens showing male catkins and leaves, no fruits, 11 Aug. 2006, *D. E. Boufford, B. Bartholomew, S. L. Kelly, R. H. Ree, H. Sun, J. P. Yue, D. C. Zhang & W. D. Zhu* 36229 (HUH, LIV). Dêgê Xian: W of road (highway 317) from Dêgê to Jinsha Jiang (upper Chang Jiang) just W of Gongya Xiang along Serqu Ge (Serqu R.), 31°41'10"N 98°35'39"E, 3100 m, dry slope above river with spinescent shrubs (c. 2 m tall) of *Rosa, Caragana* and *Berberis*, on steep dry slope above river, shrub 1.5–2 m tall, fruit green, 18 Aug. 2006, *D. E. Boufford, S. L. Kelly, R. H. Ree, H. Sun, J. P. Yue, L. L. Yue, D. C. Zhang & W. D. Zhu.* 36417 (HUH, LIV). Litang Xian: road (highway 217) from Litang to Xilong, S of Junba Xiang (Orxoi), 30°27'28"N 100°18'6"E, 3120 m, dry slopes with spiny leaved *Quercus, Abies* and mixed spinescent shrubs, steep slope among shrubs above river. Open, shrub c. 2 m tall, 10 Aug. 2006, *D. E. Boufford, Bartholomew, S. L. Kelly, R. H. Ree, H. Sun, J. P. Yue, D. C. Zhang & W. D. Zhu* 36229 (HUH, LIV).

Fig. 93 (right). *Betula skvortsovii* among boulders in open pine forest, near Yading, W Sichuan (MF).

Subgenus ASPERAE section ASPERAE subsection CHINENSES

Fruiting catkins opening and shedding seed on drying, remaining on branches for some time before catkin disintegrates; catkins scale lobes oblong or parallel sided.

10. BETULA CHINENSIS

HISTORY, ECOLOGY AND STATUS IN THE WILD

Betula chinensis was named by Carl Maximowicz in 1879 from specimens collected by Dr Bretschneider, near Beijing. Emil Bretschneider (1833–1901) is best known for his monumental *History of European Botanical Discoveries in China* (1898). He spent the years 1866 to 1883 in Beijing, as physician to the Russian legation.

At present *Betula chinensis* is known from north-eastern China, where it is found near Beijing, possibly as far west as Gansu, and in Korea. According to *Flora of China* (Li & Skvortsov 1999), it grows in 'broad-leaved forests in mountain valleys, shaded, rocky mountain slopes', the first perhaps an unusual habitat for a light-demanding species; they also state that it is one of the most valuable timber trees in north China, and describe the wood as 'hard and dense, close-grained, very fine textured, used for making pestels and wagon axels'. In Korea, Prof. Chang tells us that it is a shrub of exposed mountain ridges.

Cytological studies by HMcA show that this species occurs in two cytotypes, the common octoploid and a rarer hexaploid, found in Korea. These are described in detail below.

TAXONOMY, RECOGNITION AND RELATIONSHIPS

With its relatively small, more or less rhomboid or diamond-shaped, coarsely toothed leaves, *Betula chinensis* most resembles *B. dahurica* or *B. gmelinii*. However, as so often in *Betula*, it is the fruiting catkins which are distinctive. Those of *B. chinensis* are more or less spherical when mature, dry, and open, resembling those of *B. costata* and very similar to, but somewhat smaller than, those of *B. globispica* or *B. fargesii*. Its fruiting catkin scales have elongated lobes, and are more or less glabrous except for marginal cilia and base of abaxial surface, but lack the apical twisted hair tuft characteristic of the *B. potaninii* species group. From its close relatives, *B. chinensis* is easily distinguished by the thin texture of its leaves, by the flexibility of the shoots, and by its twigs, which are thinner than the broad buds. Added to this is the unmistakable crowding of the lenticels on trunks and branches, only rivalled in *B. schmidtii*, but that is usually a substantial tree.

An unusual feature of this species, linking it to the related *Betula globispica*, is that only one seed is produced in the axil of each fruiting catkin scale (De Jong 1993) (Fig. 96).

CYTOLOGY

Octoploid and rare hexaploid populations of *Betula chinensis* have been found. North Korean and most South Korean populations are octoploid with 2n=112. Judging by the herbarium specimens seen, populations from mainland China are also octoploid; hexaploid plants (2n=84), distinguishable by their more deeply toothed leaves with more numerous veins, have been raised from seed from South Korea.

Following the discovery of the hexaploid cytotype in the 1970s, we tried to get seed of wild source *Betula chinensis*. One of us (KBA) spent a frustrating day in 1981 looking unsuccessfully for *B. chinensis* in Korea, and a subsequent expedition from Kew, one of whose aims was to obtain seed of *B. chinensis*, was equally unsuccessful. However, through the botanic gardens seed exchange system,

PLATE 8. *Betula chinensis*
Painted by Josephine Hague

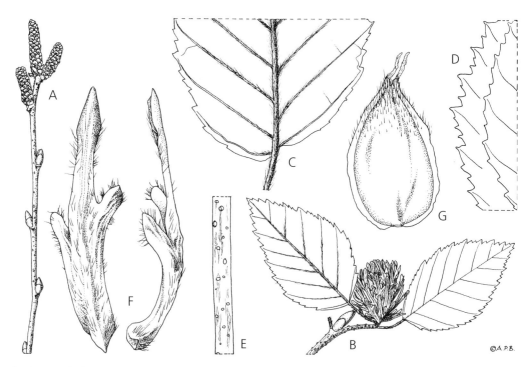

Fig. 94. *Betula chinensis*. **A** twig with buds and immature male catkins × 1; **B** fruiting shoot with mature fruiting catkin × 1; **C** underside of leaf base × 2; **D** hexaploid (left) and octoploid (right) margins of long shoot leaves showing difference in toothing × 2; **E** bark of twig × 2; **F** female catkin scale, back and side views × 6; **G** seed × 10. **A, E** *Beyer, Erskine & Cowley* 187 (Kwangumsong, Mt Sorak, S Korea, 6 Oct. 1982); **B, F, G** from *E. H. Wilson* 10627 (Mts behind Kwachoji, Keiki Province, Korea, 20 Aug. 1918); **C** from *H. McAllister* (Ness BG). DRAWN BY ANDREW BROWN.

several different collections were grown at Ness. One, distributed by Suwon Arboretum of Seoul University, from Chiak-san in Gangwondo in central South Korea, collected by *Hui Kim* (no. 503) *Chang, Choi & Min* on 30 Sept. 2000, gave rise to distinctive-looking seedlings with sharply toothed leaves and proved to be the hexaploid, a further collection to the earlier hexaploid population found on Mt Jiri (presumably Chi-ri) by staff of the Kwanak Arboretum.

From the first two samples studied, one of each cytotype, it seemed that there are consistent differences in fruiting catkin scale and nutlet size, something which might be expected where there is a difference in ploidy level. However, when Professor Chang of Seoul kindly sent a number of samples from Surak-san near Seoul, all proved to be octoploid and the variation in scale and nutlet size was very great, overlapping with that of the cultivated hexaploids. Such variation is likely to be largely the result of environmental factors, but makes the use of their size of little use for separating the cytotypes.

Several plants from each of a number of populations have now been counted, so we can be fairly certain that whole populations are either hexaploid or octoploid. Seedlings raised from each cytotype are similar in appearance and can be distinguished morphologically from populations of the other cytotype. Judging from chromosome counts the octoploid is probably more common and all the specimens in E, BM, and K appear to be the octoploid. However, until more collections of both have been checked and compared, and a geographical difference has emerged, we do not consider that the two cytotypes merit different names.

From our present state of knowledge, the cytotypes can be distinguished as follows:

Bud blunt; leaf tooth mucro about half tooth length; leaves appearing almost double toothed; fruiting catkin scale to 6.5 × 2 mm, hairy abaxially. . . . . 6x *chinensis* (2n=84)
Bud acute; leaf tooth mucro less than half (about ¹/₃) tooth length; leaves very evenly toothed; fruiting catkin scale to 8 × 3.5 mm, glabrous abaxially . . . . 8x *chinensis* (2n=112)

Although it is not easy to explain how a hexaploid could be derived from an octoploid or vice versa, such chromosome number variation within and between populations of a species is known: it can arise within a species or as a result of hybridisation with a related species of different ploidy level. In this case the tetraploid *Betula ermanii* is the only possible parent, as a hybrid between it and octoploid *B. chinensis* would be expected to be hexaploid. However, hexaploid *chinensis* shows no trace of the much larger, very different fruiting catkins and longer leaf stalk of *B. ermanii*.

CULTIVATION

In cultivation *Betula chinensis* makes a shrub or small tree with much variation in habit. Some individuals form low, spreading shrubs of no more than about two metres while others are more upright. The breaking buds in spring and young shoots are frost sensitive and the thin twigs and buds seem susceptible to winter desiccation. In many parts of Britain young shrubs are often reduced to a tangle of dead twigs. The species might thrive better in the more continental climates of central Europe or North America. There is a fine specimen in the Sir Harold Hillier Arboretum and other more recent introductions from Korea are now growing well in a number of gardens.

Fig. 95. Autumn colour, fruiting catkin and male catkins on hexaploid *Betula chinensis* in cultivation at Ness from Mt Chiak-san, Wonju city, Gangwondo, South Korea, *Hui Kim* 503 (Chin-Sung Chang, Do-yol Choi, Woong-Gi Min. 30 Sept. 2000) (HMcA).

Young plants from Chiak-san put out on a dry sunny slope in sandy soil have thrived and develop attractive orange-yellow autumn colour as illustrated in Plate 8. In 2007, in their sixth year, they bore their first fruiting catkins. Once established, older plants show less dieback in winter and bud break seems to be sufficiently delayed to avoid damage by late spring frosts. On these plants only the male catkins appear to be seriously damaged by desiccating winter wind. Young plants are fairly shade tolerant but mature trees are less so and shaded branches die; the wood of these dead branches remains fairly tough unlike that of most other birches.

Like its relatives, *Betula chinensis* has a very vigorous, deep root system, perhaps more so than any other birch. From the habitat in which they grow in the wild, established bushes on deep soils may be expected to be drought tolerant and, given adequate summer warmth and a long growing season, tolerant of exposure as well. In the nursery it is best to transplant them every year to promote the development of a compact root system.

Propagation by cuttings is relatively easy. Both cytotypes appear to be almost totally self-incompatible so seed can only be produced from stands of more than one clone.

Betula chinensis Maxim., *Bull. Soc. Imp. Naturalistes Moscou* 54 (1): 47 (1879). Type: Non procul a Pekino, prope cacumen montis Conolly dicti, alt. 5–6000 ped. frequens, medio Julio 1877 fructif., *Dr Bretschneider* (holotype LE).

B. *ceratoptera* G. H. Liu & Y. C. Ma, *Bull. Bot. Res.*, Harbin 9 (4): 55 (1989). Neimenggu, Ningcheng Xian Jiangshuan, alt. 1000 m, 21 July 1981, *Lei Xi-ting et al.* 216 (NMU).

B. *exaltata* S. Moore, *J. Linn. Soc., Bot.* 17: 386, t. 16 (1879) [1880 publ. 1879].

B. *jiadongensis* S. B. Liang, *Bull. Bot. Res.*, Harbin 4 (2): 155 (1984).

B. *liaotungensis* A. I. Baranov in T. N. Liou, *Ill. Fl. Lign. Pl. NE. China*: 559 (1958).

DESCRIPTION. *Shrubs or small trees* to 5 m; trunk grey-brown with crowded prominent raised lenticels, becoming black-grey with age. *Twigs* to 4 mm in diam.; thinner than prominent buds. *Buds* brown, silky hairy, about 6 × 3 mm. *Shoot* densely hairy when very young, silky hairs persistent throughout first year. *Leaves* decreasing noticeably in size from base to apex of shoot as later formed leaves are conspicuously smaller than the basal leaves which were preformed in the bud; petiole (2)–20 mm less than 1 mm in diam.; petiole:blade ratio very variable from 1:5 to 1:7, blade ovate to rhombic, cuneate or rounded at base with characteristic twist towards acute or acuminate apex, 15–69 × 10–50 mm, with 7–10 pairs of lateral veins, serrate with 2(–3) teeth between major teeth, initially hairy abaxially, especially along veins, axillary hair tufts present, sometimes with conspicuous red glands on the leaf underside, but sometimes totally eglandular. *Male catkins* dark brown, about 18 × 4 mm; scales peltate, shield-shaped, about 2.5 × 2 mm. *Fruiting catkins* 10–22 × 6–15 mm, very variable in size (as are scales and nutlets, although this does not appear to be correlated with ploidy level, both cytotypes being very variable). *Scales* about 5–9 × 3.5 mm, more or less glabrous except for marginal cilia and base of abaxial surface; central lobe about 4 mm, oblong; lateral lobes about 2 mm, oblong. *Seeds* 1 per catkin scale (instead of the usual 3), 2.5–4 × 1.75–2.5 mm; style about 1 mm, broad, not filiform; wing 0.2 mm, about 1/5 width of nutlet, opaque, narrow rim more than one cell thick.

DISTRIBUTION. China (Gansu, Hebei, Henan, Liaoning, Nei Mongol, Shaanxi, Shandong, Shanxi), North and South Korea.

HABITAT. Broad-leaved forests in mountain valleys, shaded rocky mountain slopes, 700–3000 m (Li & Skvortsov 1999); in Korea, a shrub of exposed mountain ridges (Chang, pers. comm.).

CHROMOSOME NUMBER. Hexaploid: 2n=84 and octoploid: 2n=112.

Fig. 96. Octoploid *Betula chinensis* North Korea, Kumgang Mts. 'Spikey' open fruiting catkins with wingless nutlets borne singly in axils of catkin scales (PW).

11. BETULA GLOBISPICA

Betula globispica forms a small, spreading tree with glossy leaves with quite deeply impressed veins, which turn golden yellow in autumn. A most attractive and distinctive feature is its persistent fruiting catkins, which are conspicuous on the leafless twigs throughout the winter and remain on the tree for over a year.

Betula globispica is rare in the wild in Japan (Ohwi 1965; Oldfield *et al.* 1998) and is very rare in cultivation. It was described in 1894, but, as far as can be determined, was first introduced to cultivation in Britain in 1981. The tree at Kew planted in 1957 (Clark in Bean 1970), was obtained from Berlin Dahlem Botanic Gardens, and was probably *B. fargesii*.

Vegetative shoots of *Betula globispica* are superficially similar to those of *B. ermanii*, differing in the broader leaves, with deeper and coarser teeth. However, the fruiting catkins are so distinctive that they could not be mistaken for those of any other birch except perhaps *B. fargesii*, being more or less spherical, and, when dry, with conspicuous spreading bract lobes; the nutlets have no wing, merely a thickened rim. The fruiting catkins do not disintegrate readily and are so persistent that those from at least two successive years are usually present on the branches at the same time; attempts to detach the scales often end with the scales breaking.

An unusual feature of this species, linking it to *Betula chinensis*, is that only one seed is produced in the axil of each fruiting catkin scale.

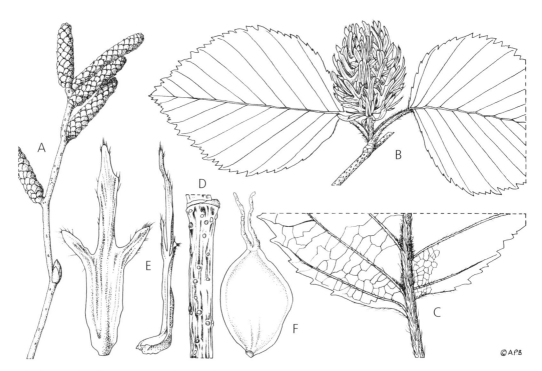

Fig. 97. *Betula globispica.* **A** twig with bud and immature male catkins × 1; **B** fruiting shoot with mature fruiting catkin × 1; **C** undersurface of leaf base × 2; **D** bark of twig × 2; **E** female catkin scale, back and side views × 4; **F** seed × 7. **A** from *H. McAllister* (Ness BG, 17 Nov. 2010); **B–F** ex herb. *Takeda* (14 Aug. 1904). DRAWN BY ANDREW BROWN.

Fig. 98. *Betula globispica* Nikko, N Honshu (KBA).

Fig. 99 (left). Trunk of *Betula globispica*, Nikko, Honshu, Japan (KBA).

Fig. 100 (above). Leaves and fruiting catkin of *Betula globispica* (HMcA).

CULTIVATION

Seed of *Betula globispica* was obtained from Honshu near San'no, Kuriyama-mura, in Tochigi Prefecture in 1981, collected by Rindo Kinenhi and sent to us by Dr H. Ohba. About twelve seedlings were raised. Six were planted out at Ness and became established; five later died of drought, although two of them had grown to good trees of two metres with several main stems. Those which died were on light sandy soil, whereas the single surviving tree is on heavy clay. It was ten years old before it first fruited and has now (2008) grown to about 6 metres and fruits freely every year, producing large quantities of viable seed. It is therefore self-compatible.

The above experience at Ness suggests that it is very sensitive to drought, even as a relatively well-established young tree. It thrives on very heavy clay which, although wet, does not become waterlogged. It is therefore likely to do well in situations where the soil never dries out. While it does not seem to be particular about soil texture, having grown well for several years on both light and heavy soils, dry conditions as can be experienced on light soils can be fatal. It is quite likely that it will be tolerant of some degree of winter waterlogging.

It can be propagated easily by seed, although seedlings are slow growing and highly susceptible to slugs and drought. Seed stored in a refrigerator for 12 years still shows excellent germination.

Betula globispica Shirai, *Bot. Mag. (Tokyo)* 8: 318 (1894). Type: I first discovered this species on 17th October 1893 at Katsuradaira Forest situated near the summit of a peak in the Mitsumine mountain at a distance of about 24 cho from the Mitsumine temple. I found the same species for the second time on 3rd April of this year in Nikko, near the tea house in front of the Kegon-waterfall (holotype ?TI).

DESCRIPTION. *Tree* said to grow to 70 feet (c. 21 m) (Bean 1970) but quoted in the Japanese flora (Satake *et al.* 1989) to 15 m. *Bark* smooth, dark grey. *Twigs* pale brown. *Bud* to 8.5 × 3.5 mm, brown, glabrous. *Leaves* petiole 5–15 mm. Lamina broadly triangular-ovate, 40–70 × 30–50 cm, apex shortly acuminate, conspicuously sharply toothed; lateral veins 8–10 pairs. *Male catkins* up to 2 terminal with 4 laterals, to 30 × 4 mm in winter, brown; expanding to 70 × 11 mm in spring; scales peltate with ciliate margins, 3 × 2.25 mm. *Fruiting catkins* very uniform in size, about 25–35 × 25 mm, more or less spherical; scales 10–15 × 6 mm, glabrous except for ciliate margin; terminal lobe c. 5 mm, oblong-spatulate; lateral lobes c. 2 mm, oblong. *Seed* 1 per catkin scale (instead of the usual 3), 4 × 2 mm, glabrous; style 1 mm with a few cilia; wing a mere rim, 0.2 mm wide, $^1/_{10}$ width of nutlet, opaque, more than one cell thick.

DISTRIBUTION. Japan, Honshu (Kanto and central district).

HABITAT. Steep mountain slopes and rock outcrops (Tabata 1966).

CHROMOSOME NUMBER. Decaploid: 2n=140.

12. BETULA FARGESII

Betula fargesii was described from specimens collected by Père Farges on the western side of the Daba shan, near present-day Chengkou, in the province of Chongqing, formerly eastern Sichuan, near the border with southern Shaanxi.

Paul Guillaume Farges was born in France in 1844 and went to China in 1867; from Chengdu he travelled to northeast Sichuan (now in Chongqing province) where he made many collections on Daba shan. In 1893 he was sent to Chongqing to be chaplain of a local hospital. He sent over 5000 herbarium specimens to the Muséum National d'Histoire naturelle, many of them new species, and, in addition, seeds to Maurice de Vilmorin for his arboretum near Paris. He died in Chongqing in 1912.

Betula fargesii is very rare in cultivation, and very different in overall appearance from almost all other species, being a spreading shrub with greyish twigs, small, greyish-green, narrow leaves, especially on the long shoots, and fruiting catkins with conspicuous spreading elongated scale lobes like those of *B. globispica* and *B. chinensis* (Plate 8; Fig. 33).

The 1980 *SABE* (Sino-American Botanical Expedition to Hubei) (Bartholemew *et al.* 1983) made two herbarium and seed collections of *Betula fargesii*, from western Hubei, around 160 km east of the type locality. *SABE* 183 was initially identified as *B. insignis* and *SABE* 904 as *B. fargesii*. Cuttings of *SABE* 183 were sent under the name *B. insignis* from the University of British Columbia (UBC) by Peter Wharton to Jim Russell at Castle Howard, Yorkshire, England who, in turn, gave cuttings to Ness. It was only after the Guizhou expedition of 1985 brought back seed of true *B. insignis* and these grew to some size that it was realised that the tree of *SABE* 183 at Ness was not *B. insignis*.

In 1997, after studying specimens in Harvard University Herbarium (HUH) and cultivated plants growing in the Arnold Arboretum, we (HMcA) decided that both *SABE* collections are *Betula fargesii*; illustrations of the isotypes of *Betula fargesii* are viewable on http://www.usna.usda.gov by using the 'search our site' facility.

CULTIVATION

The shrubs at Ness derived from the cuttings of *SABE* 183 have grown to about three metres with weeping terminal shoots, but the trees from which *SABE* 904 was collected were 40 feet tall. The only other herbarium specimen seen, *Chu* 3910, is also recorded as being from a tree of 40 feet (12 m) tall. It will be interesting to see what height the seedlings of the *SABE* collections will reach, since like its closest relative *Betula globispica*, *B. fargesii* is self-compatible and the single clone in cultivation at Ness produces a high proportion of viable seed.

PLATE 9. *Betula fargesii*
Painted by Josephine Hague

Betula fargesii Franch., *J. Bot (Morot)* 13: 205 (1899). Type: China occidentalis, Provincia Sutchuen, ad Héoupin in vicinitate Tchen keou tin, alt. 2200 m, *P. Farges* 1012 (holotype P, isotype NA). *Betula chinensis* Maxim. var. *fargesii* (Franch.) P. C. Li.

DESCRIPTION. A multi-stemmed *shrub* or spreading *tree* to 15 m. *Young shoots* and *leaves* quite densely covered with silky hairs, and with small patent velvety puberulent hairs present mainly in leaf axils and on the peduncle, but also present elsewhere on the shoot but not on leaves; small red glands present on shoot and leaf veins on both surfaces and to some extent on the lamina underside. *Twigs* about 2 mm in diam., greyish due to the epidermis having lifted off the underlying tissues. Thicker, more vigorous twigs have a rich brown colouration where the expanding twig is no longer covered by original epidermis; at least some short puberulence present on grey areas of twig where the epidermis is still present, especially in leaf axils. Silky hairs absent as they have fallen off. *Lenticels* not conspicuous; thicker and older twigs are light brown and glabrous. *Leaf petiole* to 10 mm; blade 45–70 × 25–45 mm, glossy somewhat grey-green above; with 8–9 pairs of veins with 2(–3) teeth between main veins in mid-leaf, teeth broader than long; reddish glands persistent on main veins on leaf underside, silky hairy on veins, especially on underside. *Bud* about 7 × 2.5 mm, brown. *Male catkins* up to 2 terminal and 4 lateral, to 20 × 3.5 mm in winter. *Fruiting catkin* indehiscent, very persistent, to 25 × 25 mm when dry, more or less globose; peduncle 1–2 mm; scales < 13.5 × 10 mm, abaxial surface ciliate-hairy, adaxial surface glabrous; terminal lobe < 7 mm, oblong-spatulate; lateral lobes < 4.5 mm, oblong. *Seed* < 3 × 3 mm; style 1 mm; wing very narrow, 0.1–0.3 mm, opaque, more than one cell thick (Fig. 101).

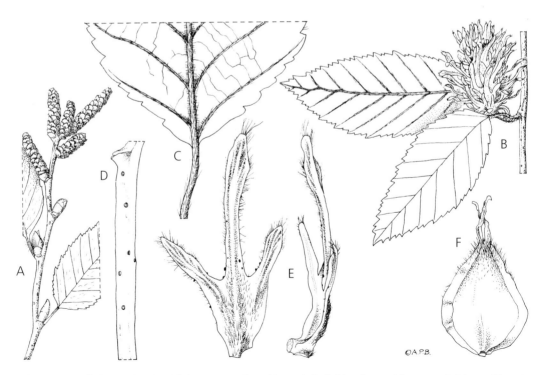

Fig. 101. *Betula fargesii.* **A** twig with immature male catkins × 1; **B** fruiting shoot with mature fruiting catkin × 1; **C** undersurface of leaf base × 2; **D** bark of twig × 2; **E** female catkin scale, back and side views (resin glands present) × **5**; **F** seed × 10. All from *H. McAllister* (Ness BG, 30 Aug. 2010). DRAWN BY ANDREW BROWN.

DISTRIBUTION. West Hubei, in Shennongjia Forest Reserve; eastern Chongqing (formerly in Sichuan), near Chengkou Xian. It seems likely that this species is rare in the wild and may be confined to a small area of western Hubei and adjacent provinces.

HABITAT. Woods and rocky hillsides.

CHROMOSOME NUMBER. 2n=140.

SPECIMENS SEEN. Hubei, Shennongjia forest distr., 31°30'N 110°30'E, along the Qiaodonggou Canyon, W of rd between Jiuhuping forest farm and Bancang, 1950 m, 27 Aug. 1980, *SABE* 183 Sino-American Botanical Expedition) (HUH!) (collected as *Betula insignis*), (checked on internet against SABE collection data). Shennongjia forest distr., 31°30'N 110°30'E vicinity of Dalongtan and Xiaolongtan on W side of road, 2300–2,600 m, 9 Sept. 1980, *SABE* 904 (HUH!) collected as *Betula fargesii*, specimen in HUH determined as *B. delavayi* by Skvortsov (1998)]. Sichuan, Pao-hsing-hsien. 2200 m, 20 Sept. 1936, among woods, shrubs 15 m high, common, *K. L. Chu* 3910 (E! K!, BM!).

Fig. 102. *Betula fargesii*. Winter twig with buds and male catkins. DRAWN BY JOSEPHINE HAGUE.

Fig. 103. *Betula fargesii* (*SABE* 183) China, Hubei. Shrub in cultivation at Ness with spreading arching habit (PW).

Fig. 104. *Betula fargesii* (*SABE* 183) China, Hubei. Foliage and fruiting catkins (PW).

Subgenus ASPERAE section LENTAE

Trees or large shrubs. Trunks never with any trace of white, but often glossy and peeling similar to the mode of the white barked birches. *Twigs* always, if faintly in the case of *Betula medwediewii* and *B. megrelica*, smelling of oil of wintergreen (methyl salicylate) when bruised. *Buds* ovoid, acute. *Leaves* ovate to lanceolate, to almost orbicular in *B. medwediewii*. *Male catkins* always relatively large and showy, terminal, and often also lateral, on the previous year's long shoots. *Female catkins* always remaining erect as they develop into the fruiting catkins. *Fruiting catkins* always more or less persistent over winter, only opening to a limited extent with scales usually remaining more or less erect and appressed to the axis so that the nutlets are only released when the fruiting catkin disintegrates; fruiting catkin scales 3-lobed with the lateral lobes always pointing forwards, never recurved; nutlets always with a distinct (wider than 0.5 mm) but narrow wing at least two cells thick and contiguous with styles, being an extension of the ovary.

13. BETULA INSIGNIS

HISTORY, ECOLOGY AND STATUS IN THE WILD

Although a most handsome birch, and discovered as long ago as 1899, *Betula insignis* was, as far as we know, not in cultivation until 1985 when seed was introduced from the Fanjing Shan in north-eastern Guizhou province, by the Simmons, Fleigner, Russell expedition (*SFRX* or *GUIZ* 82). It had been described by Franchet from a collection made by Père Farges in the same area as *B. fargesii*, in eastern Sichuan, now in Chongqing province near the Shaanxi border. An illustration of the isotype of *B. insignis* can be seen on http://www.usna.usda.gov by using the search-our-site facility. A very similar birch from North Vietnam and south Yunnan, is here described as *Betula insignis* subsp. *fansipanensis* Ashburner & McAll.

   Betula insignis is very similar to a fossil species, *B. leopoldae*, described from the Eocene in British Columbia (c. 50 million years bp) (Crane & Stockey 1987; McAllister 2005a). This is the earliest known fossil certainly referable to *Betula* and consists of many well-preserved leaves, fruiting catkins, catkin scales and seeds — all of which are illustrated in the first paper quoted. The most significant difference between the fossil and living material is perhaps the wide variation in leaf size and shape in *B. leopoldae*. Many of the characteristics of these two species (see Chapter 1) are those predicted for the ancestral form of the genus *Betula*, with leaves not unlike those of some species of *Alnus* or *Carpinus*. But there is no possibility that living *B. insignis* itself is the ancestor of other birch species, partly because of its high chromosome number, decaploid with 2n=140, while many of the more recently evolved common species are diploid with 2n=28.

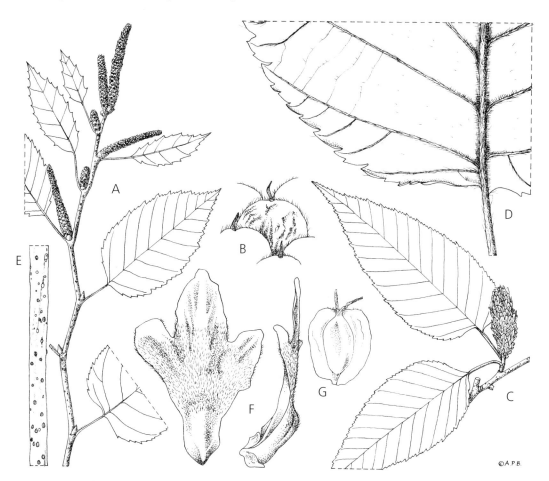

Fig. 105. *Betula insignis.* **A** twig with buds and immature male catkins × ¾; **B** male catkin scale × 10; **C** fruiting shoot with mature fruiting catkin × ½; **D** underside of leaf base × 2; **E** bark of twig × 2; **F** female catkin scale, back and side views × 7; **G** seed × 7. All from live material RBG Kew, *GUIZ* 1993-1193, 27 Aug. 2010. DRAWN BY ANDREW BROWN.

Fig. 106 (left). *Betula insignis* subsp. *insignis* (*GUIZ* 82). Fruiting catkins on shoots bearing small leaves with no buds in their axils. End of shoot (dead) towards right bore male catkins in the spring. Leaf subtending central fruiting catkin shows basal two veins curling away from margin and not running into tooth (camptodromous state) (PW).

Fig. 107 (right). *Betula insignis* subsp. *insignis* (*GUIZ* 82). Trunk of young tree with lenticels (PW).

Jim Russell's notes describe the excitement of finding this species (*Betula insignis*, here referred to as an *Alnus*, which perhaps reveals how little-known it was, even to Chinese botanists; in cultivation this collection was initially distributed as *Betula austrosinensis*):

'First collected from a spray fallen from above in the area where *Fagus longipetiolata* was the dominant tree. Professor Huang stated that it was an *Alnus* but definitely not *A. nepalensis*, and that the fruits seemed consistent with *A. trabeculosa*. The tree grows in high forest at 1200 m. The ascent here (between Tong Kuanchan and Wen Bao Yen) was still very steep but there was a wide shoulder of mountain to the right of the track. With the *Fagus* grew the *Alnus* (*sic.*) and a few *Tsuga chinensis*. It was near here that we saw the two trees of *Stewartia sinensis*.

Coming down three days later, we had more time and it became clear that this alder (*sic*) was fairly common, a rough-barked tree of some seventy to eighty feet. The strobiles are upright, in clusters on ascending twigs and, from below, look curiously like beech mast.'

The mention here of *Fagus longipetiolata* recalls the phytogeographic association of birches of section *Lentae* with other species of beech — for example *Betula alleghaniensis* with *Fagus grandifolia* in North America and *B. grossa* with *Fagus crenata* in Japan. The ecological role of *B. insignis* corresponds to that of other members of section *Lentae*, i.e. it does not form pure stands, but grows scattered in mixed forest.

While we have not seen the type specimen of *Betula kwangsiensis*, a specimen annotated *kwangsiensis* by Metcalf himself and dated 17 May 1941, is in Harvard (HUH!) and in Kew (K!); this is *Steward, Chiao & Cheo* 735, from Ta Ho Yen on the Fanjing Shan at 1200 m. It is very similar to the cultivated material of *B. insignis GUIZ* 82, and from the same altitude.

TAXONOMY, RECOGNITION AND RELATIONSHIPS

Betula insignis most closely resembles *B. alleghaniensis* or *B. lenta*, but its handsome leaves are much larger than those of either species. Its fruiting catkins are more bulky and differ in appearance because of the more elongated central lobe of each scale. Its leaves are strikingly similar to those of the distantly-related *B. cylindrostachya*, but *B. insignis* lacks attenuate tips to the leaf teeth, and differs in leaf venation. In *B. insignis* venation is semi-craspedromous, i.e. the ends of one or two basal pairs of secondary veins are barely distinguishable from the tertiary venation; they curl back at the tooth base and send a branch into the tooth (Fig. 106). In contrast, the venation in *B. cylindrostachya* is craspedodomous, i.e. the secondary veins are significantly thicker than the third order venation and run straight into the teeth as in almost all other species.

PLATE 10. *Betula insignis*
Painted by Josephine Hague

Fig. 108. *Betula insignis* subsp. *insignis* (*GUIZ* 82). Fruiting catkins crowded on twigs which will die after fruiting (PW).

CULTIVATION

Trees of *Betula insignis* grow best in heavy wet clay soils. Their roots are evidently tolerant of poor aeration. In drier soils they survive but grow much more slowly and reach a lesser height (Albertson & Weaver 1945). Other early-evolved angiosperm trees such as planes (*Platanus*), alders (*Alnus*), *Liquidambar* and wingnuts (*Pterocarya*) show a similar response and, like them, *B. insignis* might be suitable for planting in poorly aerated and heavy soils in cities.

The large clusters of male catkins, the showiest of all birches, are very conspicuous in April (Plate 10; Fig. 109), and can reach 16 cm when fully open. The fruiting catkins are slow to break up, and the scales remain attached for up to a year, hanging down on a thread (Fig. 110), so they appear very different from those of *Betula globispica* and *B. fargesii* in which the scales separate from one another but remain firmly attached to the axis, releasing the seeds in a similar way to a pine cone.

Trees in sheltered situations can develop wonderful yellow-orange autumn colour, and Andrew Jackson, head of Wakehurst Place in Sussex, reports that it is one of the best autumn-colouring trees there (McAllister 2005a).

Betula insignis is self-incompatible, and therefore more than one clone must be grown to obtain viable seed. Cuttings from young trees root relatively easily, but those from older and fruiting trees are more difficult.

Betula insignis Franch., *J. Bot. (Morot)* 13: 206 (1899). Type: China occidentalis, provincia Sutchuen, in vicinitate Tchen Keou Tin, alt. 1400 m, *R. P. Farges* 83 (holotype P, isotype USNA).

subsp. **insignis**

Betula kwangsiensis Metcalf, *Lingnan Sci. J.* 20: 216 (1942). Type: Kwangsi, Dar Yeung Kiang, Luchen, border of Kweichou, 3900 ft, 27 June 1928, tree 40 ft, 1.2 ft diam., *R. C. Ching* 6279. (holotype ?, isotype HUH).

Betula austrosinensis Chun ex P. C. Li, *Acta Phytotax. Sin.* 17 (1): 89 (1979). Type: Kwangsi: Lin-kuei Hsien, Chi-fen shan, in valley and mountain top, 9 Oct. 1956, *Liang Heng* 100231 (holotype PE).

DESCRIPTION. *Tree* to 25 m. *Bark* brown, greyish-brown to greyish-black, fissured. *Twigs* about 2.25 mm in diam., mostly rich brown with greyish patches, glabrescent; lenticels conspicuous, white; shoot when young with silky hairs, which mostly soon fall, puberulent hairs and glands absent. *Petiole* up to 25 mm. *Leaf* lanceolate, blade to 160 × 68 mm, but much smaller on female and fruiting catkin bearing shoots, with 12–13 pairs of veins and 3–4 teeth between main vein ends; when young with silky hairs, which mostly soon fall except on the leaf veins and petiole, lower (abaxial) surface of lamina with small reddish glands; upper surface not glossy, with glands restricted to veins. *Male catkins* from 27– c. 40 × 5 mm in winter, expanding to 110–160 × 9 mm at anthesis, up to 9 clustered at the ends of strong shoots of the current year's growth, mostly singly, rarely in 2s or 3s, or up to 5 in the axils of leaves, green flecked with brown. *Fruiting catkins* to about 30–50 × 20 mm, cylindric-ovoid, slow to break up; seeds dispersed as the bracts separate from one another. *Scales* to 11.5 × 4 mm, hairy on both surfaces and margins, persistent and remaining attached to the axis, or hanging down from the axis on a thin thread for at least a year after ripening; central lobe to 5 mm, oblong; lateral lobes to 3 mm, oblong. *Seeds* to 3 × 3 mm, wing c. 0.6 mm, c. $^{1}/_{3}$ width of nutlet, opaque, more than one cell thick; style 1.8 mm.

DISTRIBUTION. China: Chongqing (formerly part of E Sichuan), Guizhou, Hunnan, W Hubei, Sichuan, Yunnan, (Fujian, Guangdong, Guangxi, Jiangxi). We have not seen specimens from the provinces in brackets which are listed in *Flora of China* (Li & Skvortsov 1999) under *Betula austrosinensis*.

HABITAT. Mountain forests.

CHROMOSOME NUMBER. Decaploid: 2n=140.

Fig. 109 (right). *Betula insignis* subsp. *insignis* (*GUIZ* 82) male catkins elongating and female catkins (green) emerging from buds (PW).

BETULA INSIGNIS subsp. FANSIPANENSIS

This subspecies is very similar to *Betula insignis* subsp. *insignis*, differing mainly in the knobbly scales on the male catkins in winter and its smaller fruiting catkins. Its young leaves are a remarkable purple in colour and most attractive (Plate 11; Figs 113, 115). It occurs in southeast Yunnan and the high mountain range of Fan-Si-Pan in the adjacent area of North Vietnam, representing a south-easterly vicariant of subsp. *insignis*. Coming from a low latitude and relatively low altitude, it is less hardy than subsp. *insignis* and appears to require a long frost free growing season in order to produce sufficiently mature twigs to survive winter desiccation. Trees have grown well at Ness Gardens since 1995, but we do not know whether it survives in any other of the gardens to which it was distributed.

CULTIVATION

Betula insignis subsp. *fansipanensis* was first introduced into cultivation by Keith Rushforth in 1991. It is probably intolerant of dry soils, but may survive in waterlogged heavy clay as a plant at Ness is thriving reasonably well with the base of its trunk standing in water for much of the winter, although

Fig. 110 (left). Knobbly male catkins and emerging female catkins of *Betula insignis* subsp. *fansipanensis* (*KR* 2344) showing probable winter desiccation damage to male catkin tips and wind or frost damage to some leaf tips (PW).

Fig. 111 (right). *Betula insignis* subsp. *fansipanensis* (*KR* 2344). Male and female young catkins and remains of previous year's fruiting catkins with scales hanging by thread from central catkin axis (PW).

PLATE 11. *Betula insignis* subsp. *fansipanensis*
Painted by Josephine Hague

Fig. 112 (left). *Betula insignis* subsp. *fansipanensis* (*KR* 2344). Trunk of young tree (PW).

Fig. 113 (right). *Betula insignis* subsp. *fansipanensis* (*KR* 1878) with purplish young leaves (PW).

Fig. 114 (left). *Betula insignis* subsp. *fansipanensis* (*KR* 1878). Male catkins clustered towards end of twig with groups of up to three in axil of each leaf (PW).

Fig. 115 (above). *Betula insignis* subsp. *fansipanensis* (*KR* 1878). Young shoot with deep purple leaves on emergence (PW).

it is only fair to say that a nearby tree in an only slightly drier situation is growing more rapidly. As with subsp. *insignis*, cuttings of young plants have rooted fairly easily, but since the trees have begun to bear catkins, cuttings have been very difficult to root.

subsp. **fansipanensis** Ashburner & McAll., **subsp. nov.** Affinis subsp. *insignis*, sed amentis fructiferis gracilioribus, 15 mm diam. et amentis masculis torulosis, squamis umbonatis differt. Typus: N Vietnam, Lao Cai prov., Sapa, hillside N of Sapa and opposite Ban Khoang, 6800 ft, 22°23'N 103°48'E, N facing aspect, 12 May 1992, (Keith Rushforth) *KR 2344*. Collected as *Carpinus pubescens*. Cultivated specimen from University of Liverpool Botanic Gardens, Ness, 13 Sept. 2006, *H. McAllister* (LIV).

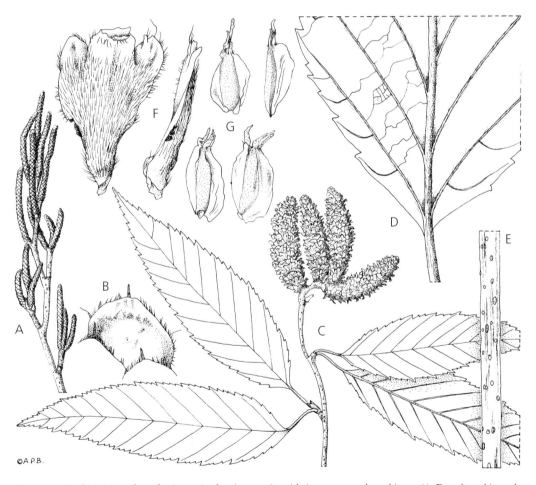

Fig. 116. *Betula insignis* subsp. *fansipanensis*. **A** winter twig with immature male catkins × ½; **B** male catkin scale showing swollen, smooth, cushioned base × 10; **C** shoot with terminal mature fruiting catkins × ¾; **D** underside of leaf base × 2; **E** bark of twig × 2; **F** female catkin scale, back and side views × 7; **G** selection of seeds showing malformation, upper pair with two styles, lower pair with three styles (frequent in this material) × 7. All from *H. McAllister* (Ness BG, Aug. 2010, ♂ catkins on *KR1878* in cultivation at Ness). DRAWN BY ANDREW BROWN.

Note abnormal fruiting twig with clustered terminal fruiting catkins — the defining character of *Betula gynoterminalis* and also seen in *B. medwediewii*.

DESCRIPTION. Differs from subsp. *insignis* in having male catkins knobbly with raised bosses on each scale and smaller fruiting catkins. Young leaves more distinctly purplish on emergence. *Male catkin* to 80 × 5.5 mm. *Fruiting catkins* 45 × 15 mm. *Scales* 10 × 6 mm, both surfaces and margin with rather stiff hairs; terminal lobe 3 mm, triangular-oblong, lateral lobes 1.5 mm, triangular-oblong. *Seed* 4 × 3 mm, with wing 0.4 mm, ⅓ width of nutlet, opaque, more than one cell thick; style 1 mm.
DISTRIBUTION. SE Yunnan; North Vietnam, Fan-Si-Pan.
HABITAT. Forest on mountain ridges, on limestone.

14. BETULA GROSSA

HISTORY

Betula grossa is a very handsome, hornbeam-like tree, found in the three southern islands of Japan. It was described by Philipp von Siebold (1796–1866) and Joseph Zuccarini from specimens collected by Siebold. At the same time he described two similar species, *B. carpinifolia* Sieb. & Zucc. and *B. ulmifolia* Sieb. & Zucc., but later authors have considered them to be the same as *B. grossa*.

ECOLOGY AND STATUS IN THE WILD

According to E. H. Wilson (in Sargent 1916), *Betula grossa* is common in mixed woods in Shinano province in central Honshu in Japan. Its range extends from north Honshu southwards to Shikoku and Kyushu, reaching higher elevations in the south. Ohwi (1965) describes this as a mountain tree; it does not occur in Hokkaido or on the mainland of Asia.

Fig. 117. *Betula grossa*. Foliage and fruiting catkins (PW).

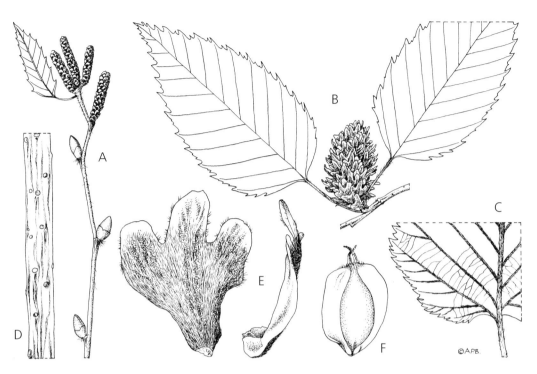

Fig. 118. *Betula grossa*. **A** twig with buds and immature male catkins × 1; **B** fruiting shoot with mature fruiting catkin × 1; **C** underside of leaf base × 1; **D** bark of twig × 2; **E** female catkin scale, back and side views × 7; **F** seed × 7. All from *Furuse* 48083 (Gifu Prefecture, Hondo, Japan, 17 Sept. 1969). DRAWN BY ANDREW BROWN.

There are many collections from Shinano province, and many from Lake Chuzenji and Nikko in Shimotsuke province. It is an early colonising species following disturbance in forests dominated by beech (*Fagus crenata*), a role not dissimilar to that played by *Betula lenta* in eastern North America (Nakashizuka 1989; Yamamoto *et al.* 1995). It can evidently reach considerable size judging from a single smooth-barked specimen seen in Miyagi prefecture in northern Honshu (KBA pers. obs.), although the maximum height is usually given as 25 m.

TAXONOMY, RECOGNITION AND RELATIONSHIPS

With its conspicuously hairy leaves and young stems, and small upright fruiting catkins, *Betula grossa* is only likely to be confused with *B. alleghaniensis* of eastern North America. The most obvious difference is that the leaf bases of *B. grossa* are usually cordate, while those of *B. alleghaniensis* are usually truncate. A more precise difference which can be seen with a hand lens: whereas both species have long (1–2 mm), silky hairs which lie more or less parallel with the leaf, only *B. alleghaniensis* also has fine (less than 0.3 mm) velvety puberulent hairs which stand out at right angles to the surface of the veins underneath the silky hairs, but their character has been checked in only relatively few specimens so its value requires confirmation.

In addition to the above, the leaves of *Betula grossa* are less double toothed, having only one order of toothing, and are more persistently hairy on their upper surface. The bark is very similar to that of *B. lenta* with its hint of flaking and, like *B. lenta*, lacks the metallic brilliance seen in *B. alleghaniensis*. There is a certain thick-twigged sturdiness about *B. grossa* when compared with *B. lenta*.

CULTIVATION

When fully grown, *Betula grossa* is likely to be too large for most gardens, but young trees grow relatively slowly, forming neat small spreading trees to about eight metres in thirty years. As long as the situation is not too exposed or the soil too dry, it will give a reliable display of autumn colour every year. It will thrive in any good, rich and well-drained soil.

Betula grossa Sieb. & Zucc., *Abh. Math.-Phys. Cl. Königl. Bayer. Akad. Wiss.* 4 (3): 228 (1846). *Betula sollennis* Koidz., *Fl. Symb. Orient.-Asiat.*: 20 (1930).

ILLUSTRATION. *Woody Plants of Japan*: 118 (1985).

DESCRIPTION. *Tree* with dark grey smooth bark, eventually very large, to 25 m. *Young shoots* yellow-green, very silky hairy on youngest internode, without short puberulent hairs, with small brown glands; lenticels pale brown and slightly raised, small, linear-ovate, less than 0.5 mm. *Twigs* light brown, with persistent more or less patent setose hairs present on most first year twigs through winter. *Buds* large, on strongest, vigorous twigs to 10 × 5.0 mm, ovoid and acute but not acuminate, projecting outwards from twig as much as c. 45°, scales shining chestnut-brown. *Petioles* hairy, 1–2.5 cm, to 2 mm broad at base, with a few glands. *Leaves* broadly ovate, blades 5 × 3 to 10 × 6 cm, cordate at base, sometimes deeply so with sinus as much as 10 mm deep, with 8–14 pairs lateral veins; leaf margins with 4 or even 5 acute teeth between major teeth; *leaves* on fruiting spur shoots smaller, distinctly cordate, the blade as little as 3.6 × 2 cm, petiole:blade ratio 1:3.5 to 1:5. *Male catkins* rather short, to 20 × 5 mm, scales pale brown and shiny with darker brown fringes, sometimes thickly encrusted with white resin, and occasionally more or less knobbly with raised bosses on scales, as many as 2 axillary below the 2 terminal. *Fruiting catkins* erect, almost sessile, ovate to oblong, to 25 × 15 mm, scales ciliate, to 7.5 × 6.5 mm, laterals more than half the length of mid-lobe. *Seed* to 4 × 3.5 mm with nut c. 2 mm broad.

DISTRIBUTION. Japan: Honshu, Shikoku, Kyushu.

HABITAT. Mountain forests with other trees, not forming pure forest.

CHROMOSOME NUMBER. Hexaploid: 2n=84.

Fig. 119. *Betula grossa.* Bark on trunk of young tree (PW).

15. BETULA MEDWEDIEWII

Betula medwediewii is a wide, spreading shrub, with large leaves, long male catkins and rich yellow autumn colour. It was described by Eduard Regel in St Petersburg from specimens collected in the Caucasus in 1886 by Jakob Medvedev (1847–1923).

Betula medwediewii is clearly distinguished from all other birches by its thick, stiff twigs with large, ovate to circular leaves, and large, upright persistent fruiting catkins with brown scales. The fruiting catkins are mostly borne on shoots on the outside of the shrub, so that they occur in groups along shoots sticking out from the shrub. These shoots usually die after fruiting, as no buds are formed in the axils of the leaves of the fruiting spur shoots. The twigs, when scraped, smell of oil of wintergreen.

Apart from the even rarer Caucasian *Betula megrelica*, (see p. 192), *B. medwediewii* is probably most closely related to *B. alleghaniensis*, the yellow birch, of eastern North America, and *B. insignis* of central China. However, these east Asian and North American species are large, moisture-requiring trees whereas *B. medwediewii* is a drought-tolerant mountain shrub.

The subfossil birch, *Betula islandica*, described from mid-Miocene deposits in Iceland, is apparently similar in appearance to *B. medwediewii* (Denk *et al.* 2005), but as it is known only from leaves and fruiting catkin scales, we don't know whether it was a tree or a shrub.

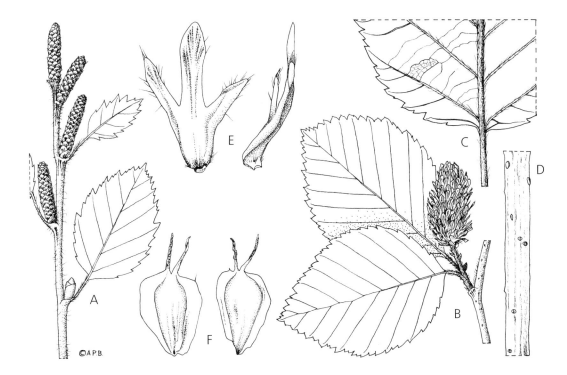

Fig. 120. *Betula medwediewii.* **A** twig with bud and immature male catkins × 1; **B** fruiting shoot with mature fruiting catkin × ¾; **C** underside of leaf base × 1½; **D** bark of twig × 2; **E** female catkin scale, back and side views × 5; **F** seed (showing vestigial hairiness on wing) × 7. **A, B, D–F** from *Davis & Hedge* D32343 (Şavval Tepe, Çoruh (Artvin), Turkey, 14 Aug. 1857); **C** from *Davis & Hedge* D29913 (Tiryal Dağ, Çoruh, Turkey, 23 June 1957). DRAWN BY ANDREW BROWN.

Betula medwediewii is one of a group of rare relict trees and shrubs found in the western Transcaucasus and the adjacent mountains of north-eastern Turkey (Melville 1939; Eminagaoğlu *et al.* 2006; Browicz 1975). It grows on the upper margins of *Picea orientalis* forest, and in subalpine scrub, and is often accompanied by such rarities as *Epigaea gaultherioides*, *Quercus pontica* and *Rhododendron ungernii*, for example in the Mtirala reserve and in the Korolistzkhali river gorge near Batumi (Dmitrieva 1960).

Seed was obtained from a collection made by Peter Davis (*D. 32343*) in Turkey, near Çoruh, which was the only example of the species growing in the Royal Botanic Garden, Edinburgh and so could be trusted to come true. Seedlings from this and from a shrub of unknown origin in Hergest Croft gave chromosome counts of 2n=140, as did six collections, supposedly from the Mt Migaria region, which were sent from Russia. From what is now known from Paul Bartlett's 2013 collections from Mt Migaria and Mt Javari, it would seem very unlikely that these collections did in fact come from that region. They confirm that *Betula medwediewii* is quite variable in leaf size and shape, but we will have to wait until the recent collections of *B. megrelica* mature to judge whether the two species are morphologically always distinct.

CULTIVATION

Betula medwediewii is occasionally grown in arboreta and a few large gardens. It forms a large shrub, or, when it is grown on a single stem, a goblet-shaped small tree, which develops beautiful golden yellow autumn colour, and the leaves remain on the tree for some time. Although slow growing as a seedling, it is easy to grow and, being a mountain plant, is tolerant of exposure and very hardy. The cultivar 'Gold Bark' has shining pale brown bark, and good yellow autumn colour.

Fig. 121. *Betula medwediewii*, male catkins and fruiting catkins from previous year, 5 April 2006 (PW).

Fig. 122. *Betula mewediewii* in autumn colour (PW).

Many gardeners and botanists have noticed that *Betula medwediewii* comes true from seed and does not hybridise with other birches, perhaps because it flowers later than other species of *Betula*; in addition, single trees bear fertile seed, so it is clearly self-compatible. Also, it is rarely grown alongside other species of its section, so hybridisation would be unlikely, and its chromosome number, 2n=140, differs widely from that of most other birches.

Betula medwediewii Regel, *Trudy Imp. S.-Peterburgsk. Bot. Sada* 10: 375 (1887). Type: in Transcaucasiae, monte Somlia jugorum Adscharo guriensium ad fines sylvarum, 6800 ft alt., Julio 1886 Cl. Medwediew legit *Medwediew* 107 (holotype LE).

DESCRIPTION. *Tree or shrub* to 5 m. *Trunk and branches* with metallic, reflective bark with conspicuous lenticels. *Twig* stout, 2–4 mm thick, long silky hairy when young, then shining brown. *Bud* ovoid, with green brown-edged scales. *Petiole* to 2 cm. *Leaf* blade 3.5–12 × 2–7 cm, usually 5–9 × 3–6 cm, ovate to orbicular, with (6–)8–10(–13) pairs of veins. *Male catkins* 4–5 cm, in groups of 4 or 5 at the ends of long shoots whose lower buds often bear the female catkins. *Female catkins* produced from successive buds below the male catkins. *Fruiting catkins* erect (20–)25–40(–50) × 12–18 mm, ovate to cylindrical-oblong. *Scales* 8–12 mm, lobes oblong, parallel sided, central to 6 mm, laterals to 4 mm. *Seeds* c. 3 mm with wing less than half nutlet width.

DISTRIBUTION. South-western Georgia, in Adjara, and north-eastern Turkey, in Rize and Çoruh.

HABITAT. Subalpine mixed forests, at the upper limit of *Picea orientalis* and in *Rhododendron* scrub, 1200–2400 m.

CHROMOSOME NUMBER. Decaploid: 2n=140.

16. BETULA MEGRELICA

Betula megrelica is often treated as a synonym of *B. medwediewii* (Rehder 1940; Govaerts & Frodin 1998). Although Bean (1970), perhaps following *Flora of the USSR* (Kuzeneva 1936), describes it as a tree, it is a mountain shrub only known from Mt Migaria northwest of Kataisi, in the Mingrelia region of the Caucasus in western Georgia; it is listed in the *Russian Red Data Book* as being very rare (Iliashenko & Iliashenko 2000). The much smaller, narrower leaves, and smaller fruiting catkins, give shrubs of *B. megrelica* a different appearance from those of *B. medwediewii*.

In 1982 Professor A. K. Skvortsov, the Russian authority on birches, kindly sent us seed from a shrub of *Betula megrelica* growing in the Main Botanic Garden in Moscow. He wrote that this plant had been grown from a seedling collected on Mt Migaria. The seedlings raised at Ness were somewhat variable, but several were very distinct from the more familiar Turkish *B. medwediewii* in having smaller, narrower, more hairy, oval leaves, greyish hairy twigs, and smaller fruiting catkins, and agreed with the descriptions and illustrations of *B. megrelica* in *Flora of the USSR*.

A chromosome count on one of these gave 2n=168, whereas chromosome counts on *Betula medwediewii* gave 2n=140. *B. megrelica* therefore appears to be a dodecaploid with twelve times the base chromosome number for *Betula*, (x=14), whereas *B. medwediewii* is decaploid with ten times the base chromosome number. The differences in leaves, twigs and catkins (shown in Plate 12; Figs 123–126), together with the difference in ploidy level, seem to justify the maintenance of *B. megrelica* as a species distinct from *B. medwediewii*.

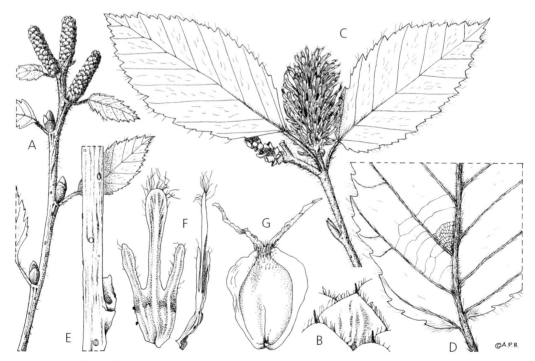

Fig. 123. *Betula megrelica*. **A** twig with buds and immature male catkins × 1; **B** male catkin scale × 10; **C** fruiting shoot with mature fruiting catkin (subterminal with remains of male catkin on left) × 1; **D** underside of leaf base × 2; **E** bark of twig × 2; **F** female catkin scale, back and side views (resin glands present) × 5; **G** seed × 10. All from *H. McAllister* (Ness BG, 30 Aug. 2010). DRAWN BY ANDREW BROWN.

PLATE 12. *Betula megrelica*
Painted by Josephine Hague

Fig. 124. A lone *Betula megrelica* shrub growing on a limestone outcrop on the north face of Mt Migaria (PB).

Paul Bartlett of Stone Lane Gardens has visited the Mt Migaria region in 2012, 2013 and 2015 in co-operation with Georgian botanists and conservation authorities and collected seed and twigs from Mt Migaria and the nearby Mt Jvari and Mt Askhi. These have all proved to be dodecaploid [2n=12x (c. 168)] as assessed by flow cytometry (Wang *et al.* in press 2016). This confirms that the shrub birches of these isolated limestone mountains are distinct cytologically and geographically from *Betula medwediewii*. Following the field studies of 2013, *B. megrelica* has been given 'endangered' status in the IUCN Red List.

CULTIVATION

As it is a smaller, denser, slower growing plant than *Betula medwediewii*, *B. megrelica* is suitable for smaller gardens, and would make an interesting addition to them. Both species are drought tolerant, deep-rooted shrubs which are very tolerant of exposure. The autumn leaf colour is a bright yellow similar to that of *B. medwediewii*.

Sizeable shrubs of *Betula megrelica*, which have reached two to three metres after about 30 years, can be seen at the University of Liverpool Botanic Gardens, at Ness in the Wirral, Cheshire, at Wakehurst Place in Sussex, and at Stone Lane Gardens near Chagford in Devon. Seed has been distributed through the botanic gardens seed exchange, but it is not known if plants are established in any other gardens.

Like *Betula medwediewii*, *B. megrelica* is self-compatible, and so even single, isolated bushes produce large quantities of viable seed every year. However, it is possible that *B. megrelica* might hybridise with *B. medwediewii*, should it be nearby. Seed germinates easily. The young seedlings are slow growing, but very stiff and sturdy and easy to raise as long as they are not allowed to be shaded by weeds. Young plants have a very extensive root system. *B. megrelica* is also easily propagated by cuttings taken as soon as the new shoots are long enough in June or early July and rooted in mist. The buds do not usually break in the year in which the cuttings are taken, so as soon as the leaves begin to colour in autumn, the rooted cuttings should be placed outside and kept cool to avoid depletion of their carbohydrate reserves during the winter.

Betula megrelica Sosn., *Trudy Tiflissk. Bot. Sada* 1: 42 (1934). Type: Hab. in Megrelia in m-tibus Migaria, Chokashi, Dzvari (2000 m), Chitazkali (1250–1400 m) in regione subalpina (holotype, ?TBI).

DESCRIPTION. *Shrub* to 4–8 m with several main stems radiating from just above soil level, in the wild often bent downwards by snow and up to 14 m long. The trunk and branches are a metallic silver to grey-brown, generally only peeling in small tatters at the base of older trunks. Raised grey-brown lenticels are sparsely scattered on the trunk. *Twig* matt grey to mid-brown, with white lenticels, stiff, 3–6 mm diam. *Bud* to 10 × 5 mm, ovoid, acute to blunt, green to greenish-brown, sometimes with a coating of white resin. *Young shoot* grey to pale brown usually with a few pale orange resin glands and scattered long silky hairs c. 2 mm long. Occasionally with dense puberulence of very short (c. 0.1 mm but variable) hairs about their own height apart, and few hairs intermediate in length between the silky and very short hairs. Shoots, twigs and stems smell faintly of wintergreen when scratched. *Petiole* c. 10–20 mm, densely covered with long (c. 2 mm) silky hairs, very short hairs absent. *Leaf* lamina 41–80 × 22–60 mm, broadly elliptical-ovate to almost as wide as long, often more or less sharply double toothed almost to base, glabrescent beneath; lateral veins 7–11 pairs, densely silky hairy on lower (abaxial) surface, on upper (adaxial) surface with scattered silky hairs which extend onto the lamina and are particularly noticeable at the margin; small reddish glands scattered over the lower surface of the lamina but confined to the main veins on the upper surface. Leaves in shade are generally

Fig. 125. *Betula megrelica* fruits on previous season's growth. High on north face of Mt Migaria (PB).

Fig. 126. *Betula megrelica* shoots with terminal withered male and current year's fruiting catkins. Smaller and narrower leaves and grey twig differentiate this species from *B. medwediewii* (PW).

thinner, rounder and with a less shiny adaxial surface. There is great variation in shape and size of leaves and fruiting catkins, amounts of pubescence and colour of shoots. *Male catkin* cylindrical, stout, appearing singly on the tip and the first (distal) 3 bud axils of current season's growth. When living c. 20–30 × 5.5 mm with green, brown tipped scales, when dry c. 20 × 3.5 mm with brown, clearly overlapping triangular scales. *Fruiting catkin* erect, with peduncle c. 5 mm, borne singly (occasionally in pairs) at lowest (proximal) leaf axils of previous season's growth, to 18–50 × 17 mm when dry; scales divergent, 6–8 × 3 mm with central lobe spatulate, 3–5 mm long, and laterals to 2.5 mm long. Tips of scales frequently bear fine hairs to 1.5 mm. Occasionally the central lobes curve out dramatically, producing a spikier fruiting catkin. *Seed* c. 2.5 × 1.5 mm with wings c. 0.25 mm broad. Seed shape can vary from almost round to being sharply pointed at one end.

DISTRIBUTION. Western Georgia, southern Caucasus: Mt Migaria, (42°62'N 42°34'E), Mt Jvari and Mt Askhi.

HABITAT. Mountain scrub.

CHROMOSOME NUMBER. 2n=168.

17. BETULA LENTA

A fine winter haze of thin twigs, flat many-veined leaves, a straight trunk with blackish bark and an elegant rounded crown mark the cherry birch or sweet birch, *Betula lenta* from eastern North America. It was named by Linnaeus in 1753.

In the wild, *Betula lenta* is one of the dominant species of the moister hardwood forests of the northern Appalachians (Furlow 1997). In the northern part of its range, in southern Maine, Vermont,

New Hampshire, New York, and just into Ontario, *B. lenta* occurs only at low levels scattered through the mixed secondary forest. In the south its range extends just into Alabama and Georgia. In North Carolina for example, it is abundant at about 300–600 m, frequently dominating open spaces and filling gaps in the forest much as the white birches do, but in this species its dark stems make it less obvious. It must surely be more shade-tolerant than the white birches, yet apparently not much more; many a lifeless pole of *B. lenta*, shorn of branches, may be seen shaded out in the mixed secondary forest.

Unlike *Betula alleghaniensis*, *B. lenta* does not vary greatly in appearance. It is easy to distinguish from the other wintergreen-bearing birches, particularly *B. alleghaniensis* and *B. grossa*, by the thin papery quality of its leaves and the upward sweep of the numerous and close veins at less than 40° to the midrib. The sharply acute conical buds, pointing outwards, are distinctive. In bark it could be confused with *B. grossa*, but certainly not with *B. alleghaniensis*, which has more metallic, yellowish bark. In *B. grossa* and *B. alleghaniensis* the shoots, twigs and leaves are coarser and generally hairier, and the lateral veins commonly more grooved or impressed in the leaf, especially on spur leaves. Its fruiting catkins are usually smaller than those of *B. alleghaniensis* and the scales are nearly always glabrous, whereas those of *B. alleghaniensis* are

always hairy. Other minor differences with *B. alleghaniensis* are that *B. lenta* is a smaller tree, reaching only 20 m, its shoots and twigs are thinner, its young-shoot leaves glossy above and lighter green than those of *B. alleghaniensis*, with the main and sharpest teeth 2–3 mm broad at base, small and fine while those of *B. alleghaniensis* are coarser, to 4 mm broad.

Betula lenta is still important as a timber tree; its wood is similar to that of *B. alleghaniensis* (Furlow 1990) and used for the same purposes. A former use for the wood, now superseded by synthesis, was the extraction of methyl salicylate by distillation. The sap is also sweet, hence the name, thus convenient for fermenting; as in all birches it pours forth freely when the trees are tapped in the spring.

CULTIVATION

It could be said that in the garden *Betula lenta* has no advantage over a *Carpinus*. Its aspect is much the same, except that generally it is not as vigorous as hornbeams like *C. betulus* and so less likely to become a problem. Its particular attributes are its delicacy, its finely veined leaves and their autumn colour — together with the always-intriguing presence of oil of wintergreen in the twigs and wood. Seedlings from the north of its range, in New England, are quicker

Fig. 127. Almost black trunks of *Betula lenta*, Vermont, USA (KBA).

growing in Britain. There appear to be no problems with dieback or early leafing with these, nor with those from the Appalachians further south, although those of southern provenance are definitely slower-growing here.

Betula lenta is not common in cultivation in Europe, but thrives in any soil not subject to drying out during the growing season. Long dry spells in summer may cause premature leaf fall and subsequent winter dieback, presumably due to desiccation of the twigs. As a species of bottomlands in the wild, it is likely to do well on poorly aerated, heavy wet clays and could be tried as a street tree on clay soils. However, in New England it is regarded as very drought tolerant and is a common coloniser of exposed dry sites as on Hemlock Hill in the Arnold Arboretum.

Betula lenta L., *Sp. Pl.* 2: 983 (1753). Type: Herb. Linn. No. 1109.8 (lectotype, LINN, designated by Reveal in Jarvis 2007: 349).

DESCRIPTION. *Tree* to 20 m; trunk straight, sometimes a little fluted, but scarcely buttressed by main roots, as in *Betula alleghaniensis*. *Bark* brown to black-brown; epidermis scarcely reflective, with little lustre or shine, tight and flaking a little, appearing dark grey or almost black when older. *Twigs* thin, not more than 3 mm diam., sometimes as little as 2 mm, glabrescent, yellow-brown to deep brown, smooth and shining, with few, obscure resinous glands. Lenticels very small, dot-like, protruding

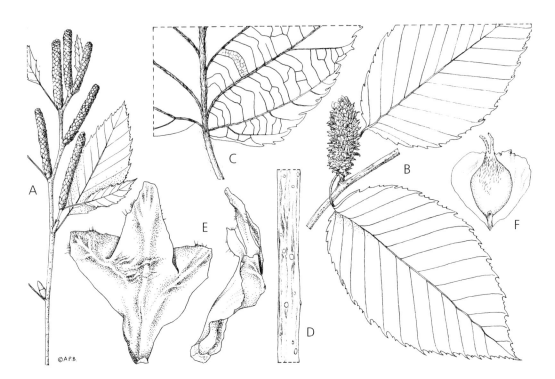

Fig. 128. *Betula lenta.* **A** twig with bud and immature male catkins × ¾; **B** fruiting shoot with mature fruiting catkin × ¾; **C** underside of leaf base × 2; **D** bark of twig × 2; **E** female catkin scale, back and side views × 7; **F** seed × 7. **A** from *Clausen* 5592 (collected as *B. alleghaniensis* but det. by K. Ashburner as *B. lenta*, Deep Gap, NC, USA, Sept. 1941); **B–F** from *Chambers et al.* NYCDPR-BBG-009 (Staten Island, NY, USA, 40°25'45.36"N 74°8'33.06"W, 11 Aug. 2006). DRAWN BY ANDREW BROWN.

Fig. 129. *Betula lenta*, male and female catkins and previous year's fruiting catkins (PW).

slightly only on two year old twig. *Buds* conical-ovoid, acute 7–11 × 2–6 mm, divergent from the twig. *Young shoots* initially slightly hairy, glabrescent, greenish-yellow and shining, with small clear resinous glands. *Petioles* to 15 mm, 1.5 mm wide or less at base. *Leaves* elliptic-ovate, 5.0–10 × 2.5–6.0 cm; basal and spur shoot leaves (preformed in bud) consistently cordate at base; lateral veins 12–18 pairs, not impressed, few silky hairs along midrib and lateral veins beneath; margins single-toothed, or more or less double-toothed on vigorous shoots, the teeth curved forwards, with main and sharpest teeth 2–3 mm broad at base; autumn colour yellow to yellow-brown. *Male catkins* in winter 11–16 × 3 mm smooth, scales green, expanding to 5–10 cm long at anthesis; catkins often 2 at tip of twig, with a 3rd axillary immediately below and frequently a 4th or even 5th below that. *Female catkins* 4–6 × 1–2 mm. *Fruiting catkins* erect, hard and persistent, ovoid, 1.5–3.5 × 1.0–2.5 cm usually solitary at the end of a short shoot; peduncle 2–6 mm. *Bracts* 5–7 × c. 4–5(–9) mm, usually glabrous but rarely ciliate on margins, lateral lobes longer and often broader than central lobe, about 2.5 mm long. Seeds to c. 3.0 × 4 mm with nutlet broadly elliptical to 3 × 2 mm and wings to 1.25 mm wide.

DISTRIBUTION. Ontario to Maine, south to North Carolina, Tennessee and N Georgia in the mountains, with outliers in Mississippi.

HABITAT. Moist, cool forests and rocky slopes, 0–1500 m.

CHROMOSOME NUMBER. Diploid: 2n=28.

BETULA LENTA forma **UBER**

When first discovered, *Betula lenta* f. *uber*, the Virginia round-leaf birch, was described as a variety of *B. lenta* (Ashe 1918). It was also considered to be a distinct species, closely related to the dwarf birches of section *Apterocaryon* (Fernald 1950; Lucas & Synge 1978), on account of the rounded shape of the leaves, even though its supposed relatives in section *Apterocaryon* grow at least 800 km further north. Johnson (1954), however, described it as a variety of *B. lenta* and Lucas & Synge (1978) stated that 'less than 1% of the seed collected from wild trees developed into *Betula uber*, the rest being F[1] hybrids with *B. lenta*'. This behaviour for a species seems very unusual. It could only happen if the seed came from an isolated tree of an almost self-incompatible species, and yet field reports state that *B. uber* trees do not grow in total isolation (Kinkhead 1976; Ogle & Mazzeo 1976). Sharik & Ford (1984) while suggesting a relationship with *B. lenta* based on morphology and the presence of oil of wintergreen, still regarded *B. uber* as a distinct species, although they admitted that it differs only in leaf shape. They deferred a decision, but only because the leaf shape variation is discontinuous; no trees with leaves intermediate in shape between those of *B. lenta* and *B. uber* exist.

After its initial discovery, it appears to have been lost until seventeen trees were rediscovered in 1974. In spite of the excitement at its rediscovery in Virginia, all arguments point to it being no more than a form of *Betula lenta*. Sharik (1984) states that 'in one trial, only 1.2% of the progeny of six *B. uber* mother trees possessed round leaves and the remainder were indistinguishable from those of *B.*

Fig. 130. *Betula lenta* (seedling from *B. lenta* f. *uber*). Twigs with spur shoots bearing two leaves (which were preformed in bud), fruiting catkins, and remains of male catkins are ends of twigs (PW).

lenta.' He goes on to conclude that biologists have postulated that the round leaf form is probably due to a single recessive gene.

We carried out our own test with seed from an isolated tree of *Betula uber* growing in Ness Gardens; although of low viability, six seedlings were raised in 1997. They were indistinguishable from *B. lenta* and their elongated leaves quite distinct from the small, round leaves of *B. uber*. There were absolutely no *B. lenta* individuals in the vicinity. It therefore seems that it is not possible to distinguish most seedlings resulting from the selfing of *B. uber* from those of *B. lenta* itself and that the total lack of any of the characteristics of *B. uber* in them must rule out the possibility of a hybrid.

Betula uber, should therefore be considered merely a form of *B. lenta* because leaf shape is the only consistent differential character (Ashburner & McAllister 2005). With only around 1% of seed coming true to f. *uber*, it appears that several genes are involved in this variety. This mutation appears to have occurred only in Virginia, and it is surprising that there is no known counterpart to the *B. uber* leaf form in the other species of section *Lentae*.

Last of all, this is no denigration of its interesting qualities, which are well worthy of conservation. But *Betula lenta* forma *uber* is of much less significance than taxa which can be unequivocally regarded as species.

Betula lenta f. uber (Ashe) McAll. & Ashburner, *Curtis's Bot. Mag.* 21(1): 58 (2004).

Betula lenta var. *uber* Ashe, *Rhodora* 20: 64 (1918). Type: Virginia, Smythe Co., Banks of Dicky creek,
    south of Rye valley station, 14 Jan. 1914, *W. W. Ashe* (isotypes GH, MICH, NY, NA,).
Betula uber (Ashe) Fernald, *Rhodora* 47: 325 (1945).

DESCRIPTION. Like f. *lenta*, but a slender *tree* to 10 m. *Petiole* 10–15 mm. *Leaves* on spur shoots and early leaves with blade suborbicular with a cordate base, 2.1–4.0 × 2.4 × 3.5 cm; later formed leaves on long shoots more elongated, 2.7 × 2.0 to 5.4 × 3.9 cm. *Fruiting catkins* erect, hard and persistent, ovoid, to 18 × 9 mm. *Bracts* c. 4.5 × 4.0 mm, more or less ciliate on margins, lateral lobes about equal to central lobe, about 2.5 mm long.
DISTRIBUTION. Virginia, in Smythe Co.
HABITAT. Stream banks and flood plain in forest, 500 m, with *Betula lenta*.
CHROMOSOME NUMBER. Diploid: 2n=28.

18. BETULA ALLEGHANIENSIS

HISTORY

The yellow birch was first named by François André Michaux (1770–1825) in his *Histoire des Arbres Forestière* published in 1811. He called it *Betula lutea*, but unfortunately this name proved to be illegitimate, and the name *alleghaniensis* published by Nathaniel Britton in 1904 for a species he considered distinct from both *B. lutea* and *B. lenta*, is now used instead. In Quebec *B. alleghaniensis* is called merisier or bouleau jaune.

ECOLOGY AND STATUS IN THE WILD

Yellow birch occurs throughout eastern North America from the southern parts of Quebec and Labrador in the north, reaching to 49°N in Quebec and as far west as Minnesota and western Iowa, south in the Appalachians to northern Georgia and Alabama. It is very common in New England and throughout its range occurs in mixed forest, in association with *Acer saccharum*, sugar maple,

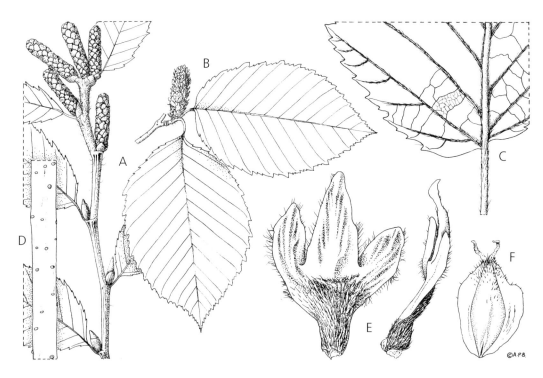

Fig. 131. *Betula alleghaniensis*. **A** twig with buds and immature male catkins × 1; **B** fruiting shoot with mature fruiting catkin × ½; **C** underside of leaf base × 2; **D** bark of twig × 2; **E** female catkin scale, back and side views × 7; **F** seed × 7. **A** from Herb. Oakes (Ipswich, Mass., USA); **B** from *Cahilly* AWC-12-92 (Millbrook, NY, USA, 19 Sept. 1992); **C** from *Rousseau* 35382 (Guysborough Lake, Nova Scotia, 8/9 Aug. 1930); **D** from *Joliceur* 3856 (St. Calixte, Québec, 20 July 1953); **E, F** from *Howick & Warner* WH55 (Mt Toby, Mass., USA, 9 Oct. 1985). DRAWN BY ANDREW BROWN.

and *Fagus grandifolia*, both in lowland high forest and, especially with the beech, at high altitudes in the southern part of its range. Its wood is hard, heavy and moderately strong and can be of large dimensions, as trees can reach 30 m. It is the most important of all *Betula* species for timber, although as a timber, it is usually not distinguished from that of *B. lenta* or the Japanese *B. maximowicziana*.

The common name yellow birch is, in a sense, a wishful euphemism, in that very often stems are brownish, but glint vividly in the sun after the canopy has been thinned by sharp frosts. Throughout the natural populations, and in garden or arboretum, it is the yellower bark of younger trees which is most attractive. Hybridisation with *Betula lenta* has been proposed to explain the darker barked forms, but there is little evidence of introgression in other characters. It seems most likely that there is more variation within *B. alleghaniensis* itself than has been generally accepted (Dancik & Barnes 1969, 1971, 1974, 1975). Hybrids between the two species do occur (Sharik & Barnes 1971) and introgression is theoretically possible, but there are few reports of tetraploids and pentaploids of intermediate appearance as are often found between white barked birches.

Introgression is much less likely from a diploid into a hexaploid than into a tetraploid and it is interesting how many cases there are in both *Betula* and *Alnus* of very similar diploid/hexaploid species pairs whereas there are no very close diploid/tetraploid species pairs.

At higher altitudes, the trees of *Betula alleghaniensis* can look very different, as on Bromley Mountain in Vermont where they are scrubby and dwarfed, and in the Appalachians further south, where for example on Mt Mitchell in North Carolina they are taller but gnarled, twisted and windswept, mostly showing the flaky and plated bark of older trees (Fig. 48). They share dominance to the tree-line (the 'balds') with beech. It is almost as if *B. alleghaniensis* has here taken the ecological rôle of the mountain birch *B. cordifolia*. In this southerly, and possibly drier environment, the more deeply rooting *B. alleghaniensis* occupies the ecological position filled further north by white barked birches.

In Newfoundland, and no doubt elsewhere in the northern part of its range, there are hybrids with *Betula papyrifera*, for example in Michigan, (Barnes *et al.* 1974). In the fens and bogs of SE Michigan *B. alleghaniensis* also hybridises with the multi-stemmed, densely downy bog birch *B. pumila* of eastern North America (Dancik & Barnes 1972; Clausen 1963). These hybrids, *B. ×purpusii*, are intermediate between the two species, *i.e.* downy shrubs with more or less ovate leaves. A backcross between this hybrid and *B. alleghaniensis* has resulted in the creation of a new species, *B. murrayana* (see below).

TAXONOMY, RECOGNITION AND RELATIONSHIPS

Betula alleghaniensis can often be recognised solely by its trunk, which has a metallic glinting bark with thick persistent flakes that curl back on stouter stems and branches (Fig. 133), and develop thick plates on older trunks, which can become almost black. These bark characteristics are unique in the genus, no other birch having similar 'metallic' flaking bark. The trunks are buttressed and have the silvery 'toes' streaked with lenticels at the base of the trunk, formed as the roots spread out near the soil surface (Fig. 134).

Fig. 132. *Betula alleghaniensis* and *B. papyrifera* (the two trees behind the figure on left) with *Fagus grandifolia* and young *Abies balsamifera* in Franconia Notch, New Hampshire, USA (HMcA).

Fig. 133. Trunk of *Betula alleghaniensis*, Vermont, USA (KBA).

The closely-related *Betula lenta* differs in its thinner, smooth, polished and almost hairless twigs, with smaller and more sharply pointed buds. Typical *B. alleghaniensis* has thicker matt hairy leaves with fewer, more impressed veins, double-toothed in characteristic 'stepped' fashion with a deep notch above each major tooth interrupting the run of secondary teeth between. The major teeth are 4–5 mm wide at their base, while *B. lenta* has teeth to 2 mm wide at the base, but this may not always be evident, especially in shade leaves; *B. alleghaniensis* also has more downy shoots and larger fruiting catkins with larger and hairy scales.

Betula grossa is not nearly so downy on twig and leaf, has bigger outward-pointing buds, a much narrower vein-gap with more impressed veins, and dark, scarcely glinting bark like that of *B. lenta*. The most useful qualitative character is that in *B. alleghaniensis*, the leaf underside and petiole have both silky and short puberulent hairs, while on the leaves of *B. grossa* puberulent hairs are totally absent.

CULTIVATION

Betula alleghaniensis is very handsome in the garden, for its bark and for the brilliant translucency of its young leaves, as well as their rich yellow autumn colour. Trees thrive where there is plenty of moisture, but appear to grow perfectly well in drier conditions, although in the British Isles they never seem to be fast-growing.

The strongest-growing trees may be the most desirable for the arboretum, but not necessarily for

Fig. 134. Trees of *Betula alleghaniensis* on rocky ground with surface roots showing similar bark to trunks. Note tree in background growing from top of rock. Franconia Notch, New Hampshire, USA. October 8th (HMcA).

Fig. 135 (left).   Rotten tree stump with seedlings of *Betula alleghaniensis* on moss to left of stump. Seedlings can only establish in such situations where not smothered by fallen leaves. Franconia Notch, New Hampshire, USA (HMcA).

Fig. 136 (right).   Close-up of seedlings of *Betula alleghaniensis* in previous photograph with fallen leaves of *Acer rubrum*, seedlings having have produced only one or two post-cotyledonary leaves. October 8th (HMcA).

the garden. Collections from the Maritime Provinces of Canada or New England seem to grow best and most rapidly in the British Isles; those from inland, for example from the north shore of Lake Huron, also grow well. Less suitable, and perhaps too slow-growing, are the southerly montane ones from North Carolina and Virginia. It would be interesting to compare the performance of trees of these provenances in central Europe.

Seed has a high viability and that harvested in garden or arboretum seems to come entirely true. Hybridisation would be most likely with *Betula grossa* as this is also hexaploid, but this species is rare in cultivation and it is not known whether seed has been raised from gardens in which these species grow near to one another.

Clones with the best yellow bark could be grafted onto seedling *Betula alleghaniensis* itself, or even onto *B. lenta*, in preference to *B. pendula* or *B. papyrifera* stocks. Cuttings of young shoots would be expected to root easily under mist.

Betula alleghaniensis Britton, *Bull. Torrey Bot. Club* 31: 116 (1904). Type: collected on the upper slopes of Mt Pisgah, western North Carolina, Sept. 21, 1897, distributed by the Biltmore herbarium, no. 1619 (holotype: ?).

B. lutea F. Michx., *Hist. Arb. Forest.* 2: 152 (1811) *nom. illeg.*

B. lenta var. *lutea* (F. Michx.) Regel in A. P. de Candolle, *Prodr.* 16 (2): 179 (1868).

B. lutea var. *alleghaniensis* (Britton) Ashe.

B. lutea f. *macrolepis* Fernald, *Rhodora* 34: 95 (1932).

DESCRIPTION. *Trees* to 30 m, though reduced to spreading stunted shrubs in exposed montane habitats; trunk cylindrical, but sometimes fluted; main roots forming buttresses. *Bark* brown or yellowish, strikingly metallic with translucent and highly reflective surface, black and peeling on old trees. *Shoots* and *twigs* very hairy when young, becoming less so, but shoots particularly hairy towards base, feeling more or less smooth and velvety, with many small dot-like brown glands; mature twigs brown above, silvery below. *Buds* ovate-conical, 6–11 × 2–6 mm, diverging from twig. *Petiole* 12–18 mm, to 2 mm at base, grooved above. *Leaves* oblong-ovate, acute with rounded or cordate base and acuminate apex, mostly (6–)9–10 × (3–)5–7 cm. Lateral veins 7–10, mostly spaced about 10 mm or just less, but sometimes to 12 mm apart, always compressed or grooved into the leaf and not protruding above; tertiary veins conspicuously proud below; leaf margins mostly coarsely double-toothed with major teeth 4–5 mm broad at base, commonly 3 minor teeth between the major. Leaves on vegetative spur shoots large, but reduced to as little as 45 × 20 mm on some fruiting spurs. *Male catkins* smooth, 3–4(–5) at twig ends, 50–100 × 5–8 mm at anthesis. *Female catkins* 6–8 × 1.5–2.5 mm. *Fruiting catkins* cylindrical-ovoid, hard, woody and persistent, 1.5–3.5(–6.5) × 1.0–2.5 cm; peduncle 2–8 mm. *Scales* hairy, very variable in size, from 5–9(–15) × 4–8 mm (Furlow & Mitchell 1990), woody, concave, hairy, the 3 lobes ± equal. *Seeds* ovate, 3–4.5 × 4–5.8 mm, with nutlet 2.3–3.4 mm broad; wing narrower than nutlet 0.8–1.3 mm, tapering from apex to base.

DISTRIBUTION. Eastern North America from Newfoundland south to Tennessee and west to Minnesota.

HABITAT. On moist soils in lowlands and mountains.

CHROMOSOME NUMBER. Hexaploid: 2n=84.

19. BETULA MURRAYANA

Betula murrayana was described as a new species in 1985 by B. V. Barnes and B. P. Dancik. It is intermediate between *B. alleghaniensis* and *B. pumila*, the two species from which it is derived, but closer to *B. alleghaniensis* (see Chapter 1). This is not surprising as it has inherited 84 chromosomes from that species and only 28 from *B. pumila*. Its current status is perhaps best given by quoting from the internet site of the Center for Plant Conservation. http://www.centerforplantconservation.org/ASP/CPC_ViewProfile.asp?CPCNum=8247 (accessed December 2010):

> Originally two trees were found, but in the early 1980s one died unexpectedly. Only one known original individual remains in the University of Michigan's demonstration forest. Several cuttings from this original have produced plants at the University of Michigan's Matthaei Botanical Gardens and at The Holden Arboretum. It would be expected to be easy to propagate by seed or cuttings.

Despite the distance in relationship and dissimilarity in appearance between its parents, the tall tree yellow birch *Betula alleghaniensis* (2n=6x=84) and the dwarf bog birch *B. pumila* (2n=4x=56), hybrids between them are well known and have the expected pentaploid chromosome number of 2n=5x=70. The primary hybrid is so frequent that it was named *B. purpusii* before its hybrid nature was realised, so it is correctly named *B. ×purpusii* (Dugle 1966). The two unusual trees more similar to *B. alleghaniensis* than *B. ×purpusii*, but also intermediate in appearance between the yellow birch and the bog birch, were noticed in southeastern Michigan and found to have a chromosome number of 2n=112. While *B. ×purpusii* is relatively infertile, the strange new trees were fertile and, having a different chromosome number, clearly had a different origin. From their appearance and chromosome number it appears that they contain a full set of chromosomes from *B. alleghaniensis* and a half set from *B. pumila*. This gives them an even number of chromosome sets (8 × 14). It was

Fig. 137. Branch of *Betula murrayana* cultivated at Ness from seed collected at St Williams, Norfolk Co., Ontario in 1987. Note smaller leaves on shoots bearing fruiting catkins (HMcA).

deduced that *B. murrayana* had arisen from the union of an unreduced gamete from the hybrid *B.* ×*purpusii* (n=70) and a normal reduced gamete from *B. alleghaniensis* (n=42). This is shown in Table 7 in which the six chromosome sets in *B. alleghaniensis* are represented by (AA)(A^{1}A^1)(A^{2}A^2) and the four in *B. pumila* by (BB)(B^{1}B^1), the superscripts indicating that there may be differences between the genomes within each species, which may themselves be of allopolyploid origin. It would appear that there is sufficient similarity between the two different 'B' genomes in *B. pumila* to allow regular pairing between them in the new species.

In 1987 Pru Barnes (KBA's niece) collected seed of *Betula pumila* and *B. alleghaniensis* near St Williams in Norfolk county in southern Ontario, adding the note that the *B. alleghaniensis* might be hybrid. Although the collected twigs of *B. pumila* are certainly of that species, the two seedlings raised are *B.* ×*purpusii*, suggesting a single or few *B. pumila* bushes surrounded by *B. alleghaniensis,* a situation conducive to the formation of *B.* ×*purpusii* and *B. murrayana*. A single seedling was raised from the possibly hybrid *B. alleghaniensis*. This was regarded as a small-leaved form of that species until 2011 when chromosome counts of its seedlings gave 2n=112. The 'possibly hybrid' *B. alleghaniensis* is clearly *B. murrayana,* the original tree (or population) at St Williams presumably representing a second, independent, origin of the species about 320 km (200 miles) east of the type locality. Such an occurrence was predicted by Barnes & Dancik (1985) when they stated 'it is highly likely that it will be found elsewhere in the Great Lakes — St Lawrence Valley area.'

As the single tree in cultivation at Ness produces viable seed which comes true, it is self-compatible. These seedlings are now being distributed so that this potentially attractive small tree can be made more widely available.

Table 7

The probable origin of *Betula murrayana*.

	B. pumila		*B. alleghaniensis*	
Tree 2n	BBB^{1}B^1	×	AAA^{1}A^{1}A^{2}A^2	
Gametes (n)	BB^1		AA^{1}A^2	
		B. × purpusii	×	*B. alleghaniensis*
Tree (2n)		AA^{1}A^{2}BB^1		AAA^{1}A^{1}A^{2}A^2
Gametes (n)		AA^{1}A^{2}BB^1 (unreduced)		AA^{1}A^2
Tree (2n)			*B. murrayana* AAA^{1}A^{1}A^{2}A^{2}BB^1	

Betula murrayana forms a tree smaller than *B. alleghaniensis*, and differs from it mainly in its reddish-brown bark and in its leaves, which are 5–11 cm long, with only 7–10 pairs of lateral veins, and ovate with an acute or slightly acuminate apex and cuneate base. The fruiting catkins are erect, and smaller than those of *B. alleghaniensis*, 2–4 cm long.

Fig. 138. Twig with fruiting catkins of *Betula murrayana* cultivated at Ness from seed collected at St. Williams, Norfolk Co., Ontario in 1987 (HMcA).

Betula murrayana B. V. Barnes & Dancik, *Canad. J. Bot.* 63: 226, figs 1D, 2, 3 (1985). Type: Michigan, Washtenau Co., south shore, Third Sister Lake; latitude 42°16'N; longitude 83°48'W; NW 1/2, SE 1/2, section 26, T 2 S R 5 E; collection no. 660004 (holotype MICH).

DESCRIPTION. *Tree* to 15 m, usually with several trunks; *bark* smooth, dark red to reddish-brown with prominent lenticels. *Twigs* covered with small resinous glands, glabrous to sparsely pubescent, smelling of oil of wintergreen. *Leaves* ovate, 5–11 × 3–6 cm; with 7–9 pairs of veins. *Fruiting catkins* ovoid, 2–4 × 1.5–3 cm. *Seed* c. 2.5–3 × 2.5–3 mm, with nutlet 1.75–2 mm broad and wing about ¼ nutlet width; scale c. 5 × 5.5 mm with lateral lobes patent or erect, often more or less square.

DISTRIBUTION. Found in only one site in Michigan, by Third Sister Lake in Saginaw Forest — the demonstration forest for the University of Michigan, USA; and a second site near St Williams, Norfolk Co., Ontario, Canada.

HABITAT. Wet swampy forest with *Betula pumila*.

CHROMOSOME NUMBER. 2n=112.

Subgenus ACUMINATAE section ACUMINATAE

Genus *Betulaster* Spach, *Ann. Sci. Nat., Bot.* sér. 2, 14: 182, 198 (1841).
"Abteiltung" *Acuminatae* Regel, *Bull. Soc. Imp. Naturalistes Moscou* 38, N4: 418 (1865).
Sect. *Betulaster* (Spach) Regel in DC., *Prodr.* 16, 2: 179 (1869).
Sect. *Acuminatae* (Regel) Schneid. in Sarg., *Pl. Wilson.* 2: 465 (1916).

Leaf teeth with elongated nipple-like tips; *male catkins* lacking secondary bracts (Furlow 1990); *female catkins* pendent, usually borne in clusters of 2–6 (except *Betula luminifera*); 4 (fewer than 4 and often only 1 in other sections) tepals in the flowers (Furlow 1990); pendent *fruiting catkins* cylindrical, ripening in summer (except *B. maximowicziana*), *fruiting catkins* having somewhat persistent bracts which have rudimentary lateral lobes (except *B. maximowicziana*) and do not enclose the wide thin wings of the seeds which are usually considerably wider than the nutlets, and are therefore very obviously visible in the fruiting catkins. The leaves are large and lanceolate to ovate to elliptical in the three Sino-Himalayan species, but broader and markedly cordate in the Japanese *B. maximowicziana*. Most of our knowledge of this subgenus, and therefore this account, is based on Skvortsov's excellent taxonomic revision (Skvortsov 1997).

20. BETULA ALNOIDES

HISTORY

This widespread Himalayan birch was described by Francis Buchanan-Hamilton (1763–1829) in David Don's introduction to the flora of Nepal (*Prodromus florae Nepalensis* 1825), and the name *alnoides* acknowledges its general similarity to an alder. David Don was librarian to the Linnean Society, and later, Professor of Botany at King's College, London. His flora was mostly based on collections made by Buchanan-Hamilton and Nathaniel Wallich, who were both in the Bengal Medical Service and, in turn, superintendents of Calcutta Botanic Garden in the early 19th century.

ECOLOGY AND STATUS IN THE WILD

Betula alnoides is a component of the tropical to subtropical forests from Nepal eastwards to Vietnam, mostly below 2000 m, although, if records are correct, perhaps reaching 2600 m in Yunnan. It is unlikely to be hardy over most of the British Isles and has probably only been introduced to cultivation very recently from Vietnam (T. Hudson, pers. comm.). Skvortsov reports trees in the wild near Kalimpong, West Bengal, developing new shoots in mid-October while still retaining their old leaves. This species is therefore probably evergreen.

Although *Betula alnoides* and the related *B. cylindrostachya* and *B. luminifera* grow in tropical to warm-temperate forest, they are also reported to behave as primary colonisers following disturbance such as land slips or road building in the same way as the white barked birches of section *Betula*, and they are commonly found along rivers, where gravels have been exposed by floods. The production of vast numbers of very small, very broadly winged seeds ensure that they are widely dispersed.

TAXONOMY, RECOGNITION AND RELATIONSHIPS

Betula alnoides has been much confused with *B. cylindrostachya*, a problem sorted out by Skvortsov (1997). In the field the most useful distinguishing characters are probably the autumn and winter (October to January) flowering time, spring fruiting (February to April), very glandular shoots, and sparse marginal toothing of *B. alnoides*; and the spring (March to May) flowering and much hairier shoots with few if any glands and closer toothing of *B. cylindrostachya*. Skvortsov (1997) also distinguished these two species by leaf shape, fruiting catkin width, and size and shape of nutlet, nutlet-wing, and bract. Our study of numerous herbarium specimens confirms that hair density and presence and density of glands on young shoots are correlated with the fruiting catkin characters used by Skvortsov. The differences may be summarised as follows (modified from Skvortsov 1997).

Young shoots conspicuously glandular and often only thinly hairy; leaves lanceolate, base usually cuneate, toothing sparse with (0)1–2(3) teeth between major teeth at 2er vein ends; flowering (September to) October to January; fruiting catkin up to 6 mm broad, usually 4 catkins per cluster; nutlet c. 1.0 × 1.0 mm: 2n=28. . . . . *Betula alnoides*

Young shoots eglandular or with a few to many glands, often densely hairy; leaves ovate to oblong-ovate, base rounded or slightly cordate, toothing serrate with up to 7 teeth between major teeth at 2er vein ends; flowering March to May; fruiting catkin usually more than 6 mm broad, usually 2 catkins per cluster; nutlet longer than broad, c. 1.5–3.2 × 1.0–1.5 mm: 2n=56 . . . . *Betula cylindrostachya*

The differences in fruit wing shape mentioned by Skvortsov, more or less symmetrical in *Betula alnoides* and forward pointing and extending well beyond nutlet in *B. cylindrostachya*, seem to be very variable even within a catkin, but observations on a number of seeds can usually be used to distinguish the two species (Fig. 33). Many fruiting catkins, assigned to either species on other characters, seemed to be about 6 mm wide, so the width difference was not as obvious as implied in Skvortsov's key (3.5–6 mm wide in *B. alnoides* and 6–8 mm wide in *B. cylindrostachya*).

Published chromosome counts for *Betula alnoides* are diploid, 2n=28 (Mehra & Sareen 1973), so it was a surprise when two collections from Bhutan by Keith Rushforth cultivated under this name, and one from the Garwhal area of NW India by KBA, were found to be tetraploid (2n=56). This chromosome number, combined with Skvortsov's key characters, e.g. their spring flowering, broader fruiting catkins and larger nutlets, showed that these collections were in fact *B. cylindrostachya*. More cytological studies on Himalayan material need to be made to confirm these findings and check that these cytological differences are consistent and correlate with the morphological distinctions used to distinguish the species. The difference in chromosome number explains why these closely related species remain distinct and why hybrids have not been reported; in addition many collectors and observers have not distinguished between the two species. Skvortsov described *B. alnoides* and *B.*

cylindrostachya as overlapping in altitudinal range and sometimes growing together in the Himalaya, although with *B. alnoides* being tropical to subtropical and autumn flowering while *B. cylindrostachya* is warm temperate and spring flowering. Their different flowering times and difference in ploidy levels greatly reduce the opportunities for interbreeding.

CULTIVATION

We have not seen living material of *Betula alnoides* in Britain, as it is of tropical or subtropical origin it may not be hardy in areas subject to significant frost.

All four species of this subgenus seem to be relatively tolerant of a range of soil texture and soil moisture. In subtropical or warm temperate climates *Betula alnoides*, *B. cylindrostachya* and *B. luminifera* are rapid-growing, attractive trees likely to be suitable for areas with adequate summer moisture, while *B. maximowicziana* is an attractive large tree for temperate climates.

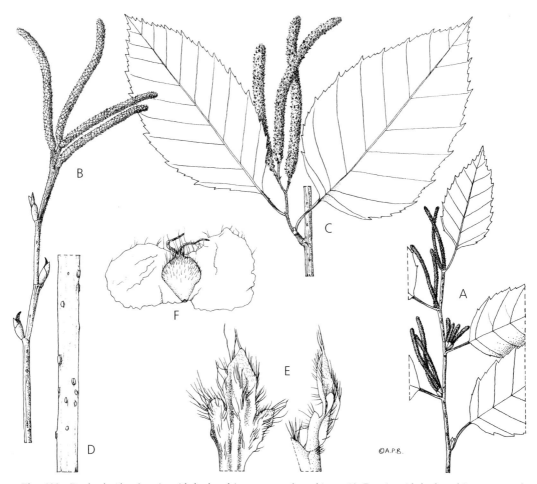

Fig. 139. *Betula alnoides*. **A** twig with bud and immature male catkins × ½; **B** twig with buds and immature male catkins × ¾; **C** fruiting shoot with mature fruiting catkins × ¾; **D** bark of twig × 2; **E** female catkin scale, back and side views × 15; **F** seed × 10. **A** from *Sørensen et al.* 5655 (Danish Thai Expedition, Doi Sutep, Thailand, 14 Oct. 1958); **B** from *Strachey & Winterbottom* 4 (Naini Tal, Kumaon); **C** from *Forrest* 26167 (Mid-West Yunnan, 25°20'N 98°30'E, Dec. 1924); D from Type 2794 (?collector, Kumaon); **E, F** from *Henry* 10437 (Yunnan, Mengei). DRAWN BY ANDREW BROWN.

Betula alnoides Buch.- Ham. ex D. Don, *Prodr. Fl. Nepal.*: 58 (1825). Type: Nepal: in sylvis ad Narainhetty, *F. Hamilton* (holotype BM).

Betula acuminata Wall., *Pl. Asiat. Rar.* 2: 7 (1830). Type: Nepal: without locality, *N. Wallich* 2793 (holotype BM).

Betula alnoides var. *acuminata* (Wall.) H. J. P. Winkl. in Engler, *Pflanzenr.* IV, 61: 89 (1904).

DESCRIPTION. *Trees* to 30 m. *Bark* dark brown. *Shoot* variably white silky-hairy, glandular over whole of current year's twig. *Petiole* 1.2–3 cm, hairy and glandular as twig. *Leaves* lanceolate to ovate-elliptic, (4–)12 × 2.5–5.5 cm, base cuneate, apex acuminate to caudate-acuminate, margin serrate, with 10–13 pairs of secondary veins ending in prominent teeth with (0)1–2(3) teeth between, hairy along veins beneath (abaxially) and in secondary vein axils, glandular beneath, glabrous and eglandular to thinly glandular on veins above (adaxially). *Fruiting catkins* with peduncle 2–3 mm, usually 4 in a raceme, hairy, pendulous, cylindrical, 50–80 × 4–6 mm. *Scales* 1.5–3 mm, hairy, lateral lobes much reduced. *Nutlet* c. 1 × 1 mm, wings about twice as wide as nutlet, often more or less orbicular and symmetrical on either side of nutlet; style c. 1.5 mm.

DISTRIBUTION. NW India, Himachal Pradesh (Beas river valley) eastwards through Nepal, NE India (Sikkim, W Bengal, Assam, Nagaland), Bhutan, Assam (Khasia Hills), N Burma (Myanmar), S China (Fujian, Guangxi, Hainan, Hubei, SW Sichuan, NW Yunnan), Thailand, Laos, Vietnam.

HABITAT. Tropical to subtropical forests, 1,000–2,000(–2,600?) m.

FLOWERING TIME. (September) October to January.

FRUITING. February to April.

CHROMOSOME NUMBER. Diploid: 2n=28 (Mehra & Sareen 1973).

21. BETULA CYLINDROSTACHYA

Betula cylindrostachya was described by John Lindley in Wallich's *Plantae Asiaticae Rariores*, from specimens collected by Robert Blinkworth, a correspondent of Wallich's. *B. nitida* had been described earlier by David Don, from a specimen collected in Srinagar, but was considered by Lindley to be intermediate between *B. cylindrostachya* and *B. alnoides*. These early collections are discussed in detail by C. Schneider in *Plantae Wilsonianae* (Sargent 1916) and Skvortsov (1997).

As mentioned above, *Betula cylindrostachya* has been much confused with *B. alnoides*, but can be most easily distinguished from that species by its spring, rather than autumn to winter, flowering time, larger (c. 2 × 1 mm) nutlets, hairy usually almost eglandular young shoots, and much more closely toothed leaf margin. The glands may be difficult to detect under the dense hair and glandular specimens might be referable to *B. alnoides* or possibly hybrids. Most recent introductions to cultivation in Britain named *B. alnoides* appear to be *B. cylindrostachya*, and have been found to be tetraploid whereas *B. alnoides* is reported to be diploid (Mehra & Sareen 1973).

Fig. 140. Trunk of *Betula cylindrostachya*, Sotul, Garwhal, Himachal Pradesh, N India (KBA).

Fig. 141 (left). Fruiting catkins and foliage of *Betula cylindrostachya*. Very strong secondary veins running straight into teeth (craspedodromous). Bhutan. (*KR* 976) (PW).

Fig. 142 (right). Trunk of *Betula cylindrostachya* in the wild in Arunachal Pradesh in north-east India showing large lenticels (PB).

Skvortsov (1997) reports that 'the two species may grow in close proximity', and their altitudinal distributions overlap; in addition, they grow in the much same areas.

The recently described *Betula fujianensis* (Zeng *et al.* 2008) would seem from its description to be synonymous with *B. cylindrostachya*, the only significant difference being the glandular shoots and the absence of hair tufts in the secondary vein axils on the abaxial leaf surface (both variable in *B. cylindrostachya*). Previously these Fujian populations had been regarded as *B. alnoides* (Li & Skvortsov 1999; Zeng *et al.* 2008) but the deciduous, spring flowering, densely hairy young shoots, often subcordate leaf base, suborbicular to ovate nutlet, and greater hardiness reported by Zeng *et al.* led these authors to deduce that they did not belong to that species and they described these populations as a new species, *B. fujianensis*. Zeng *et al.* compare their new species with *B. alnoides* and *B. luminifera* but make no comparison with *B. cylindrostachya*.

CULTIVATION

Betula cylindrostachya is probably rare in cultivation outside its native area but promises to be an attractive species for warm temperate climates, although it might naturalise itself and become invasive. Young trees grow rapidly to form large spreading trees.

A tree labelled *Betula cylindrostachya* has been in cultivation at Kew since 1971 and another since 1996. The collections of *B. cylindrostachya* from Bhutan (*Keith Rushforth* 876, 976), from Garhwal collected by Kenneth Ashburner, and Tibet (*Keith Rushforth* 6000) have survived without significant damage for more than ten years at Ness and have grown vigorously to form spreading trees which look as if they may eventually grow large and attractive; they are now bearing catkins and fruiting freely. These collections have been found to be tetraploid (2n=56).

The main difficulty in growing *Betula cylindrostachya* and *B. luminifera* in the British Isles appears to be the risk of damage from late spring frosts. All the buds on young trees break at the same time, and if these are killed by frost the trees appear to be incapable of generating new buds from the base of their trunks. In this they differ from other subgenera of *Betula* in which at least young trees regenerate easily from basal buds on the trunk. Due to early bud break, the delicate young leaves are often damaged by desiccating winds. In any batch of seedlings there is variation in time of bud break, and the plants with later breaking buds may be the only ones which survive. As Hacke & Sauter (1996) mentioned in a discussion on time of bud break, epigenetic (inherited characteristics affecting gene expession but not involving changes in DNA) factors may result in trees grown from seed from cultivated trees being less susceptible to frost damage than trees grown in the same area from seed of wild origin; this is in addition to the effects of genetic selection by which the only seed parents are those trees which survive to seed-bearing age.

Thus, now that naturally selected trees of Garhwal origin (many having died at the seedling stage from late frosts) are fruiting at Ness, it may be possible to produce seedlings adapted to higher latitude temperate climates.

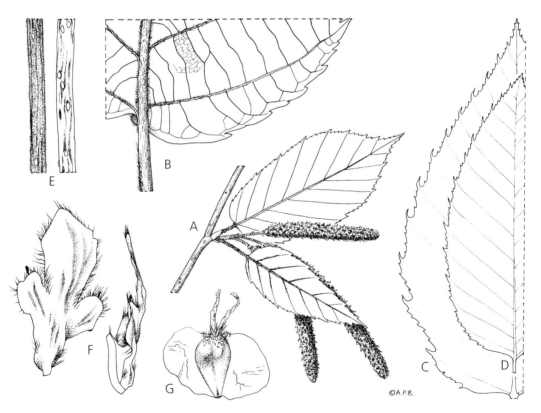

Fig. 143. *Betula cylindrostachya.* **A** fruiting shoot with mature fruiting catkins × ¾; **B** underside of leaf base × 2; **C, D** leaf outlines, mature stem leaves, largest on specimen, showing variation of marginal dentation × ¾; **E** bark of twig (left, hairy specimen; right, glabrous) × 2; **F** female catkin scale, back and side views × 15; **G** seed × 10. **A, E** right, **F, G** from *Forrest* 17657 (Yunnan, 1917–19); **B** from *McLaren* D221 (Western Yunnan, 1933); **C** from Type 2794 (?collector, Kumaon); **D** from *Grierson & Long* 4667 (Bhutan, Mendegang, 27°31'N 89°49'E, 24 April 1982); **E** left from *Forrest* 9616. DRAWN BY ANDREW BROWN.

Betula cylindrostachya Lindl. in Wall., *Plantae Asiaticae Rariores* 2: 7 (1831). Type: Kumaon, *R. Blinkworth*. Wall. cat. 2794 (holotype BM).

B. alnoides Wall. var. *cylindrostachya* (Lindl. ex Wall.) Regel, *Nouv. Mém. Soc. Imp. Naturalistes Moscou* 13 (2): 129 (1861).

B. alnoides Buch.-Ham. var. *cylindrostachya* (Lindl. ex Wall.) H. J. P. Winkl. in Engler, *Pflanzenr.* IV, 61: 91 (1904).

B. rhombibracteata P. C. Li, *Acta Phytotax. Sin.* 17 (2): 88 (1979). Type: Yunnan, The-chin Hsien, Tse-Ching, (Dêqên Xian, Weixi Xian), 20 July 1940, *Feng Kuo-mei* 5675 (PE).

B. nitida Lindl. ex Wall., *Plantae Asiaticae Rariores* 2: 7 (1831), haud D. Don (1825).

Betulaster cylindrostachya (Lindl.) Spach, *Ann. Sci. Nat., Bot.* sér. 2, 15: 198 (1841).

B. fujianensis J. Zeng, Jian H. Li & Z. D. Chen, *Bot. J. Linn. Soc.* 156: 523–528 (2008). Type: China, Fujian, Shaxian, Luoboyan Reserve, evergreen and deciduous broad-leaved forest, 500–600 m, 5 May 2004, *Zeng* 2004050501 (holotype PE); ibid., *Zeng* 2004050502 (paratype, PE).

DESCRIPTION. *Tree* to 30 m. *Bark* dark brown. *Twig* light brown, to 3 mm in diam., glabrescent. *Buds* c. 9–11 × 3.5 mm, ovate, acuminate, incurved. *Shoots* mostly densely white hairy, eglandular or with some glands on the very youngest part, rarely very glandular. *Petioles* 0.8–1.5 cm, hairy, glandular; petiole:blade ratio c. 1:10. *Leaves* ovate-elliptic, oblong to ovate-lanceolate, (5)–14 × 2–8 cm, base rounded or sub-cordate, apex acuminate, margin serrate, with 13–14(–15) pairs of secondary veins ending in prominent teeth with up to 7 teeth between these, hairy when young on both surfaces and glandular below (abaxially), persistently hairy in secondary vein axils. *Male catkins* to 60 × 3 mm, scales peltate, often apiculate, c. 2 × 2 mm. *Fruiting catkins* usually in groups of 2 or 3, or solitary; peduncle to 20 mm, hairy; catkins to 100 × 5–7(–10) mm; transitional region between peduncle and catkin proper c. 5 mm, with small poorly developed scales and fruitlets. *Scales* 2–3.25 × 1.5–1.7 mm, base very thick, lobes very thin and acute, laterals much reduced c. 1/3 length of midlobe. *Seeds* to 2.5 × 4.5 mm usually narrowing towards base, nutlet 2–2.25 × 1–1.5 mm, usually broadest towards apex; style c. 1.75 mm.

DISTRIBUTION. Pakistan, Chitral: India, Himachal Pradesh (Beas river valley), through Nepal, Sikkim, Bhutan, Darjeeling district, Assam (Khasia Hills), N Burma, China (SW Sichuan, NW Yunnan, Fujian). 1,000–3,000 m.

HABITAT. River valleys, often in disturbed ground.

FLOWERING TIME. March to May.

FRUITING. May to June.

CHROMOSOME NUMBER. Tetraploid: 2n=56.

22. BETULA LUMINIFERA

Betula luminifera is a common species in western China, from Yunnan as far north as Shaanxi and Gansu, and east to Jiangsu; it was collected by both Père Delavay and Père Farges, but not described as a distinct species until Winkler's account of *Betula*, published in Das Pflanzenreich in 1903. It forms a slender tree, and grows in forest in the valleys, at no great altitude, between 1000 and 2500 m in western Sichuan, according to Wilson (in Sargent 1916).

Flowering is given as May to June in *Flora of China*, but in cultivation at Ness the Rix provenance (*EMR* 4121) from Baoxing in Sichuan, is one of the earliest birches to break bud, and expands its young leaves and catkins in early April (Fig. 59).

Betula luminifera is quite variable, both in leaf shape and in catkin length. The Farges specimens have broadly ovate and almost cordate leaves. *B. wilsoniana* C. K. Schneid. from near Ichang is tomentose

PLATE 13. *Betula luminifera*
Painted by Josephine Hague

with very long catkins, while *B. hupehensis* C. K. Schneid. is nearer glabrous, with shorter catkins.

Skvortsov regarded *Betula cylindrostachya* as being most similar and closely related to *B. luminifera*, despite the female catkins being borne singly in *B. luminifera* and in clusters of two to six in *B. cylindrostachya*. We agree with his comment that this is probably not a totally reliable or stable character as weak shoots of *B. cylindrostachya* often bear single fruiting catkins. The few chromosome counts we have made suggest that *B. luminifera* is diploid while *B. cylindrostachya* is tetraploid, so that regular interbreeding is unlikely. Some gene flow, especially from the diploids *B. alnoides* and *B. luminifera* to the tetraploid *B. cylindrostachya*, might be expected.

CULTIVATION

From the horticultural point of view, *Betula luminifera* has a very attractive dark, glossy bark. In spring the very long thin male catkins give a delicate appearance to the tree (Plate 13; Fig. 145). The leaves tend to become more battered than those of KBA's collection of *B. cylindrostachya* from Garhwal, but in warmer climates these species could thrive and their catkins provide an attractive display very early in spring.

Seedlings and saplings of this species have very densely hairy young shoots and leaves, but older plants have shoots which are almost glabrous. The sole surviving clone of *EMR* 4121 was the only one which retained the hairy juvenile shoots until it reached several metres tall and is presumably hardier and more suited to the British climate than its sibling seedlings.

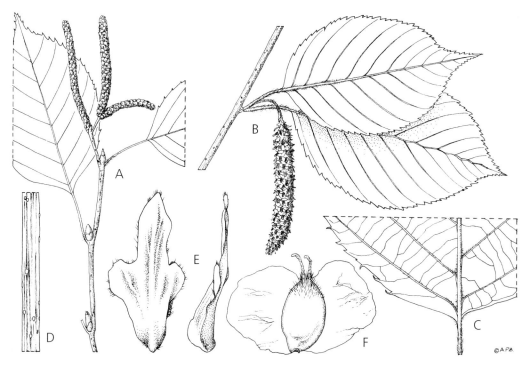

Fig. 144. *Betula luminifera*. **A** twig with buds and immature male catkins × ¾; **B** fruiting shoot with mature fruiting catkin × ¾; **C** underside of leaf base × 2; **D** bark of twig × 2; **E** female catkin scale, back and side views × 10; **F** seed × 10. **A** from *Kirkham et al*. SICH 2027 (30 Sept. 1999, 32°39'53"N 107°3'57"E); **B–F** *E. H. Wilson* 17 (Western Hupeh, June 1907). DRAWN BY ANDREW BROWN.

Fig. 145 (left). *Betula luminifera* (*EMR* 4121). Fruiting catkins borne singly (PW).

Fig. 146 (right). *Betula luminifera* (*EMR* 4121). Trunk of young tree showing lenticels (PW).

At least one clone of *Betula luminifera* has been distributed as grafted plants by Hilliers (of Winchester, England) for many years and other collections have recently been introduced to cultivation.

Betula luminifera H. J. P. Winkl. in Engl., *Pflanzenr*. IV. (Heft 19): 61 (1904). Type: E Sichuan (now Chongqing), Heoupin prope Tchen-Keou, 1400 m, *Farges* s.n. (holotype P.)

B. cylindrostachys Diels 1900 (non Lindl. ex Wall. 1830). Type: *Bot. Jahrb. Syst.* 29(2): 281–282 (1900).

B. baeumkeri H. J. P. Winkl. in Engl., *Pflanzenr*. IV. (Heft 19): 61 (1904). Type: Yunnan, *J. Delavay* s.n.

B. wilsoniana C. K. Schneid., *Ill. Handb. Laubholzk*. 2: 882 (1912). Type: Western Hubei, north and south of Ichang, 1000–1700 m, woodlands, abundant, April & July 1900, *Wilson* 48.

B. hupehensis C. K. Schneid., *Ill. Handb. Laubholzk*. 2: 882 (1912). Type: Western Hubei, Fang Hsien, alt. 1000–2000 m, woodlands, June 1901, *Wilson* 2800.

B. riparia W. W. Sm., *Notes Roy. Bot. Gard. Edinburgh* 13: 155 (1921). Type: W China, Li-ti-ping, Yangtze–Mekong divide, Yunnan, forming thickets by streams, lat. 27°12'N, alt. 10,000 ft, shrub of 20–30 ft, June 1918, *G. Forrest* 16334 (holotype E).

DESCRIPTION. *Tree* to 25 m. *Bark* dark brown, not peeling, becoming reddish-brown or dull grey on old trees. *Shoots* silky hairy, sparsely glandular. *Twigs* orange-brown, to 3 mm in diam., glabrescent, smooth, lenticels longitudinally elongated. *Petiole* c. 12 mm, densely hairy. *Leaves* ovate, broadly oblong to oblong-lanceolate, rarely elliptic, base broadly cuneate to sub-cordate, apex mucronate to caudate, margin double serrate, 4.5–10 × 2.5–6 cm, secondary veins 12–14 pairs, hairy above and below when young, glandular below, secondary vein axils hairy. *Male catkins* to 2 terminal with up to 2 laterals, light brown, very slender in both winter and expanded spring state, to 20 × 2.5 mm, expanding to 100 × 4.5 mm; scales peltate, shield-shaped, about 1.5 × 1.5 mm, margins ciliate, stamens yellow. *Fruiting catkins* borne singly, rarely in pairs, pubescent, 3–9 cm × 6–10 mm, long cylindrical, with peduncle 1–2 mm. *Scales* 2–4 × 1.5 mm, margin ciliate, surfaces more or less glabrous; central lobe 2.5 mm, ovate-triangular-oblong; lateral lobes to 2 mm. *Seed* with nutlet 2–2.5 × 1.75 mm, wing c. 2 mm each side, translucent, one cell thick, styles 0.6 mm.

DISTRIBUTION. China: Anhui, Fujian, Gansu, N Guangdong, Guangxi, Guizhou, Henan, Hubei, Hunan, Jiangsu, Jiangxi, Shaanxi, Sichuan, Yunnan, Zhejiang (Li & Skvortsov 1999).

HABITAT. Broad-leaved forest in mountain areas, 200–2900 m.

FLOWERING & AND FRUITING TIME. April–June; fruiting June–August.

CHROMOSOME NUMBER. Diploid: 2n=28.

23. BETULA MAXIMOWICZIANA

HISTORY

Betula maximowicziana was collected first by Carl J. Maximowicz (1827–1891) during his expedition to Japan in 1860–1864. During this time he collected extensively from south-western Hokkaido to southern Honshu. In later years, while based at St Petersburg, he continued working on the Japanese flora, receiving specimens from many Japanese botanists, and purchasing part of Siebold's collections from his widow (Ohwi 1965; Blunt & Stearn 1994). Eduard Regel was also working at St Petersburg at the same period, and described this species in his monograph of *Betula*, as *B. maximowiczii*, but this name was illegitimate because Ruprecht had already used the name for *B. dahurica*, and three years later Regel published *B. maximowicziana*.

ECOLOGY AND STATUS IN THE WILD

Betula maximowicziana is a major timber-producing tree in Japan where it forms a tall tree with large, ovate, finely hairy leaves. It occurs naturally at lower altitudes than *B. pendula* subsp. *platyphylla* (Osumi 2005; Tabata 1966). It would appear to be more drought tolerant than many other birches, Coder (1999) mentioning only this species and *B. nigra* in a list of drought tolerant species for south-east USA. Tsuda & Ide (2005) describe it as a long-lived pioneer species of cool temperate forests with relatively little genetic diversity, but some differentiation among populations. This suggests that the populations have been through significant bottlenecks in the past with episodes of small population size.

Natural regeneration of *Betula maximowicziana* depends mainly on the previous year's seed production, but the species has been shown to have a persistent seed bank with over 16% of buried seed being viable after six years (Osumi & Sakurai 1997). This means that there is always seed present in the soil, even in years when no seeds are produced, to take advantage of any disturbance which provides favourable conditions for seed germination (Tsuyuzaki & Goto 2001).

Fig. 147. *Betula maximowicziana* AGS/J, male catkins and young leaves showing elongated teeth, 24 May 2006 (PW).

TAXONOMY, RECOGNITION AND RELATIONSHIPS

Although clearly belonging to subgenus *Acuminatae* because of its elongated leaf teeth and long, cylindrical, clustered fruiting catkins, *Betula maximowicziana* differs from the other species of the subgenus in its autumn, rather than summer, fruiting and its much thicker male catkins, similar to those of most other birch species (Fig. 62). It has therefore presumably been isolated from its close relatives, probably in the Japanese archipelago, at least since the Tertiary. *B. maximowicziana* belongs to the north-eastern group of tree species of Japan and north-east Asia, which have few species in common with the Sino-Himalayan region (Ran *et al.* 2006).

CULTIVATION

Betula maximowicziana was introduced to cultivation in Britain in 1888, when J. H. Veitch sent seed from Yezo, and a flowering specimen was illustrated in *Curtis's Botanical Magazine* in 1910. It is fairly

Fig. 148. *Betula maximowicziana* AGS/J, bark (PW).

common in cultivation and seems to be easy to grow in temperate areas of the British Isles and North America (Clarke in Bean 1970; Santamour & Lundgren 1996). It makes an attractive tree with striking large leaves and clusters of long fruiting catkins in spring. It is easily propagated from seed and is quick growing. It appears to be almost totally self-incompatible and only a single hybrid tree is known (see Chapter 2, *B.* ×*dosmannnii*). There have been earlier suggestions that hybrids might exist elsewhere (Santamour & Lundgren 1996), but, in the present authors' opinion, these are likely to be misidentifications.

The single tree of *Betula maximowicziana* in the Hergest Croft Arboretum in Herefordshire, produces large quantities of non-viable seed when growing in the presence of many other birch species, which suggests that hybridisation with species of other subgenera is very uncommon. At Ness, fruiting *B. maximowicziana* is now growing alongside fruiting trees of *B. luminifera* and *B. cylindrostachya*, so the possibility of hybridisation within subgenus *Acuminatae* could be tested, although *B. luminifera* and *B. cylindrostachya* tend to flower much earlier than *B. maximowicziana*.

Betula maximowicziana Regel in A. P. de Candolle, *Prodr.* 16 (2): 180 (1868). Type: Auf der insel Gesso (Yezo), von *C. Maxim.* entdeckt (holotype MW?).

> *B. maximowiczii* Regel, *Bull. Soc. Imp. Naturalistes Moscou* 38 (3): 26 (1865). nom. illeg. Type: Auf der insel Gesso, von *C. Maxim.* entdeckt.
>
> *B. candelae* Koidz., *Bot. Mag.* (Tokyo) 27: 147 (1913).

ILLUSTRATION. *Woody Plants of Japan*: 119 (1985).

DESCRIPTION. *Tree* to 30 m. *Bark* dark greyish, peeling in thin strips. *Twigs* c. 3 mm in diam., orange-brown, with small, dried, reddish glands, more or less persistently silky hairy, puberulent in branch and bud axils; lenticels oval, vertically elongated, c. 0.75 mm. *Bud scales* ciliate, otherwise glabrous.

Shoots silky hairy. *Petioles* 25–35 mm. *Leaves* 8–14 × 6–10 cm, broadly ovate to almost ovate-orbicular, deeply cordate at base, shortly acuminate at apex, serrate with gland tipped teeth, silky-hairy on both surfaces when young, glabrescent, with 10–12 pairs of secondary veins. *Male catkins* c. 35 × 5 mm in winter, expanding on opening; scales c. 3 × 2.5 mm, shortly ciliate, chestnut brown; raceme axis persistently hairy and glandular as twig. *Fruiting catkins* borne in groups of 2–4, (2–)3–7 cm × 6–7 mm; scales glabrous, 3–6 mm, 3-lobed with lateral lobes mostly about half the length of the central. *Seeds* c. 3 mm, with wings 2–3 times as wide as nutlet.

DISTRIBUTION. Japan: central and north Honshu, Hokkaido; Russia: Kurile Islands, Iturup.

HABITAT. Forest at low altitude, often dominant.

FLOWERING AND FRUITING TIME. May and June; fruiting Autumn.

CHROMOSOME NUMBER. Diploid: 2n=28.

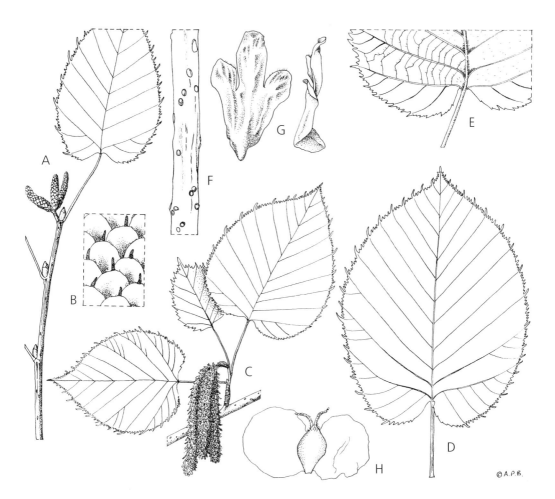

Fig. 149. *Betula maximowicziana*. **A** twig with buds and immature male catkins × ¾; **B** male catkin scales × 10; **C** fruiting shoot with mature fruiting catkins × ½; **D** large leaf × ½; **E** underside of leaf base × 1; **F** bark of twig × 2; **G** female catkin scale, back and side views × 7; **H** seed × 7. **A, B** from Stone Lane Gardens, cult. *Ashburner*, collected 19 Aug. 2010; **C** from *Murata et al.* 38018 (Abashiri, Hokkaido, 18 June 1978); **D–H** from *Maximowicz* (Hakodati, 1861, type). DRAWN BY ANDREW BROWN.

Subgenus BETULA

Trunks mostly with characteristic peeling, usually white, bark; axillary buds present on at least one leaf on the fruiting spur shoots; *leaves* with relatively long petioles in the tree species; *seeds* with thin, one cell thick, wings not attached to the styles.

Subgenus BETULA section DAHURICAE

Old trunks often very dark; *bark* initially white, peeling in small flakes and so 'shaggy'; *fruiting catkins* erect (*Betula nigra*) or horizontal more or less in the plane of the shoots, only pendent when dry. East Asia, Caucasus, South-east North America.

24. BETULA NIGRA RIVER BIRCH, RED BIRCH

HISTORY

Betula nigra was named in Linnaeus' *Species Plantarum* in 1753, based on the plant named as *Betula nigra virginiana* in Plukenet's *Historia Plantarum* (1686–1704), and *Betula foliis ovatis oblongis acuminatis serratis* of Gronovius' *Flora Virginica, exhibens plantas quas v.c. Johannes Clayton in Virginia observavit atque collegit*, published in 1743; Clayton's specimen is in the Herbarium of the Natural History Museum, viewable on the NHM website, and actually shows a specimen of *B. alleghaniensis*. John Bartram was probably responsible for the introduction of *B. nigra* to England, as he sent quantities of tree seed from Virginia to Peter Collinson, who distributed them around a group of subscribers.

This species is called river birch, or red birch, or sometimes black birch because of the often dark or reddish, conspicuously flaking bark. Cultivars with pale bark are commonly planted. There is an illustration after a painting by H. J. Redouté under the name *Betula rubra*, in François André Michaux's *The North American Sylva* (1819). A specimen in the Linnaean Herbariun, annotated by J. E. Smith has been chosen as the type.

ECOLOGY AND STATUS IN THE WILD

The north-south range of *Betula nigra* in the east of North America is very great, stretching continuously from Massachusetts in the north to Florida in the south and as far west as Texas and Ohio. It seems to be scarce in the Appalachians but can be seen by riversides in West Virginia. Otherwise it is common between the Appalachians and the sea, and also on the western side of the mountains. There are significant groves of the trees in bottomlands and streamsides in southeast Ohio, for example at Conkles Hollow in the Hocking Hills, where superb specimens of *Platanus occidentalis* are also common; the bark on the stems of river birches have tightly-knit fissures, a hazardous covering of poison ivy *Toxicodendron radicans* here and there, and sharply-toothed leaves in yellow fall colour.

Fig. 150. *Betula nigra* 'Wakehurst'. Trunk of young tree with characteristic peeling giving 'shaggy' bark (PW).

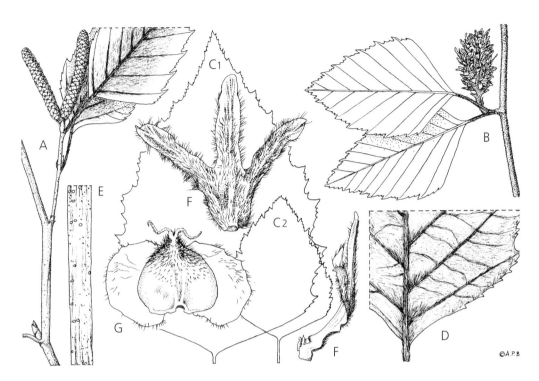

Fig. 151. *Betula nigra.* **A** immature male catkins terminally on lateral shoot from node 5 × ¾; **B** fruiting shoot with mature fruiting catkin× 1; **C1, C2** size variants of main stem leaves × 1; **D** underside of leaf base × 1½; **E** bark of twig × 2; **F** female catkin scales, back and side views × 7; **G** seed × 7. **A, C1** from *Palmer* 8930 (Monroe, Ouachita Parish, Louisiana, 14 Oct. 1915); **B, E–G** from *Palmer* 5569 (White R., Beaver, Caroll County, Arkansas, 13 May 1914); **C2** from *Beattie* (Lowell, Mass., June); **D** from from *Palmer* 7683 (Calcasieu Parish, Lake Charles, Louisiana, 18 May 1915). DRAWN BY ANDREW BROWN.

Everywhere in the wild *Betula nigra* grows on streambanks and damp ground, although in cultivation it will grow perfectly well in drier habitats. This preference for riparian habitats is probably linked to the ripening and dispersal of its seed in June. Thus, like species of willow and *Populus* (poplar, aspen, cottonwood), it is the ecological requirement for moist soil for the germinating seedlings during the summer which determines where they can establish naturally. *The Flora of North America* (Furlow 1997) points out that dispersal in June is advantageous for riverbank establishment as the habitat is then unlikely to be flooded. The fruits are winged but also more or less corky, and float easily.

The fact that *Betula nigra* is a warm temperate species and one of the most southerly in distribution, together with its riparian habitat and association with *Platanus* (a known relict), suggests that it may also be a relict species, like the Caucasian *B. raddeana* (Wolfe 1969; Collinson 1990).

TAXONOMY, RECOGNITION AND RELATIONSHIPS

This birch can be recognised instantly by many characters: firstly the peculiar bark, pink on young trees and shedding a pink dust, with clinging flakes turning back and becoming tighter as the stems increase in size, and eventually scored with close fissures; then the leaves, invariably cuneate, tapering into the petiole, more or less rhombic and more or less glaucous beneath; thirdly its thin, smooth, brown or yellow-brown twigs with their very small resinous glands and very small buds are distinctive.

Fig. 152. *Betula nigra.* Fruiting shoots with more or less horizontal fruiting catkins and conspicuously double toothed rhombic leaves (PW).

Fig. 153. Old trunks of *Betula nigra* with seedlings in the foreground on bank of Charles River, Cutler Park, Boston, Massachusetts. 19th October (HMcA).

Fig. 154. Close up of seedlings in previous photograph of *Betula nigra* with seedlings in the foreground on bank of Charles River in Cutler Park, Boston, Massachusetts. 19th Oct. (HMcA).

CULTIVATION

Seedlings from Illinois, for example, grow well in the British Isles in drier habitats, as well as wet and streamside situations. In these, the young bark is a deep orange-pink, shedding the pink dust, but they soon become untidy with persistent flakes and readily sprout epicormic 'water shoots' from the main trunks. Later, the bark is dark, firm and narrowly fissured as it is on the trees seen in Ohio.

Many specimens in arboreta, for example the cultivars 'Heritage' and 'Wakehurst', are markedly different, whiter in bark when young and with bigger, shinier, more rhombic leaves, blue-green rather than pale green, and with fewer and more widely-spaced lateral veins.

Softwood cuttings strike easily in mist. Branches always seem brittle and easily snap off in high wind. There may be considerable winter die-back in areas where the summers are too cool for the twigs to ripen properly.

Betula nigra L., *Sp. Pl.*: 982 (1753). Type: Habitat in Virginia, Canada (lectotype: *Herb. Linn.* 1109.7).
Betula rubra F. Michx., *Hist. Arb.* II: 143. t. 3 (1811).

DESCRIPTION. *Large trees* to 25 m, often with several trunks. *Bark* shaggy, peeling off in numerous small curls, whitish to quite dark brown in thicker trunks. *Twigs* hairy to glabrous. *Buds* small and narrower than adjoining internode, to 5.0 × 2.0 mm. *Scales* shining pale brown or green, more or less hairy. *Shoots* glabrous to densely hairy, smooth, with small lenticels and glands; petioles glandular. *Leaves* on long-shoots 50–90 × 30–60 mm, matt pale green and only shiny on youngest leaves, always matt and glaucous beneath, rhombic-ovate, cuneate to truncate at base; lateral veins (5–)7–10

(−12) and impressed when young, brown or crimson-brown beneath and densely hairy; leaf margin with lobes up to 1 cm deep, sharply double serrate; lateral veins ending in large forward-curving teeth, each with slender tip. Basal long shoot and spur shoot leaves (i.e. those preformed in bud) often smaller, ovate, cuneate, acute rather than acuminate, usually 38–40 × 24–30 mm, sometimes greener rather than glaucous beneath, serrations less deep, with sinus as little as 1–2 mm deep, margin sometimes scarcely double-toothed, with only 2 or 3 minor teeth between the major, with c. 6 veins. Blade:petiole ratio usually 1:5 to 1:6, to 1:3 in larger, more rhombic leaves. *Fruiting catkins* more or less erect, horizontal or pendent, more or less cylindrical but mostly tapering at base, about 15–30 × 10–25 mm. Seed almost circular, c. 4 × 4 mm with wing about half nutlet width.

DISTRIBUTION. Eastern North America, from Massachusetts in north to Florida in the south and west to Texas and Ohio.

HABITAT. Riversides and streambanks, near permanent water.

FLOWERING AND FRUITING TIME. March–April; fruiting June.

CHROMOSOME NUMBER. 2n=28.

25. BETULA DAHURICA

HISTORY

Betula dahurica was first collected by Peter Simon Pallas (1741–1811) on his expedition across Russia, which lasted from 1768 to 1774. He was both zoologist and botanist, born in Berlin and studied at the Universities of Halle and Göttingen, before receiving his doctorate at Leiden at the age of 19. Catherine the Great invited him to be a professor at the Academy of Sciences in St Petersburg, and from there he travelled to the Altai, Lake Baikal, and further east into the upper Amur basin, where he collected *B. dahurica*. His travels were published almost as soon as he returned in *Reise durch verschiedene Provinzen des Russischen Reichs*, and later many of the plants, including *B. dahurica*, were beautifully illustrated in *Flora Rossica* which was published in parts between 1784 and 1815.

ECOLOGY AND STATUS IN THE WILD

Li & Skvortsov (1999) describe this species as occurring in mixed or coniferous forests, on dry or moist soils and exposed rocky places. In some situations it can be the dominant or co-dominant species in the forest (Zhang *et al*. 2007; Hou *et al*. 2006).

From the study of herbarium specimens and limited field experience, *Betula dahurica* appears to have a fairly distinctive appearance as the few veined, sharply toothed, often rhomboid leaves give a characteristic form to its canopy; the variation in hairiness and other detailed features has little effect on the overall appearance of the trees. It shares with the eastern North American *B. nigra* the shaggy peeling bark (Figs 159, 160) which distinguishes section *Dahuricae* (*B. nigra*, *B. dahurica* and *B. raddeana*) from other birches, or indeed from any other tree.

The distribution of *Betula dahurica* is similar to that of *B. schmidtii* and *B. costata* and to some extent *B. chinensis*, all four being confined to the region of north east Asia rich in relict species. Like *B. fruticosa*/*B. tatewakiana* and the enigmatic *B. gmelinii*/*B. apoiensis*, *B. dahurica* has outlying small rare populations on the Japanese islands.

TAXONOMY, RECOGNITION AND RELATIONSHIPS

Apart from its characteristic flaking bark, the leaf, twig and fruiting catkins of *Betula dahurica* are similar to those of many other white barked birches with which they can be confused, especially on

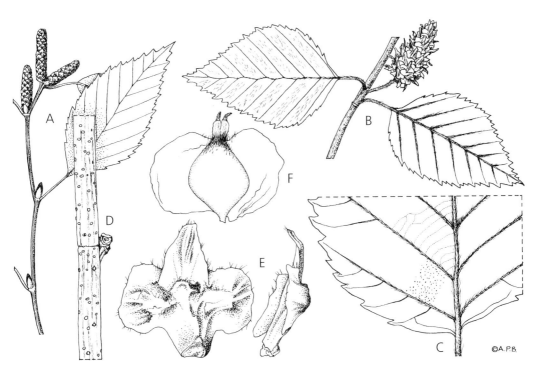

Fig. 155. *Betula dahurica.* **A** twig with buds and immature male catkins × 1; **B** fruiting shoot with mature fruiting catkin × 1; **C** underside of leaf base (showing very faint tertiary veins and gland pattern) × 2; **D** bark of twig × 2; **E** female catkin scale, back and side views × **7**; **F** seed × 10. **A** from *E. H. Wilson* 8971 (Yukyo-to-Sohyo, N Hankyo province, Korea, 19 August 1917); **C, D** from *E. H. Wilson* 8909 (Funei-to-S-ha-yurei province, Korea, 15 Aug. 1917); **B, E, F** from *Furuse* 20899 (Kanayama, Mt Midzukaki, Hondo, 21 July 1949). DRAWN BY ANDREW BROWN.

the herbarium sheet. The more or less diamond-shaped leaves with coarse toothing and the relatively few veins (up to 8 pairs) help to distinguish leafy shoots of the three species of *Dahuricae* from other species of *Betula*.

Within section *Dahuricae*, *Betula dahurica* can be distinguished by its non-spiky later-ripening fruiting catkins, leaves less distinctly double-toothed than in *B. nigra*, and from *B. raddeana* by usually more than five pairs of leaf veins. Early in our study of birches we acquired wild source seed from Korea and of Japanese and Kurile Island sources from cultivation in Hokkaido and Tallin (Estonia) respectively, and, as far as we could tell from the juvenile plants, these appeared to be true to the species. Korean origins all gave chromosome counts of 2n=112, but the Kurile Island material gave 2n=84 and is now thought to be derived from hybridisation with a tetraploid species, possibly *B. ermanii* which is quite similar morphologically in twig and leaf. Later, seed was obtained from the Japanese Honshu population, but chromosome counts were not made before the trees were planted out in the garden so seedlings of these were raised (which appeared to be coming 'true') and counted giving 2n=84. However, over the last two years, the flow cytometry work of Drs Nian Wang and Richard Buggs at Queen Mary University, and a chromosome count on a more recent collection from the wild in Japan have confirmed that the Japanese populations are octoploid like those of the Asiatic mainland.

The Hokkaido population had already been distinguished morphologically (Ohwi 1965), but we had great difficulty attempting to find differential characters between what we thought were distinct cytotypes. On the other hand, from the limited number of specimens we have examined, there do seem to be differences between the Japanese and mainland Asian populations, and the Japanese populations certainly do perform much better in cultivation in the British Isles, indicating physiological differences. However, whether the morphological distinctions between the continental and Japanese populations will hold following further study of a very variable species is uncertain.

Most distinct is the Honshu population from Nobeyama which has uniformly small leaves with conspicuous double toothing and usually seven pairs of veins which are closely and evenly spaced, giving the leaf a more strongly veined appearance than seen in any other population. We therefore propose that the continental populations are recognised as var. *dahurica*, the Kurile Island and Hokkaido populations as var. *okuboi*, and the more distinct Honshu populations as var. *parvifolia*.

CULTIVATION

In the wild *Betula dahurica* can grow into a tall tree, but in cultivation, at least in Europe, individuals of most provenances make relatively small, often several stemmed trees, due probably to winter dieback of the leading shoots. Young trees grow relatively slowly in comparison with other white barked birches, at least in cultivation in the British Isles, and are slower to reach maturity and produce catkins. For many years seedlings only produce long shoots, and it is only after about ten years that spur shoots are formed and catkins produced. This species is rare in cultivation in the British Isles and Bean (1970) dismisses it for amenity use because of its susceptibility to late frosts. As so often, this may be a problem with the very few collections in cultivation at that time.

Experience with recent introductions suggests that the Japanese Honshu population and the Korean introduction (*KFBX* 43) may be better suited to the British climate, as trees of these provenances are growing well in several places within the British Isles, the Honshu one at Ness, Dawyck near Peebles in the Scottish borders (and the Arnold Arboretum in Boston), and the Korean introduction at Kew and Ness.

Trees of the Honshu provenance form narrow, attractive, erect trees often with a single trunk and relatively short branches (Fig. 158). They have smaller leaves than other collections, which gives a lighter canopy. Their peeling bark is particularly attractive (Fig. 159), and they promise to be a valuable addition to the range of trees in cultivation. They fruit freely every year.

Fig. 156. Trunk of *Betula dahurica* in nature reserve near Khabarovsk, Russian Far East (HMcA).

Fig. 157 (left). *Betula dahurica* var. *dahurica* in Ussuri forest nature reserve north of Vladivostock (HMcA).

Fig. 158 (right). *Betula dahurica* var. *parvifolia* Japan, Nagano Pref, Minamisaku-gun, Minamimaki-mura, Nobeyama 35°55'N 138°30'E, 1400 m. Deciduous forest with *Quercus grosseserrata* (Makoto Amano 3 Sept. 1991). Three trees at Ness of the hexaploid Japanese subspecies from Honshu thrive much better in cultivation in the British Isles than provenances of other origins (PW).

Trees originating in Hokkaido grow well in Devon, but further north at Ness near Liverpool they suffer from late spring frosts and form twisted trunks due to dieback of leading shoots. The form of these trees is however interesting and attractive. At Ness, the Kurile Island population from Okeanskiy on the island of Iturup suffers severely from winter dieback, especially of the lower twigs, and forms a rather open, leggy, unattractive tree. We had some concerns over the origin of our plants of this population, as they were grown from seed collected in Tallinn Botanic Garden, in Estonia.

The trees of Korean provenance are thriving and making attractive small trees with the characteristic flaky bark. Their relatively dense, early-developing young foliage is bright green and one of the first to expand in spring, an attractive feature but one which makes them sensitive to late frosts. Their fruiting behaviour, like that of other species from eastern continental Siberia, is often very erratic between years and trees, at least in cultivation. Collections from further north in Far Eastern Russia are not doing so well, showing considerable dieback of the thinner, weaker twigs in winter.

As mentioned in the section on pests and diseases (Chapter 3, p. 87), this species initially appeared to be resistant to the bronze birch borer, but recent reports of attacks indicate that any resistance is not complete (Dosmann, pers comm.).

Fig. 159 (left). *Betula dahurica* var. *parvifolia*. 'Shaggy' bark of young branch showing rich red-brown colour of inner surface of bark curls (PW).

Fig. 160 (right). *Betula dahurica* var. *parvifolia* trunk (PW).

Betula dahurica Pall., *Reise Russ. Reich.* 3: 224 (1776).

B. dioica Pall., *Reise Russ. Reich.* 3: 321 (1776).

B. davurica Pall., *Fl. Ross.* 1: 60 (1784).

B. maximowiczii Rupr., *Bull. Cl. Phys.-Math. Acad. Imp. Sci. Saint-Pétersbourg* 15: 139 (1856).

B. maackii Rupr., *Bull. Cl. Phys.-Math. Acad. Imp. Sci. Saint-Pétersbourg* 15: 380 (1857).

B. wutaica Mayr, *Fremdl. Wald-u. Parkbaume Eur.*: 450 (1906).

ILLUSTRATIONS. P. S. Pallas (1784). *Fl. Ross.* 1: 60 t. 39, fig. A. *Woody Plants of Japan*: 117 (1985).

DESCRIPTION. *Tree* to 20 m. *Bark* initially creamy white on young branches, becoming irregularly covered with curling flakes of white peeling bark giving a very distinctive appearance as branches increase in diameter; eventually thick and corky and black-brown in old trunks (Figs 156, 157). *Shoots* very warty with copious resinous secretions, with variable amounts of longer silky hairs and short puberulent hairs, especially in axils of buds, becoming a rich brown colour, relatively thin, to about 3 mm diam.; vigorous and epicormic shoots of current year often branching freely. *Twigs* persistently hairy with both long ciliate and short puberulent hairs persisting on 2nd year twigs, especially near buds. *Leaves* with blade usually more or less rhombic to ovate but varying from broadly ovate to elliptic, cuneate at base, 4–8 × 3.5–5 cm; with 6–8 pairs of veins. Reddish glands mainly restricted to veins on upper surface and lamina of lower surface. *Male catkins* about 18 × 4 mm. *Fruiting catkins* more or less erect to horizontal to pendent, to about 30 × 10 mm, only sometimes 'spikey' due to patent or recurved scale midlobes even in unripe catkin; scales hairy to glabrous, glandular; style c. 0.75 mm, pyriform. *Seeds* with wings as wide as or narrower than nutlet, more or less adnate to style.

DISTRIBUTION. East Asia east of Lake Baikal south to eastern Mongolia, China and Korea, Japan, and the Kurile Islands.

HABITAT. Mixed broadleaved/coniferous and coniferous forests on dry or moist soils and exposed rocky places; 400–1300 m.

Key to the varieties of *Betula dahurica*

1. Male catkin scales appearing circular with margins mostly fully exposed; leaves small (less than 65 × 50 mm), double toothed with fine toothing (mostly more than 5 teeth per 10 mm of mid-leaf margin); fruiting catkin scales 5–6 mm (Honshu, Japan) . . . . **dahurica** var. **parvifolia**
1a. Male catkin scales overlapping apically; leaves often larger, single or double toothed with coarser toothing (mostly fewer than 5 teeth/10 mm of mid-leaf margin) **2**

2. Fruiting catkins scales 5–6 mm (Hokkaido, Japan; Iturup, Kurile Islands) . . . .
. . . . **dahurica** var. **okuboi**
2a. Fruiting catkins scales (5.5) 6.5–7 mm (mainland Asia) . . . . **dahurica** var. **dahurica**

Betula dahurica var. dahurica

DESCRIPTION. *Fruiting catkin scales* usually more than 6 mm long. *Seed* usually more than 2 mm long. Chromosome number 2n (8x)=112.

DISTRIBUTION. China: Hebei, Heilongjiang, E Jilin, N Liaoning, Nei Mongol, Shaanxi, Shanxi. E Mongolia. Korea. Russia: E Siberia: Dahuria. Far East: Ze, Buryatia, S Uda, Ussuri (from Kuzeneva 1936).

HABITAT. In mixed broad-leaved and coniferous forests on dry or moist soils and exposed rocky places, sometimes the dominant or co-dominant species in the forest.

BETULA DAHURICA var. OKUBOI

This variety was described originally from Hokkaido, and differs from var. *dahurica* in its chromosome number and in its shorter fruiting catkin bracts (< 6 mm) and shorter nutlets (< 2.0 mm). According to Ohwi (1965) it has leaves with up to nine pairs of lateral veins and fruiting catkin bracts with the central lobe longer than the lateral lobes.

Betula dahurica var. **okuboi** Miyabe & Tatew., *Trans. Sapporo Nat. Hist. Soc.* 17: 49 (1941).

DESCRIPTION. *Fruiting catkin scales* less than 6 mm long. *Seed* usually less than 2 mm long. Chromosome number 8x (flow cytometry).

HABITAT. Unknown.

DISTRIBUTION. A few populations in Japan on Hokkaido and in the Kurile Islands on Iturup.

BETULA DAHURICA var. PARVIFOLIA

The trees in cultivation at Ness originated from Japan: Honshu, Nagano Prefecture, Minamisaku-gun (county), Minamimaki-mura (village), Nobeyama, 35°55'N 138°30'E, deciduous forest with *Quercus grosseserrata*, c. 1,400 m, 3 Sept. 1991, *Makoto Amano*. Dr Amano reported 'The species was not so rare in the locality, however, I found only one seed bearing tree.' This does not mean that it is self-compatible as other trees may have borne male catkins. The other cultivated living collections and all the herbarium specimens seen from Honshu seem to be from the same locality, so as far as is known, it is very restricted in distribution. All the specimens are very similar and immediately recognisable by the characters mentioned below.

Betula dahurica var. **parvifolia** Ashburner & McAll. **var. nov.** Type: Cultivated specimen from tree at Ness (LIV) grown from seed collected by Dr M. Amano, 3 Sept. 1991, Nobeyama, Minamimaki-maura, Minamisaku-gun, Nagano Prefecture, Honshu, Japan. 35°55'N 138°30'E, c. 1,400 m, deciduous forest with *Quercus grosseserrata*.

DESCRIPTION. *Male catkin scales* appearing circular with margins mostly fully exposed. *Leaves* small (less than 65 × 50 mm), double toothed with fine toothing, mostly more than 5 teeth per 10 mm of mid-leaf margin; *Fruiting catkin scales* less than 6 mm long. *Seed* usually less than 2 mm long. Chromosome number 2n (8x)=112.

DISTRIBUTION. Japan, Honshu, Nagano Prefecture.

HABITAT. Deciduous forest.

26. BETULA RADDEANA

HISTORY

Betula raddeana was described in 1887 by Ernst Rudolph Trautvetter and named after Gustav Ferdinand Richard von Radde (1831–1903). Radde was born in Danzig (Gdansk), but like many German biologists in the early 19th century, found employment in Russia. He was largely self-taught, but became an expert on beetles, as well as being a plant collector, first in the Crimea, then on an expedition across Siberia; from 1864 he lived in Tbilisi, studying the birds, insects and flowers of the Caucasus, the Black sea coast and the shores of the Caspian as far east as Turkmenia.

Until 1981 *Betula raddeana* was probably represented in Britain solely by herbarium specimens, but since that time, a few living plants have come into cultivation at Ness, so we have been able to study them in more detail. From studying herbarium specimens, we thought that *B. raddeana* was only slightly different from, and closely related to, *B. utilis*, and could be regarded as an outlier of that Himalayan species in the Caucasus.

In 1979 Dr Nora Gabrielian of Erevan in Armenia kindly sent some herbarium specimens of Armenian species, including a birch specimen, which appeared almost identical to *Betula nigra* in its rhomboidal leaf shape and the dense creamy-white hairiness of its young shoots and leaf undersides. We realised that this specimen could only be *B. raddeana*, and suggested a relationship between this species and *B. nigra*.

In 1980 an expedition from the Cary Arboretum, New York State, collected seed of *Betula raddeana* from Karachayevsk in the Karachai-Cherkesk Autonomous region of the northern Caucasus and distributed some via the Botanic Gardens seed exchange. The single seedling raised at Ness grew into a rather thick-twigged, fastigiate young tree which proved to be hexaploid (2n=84), suggesting that it was distinct from the tetraploid *B. utilis* (2n=56) from the Himalaya. Morphologically the tree, like most herbarium specimens of *B. raddeana*, was unremarkable, but the rhombic leaves with only five or six pairs of veins suggested a relationship with *B. nigra* and *B. dahurica*. The chromosome number of 2n=84 is the same as that of Japanese material of *B. dahurica*. Another similarity which was noticed was that young trees of these two species produce hardly any short spur shoots in their early years, all shoots showing some degree of internodal extension so that even the weakest shoots are elongated.

As only a single seedling had been obtained, propagation by cuttings was attempted, although experience with other white barked birches suggested that they would fail. It was therefore a great surprise when almost every cutting rooted within three weeks and grew on vigorously, with buds of most of the cuttings breaking to produce new shoots. This physiological character supported the proposed relationship with *Betula nigra*, which also roots easily from cuttings, and with *B. dahurica*, which can usually be propagated by cuttings, although neither of these root with the ease of this seedling of *B. raddeana*.

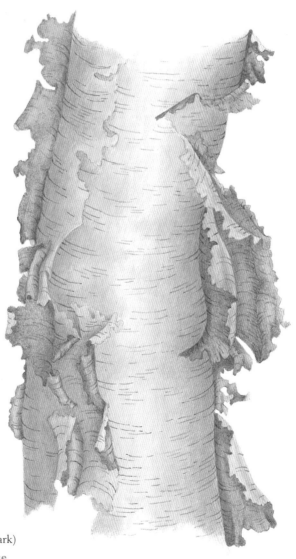

PLATE 14. *Betula raddeana* (bark)
Painted by Josephine Hague

When it came to transplanting from the nursery, another feature became evident. As most white barked birches do not transplant easily it is important to dig them up carefully, retaining as much as possible of the surface root system with soil attached. It is usual to cut the soil a good distance from the base of the tree and lift so that the tree comes up with a shallow plate of soil and roots intact. On attempting to lift the *Betula raddeana* in this way nothing happened, the tree remaining fast in the ground. Further excavation revealed a number of strong roots growing vertically downwards, firmly anchoring the tree. This provided yet another link to section *Dahuricae* which could not have been detected from herbarium specimens (McAllister 1999). *B. raddeana*, *B. dahurica*, and *B. nigra* are all easy to transplant, surviving even when considerable damage is done to the roots and little soil remains attached.

In 1988, through the kindness of Dr Skvortsov of Moscow University Botanic Gardens, seed of *Betula raddeana* was obtained from Gunib in Dagestan in the eastern Caucasus (approx. 46°57'E 42°23'N). There is a herbarium specimen in Kew presumably collected at the same time and annotated 'A. K. Skvortsov. 22/23-7-1987, Dagestan, Gunib District, supra pagum Gunib, 1700–1800 m' (K!). The two seedlings raised from this collection were different from one another in degree of hairiness and had a spreading branching pattern unlike the fastigiate earlier introduction. They also differ in forming single-stemmed trees in contrast to the multi-stemmed, basal sprouting habit of the other two collections in cultivation (see below). However, in other ways they behaved

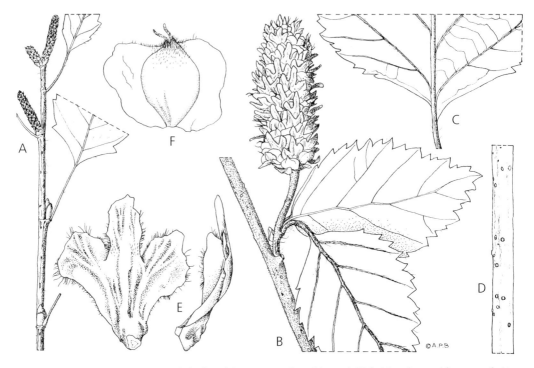

Fig. 161. *Betula raddeana.* **A** twig with buds and immature male catkins × 1; **B** fruiting shoot with mature fruiting catkin × 2; **C** underside of leaf base × 2; **D** bark of twig × 2; **E** female catkin scale, back and side views × 8; **F** seed × 8. **A** from freshly collected shoot (Stone Lane Gardens, 19 Aug. 2010); **B, D** from *Skvortsov* (tree A Daghestan, 25 km from Botlikh, 17 July 1987); **C** from *Grossheim* s.n. (Daghestan near Gunib, 12 June 1925); **E, F** from *Skvortsov* (tree T Daghestan, Gunib, 22/23 Aug. 1987). DRAWN BY ANDREW BROWN.

Fig. 162. *Betula raddeana* female catkins typically borne on relatively elongated leafy shoots and variable (erect to horizontal) in attitude (PW).

similarly, rooting easily from cuttings, being deep rooted, easy to transplant, vigorous, and drought tolerant. By 2004 the base of the trunk of one of these had developed somewhat shaggy bark, typical of *B. nigra* and *B. dahurica*. No mention of this character has been found in the literature and the development of a shaggy bark by at least one tree of *B. raddeana* adds support to the proposed relationship with other species in the *Dahuricae*; its absence, indeed, might have thrown some doubt on this relationship.

In 1993, seed of *Betula raddeana* of another provenance was obtained through the Bergen Botanic Gardens seed list (under the name *B. litwinowii*) from Chewi, east of Kazbegi village, Georgia, at 1,800–2,000 m. This accession is more slender in habit, less vigorous in growth rate, and multi-stemmed from a very early age. As with other birch species, it may be that trees from a higher altitude tend to be multi-stemmed and more shrubby in habit. The four surviving trees of this provenance are developing attractive habits, three spreading and one fastigiate, and with whiter bark than the other two collections. Their eventual size in cultivation is uncertain, but they may prove attractive, small, multi-stemmed white barked birches suitable for small gardens and confined spaces.

TAXONOMY, RECOGNITION AND RELATIONSHIPS

The close relationship of *Betula raddeana* and *B. dahurica* is another example of pairs of similar species found in the Caucasus and Eastern Asia, such as is seen in genera such as *Pterocarya*, *Parrotia*, and *Zelkova*. It is thought that originally continuous populations were split by the raising of the Himalaya, which began in the Miocene period about 30 million years ago and caused the drying out of much of central Asia which had formerly been the northern shore of the Tethys Sea.

Herbarium specimens of *Betula raddeana* can be difficult to differentiate for certain from some forms of *B. pubescens*, but they are not hard to identify in the living state, with their immature fruiting catkins held upright to horizontally, appearing 'spikey' because of the spreading scales; these ripen and disperse their fruits earlier than most other birches.

The following table was constructed using the three living accessions of *Betula raddeana* and a number of specimens of *B. pubescens*.

Table 8

Distinguishing characters of *Betula pubescens* and *B. raddeana*.

Character	*Betula pubescens*	*Betula raddeana*
Leaf shape	Ovate–deltoid–rhombic, broadest in lower half in all leaves	Elliptic–ovate, broadest around middle, especially in long-shoot leaves
Leaf toothing	Teeth ending main veins usually larger than other teeth	All teeth of much the same size
Lower leaf margin where joining petiole	Usually convex, rarely concave, at junction	Concave with blade margin more or less decurrent
Colour of lower (abaxial) leaf surface	Distinctly whiter	Greener (obvious with comparison)
Tertiary venation (veins joining 2er veins)	Little thicker that 4er veins, hardly forming a distinct order of venation	Distinctly thicker than 4er veins, forming 'chevrons' joining 2er veins
Number of male catkins	1–3, terminal and in axil of most distal leaf	(1–)3–5, terminal and in axils of up to 3 distal leaves
Fruiting catkin (immature)	Scales appressed, nutlets not visible between scales	Scales patent or recurved, giving 'spikey' appearance with nutlets visible between scales
Catkin scale	Central lobe about equal to lateral	Central lobe longer than lateral

ECOLOGY AND STATUS IN THE WILD

Little appears to have been written on the ecology of the Caucasian *Betula raddeana*. Seifriz (1931) describes it as generally occurring below *B. pubescens* and co-occurring with the enigmatic *B. pubescens* var. *raddeana*, which might be the hybrid between these species as this hybrid is suggested to occur (Kuzeneva 1936). Doluchanov (1939) discusses *B. raddeana* and describes *B. litwinowii* Doluch.

Betula raddeana is included in the *Red Data Book of the Russian Federation* (Iliashenko & Iliashenko 2000). Although many populations are recorded, they are scattered through the valleys of the Caucasus and many may be small. It is also very rare in cultivation and further collections may not be easy to obtain in the near future given the political instability in the region. For a rare species, it is surprisingly variable, apparently even within a population if the two individuals from Gunib are typical. However, it is likely that individual trees are long lived; they are able to produce stool shoots, and great variability has been reported in other tree species with this capacity for long life (Rogers 2000).

CULTIVATION

It is now thirty years since *Betula raddeana* was first introduced from Karachayevsk in 1980. It is an attractive, broadly fastigiate tree with a cluster of white barked trunks and is in the trade as *B. raddeana* 'Hugh McAllister' as material originated from Ness. The Gunib trees are beginning to develop a slightly shaggy bark. Both these accessions are growing strongly with annual apical growth of between 0.5 and 1 m and are drought tolerant and deep rooted, allowing other plants to be grown beneath them in a way difficult under the shallow rooted *B. pendula*. These characteristics make it likely that the species, and perhaps particularly fastigiate forms, could be useful in horticulture, perhaps even as a street tree where the shallow rooted silver birches grow poorly. In this context the species' ability to root from cuttings is critical. As many of the desirable characteristics of *B. raddeana* are of the root system, it might even be of use as an understock to produce other birches suitable for street planting.

Fig. 163. *Betula raddeana*, Russia, N Caucasus, Dagestan, Gunib. Flaking bark showing some similarity with that of its presumed relatives *B. nigra* and *B. dahurica* (PW).

Betula raddeana Trautv., *Trudy Imp. S.-Peterburgsk. Bot. Sada* 10: 129 (1887). Type: 'Described from Dagestan' (holotype TBI; co-type LE).

Betula aischatiae Husseinov, *Novosti Sist. Vyssh. Rast.* 8: 126 (1971). Type: Caucasus, Dagestan, distr. Akuscha, nr Gapschima, 24 June 1968 (holotype LE).

Betula victoris Husseinov, *Novosti Sist. Vyssh. Rast.* 8: 128 (1971). Type: Caucasus, Dagestan, nr Schukta, 24 June 1968 (holotype LE).

Betula maarensis V. N. Vassil. & Husseinov, *Novosti Sist. Vyssh. Rast.* 8: 129 (1971). Type: Caucasus, Dagestan, Akushinsky region, 4 km from village Akusha, Maara estate, 27 Aug. 1968 (holotype LE)

DESCRIPTION. *Tree* of small to medium size, possibly sometimes a shrub. *Bark* white or whitish, peeling. *Twigs* brown to grey, to 2.5–3.5 mm in diam., slightly rough with whitish lenticels about 0.75–1 mm diam.; even in 2nd year usually with some persistent puberulent hairs. *Buds* grey-brown and hoary grey hairy, ovoid, to 6 × 3.5 mm. *Young shoots* thinly or densely covered with silky hairs, especially near nodes, over a dense layer of short puberulent hairs, few clear glands present. *Leaves* with a puberulent petiole; lamina rhombic, base cuneate to truncate (especially on spur shoots), with about 5 pairs of veins. Upper leaf surface glabrescent with thin red colleters on the midrib, which produce gum which sticks to silky hairs; lower (abaxial) surface glabrescent except in vein axils, without glands, lamina not puberulent. *Male catkins* 1–3 terminal with up to 2 lateral, up to 20–25 × 3–4 mm in winter, expanding to 45–70 × 7–8 mm in spring; scales peltate, c. 2.5 × 2 mm, shield-shaped with ciliate margin. *Female catkins* usually solitary. *Fruiting catkins* c. 25 × 12 mm, cylindrical. *Scales* c. 6.5 × 6 mm, eglandular, with abaxial surface densely velvety puberulent hairy, adaxial surface glabrous or puberulent, margins ciliate; central lobe 2.5 mm, triangular; lateral lobe 2 mm, more or less square. *Seed* c. 2.5 × 4.5 mm, hairy; style 0.75–1 mm; wing 1.5 mm, ½ to equal to nutlet in width, translucent, one cell thick, with resinous glands on wing; wing more or less adnate to styles.

DISTRIBUTION. Caucasus: Georgia. Russia (Dagestan, Karachayevsk Autonomous Region) Armenia, Azerbaijan).

HABITAT. Mountain forests; scattered throughout pine, mixed or beech forest in the subalpine belt, and at the treeline.

FLOWERING TIME. Flowers April to May; fruits ripening July.

CHROMOSOME NUMBER. 2n=84.

Subgenus BETULA section COSTATAE

Bark often peeling in large sheets; *leaves* usually with 8 or more veins. *Seed* wings narrower than the nutlet. Sino Himalaya and East Asia.

27. BETULA COSTATA

HISTORY

Betula costata is a beautiful tree with delicate, handsomely ribbed leaves with numerous veins, and a long drawn out apex. It was named by Trautvetter from specimens collected by Carl Maximowicz on his expedition to the Amur river area of eastern Siberia in July and August 1856. Maximowicz set out from St Petersburg in 1852, and sent back reports and specimens, which were compiled into *Primatiae Florae Amurensis*, published in 1859. By 1860, he was back in the Far East, and visited Japan (see *B. maximowicziana*, p. 219).

ECOLOGY AND STATUS IN THE WILD

Mature trees in the wild are very striking in their fine haze of winter twigs, which, in their slenderness, distinguish this species from *Betula ermanii*. Old trees have magnificent rounded crowns. The smooth bark is described as grey-brown (Rehder 1940; Nakai 1915), but HMcA was shown a grove in the Ussuri forests with beautiful creamy-white bark (Fig. 166). KBA saw magnificent fine-twigged mature trees in early autumn in mixed forest in the Odae-San National Park in central South Korea (Fig. 165), and also plenty of younger, comparatively smooth barked trees in the same area. The bark of the older trees was plated and strongly reminiscent of mature trees of *B. ermanii* in Hokkaido.

TAXONOMY, RECOGNITION AND RELATIONSHIPS

Betula costata is a taxonomically isolated species, bearing some resemblance to *B. ermanii* in vegetative characteristics, but is much more delicate in appearance with thinner twigs, a much shorter petiole relative to the leaf blade, a more elongated leaf with usually more numerous veins, and a single order of toothing with much less sharply toothed leaves. It is usually classified with *B. ermanii* and species of the *B. utilis* aggregate but is aberrant in this group in its rigidly erect small fruiting catkins and short petioles.

Betula costata is here treated as the type species of section *Costatae*, which in this book is taken to include only the *B. utilis* group and *B. ermanii.* [In most previous classifications (e.g. Rehder 1940) all the species here placed in subgenus *Asperae* were placed in section *Costatae.*] However, *B. costata* is similar to species of subgenus *Asperae* in its erect catkins and relatively short petioles, and of course, the large number of leaf veins, but it differs from all of these in its thin winged nutlets and white bark; from section *Lentae* it differs in the lack of oil of wintergreen in its twigs; the transparent wings of the nutlets differentiate it from species of subgenus *Asperae.*

Being a diploid and with its mixture of characters of different species groups, *Betula costata* is a key species in any study of relationships within the genus. There is some, possibly superficial, similarity with *B. schmidtii,* another diploid species with no obvious very close relatives. It is interesting that it is the only taxonomically isolated, and therefore presumably relict, species in the north-east Asian refugium (Ran *et al.* 2006) which does not occur in Japan also.

CULTIVATION

Unfortunately, like many probable Tertiary relics, *Betula costata* appears to be drought sensitive, and well-established 4 m trees have died of drought at Ness. In Devon, at Stone Lane Gardens, KBA found it impossible to cultivate young trees without annual die back, and observed at Westonbirt that a group of apparently thriving trees at least 2 m tall a few years ago, are now becoming threadbare

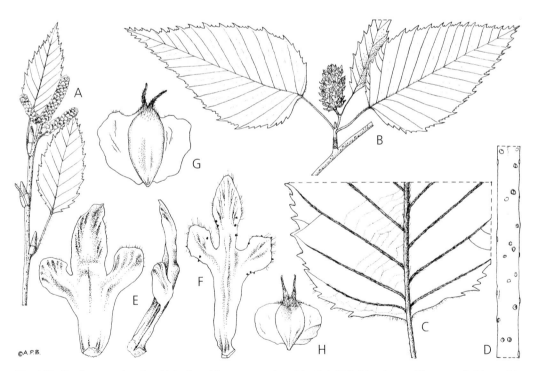

Fig. 164. *Betula costata.* **A** twig with buds and immature male catkins × 1; **B** fruiting shoot with mature fruiting catkin × ¾; **C** underside of leaf base × 2; **D** bark of twig × 2; **E** female catkin scale, back and side views × 7; **F** female catkin scale back view (hairier, longer and flatter than **E**) × 7; **G** nutlet (larger than **H** and with hairy styles) × 7; **H** seed × 7. **A** from *T. N. Liou* 1874 (Hopei, Wulingshan, 17 Sept. 1930); **B, F, H** from *Wilson* 8629 (French Mine, Taiyudo Province, N Heian, Korea, 18 June 1917); **C** from *Wilson* 10483 (Kongo-San, Kogan Province, July 1918); **D, E, G** from *Kirkham et al.* KFBX 93 (Chiri-San, S Korea, 7 Oct. 1989). DRAWN BY ANDREW BROWN.

Fig. 165. Large old tree of *Betula costata*, Odae-San, South Korea (KBA).

Fig. 166. *Betula costata* in Ussuri forest nature reserve north of Vladivostock (HMcA).

with die-back, probably due to late spring frosts or winter dieback due to inadequate cuticle thickness as a result of what is often called insufficient 'ripening' of the new twigs, caused by too short or too cold a growing season (OECD 2003; Barclay & Crawford 1982).

It is worth mentioning here that many older trees in arboreta named *Betula costata* are really *B. ermannii* 'Grayswood' (q.v.).

Betula costata Trautv. in Maxim., *Prim. Fl. Amur.*: 253 (1859). Type: Amur, 1859, *C. Maximowicz* (holotype LE).

DESCRIPTION. *Tree* to 30 m. *Bark* grey-brown, peeling in moderately thick flakes on young trees, becoming plated and rough on old trees. *Twigs* with very small glands; lenticels pale buff, to 1 mm. *Buds* small, acute and slender, with purplish or maroon scales. *Leaves* oblong-ovate to broadly lanceolate, acuminate, 50–100 mm, with 10–16 pairs of veins which are impressed above, finely double-serrate; petioles 8–15 mm, short in proportion to the leaf blade with a petiole:blade ratio of 1:6 to 1:10. *Fruiting catkins* globose to oblong, c. 20 mm, short stalked to sub-sessile; bracts 6–9 mm. *Seeds* longer than broad but almost circular, c. 2 mm, wing up to half nutlet width.

DISTRIBUTION. Eastern Siberia: Amur and Ussuri valleys, Korea.

HABITAT. Temperate rainforest.

28. BETULA ASHBURNERI

HISTORY

In south-east Tibet in October 1997, Hugh McAllister and Keith Rushforth noticed that there seemed to be a significant difference between the lower altitude, single trunked, large leaved, typical *Betula utilis*, and a shrubby birch in the subalpine scrub around and above the tree-line. This shrubby birch, although it had few leaves left at that time of year, looked different in many ways. It formed multi-stemmed bushes to about 4 m (Fig. 50), occasionally to 8–10 m, with dark-grey and peeling bark, had smaller leaves than *Betula utilis* with fewer veins, and small, erect fruiting catkins (Figs 27, 167). From observations in the field HMcA still thought that it might simply be an odd form of *B. utilis*, but KR considered that it was probably a distinct species.

Seedlings raised from these shrubby birches from several localities in the area were diploid, 2n=26, whereas all reported counts for *Betula utilis*, including some from this area, are tetraploid, 2n=56. This indicated that there are two genetically distinct taxa present, and the shrubby taxon is here treated as a new species, *Betula ashburneri*, in memory of Kenneth Ashburner who died in July 2010.

Betula ashburneri can be distinguished from *B. utilis* by its multi-stemmed shrubby habit, more elongated buds, persistent stipules, small glossy leaves with fewer, 6–10, pairs of veins (Fig. 169), and more or less erect fruiting catkins with the central lobes of the scales patent or reflexed (Figs 27, 167), and small nutlets less than 2 × 1 mm (minus wing).

Betula utilis and *B. ashburneri* are also easily distinguishable as seedlings. In seedlings of *B. utilis* successive leaves increase in size, the seedlings soon producing relatively large leaves with at least 8 pairs of veins, and the seedlings grow relatively slowly. In contrast, seedlings of *B. ashburneri* grew rapidly, producing numerous neat, small, glossy leaves of a very uniform size regularly arranged along the shoots (Plate 15; Fig. 169). These were 42 × 28 mm to 50 × 35 mm with 6–8 pairs of veins.

We examined herbarium specimens of all holdings of *Betula utilis* in the Royal Botanic Gardens, Edinburgh (E); the Natural History Museum, London (BM), Kew (K), and Harvard University Herbaria (HUH), finding sixteen earlier collections referable to *B. ashburneri* (listed below).

PLATE 15. *Betula ashburneri*
Painted by Josephine Hague

In 2009, several mature small trees of *Betula ashburneri* were found in Benmore, an annex of the Royal Botanic Gardens Edinburgh, grown from seed collected in Bhutan in 1984 by Ian Sinclair and David Long (*S & L* 5297). Several trees of this collection were clearly the new species, but one more closely resembling *B. utilis* is likely to be a hybrid. This is, therefore, the first record of a hybrid of *B. ashburneri* and the first recorded introduction of the species to cultivation.

It may seem surprising that earlier observations on herbarium specimens failed to detect this taxon, but it has to be said that they would have been easy to pass over as a somewhat 'odd' form of *Betula utilis* — as HMcA had done on pre-1997 visits to herbaria. However, having worked out the differential characters from the clearly distinct living trees, it became easy to spot herbarium specimens of the new taxon (McAllister 1999). All have the persistent stipules, very warty twigs (Fig. 170) and small nutlets so characteristic of *B. ashburneri*.

HABITAT AND ECOLOGY

In the Yarlung Tsangpo drainage of south-east Tibet, *Betula ashburneri* has been found as far west as the Gyamda chu at 29.52°24.6'N 92.31°43.8'E at c. 4260 m where there are coppiced trees to 6 m with attractive mahogany peeling bark; on the Potrang or Gyetsa La at 29.02°59.3'N 92.23°18.7'E, at 4180 m where it grows in the *Juniperus*-dominated zone.

At Pasum Tso it occurs from the shores of the lake at 30.00°19.3'N 93.57°31.3'E, at 3560 m (where it grows intermingled with *Betula utilis*), east to the west flank of the Nambu La, 29.55°09.5'N 94.13°53.4'E, 4130 m. At Pe it occurs on the valley leading to the Doshong La at 3400 m. Between Kyikar and the Nam La it is found at c. 3550 m, at 29.35°09.0'N 94.57°34.0'E. In the Rong chu valley it occurs on the south side on the approach to the peak named Makendro at 29.49°48.7'N 94.45°18.6'E, 3350 m) and on the route to Temo La between 3700 and 3900 m on a south aspect where there are old trees to 10 m with black bark.

Fig. 167. Fruiting catkins on tree of *Betula ashburneri* in the wild in Arunachal Pradesh showing thin brown reflexed tips of catkin scales (PB).

In Pome on the Showa La at c. 29.52°09.7'N 95.22°34.2'E, 3800 m on the north side of valley, it forms a shrub 5–6 m in snow runnels and in open meadow where snow lingers. Here the bark is black, scarcely peeling, the leaves 4–5 cm, ovate, finely doubly toothed or simply toothed, 7–10 pairs of veins, the catkins 1.5–2 cm, and it grows with *Betula utilis* in the woodland zone on the sides of the valley where snow does not linger.

RELATIONSHIPS

Although this species is clearly most similar to *Betula utilis* because of the shape and somewhat leathery texture of the leaf and the exfoliating bark and the usually knobbly male catkins, its relationships are rather obscure, with similarities to species of several other groups.

In subgenus *Betula* persistent stipules are otherwise only found in the dwarf birches of section *Apterocaryon*, and the small nutlets might also suggest a relationship with this group. The fruiting catkins are similar to those of the *Acuminatae* in the persistence of the scales on the fruiting catkin

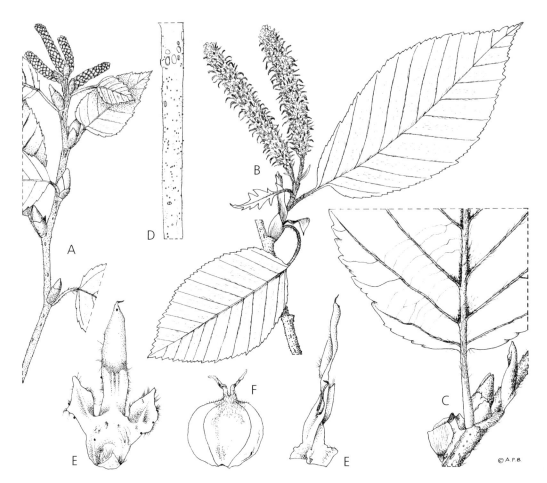

Fig. 168. *Betula ashburneri.* **A** mature twig showing dormant buds and immature male catkins × 1; **B** mature twig showing leaves and mature fruiting catkins × 1; **C** underside of leaf base × 2; **D** bark of twig × 2; **E** female catkin scale, front and side views × 10; **F** seed × 10. DRAWN BY ANDREW BROWN from isotype at K.

Fig. 169. Leafy shoot and buds of *Betula ashburneri* in the wild in Arunachal Pradesh (PB).

Fig. 170. Twig of *Betula ashburneri* in the wild in Arunachal Pradesh showing reddish translucent glands on current year's shoot (top right) becoming white on second year shoot (PB).

axes and the scales having very thick bases and thin lobes, although this state is even more extreme in the *Acuminatae*. The flattened ribbon-like hairs found in the vein axils on the leaf underside have not been detected in any other species.

Betula ashburneri is here classified in section *Costatae* on the basis of its overall appearance, but it is very distinct from any other diploid in the section, and is most similar to the tetraploid *Betula utilis*.

CULTIVATION

From the appearance of the shrubs in the wild we would not have predicted that this species would be of any great horticultural merit. However, young plants have an attractive upright habit and dark green, neat, glossy foliage (Fig. 172). The largest plants in cultivation have now reached 6 m and still have a main trunk, although some have strong basal branches. The bark on these young specimens is a shiny greenish-brown and is peeling (Fig. 173).

In the wild it might be expected to be able to hybridise with *Betula pendula* subsp. *szechuanica* should these two diploid species occur in the same habitat.

Betula ashburneri McAll. & Rushforth, *Curtis's Bot. Mag.* 28 (2): 116 (2011). Type: China, Tibet, Kongbo, on the ridge towards Namche Barwa, above Gyala (29.41°23.3'N 94.54°42.7'E), from 4065 m, 23 Oct. 1997, *Rushforth KR* 5161q (from plants in cultivation) (holotype LIV; isotypes BM, E, K).

DESCRIPTION. In the wild, a multi-stemmed *shrub* or shrubby tree from 3 m to possibly 10 m, usually with a number of trunks diverging more or less horizontally from soil level. In cultivation a *small tree* initially with a strong leader. *Buds* 8–12 × 2.5–4 mm, slender pointed, acute; scales and elongate-triangular stipules persistent; stipules c. 8–10 × 2.5 mm; young shoots reddish green, with silky hairs up to about 0.5 mm lying parallel with the twig axis, and an under layer of very fine, short (0.3 mm), erect, puberulent hairs, densely glandular, with glands initially translucent, becoming translucent reddish and then opaque white as the resin dries. *Twigs* rough with few, reddish translucent and numerous white glands persisting on 2nd year and older twigs, some hairs also persisting (Fig. 170). *Leaves* glossy, dark green, slightly cordate at base, with short petiole 5–9 mm, fully formed blade up to 5 times petiole, to (35) 42–50 × 28–35 mm, but often smaller, especially in the wild, with 6–10 pairs of closely inserted lateral veins, 2–5.5 mm apart, giving a heavily veined appearance to the leaf, with conspicuous pleating of lamina between the veins; adaxial surface ± glabrous; abaxial surface with small reddish glands and veins silky hairy with axillary hair tufts containing numerous flattened, ribbon-like hairs to 1 mm long by 0.036–0.050 mm wide. *Male catkins* c. 14 × 3.5 mm, scales square to circular, ciliate on margins. *Fruiting catkins* usually single but sometimes in 2s or 3s with peduncle 5–13 mm, ± erect, to 15(–27) × 8 mm; scales (Fig. 167), with mid-lobes protruding and patent or recurved, (2.5) 3.5–4 (5) × 2.5–4.5 mm, margins ciliate, abaxially hairy, base thick, lobes thin, central lobe oblong, 2–2.5 mm, often acutely pointed and apex very brittle, lateral lobes broadening from base, 1.5–2.0 mm long, with prominent

Fig. 171. Fruiting catkins of *Betula ashburneri* (PW).

Fig. 172 (left). Young tree of *Betula ashburneri* in cultivation at Ness (PW).
Fig. 173 (right). *Betula ashburneri*. Trunk of young (12 year old) tree (PW).

tooth-like projection on adaxial surface between points of insertion of seeds. *Seeds* 1.25–2 × 2 mm with nutlet c. 1 mm broad and wings c. 0.5–1 mm, narrower than nutlet but very variable; nutlet hairy towards apex, styles c. 0.75 mm.

DISTRIBUTION. Bhutan. China: SE Tibet, NW Yunnan, SW Sichuan. Well known from the Yarlung Tsangpo drainage from Medrogonggar (Maizhokunggar) and the Potrang La near Chusom (Qusum) to Showa. Less well known from the Salween /Kiu Kiang divide in north west Yunnan in the east. However, it is likely to occur in the little explored regions between these two centres which are c. 200 miles apart.

HABITAT. On steep mountain slopes, at the upper limit of the forest, sometimes with *Betula utilis* or with *Juniperus* sp., and in meadows by snow patches.

CHROMOSOME NUMBER. Diploid: 2n=26.

OTHER SPECIMENS SEEN. *TIBT* 2, a 1995 Tibetan collection from east of Lhasa in 1995 made by a joint Kew-Howick expedition and represented in cultivation; *Cooper* 4523 (Bhutan); *Kingdon-Ward* 13193, 13198, *LSE (Ludlow, Sherriff & Elliott)* 3845, 4824, 15251 and *LST (Ludlow, Sherriff & Taylor)* 3845, 6372 (Tibet); *Forrest* 8876, 21595 and 22748, *Yü* 19687, 20103 and 20934, *Rock* 22552 — all from what is now north-west Yunnan. The earliest *Forrest*, *Yü* and *Rock* collections date from 1912 and 1938.

Recent (2005) collections from the Gaoligong Shan near the Yunnan/N Burmese border (*GSBS* 26507, 26669, 26862) are also *Betula ashburneri*. A 2010 collection by Martyn Rix from Zheduo in Sichuan south-west of Chengdu is the most north-easterly occurrence so far recorded.

29. BETULA UTILIS

HISTORY AND NOMENCLATURE

The birches which are common throughout the Himalayas, from Afghanistan and Kashmir to western China have usually been referred to under the names *Betula utilis*, *B. jacquemontii* and *B. albosinensis*; they are clearly closely related, and have been recognised in the past as separate species, subspecies or varieties. All are tetraploid and, although the extremes are very different, it is often difficult to distinguish between them. Here they are treated as subspecies of *B. utilis*: subsp. *utilis*, subsp. *occidentalis* Kitam., subsp. *jacquemontii* (Spach) Ashburner & A. D. Schill., and subsp. *albosinensis* (Burkill) Ashburner & McAll.

Betula utilis was originally described by David Don in 1825 from Gosainthan in central Nepal. Don was then librarian to the Linnean Society and professor of Botany at King's College, London, and his *Prodromus Florae Nepalensis* described the collections of Nathanial Wallich and others, which had been sent back to London. Tony Schilling, who has studied these birches in the wild in the Himalaya, says that the type locality was actually the Gosainkund mountains which form the southern side of the Langtang valley, and not Gossainthan, which is a high peak on the Tibet-Nepal border (Schilling 1984).

In 1830 John Lindley described *Betula bhojpattra* in Wallich's *Plantae Asiaticae Rariores*. Bhojpatra is the Hindi word for *Betula utilis*, and parts of the plant are still used in Hindu ritual, as well as for medicine, and as a substitute for paper, especially for the preparation of yantras, for various purposes. As the website www.astrojyotishi.com/bhoj_patra.htm explains: 'A general looking bhojpatra Yantra is actually a powerful mean, which powersign it's diety having abnormal effects. The people of India are directed by rishis and munis to pray their Lords and dieties in the form of different yantras to get protection from any kind of troubles like diseases, loans, business, service, job, marriage, friendship, children and black magic etc. It is said in the Indian (Hindu) shastras that the yantras written and energized (praan pratishtha) on Bhoj patra are the best yantra to overcome these troubles and problems'. [sic]

In Europe and North America, *Betula utilis* is grown mainly for the beauty of its bark, and there is a general trend in the bark colour of wild *Betula utilis*, from pure white in the western Himalaya, to brown and orange-red in the east and in western China.

Fig. 174. *Betula utilis* subsp. *utilis* central Bhutan. Coppery-brown young trunk with conspicuous lenticels (MF).

Fig. 175. *Betula utilis* subsp. *utilis* with young female catkins and leaves with slightly recurved margins (PW).

ECOLOGY AND STATUS IN THE WILD

Birches of the *Betula utilis* aggregate are more or less continuously distributed from Afghanistan and northern Pakistan in the west, along the Himalaya and through the western Chinese Heng Duan mountains in north-western Yunnan and south-eastern Yunnan (as *B. jinpingensis*) in the south, to Sichuan, Hubei, and east to Hebei province (which surrounds Beijing) in northern China (see Map 10). Like the silver birches of the *B. pendula* aggregate, they all have the same chromosome number, in this case tetraploid with 2n=56.

The *Betula utilis* aggregate only comes into contact with related tetraploid species at the extreme western and north-eastern ends of its distribution; the only other tetraploid in the Himalaya and southwest China is the very distantly related Himalayan *B. cylindrostachya*. To the west of the Himalaya, in the central Asian mountains *B. utilis* meets the tetraploid *B. tianshanica*. In the northeast there may be a discontinuity between subsp. *albosinensis* in Hebei province around Beijing, and *B. ermanii* at the south-western end of its distribution in the adjacent province of Liaoning (Li & Skvortsov 1999). As we have seen very few specimens from this area, we do not know whether this break is real or whether there is some introgression. The *B. utilis* aggregate therefore shows a classic Sino-Himalayan distribution (although extending rather further north and east than is usual) and is absent from the northerly refugia of Japan and the adjacent Amur-Ussuri region on the Asiatic mainland (Ran *et al.* 2006).

Fig. 176 (left). *Betula utilis* subsp. *utilis*, with inverted V-shaped mark on trunk centred on old branch scar (PW).

Fig. 177 (right). *Betula utilis* subsp. *utilis* with much thicker lenticels than in subsp. *albosinensis* (PW).

TAXONOMY, RECOGNITION AND RELATIONSHIPS

The most useful character to identify any mature trees of the *Betula utilis* aggregate, at least in the fresh state, is the presence of the conspicuously 'knobbly' male catkins which have a prominent raised boss on each scale (Plate 16; Fig. 62). These are visible from late July, when the new catkins begin to mature, until flowering time the following spring. *B. ermanii* is closely related and, apart from the lack of bosses on the male catkins, is distinguishable by the deltoid-ovate leaf-shape, the very sharp teeth with the teeth ending the main veins more than twice as long as the neighbouring teeth, and the usually erect fruiting catkins.

Although not closely related, *Betula papyrifera* is perhaps the species most likely to be confused with *B. utilis*. Again, the knobbly male catkins are the most useful distinguishing character, and the seed wings are narrower than the nutlet in *B. utilis* but broader in *B. papyrifera*. Also *B. papyrifera* never has leaves as glossy or leathery as in some *B. utilis*.

CULTIVATION

From the decorative point of view the different barks of *Betula utilis* are among the finest in the whole genus. Bark colour ranges from the purest white in subsp. *jacquemontii* from the west Himalaya, through copper-coloured to chocolate with or without a white bloom in subsp. *utilis*, to pure copper in some subsp. *albosinensis* from northern China. Those with a white bloom of betulin over a darker colour are particularly beautiful, and the lenticel pattern can add to the effect. Trees of *B. utilis* subsp. *jacquemontii* are mostly white barked with pink protruding lenticels, with a few pink or brown barked individuals here and there, and subsp. *utilis* is mostly rosy-pink or brown barked, in which case the lenticels stand out as white or buff streaks.

The bark always peels in spectacular fashion, in scrolls and sheets. The male catkins are large and showy, often expanding much later than those of other species. The larger leaves, particularly of subsp. *jacquemontii*, give a heavier and denser appearance to the canopy than in *Betula pendula*. All variants of *B. utilis* are slower growing than *B. pendula* or *B. papyrifera* but eventually make large trees. Trees from the western end of the distribution, subsp. *occidentalis* and subsp. *jacquemontii*, often suffer from late spring frosts and severe dieback, resulting in trees of a very poor shape which may fail to fruit.

The distinctive collections of subsp. *albosinensis* in cultivation (e.g. *Wilson* 4106, *Purdom* 752 var. *septentrionalis*) owe their characteristic appearance to their elongated leaves with up to 14 pairs of veins, and thinner matt leaves which are of a softer green colour than those of any subsp. *utilis*.

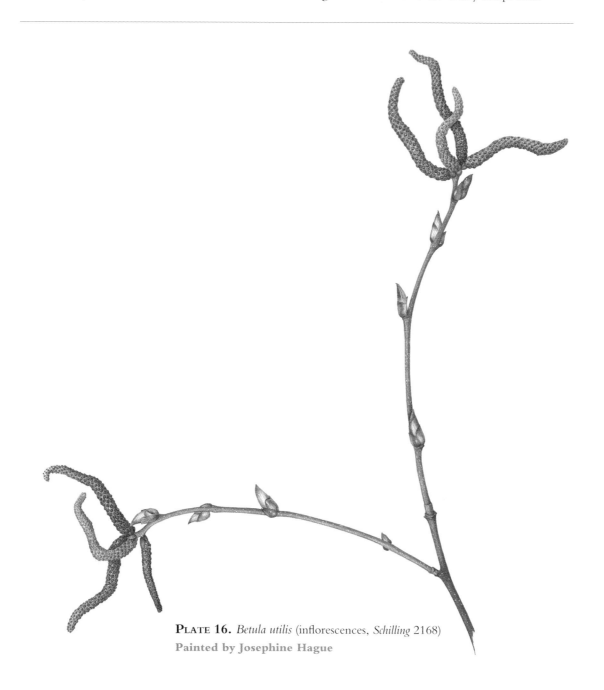

PLATE 16. *Betula utilis* (inflorescences, *Schilling* 2168)
Painted by Josephine Hague

Betula utilis agg.

DESCRIPTION. *Tree* to 35 m. *Bark* varying from pure white through copper with a bloom of white to almost black, peeling in sheets of varying size. *Twigs* variably persistently hairy with both long tangled and short puberulent hairs persisting on at least one-year old twigs, and small resinous glands, mostly opaque white but sometimes some still translucent reddish, densely crowded to scattered or sometimes very few (e.g. in *B. utilis* 'Inverleith'); epidermis grey, splitting to form a reticulate pattern on surface of twig and gradually sloughing off, taking resinous glands and hairs with it, to reveal maroon–brown shiny bark beneath; bark may begin peeling as early as the base of strong first year shoots, common on second and third year shoots, but very dependent on vigour of twig. *Buds*

PLATE 17. *Betula utilis* (bark, *Schilling* 2168)
Painted by Josephine Hague

ovoid, acute, often large, to 14 × 4.5 mm in central Nepalese subsp. *utilis*, but only to 11 × 4 mm in Gansu (subsp. *albosinensis*), more or less straight or curved, but never consistently incurved, or curved outwards as in species of the *Acuminatae*; bud scales brown to deep maroon above, greenish below, becoming wholly maroon to brown with exposure to winter sun; margins always initially ciliate, but ciliate hairs often becoming lost due to weathering. *Shoots* green, to maroon where exposed to sun, with variable amounts of long tangled hairs and short puberulent hairs and few to many small, crowded resinous glands, at first transparent brown to reddish, but usually soon becoming opaque-white, even by the third internode; lenticels inconspicuous and few, linear, 0.5–1 mm, pale brown to white and slightly protruding. *Petiole* 15–35 mm; often very glandular. *Leaves* when very young glossy as if lacquered, much crimped, maroon along the top of exposed pleats where exposed to sun, but green within pleats; ovate and commonly cordate at base, but sometimes rounded or cuneate, acuminate; lateral veins (6–)9–14 pairs, impressed and pleating the leaf in the central area (east Nepal, Yunnan, Sichuan), but leaves almost flat at western (subsp. *occidentalis*) and eastern (subsp. *albosinensis*) ends of the distribution, to c. 120 × 80 mm, profusely spotted beneath with clear glands, ± simply toothed, with long silky pale brown hairs along midrib above, and, beneath, midrib, petiole and laterals very hairy with silky white hairs, with conspicuous hair tufts in vein axils on leaf underside (Fig. 56); reddish glands dense on lower (abaxial) surface of lamina, becoming white resinous spots later in season especially towards margins, and often falling off, so that leaf underside appears glandless, upper (adaxial) surface with no glands but a few colleters on midrib, prostrate silky hairs lie on upper surface between the veins; many small brown glands on petiole; petiole:blade ratio 1:4 to 1:6. *Male catkins* always conspicuously knobbly, to 30 × 5 mm. *Female catkins* erect at flowering. *Fruiting catkins* pendent, often single, but in groups of up to 3 from single bud; peduncle 5–15 mm; catkin 30–50 × 7–12 mm; scale 5–8 mm, lobes similar in shape, mid lobe longest, often c. 3 mm, puberulent with ciliate margins. *Seed* obovate, 2–3 × 1.5–2 mm with wing to about as wide as nutlet (to wider in subsp. *occidentalis*).

DISTRIBUTION. From Afghanistan in the west, along the Himalaya to Yunnan and Sichuan, Gansu and Shaanxi in western China; also from south-east Yunnan and eastern Sichuan northwards to Hubei and Hebei around Beijing.

HABITAT. Forests, stony river beds, screes and rocky places in the mountains.

CHROMOSOME NUMBER. 2n=56.

29A. BETULA UTILIS subsp. UTILIS

Typical *Betula utilis* subsp. *utilis* is found mainly from the central Himalaya through western China to western Sichuan, east of the area occupied by subsp. *jacquemontii* and southwest of the area occupied by subsp. *albosinensis*. The bark of trees of subsp. *utilis* is predominantly, but not invariably, brown or pink, compared with that of subsp. *jacquemontii*. Somewhat white barked individuals or populations occur even as far east as Sichuan, as in *Forrest* 19505, and many individual trees, particularly in central Nepal may be impossible to assign to either subspecies, as there appears to be no sharp dividing line between the two.

To the east of the Kali Gandaki gorge in Nepal (84°35'E), trees of *Betula utilis* tend to become darker barked with more leathery leaves of a firmer texture with more numerous, more deeply indented, veins and more cordate leaf bases (Ashburner & Schilling 1985). However, there appears to be considerable overlap in these features. Both west and east of the Kali Gandaki gorge there are whiter barked trees on the drier northern slopes and darker barked populations in the wetter forest

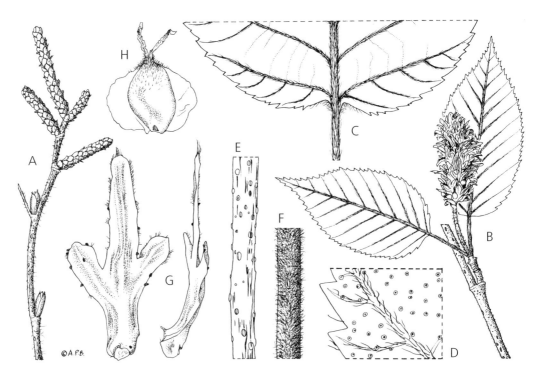

Fig. 178. *Betula utilis* subsp. *utilis*. **A** twig with buds and immature male catkins × 1; **B** fruiting shoot with mature fruiting catkin × 1; **C** underside of leaf base × 2; **D** margin of leaf showing distinct raised resin glands × 20; **E** bark of 3-year old twig × 2; **F** first-year section of twig in **E** showing dense indumentum × 4; **G** female catkin scale, back and side views (note resin deposits) × 7; **H** seed × 7. **A** from *Alpine Garden Society Expedition to Sikkim* (*AGSES*) 689 (Sikkim, below Phedang, 10,500', 25 Sept. 1983); **B–D, G, H** *AGSES* 544 from (Sikkim, below Thangshing, 12,100', 18 Sept. 1983); **E, F** from Type, HI 2792 (Kamourn). DRAWN BY ANDREW BROWN.

on south-facing slopes. Unlike the white barked trees which often form tree-line forest, the darker barked trees occur scattered through the species-rich deciduous and mixed deciduous-coniferous temperate rainforest growing alongside species of *Acer*, *Sorbus*, *Rhododendron*, *Viburnum* and *Abies*, never forming pure birch forest. Further east into south-east Tibet, Yunnan and as far as western Sichuan, this pattern continues with these birches mostly scattered through the forest. Ashburner & Schilling (1985) concluded that the distinction between *B. jacquemontii* and *B. utilis* could not be maintained at species level, and recognised *jacquemontii* as a variety of *B. utilis*.

In south-east Tibet, typical *Betula utilis* occurs rather infrequently in dense mixed forest from about 3000 m to the tree-line at about 3800 m. It grows with *Picea linzhiensis*, *Abies fordei*, and *Acer* spp. with an understorey of bamboos and a range of shrubby genera such as *Viburnum* and *Rhododendron*. At lower altitudes the *B. utilis* trees are usually very tall, up to about 20 m, and sometimes very large with trunks to 80 cm in diameter. Fruiting catkins are only borne at the very tops of the trees and can be collected only from the occasional fallen tree. These trees have very attractive, copper-orange-coloured bark flushed with white and large leaves with 11–14 pairs of veins impressed above. This combination of characters is typical for the species east of the Kali Gandaki; towards the tree-line the trees became smaller but are otherwise similar.

Betula jinpingensis was described by Li (1979) from the Jin Ping Shan in south-east Yunnan adjacent to North Vietnam; it is said to differ from *B. utilis* primarily in its shorter peduncle, wider nutlet wing, wider fruiting catkin, and the absence of glands on the branchlets. The illustration provided in the *Flora of China* suggests, as is proposed by Skvortsov (footnote in Li & Skvortsov 1999), that it should be treated as a synonym of *B. utilis* differing only in minor characters which could apply equally to other variants of subsp. *utilis*.

Betula utilis var. *prattii* was described by Burkill from specimens collected by A. E. Pratt near Tachienlu, today Kanding. The original description covered trees with orange-brown to grey or orange-grey bark, but the name has usually been attached to forms which have very dark, shining bark, often overlaid with numerous white lenticels. These dark barked forms tend to occur as large,

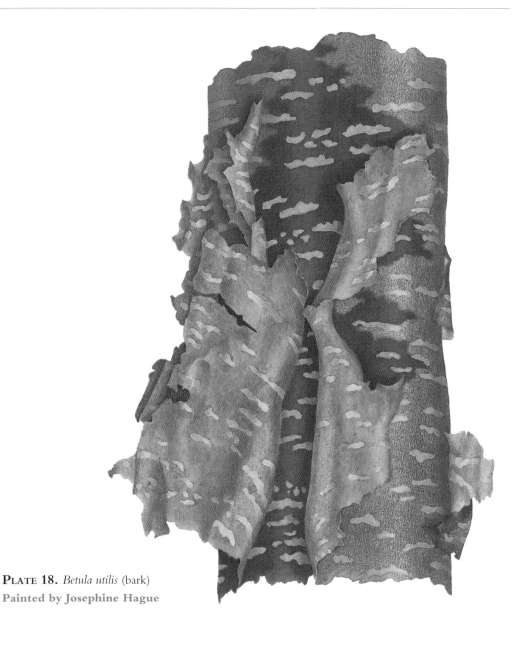

PLATE 18. *Betula utilis* (bark)
Painted by Josephine Hague

single specimens in mixed forest, at lower altitudes than most *B. utilis* in the same area, and are found from central Bhutan (Plate 18; Fig. 174) (Schilling 1993), where there is an illustration of a typical tree, eastwards to western Sichuan. Roy Lancaster (2008: 254) illustrates a tree with similar bark at Huadianba, in the Cangshan near Dali, and similar trees are common in the wet forests of Gongga Shan in western Sichuan (Fig. 179) (M. Rix pers. comm.).

Although *Betula utilis* and *B. pendula* subsp. *szechuanica* grow close to one another in north-western Sichuan, no hybrids have been recorded.

CULTIVATION

Collections of subsp. *utilis* from the wet climates of the eastern Himalayas seem to thrive in the British Isles. They are often slow to break bud, and so do not suffer from late frost. Experience suggests that they are more drought sensitive than *B. pendula* in cultivation, but they do not seem to suffer significantly whether on their own roots or grafted. A wide range of recently introduced material is available from which many selections have been made (see p. 78 in introduction, and the list of cultivars on p. 386).

Fig. 179. *Betula utilis* subsp. *utilis*. trunk of young tree with very dark bark (*Roy Lancaster* no. L 947 Sichuan, Minya Konkka, Liuba) (PW).

Betula utilis D. Don, *Prodr. Fl. Nepal.*: 58 (1825), subsp. **utilis**. Type: Hab. in Gosaingsthan nepalensium, *Wallich*.

B. bhojpattra Lindl. ex Wall., *Pl. Asiat. Rar.* (Wallich). 2: 7 (1830).

B. utilis var. *prattii* Burkill, *J. Linn. Soc., Bot.* 26: 499 (1899). Type: near Tachienlu, 13,500 ft, *Pratt 236* (holotype BM, isotypes K, P).

B. jinpingensis P. C. Li, *Acta Phytotax. Sin.* 17 (1): 89 (1979). Type: Yunnan: Jin-ping, at the margin of forest and in young forest on mountain spur, 2200 m, 4 Aug. 1958, *Ngo Yu-chih no. Chuan 10* (holotype KUN).

DESCRIPTION. *Twigs* more or less hairy with both longer silky and short puberulent hairs persisting on at least one-year old twig, small brown resinous glands, scattered or sometimes densely crowded, mostly broken down to opaque white, lenticels round to round-oval, white or buff, some nearly to 2 mm, but most about 1.5 mm, not protruding. *Buds* acute, about 8–10 × 3–4.5 mm, with brown scales. *Shoots* with variable, but often large, amounts of silky tangled hairs which are particularly thick at internodes, short puberulent hairs present; usually covered with many small, crowded resinous glands, at first transparent brown to reddish, but mostly soon becoming opaque-white, even on the third internode, lenticels difficult to see on shoots and seem few, but are linear, pale brown and protrude slightly, less than 1 mm long. *Petioles* with dense white hairs and many small brown glands; petiole:blade ratio 1:6 to 1:7. *Leaves* around 75 × 40 mm, at first light green with maroon barring, becoming dark green and rather glossy above, paler beneath, ovate, usually cordate, but sometimes rounded or cuneate; sometimes quite leathery and convex with somewhat revolute margins, acuminate with 11–13 pairs of lateral veins which are impressed on upper (adaxial) surface; toothing more or less single, with major teeth never more than 2 mm broad at base with 5 (–8) minor teeth between the major; midrib with long silky pale brown hairs above, and below, midrib and lateral veins with dense white hairs and conspicuous hair tufts in vein axils; lower (abaxial) surface of lamina with dense reddish glands, becoming white resinous spots later in season, especially towards margins, often soon falling off; upper (adaxial) surface with no glands but a few colleters on midrib and appressed hairs on the upper surface between the veins. Male and female catkins as in the species description.

DISTRIBUTION. Nepal, especially east of the Kali Gandaki gorge, eastwards through SE Tibet to Yunnan and west Sichuan where it merges with subsp. *albosinensis*.

HABITAT. Forests, stony river beds, screes and rocky places in the mountains, 2700–4400 m.

29B. BETULA UTILIS subsp. JACQUEMONTII

Betula jacquemontii was described from specimens collected by Victor Jacquemont in Kashmir. Before Jacquemont left France to escape from a sad love affair, he was asked to collect for the Jardin des Plantes, in any country of his choice. He chose India and in 1828, visited Royle in Saharanpur, before going to Simla and then up the Sutlej valley into Tibet. He spent 1831 in Kashmir, at the invitation of the Maharajah, where he collected many new species, but died of cholera in Bombay in 1832 before he could get home; his collections were returned to Paris. His birch is labelled 'Emodo', from the classical name for the Himalayas.

Fig. 180. *Betula utilis* subsp. *jacquemontii*, Sonamarg, Kashmir, India (KBA).

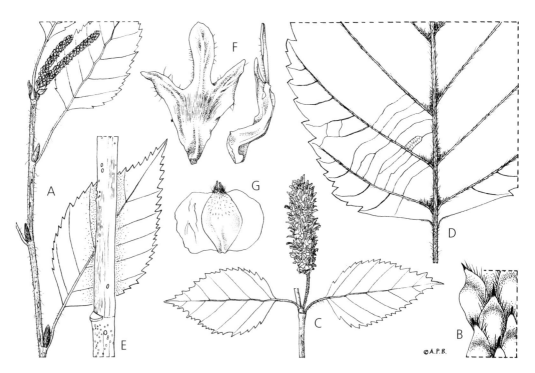

Fig. 181. *Betula utilis* subsp. *jacquemontii*. **A** twig with buds and immature male catkins × 1; **B** detail of immature male catkin scales showing distinct boss on each scale × 10; **C** fruiting shoot with mature fruiting catkin × 1; **D** underside of leaf base × 2; **E** bark of twig × 2; **F** female catkin scale, back and side views × 8; **G** seed × 8. **A, B** from *Rodin* 5640 (Swat, Pakistan, Mts W of Kalan near Bahrein 9000', 20 Aug. 1952); **C, F, G** from *N. L. Bor* 14860 (Pricken, Lahul [=Kangra], 18 June 1961); **D, E** from *Ellis* 327 (Burj near Jindi, Paugi Valley, NW Himalaya 9000', 16 Aug. 1879). DRAWN BY ANDREW BROWN.

This subspecies extends from Kashmir through west Nepal to the area of the deep gorge of the Kali Gandaki rivers at about latitude 83°28'E. Ashburner & Schilling (1985) looked for a sharp geographical division between subsp. *jacquemontii* and subsp. *utilis* where they meet in Nepal, and suggested the great cleft of the Kali Gandaki River in central Nepal might provide this, but they concluded that this huge valley represented no cut off point, but merely an acceleration in rate of change.

The young leaves of *Betula utilis* subsp. *jacquemontii* shine brilliantly, as if lacquered, and have few, widely-spaced lateral veins which scarcely pleat them. There is a limp and thin texture to the leaves, and a coarseness of the teeth (Fig. 186). This is in contrast to the less lustrous, corrugated and more leathery leaves of most trees of subsp. *utilis*. Leaves of subsp. *utilis* are more frequently cordate and generally with revolute, much more finely-toothed margins (Fig. 175). On twigs the transparent brown resinous glands of subsp. *jacquemontii* are larger and protrude more than those on subsp. *utilis*, the lenticels also protrude more, so there is a rougher feel to them. The buds are generally more cylindrical than those of subsp. *utilis*, longer and more acute.

Older trees develop thick plates of bark and form gnarled trees (Fig. 180) very much reminiscent of *Betula ermanii* on the mountains of Hokkaido or north China. At lower levels they form an open forest of scattered trees; in the valleys they are likely to be raided for fodder and firewood, or stripped of bark for roofing. Most appear white barked, but there are brown or pink barked individuals here and there, particularly at lower altitudes. There is a well-known photograph of old trees near Sonamarg in Kashmir, in Bean (1970), the trunks scored with black fissures (not quite in the same manner as *B. pendula*), and old trunks blotched with black patches. Nearby are thickets of *Rhododendron campanulatum*, and this seems to accompany subsp. *jacquemontii* throughout its range.

CULTIVATION

Betula utilis subsp. *jacquemontii* has long been a favourite in the garden and arboretum (Figs 1, 10, 182, 184, 185). There is no bark of any tree quite like it. It is not quite as pure a white, perhaps, as that of *B. pendula*, but warmer with a touch of cream. The texture is remarkable. To some the uniform white is perhaps too stark, but the effect is astonishing on a dull and drizzly winter's day, for example. The bark with its salmon-pink lenticels, rather like that of *B. ermanii* in this respect, becomes tighter on older trunks.

Many seedlings of wild origin from Uttar Pradesh, Himachal Pradesh (Fig. 182) and Kashmir (Figs 184, 185) have been introduced in recent years; those from the east of the range usually perform better in cultivation. Several named clones are now well known, and are described in the list of cultivars on p. 371.

Fig. 182 (left). *Betula utilis* subsp. *jacquemontii* (*Howick & McNamara* 1811). India, Himachal Pradesh. Ness (PW).

Fig. 183 (right). *Betula utilis* subsp. *jacquemontii*, Sonamarg, Kashmir, India (KBA).

Fig. 184 (top left). *Betula utilis* subsp. *jacquemontii*. Close up of base of naturally multi-stemmed tree grown from unpruned seedling (PW).

Fig. 185 (top right). *Betula utilis* subsp. *jacquemontii*. Whole tree of unpruned seedling growing naturally multi-stemmed (PW).

Fig. 186 (above). *Betula utilis* subsp. *jacquemontii* (*Howick & McNamara* 1846). India, Himachal Pradesh. Large (immature) fruiting catkins and large, relatively flat leaves (in comparison with those of subsp. *utilis*) (PW).

Fig. 187 (right). *Betula utilis* subsp. *jacquemontii* 'Inverleith' (PW).

Betula utilis subsp. **jacquemontii** (Spach) Ashburner & McAll., **stat. nov.** (1985). Basionym: *Betula jacquemontii* Spach, *Ann. Sci. Nat., Bot.* sér. 2, 15: 189 (1841). Type: Emodo, *V. Jacquemont* (1840) (P).

B. utilis var. *jacquemontii* (Spach) A. Henry in Elwes & Henry, *Trees of Great Britain and Ireland* 4: 981, t. 270 fig. 15 (1909).

DESCRIPTION. Subsp. *jacquemontii* may be distinguished from typical forms of the other subspecies by the following characters: *Bud* scales deep maroon but appearing lilac because of surface resin. *Petioles* 1.5–2 mm in diam. at base. *Leaves* with lamina glabrescent except for a few silky hairs on the margins and tufts of silky hairs in axils beneath, extending both along the midrib and the laterals; lateral veins 8–10(12), only pleating leaves when very young, otherwise scarcely impressed; less leathery than in subsp. *utilis* and veins more distant, the gap between major secondary veins sometimes as wide as 15 mm; tertiary veins conspicuous below but not raised; margins double-toothed with coarse teeth; on really vigorous shoots leaves may reach 120 × 80 mm.

DISTRIBUTION. India: Kashmir, Himachal Pradesh and Uttar Pradesh east to around the Kali Gandaki gorge in Nepal.

HABITAT. Forests, stony river beds, screes and rocky places in the mountains.

29C. BETULA UTILIS subsp. ALBOSINENSIS

From western Sichuan, northwards into Gansu and Shaanxi and eastwards into Hubei, and in provinces to the east and north, the predominant bark colour in trees of *Betula utilis* agg. is orange-copper, often with a bloom of white; the leaves are often (but not always) longer, thinner, more delicate looking, and with a matt rather than a glossy upper surface, and the young shoots soon become glabrous. This is *Betula utilis* subsp. *albosinensis*, which occasionally forms pure birch forests, but is usually seen as scattered individuals in mixed forest (Wilson in Sargent 1916).

Betula albosinensis was described by Burkill, citing specimens collected by Père Farges in eastern Sichuan, near Tchenkeoutin. In the same year and from the same collection, it was described as var. *sinensis* of *B. bhojpattra* by Franchet. It is interesting that Tchenkeoutin (today Chengkou), on the western side of the Daba shan, is also the type locality for *B. insignis* and *B. fargesii*.

In the *Flora of China*, Li & Skvortsov (1999) accepted *Betula albosinensis* as a separate species, based on its orange bark colour and the absence of hairs in the vein axils on the leaf underside. However, the bark character is very inconsistent, many central Nepalese populations of *B. utilis* having quite coppery bark with a bloom of white, not dissimilar to populations of *B. albosinensis* in northern China (Shaanxi). Similarly, the axillary hair character is not reliable. Study of a large number of specimens has shown that hair tufts are particularly well-developed and persistent on leaves on long shoots, although they may be inconspicuous or absent on spur shoot leaves; a Nepalese specimen clearly referable to *B. utilis* (*Beer* 625 LIV!) has no axillary hair tufts. Thus neither the bark colour nor the axillary hair character show clear geographical correlation and they cannot be used to distinguish *B. albosinensis* from *B. utilis*.

The extensive collections of trees of wild origin at the Royal Botanic Gardens at Kew and Edinburgh and especially at Wakehurst and Benmore, at Ness Gardens, Cheshire and at Stone Lane Gardens, Devon include many intermediates which cannot be assigned to either *Betula utilis* or *B. albosinensis*, so we believe it is most appropriate to treat these north-eastern variants as subspecies *albosinensis* of *B. utilis*.

Examination of recent living collections has shown that there are many populations in north-west Yunnan and Sichuan which cannot be placed in either subsp. *utilis* or subsp. *albosinensis* because they either have intermediate characters or various combinations of the different characters. *Howick &*

McNamara 1480 from Mt Luoji, 30 km S of Xichang in central Sichuan has typical subsp. *albosinensis* bark, but its leaves are almost exactly intermediate in shape between those of subsp. *utilis* and subsp. *albosinensis*. Other specimens which could not be easily assigned to either subsp. *utilis* or subsp. *albosinensis* include *KR* 1364, *SICH* 2296, *BLRM* 219, and yet others have a mixture of characters, e.g. subsp. *utilis* shaped leaf but matt: *W* 990, *KA* 1994-1938 (small leaves with only 6 or 7 pairs veins). *SBEC* 816 from the Cangshan in western Yunnan has elongated leaves similar to those of trees from much further north and east. In the Lijiang area in north west Yunnan there are both trees with long narrow matt leaves of the subsp. *albosinensis* type (*SBLE* 396, *KEG* 1644 E!) and many others with the appearance of typical subsp. *utilis*. Specimens from Hubei (*SABE* 31, 897) have spur shoots with relatively short (52 × 35 mm) leaves with few (8) pairs of veins (very similar to those of subsp. *occidentalis*) and some axillary hairs, but long shoot leaves with up to 13 pairs of veins and axils heavily bearded.

Betula albosinensis var. *septentrionalis* was described by C. K. Schneider from specimens collected by E. H. Wilson near Tachienlu (Kanding) in western Sichuan. It was said to differ from the type by its glandular twigs, leaves silky-hairy beneath, especially on the midrib, veins and vein axils, and deeply cleft bracts (catkin scales). A clone of var. *septentrionalis* has been widely cultivated, probably derived from a seedling of *Purdom* 752 from western Gansu. In the account of *Betula* in *Flora of China*, Li put *Betula albosinensis* var. *septentrionalis* as a synonym of *B. utilis*, while keeping *B. albosinensis* as a distinct species.

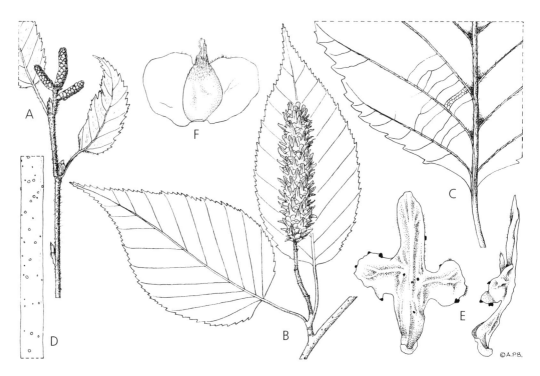

Fig. 188. *Betula utilis* subsp. *albosinensis*. **A** twig with buds and immature male catkins × 1; **B** fruiting shoot with mature fruiting catkin × 1; **C** underside of leaf base × 2; **D** bark of twig × 2; **E** female catkin scale, back and side views (resin deposits present) × 8; **F** seed × 8. **A** from *Licent* 4742 (Louo Kia Wa Sseu, Kansou, N China, 27 Aug. 1918); **B, E, F** from *E. H. Wilson* 4438 (Hupeh, June 1910); **C, D** from *E. H. Wilson* 1890 (Fang, Central China, 5 July). DRAWN BY ANDREW BROWN.

Fig. 189. *Betula utilis* subsp. *albosinensis* in birch Arboretum, Stone Lane Gardens, Devon, England (PW).

The commonly cultivated clones and a recent introduction from near Lanchow in Gansu province illustrate the extreme form of this state and have male catkins which are thinner and less knobbly than in most *B. utilis*. However, herbarium specimens of trees from Gansu and living trees from the Qinling Mountains in Shaanxi province have strikingly thick and knobbly male catkins up to 5 mm wide in the winter dormant state, much wider than the comparable catkins of *B. utilis* from any other area. In their general appearance these are unquestionably subsp. *albosinensis* with the characteristic elongated, matt leaves, but they are rather different in overall appearance from the usual concept of *albosinensis* in cultivation.

CULTIVATION

Cultivated trees of this subspecies are usually grown under the names *Betula albosinensis* and *B. albosinensis* var. *septentrionalis*, names given when very few collections were available for study in the west. The distinctive collections of subsp. *albosinensis* in cultivation (e.g. *Wilson* 4106) and var. *septentrionalis* (*Purdom* 752) owe their characteristic appearance to their elongated and thinner matt leaves which are of a softer green colour that those of any subsp. *utilis*, and have up to 14 pairs of veins,

The bark of the commonly cultivated clones of subsp. *albosinensis* is a most attractive coppery colour overlain with a bloom of white. The lenticel pattern is always very bold and striking, as white gashes and streaks on a shining dark background, and often coalescing, with quite extraordinary effect (Fig. 192). The origin of the clone distributed by Hillier's is uncertain, though it may be from *Wilson* 4106 from north-western Sichuan (near Wenchuan in the Min valley) or *Purdom* 752 from western Gansu, both of which have beautiful bark. The trees in Edinburgh of *Yu* 14547, *Rock* 13648 and *Wilson*

Fig. 190. *Betula utilis* subsp. *albosinensis*. Young stem with outer bark peeling off in large sheet (PW).

900 (var. *septentrionalis*) are rather different and most unattractive with dirty-looking, dark greyish-black bark. It is surprising that the good clones have not become more popular and more widely planted as has happened with some white barked forms of subsp. *jacquemontii*.

Betula utilis subsp. **albosinensis** (Burkill) Ashburner & McAll., **comb. et stat. nov.**
Basionym: *Betula albosinensis* Burkill in F. B. Forbes & W. B. Hemsley, *J. Linn. Soc., Bot.* 26: 497 (1899). Type: Sichuan: Tchenkeoutin, *Farges* (holotype P).
Betula bhojpattra var. *sinensis* Franch., *J. Bot. (Morot)* 13: 207 (1899). Type: in vicinitate Tchenkeoutin, *Farges* (holotype P).
Betula utilis var. *sinensis* (Franch.) H. J. P. Winkl. in Engler, *Pflanzenr.* 4: 61, 62 (1904).
Betula albosinensis var. *septentrionalis* C. K. Schneid. ex Wilson, *Plantae Wilsonianae* 2: 458 (1916).
Type: Western Sichuan: northeast of Tachienlu, Ta-p'ao-shan, alt. 3000–3600 m, July 4th 1908, *Wilson* 900 (holotype BM, isotypes K, P.)

Fig. 191 (left). *Betula utilis* subsp. *albosinensis*. Fruiting catkins and foliage showing elongated leaves of 'typical', commonly cultivated forms of this subspecies (PW).

Fig. 192 (right). *Betula utilis* subsp. *albosinensis* (Lanzhou, Gansu, China). Lenticels thin in comparison to those of other subspecies (PW).

Fig. 193. *Betula utilis* subsp. *albosinensis* with male and female catkins and remains of previous year's fruiting catkins (PW).

Fig. 194 (left). *Betula utilis* subsp. *albosinensis* (Lanzhou, Gansu, China). Variation in bark colour due to differences in time since peeling off of outer layers — darkest most recently peeled, lightest longest exposed (PW).

Fig. 195 (right). *Betula utilis* subsp. *albosinensis*, (Lanzhou, Gansu, China). Peeling bark sunlit from behind (PW).

DESCRIPTION. *Trees* to 35 m. *Shoots* soon glabrescent with a few colleters or glands near the base of the lowermost stipules of the year's growth. *Twig* c. 2.5 mm in diam., brown, glabrous; lenticels more or less circular, less that 0.5 mm. *Bud* c. 8 × 3.5 mm, chestnut brown. *Leaf* with 9–11(–14) pairs of veins, glabrescent except for veins on both surfaces, with distinctive orange colleters on veins of upper surface. *Male catkins* up to 2 terminal plus 1 lateral, only slightly knobbly (specimens from Lanchow; *Purdom* 752) to as knobbly as any *B. utilis*, < 30 × 3.25(–5) mm; scales peltate, c. 1.75 × 1.5 mm, ciliate, brown.

DISTRIBUTION. China: N Sichuan, Hubei, S Gansu, S Ningxia (Liupan Shan), S Shaanxi (Hua Shan, Taibai Shan), Shanxi, Henan, to Hebei.

HABITAT. Openings in forest, screes and rocky places in the mountains, 1000–4400 m.

29D. BETULA UTILIS subsp. OCCIDENTALIS

HISTORY AND NOMENCLATURE

Ashburner & Schilling (1985) quote the type specimen as an *S. Kitamura* collection (no number given) from Afghanistan, Nuristan, between Pushuki and Kushitaki; we have been unable to trace an image of this specimen. However, the isotype of *Betula utilis* subsp. *occidentalis* var. *parva* Kitam. is available at (http://taxa.soken.ac.jp/MakinoDB/makino/ prep_e/MAK092648.html) with the annotation "*S. Nakao* 324, Karakoram, Puiju near Baltso glacier, 3,300 m, 7 July 1955". We have been unable to trace this locality, but it may be the Baltoro Glacier in NW Pakistan, the 's' in 'Baltso' in the original herbarium label could equally be interpreted as an 'r', making this locality the most likely one. This specimen is clearly subsp. *occidentalis* var. *parva* with small, 5-veined leaves. A collection from this area by Maurice Foster (*MF* 2016) (Figs 196, 198) has been determined to be tetraploid.

ECOLOGY AND STATUS IN THE WILD

Betula utilis subsp. *occidentalis* is found at the extreme western end of the Himalaya where these very white barked trees form forest to the tree-line. It extends from western Kashmir to Pakistan, possibly Afghanistan, Tajikistan, the Pamir, and as far as west as Uzbekistan.

Specimens intermediate between subsp. *occidentalis* and subsp. *jacquemontii* have not yet been found, but the area where they might occur has not been well collected and is currently difficult to access due to political instability and the presence of land mines. From Kashmir eastwards, while still having the characteristic very white bark and white resin encrusted buds of *Betula utilis* subsp. *occidentalis*, the birches tend to have much larger, more oblong-ovate, deep green, glossy leaves and thicker twigs and are typical subsp. *jacquemontii*. The two collections from Sonamarg, Kashmir, quoted in Ashburner & Schilling (1985) as subsp. *occidentalis* are large leaved and are closer to subsp. *jacquemontii*.

Fig. 196. Grove of *Betula utilis* subsp. *occidentalis* in the Karakoram range, NW Pakistan, 12,000 ft, N aspect with *Picea smithiana*. Note birches 'combed' down the steep slope by deep annual snows at this altitude (MF).

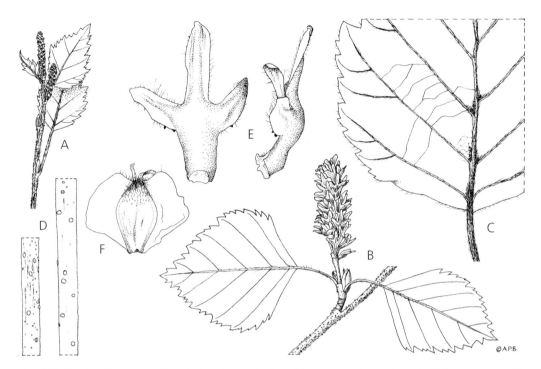

Fig. 197. *Betula utilis* subsp. *occidentalis*. **A** twig with buds and immature male catkins × 1; **B** fruiting shoot with mature fruiting catkin (fragmented towards base) × 1; **C** underside of leaf base × 2; **D** bark of twigs with more (left) or fewer (right) glands × 2; **E** female catkin scale, back and side views (resin deposits present) × 7; **F** seed × 7. All from *Siddiqi & Zaffar Ali* 4134 (Upper Saltpura, Baltistan, Pakistan, 12,000+ ft, 16 Aug. 1966). DRAWN BY ANDREW BROWN.

TAXONOMY, RECOGNITION AND RELATIONSHIPS

In overall appearance *Betula utilis* subsp. *occidentalis* is intermediate between subsp. *jacquemontii* and the central Asian *B. tianshanica*. Like subsp. *jacquemontii* it has bark peeling in large sheets, buds covered in white resin, and at least somewhat knobbly male catkins, while like *B. tianshanica* it has quite small, triangular-ovate leaves which are relatively thin, flat, with relatively few, (5–7), pairs of weakly veined leaves with a tendency to double toothing and, in the extreme west, rough warty twigs with persistent white glands.

Some SABE collections of subsp. *albosinensis* from Hubei are in fact similar in general appearance, leaf size and shape to subsp. *occidentalis*, but with a different bark colour.

Maurice Foster's introduction from northern Pakistan (*MF* (P) 2016) has not fared well in cultivation at Wakehurst (nor in North Devon, Martyn Rix pers. comm.), but is growing reasonably well if slowly at Ness and is reported to have been very white barked in the wild (Figs 198, 199, 201). Surprisingly, collections from further northwest in Tajikistan, made by Géza Kósa and colleagues from Vacratot, the Hungarian national Botanic Garden, seem to be thriving and, judging by the photographs from the wild, promise to be most attractive (Fig. 200).

Many of the isolated birch populations from the mountains of central Asia have been described as separate species: they are discussed and listed below under *Betula tianshanica*.

Fig. 198. Grove of *Betula utilis* subsp. *occidentalis* with prostrate *Juniperus* species and *Sorbus tianschanica*, stabilising scree, Hindu Kush, NW Pakistan (MF).

Fig. 199. Grove of *Betula utilis* subsp. *occidentalis* in the Karakoram range, NW Pakistan, 12,000ft (MF).

Betula utilis subsp. **occidentalis** Kitam., *Plants of West Pakistan and Afghanistan*: 37 (1964). *Betula chitralica* Browicz, *Flora Iranica* 96: 3 (1972). Type: Pakistan, Chitral, Chatiboigletscher, Moräne, 3400 m, 4 Aug. 1968, *A. Stamm & G. Wöhrl* 90 (W).

DESCRIPTION. *Twig* with many resinous glands, very hairy to glabrescent. *Bud* surface covered with resin. *Leaves* ovate; small leaf blade 48–65 × 30–50 cm, with 5–7 pairs of veins, toothing coarse, often more or less double, axillary hair tufts present and extending up midrib. *Petioles* 10–15 mm. *Fruiting catkin* with peduncle 8–10 mm, 20–28 × 8–10 mm, bracts 5.5–6.5 × 3.75–4.5 mm, squarish lateral lobes different in form from centre lobe. *Seed* 2.25–3 × 3.5–5 mm with nutlet 2.25–3 × 1.5–1.75 mm and wing 1–1.75 mm broad: style short to 0.5 mm.

DISTRIBUTION. NW Pakistan, Afghanistan, Tajikistan, Kyrghistan, Uzbekistan and Kazakhstan, from the Karakoram Range east into the Hindu Kush and north eastwards into the Pamir–Alai ranges and southern Tian Shan where it may intergrade with *Betula tianshanica*.

HABITAT. Rocky slopes at high altitude.

Fig. 200 (above). *Betula utilis* subsp. *occidentalis*. Peeling trunk. Tajikistan, Hissar mts, Karakul valley, 25–2800 m. Open *Juniperus seravschanica* forest Géza Kósa, Erzsébet Fráter, Elmér Barabits Vacratot, Hungary (GK).

Fig. 201 (right). *Betula utilis* subsp. *occidentalis*. Trunk of old tree above Bara Gah river, Ghotulti NW Pakistan. In a small population of old trees, up to 11 m in height, with one specimen 1 m in diameter. (MF).

Fig. 202. *Betula utilis* subsp. *occidentalis*. Afghanistan, Prov. Parvan, Panshir Valley, Mukeni, streamside, from 2,500 up to 3,200 m, to 5 m tall, 16 July 1962 (local name 'khadang') *Hedge & Wendlebo* 5134 (RBGE). Details from Herbarium specimen in RBGE (IH).

30. BETULA ERMANII

HISTORY

Betula ermanii, the STONE BIRCH, has been cultivated in Britain since it was introduced in around 1900. Old specimens, probably of Japanese provenance, make wonderfully beautiful large and spreading trees; upright specimens 30 m tall are reported in the wild.

This species was first described by Chamisso in 1831, from specimens collected by Erman in 1830. Adolf Erman, after whom this birch was named, was born in Berlin in 1806. Although little known today, he was a scientist and intrepid traveller in Siberia; a physicist by training, in 1828 he decided to travel around the world, departing with the expedition led by Christopher Hantseen, which was to carry out magnetic measurements in Siberia. They travelled eastwards across Russia, and when the Hantseen expedition turned back at Kyakhta on the Mongolian border, south of Lake Baikal, he proceeded onward at his own expense to Yakutsk and Okhotsk. He crossed the Sea of Okhotsk to Kamchatka and then over the peninsula to the port of Petropavlovsk, where he met the ship of Leontii Hagemeister, who was supplying Russian colonies in Alaska. Erman travelled with Hagemeister to Sitka (where he saw sea otters) and then on to San Francisco, Tahiti, the coast of South America, and finally around the Horn to Europe. On his return to Germany, he wrote up his discoveries, which were published in five volumes from 1833 to 1838.

Fig. 203. *Betula ermanii* with understorey of *Sasa*, Daisetsu National Park, Hokkaido, Japan (KBA).

ECOLOGY AND STATUS IN THE WILD

Betula ermanii is one of the finest and most distinct of birches, associated with superb landscapes. Those fortunate enough to have seen it in the wild in Japan, and perhaps climbed through the open forest which clothes the slopes of the mountains of Hokkaido, have found the 'limby' shapes of the spreading trees flattened by snow and, near the tree-line reduced to elfin proportions. Their autumn colour is a superb yellow, highlighted against dark *Pinus pumila* (Fig. 204), and contrasted with the vivid reds and oranges of suckering *Sorbus matsumurana*. The freshness of its young foliage in spring, for which *Betula ermanii* is renowned, would here not only match, but would far surpass that of cultivated plants in this setting, where the ubiquitous *Sasa* and *Pinus* thickets make walking difficult.

Betula ermanii is primarily a mountain tree which is a major if not the sole component of tree-line forests in Japan and Korea and mainland eastern Asia as far west as Lake Baikal. Like the *B. utilis* aggregate and *B. pubescens*, it is tetraploid (2n=56). It occurs to the east of *B. pubescens* and to the north and east of *B. utilis*. Its distribution does not appear to overlap with that of *B. utilis* subsp. *albosinensis* in north central China; there are no hybrids known and no indication of any introgression, but this may be due to lack of collections from the area. *Flora of China* (Li & Skvortsov 1999) gives subsp. *albosinensis* occurring north and east to Hebei Province around Beijing, with *B. ermanii* in the adjacent province of Liaoning to the north.

In the west, by contrast, *Betula ermanii* in the form of var. *lanata* (see below) appears to hybridise extensively with *B. pubescens* in the hills immediately to the east of Lake Baikal. Several hybrid populations have been raised from seed received from the Barguzin Ridge (Molozhnikov, pers.

comm.). Young trees grown from seed from a tree of *B. ermanii* in Tallin Botanic Garden (original source Mt Chekhov, Sakhalin Island) yielded about nineteen hybrids with *B. pubescens* and a single tree of *B. ermanii*.

A second, less clearly distinguishable variety, var. *subcordata*, was described from Japan. It occurs sporadically throughout Japan, and was collected from Sakhalin by E. H. Wilson, *W.* 7338; var. *japonica* is hardly distinct.

Geographically, if not ecologically, the distribution of *Betula ermanii* overlaps extensively with that of the tetraploid shrubby birches *B. fruticosa* (usually referred to as *B. ovalifolia*) and *B. gmelinii*, and Kuzeneva (1936) reports hybrids of *B. ermanii* with *B. middendorffii*; it is likely that *B. gmelinii* is a hybrid complex between *B. fruticosa* and *B. ermanii* (see below).

From Japan there is molecular evidence of hybridisation between *Betula ermanii* and *B. apoiensis* (see p. 366) although this is described in the context of introgression of *B. ermanii* into *B. apoiensis*, which itself is now considered to be a stabilised hybrid between *B. fruticosa* and *B. ermanii* (Nagamitsu *et al.* 2006a).

The frequent hybridisation between *Betula pendula* and *B. pubescens* in Europe suggests that there might be similar hybridisation between *B. ermanii* and *B. pendula* subsp. *mandshurica* in eastern Asia, i.e. primarily one way gene flow into *B. ermanii*. The only references we can find to this hybrid are Kuzeneva's (1936) reference to its occurrence in Kamtschatka (*B. avatschensis*, *Repert. Spec. Nov. Regni Veg.* no 355: 166 (1914)) and in Hultén's *Flora of Kamtchatka* (1927) in which he describes extensive hybridisation, especially towards the north of Kamchatka where *B. ermanii* occurs at low elevations. Hultén complained of this confusion in his account of a visit to the peninsula in the 1920s.

Fig. 204. Shrubby *Betula ermanii* in *Pinus pumila* scrub, Oblachnaya, Primorye, Russian Far East (KBA).

Fig. 205. *Betula ermanii* in Stone Lane Gardens, Chagford, Devon (PW).

CULTIVATION

In cultivation in Britain, *Betula ermanii*, presumably of Japanese and Korean origin, forms beautiful large trees with creamy- to pinkish-white bark peeling off in large sheets. Experience at Ness suggests that the species may be less tolerant of drought than many others, at least on its own roots. Trees of Korean and Japanese provenance have relatively warty twigs with few long silky hairs and resinous buds, these giving a characteristic appearance to the leafy twigs. Collections from further north in China and the Russian Far East are often referred to var. *lanata* and have conspicuous large, grey-hairy buds, more hairy, less resinous twigs, and less sharply toothed leaves. These collections from a more continental climate rarely grow well in the British Isles, due to early bud break and subsequent frost damage, but a collection from the Russian Far East, Prymorsky Kraj, Olchovaja Mountain has reached about four metres at Ness and has been propagated for planting elsewhere.

Several cultivars are available: 'Grayswood Hill' is presumably raised from a particularly splendid tree in Grayswood Hill, Surrey, which had reached 63 feet high and 11 feet in girth in 1966 (Bean 1970). It has peeling, creamy-white or pinkish bark and chestnut-coloured, elongated lenticels. 'Hakkoda Orange' from Mount Hakkoda has an orange cast to the bark. 'Mount Zao', selected at Stone Lane Gardens, has purplish bark.

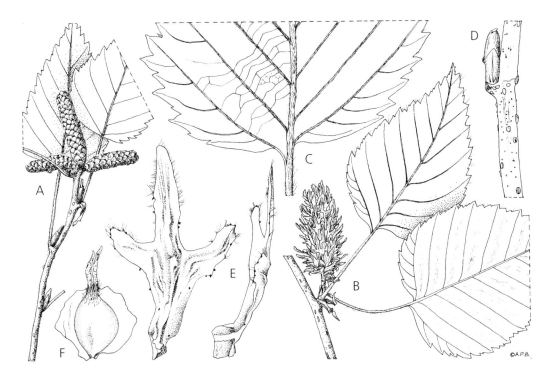

Fig. 206. *Betula ermanii* var. *ermanii*. **A** twig with buds and immature male catkins × 1; **B** fruiting shoot with mature fruiting catkin × 1; **C** underside of leaf base × 2; **D** bark of twig × 2; **E** female catkin scale, back and side views (resin deposits present) × 7; **F** seed × 7. All from *H. McAllister* (Ness BG, 30 Aug. 2010). DRAWN BY ANDREW BROWN.

Betula ermanii Cham., *Linnaea* 6: 537 (1831). Type: described from Kamtchatka (holotype LE).

Var. **ermanii**

Betula ermanii var. *subcordata* Koidz., *Bot. Mag.* (*Tokyo*) 27: 148 (1913).

Betula ermanii var. *japonica* (Shirai) Koidz., *Bot. Mag.* (*Tokyo*) 27: 148 (1913). Type: Nikko, 8 Aug. 1904 and 12 Aug. 1909' herb. Sakurai (co-types).

DESCRIPTION. *Tree* to 20 m. *Bark* grey-white to creamy-white, peeling. *Twigs* brownish-purple, variably resinous warty and hairy. *Buds* conic-ovoid, c. 8 × 3.5 mm, variably sticky-resinous and hairy to densely silky hairy. *Petioles* 10–35 mm. *Leaf blades* ovate to triangular ovate, base broadly cuneate to sub-cordate, margin usually sharply serrate, often more or less double toothed, 8–12 pairs of veins, 20–100 × 12–70 mm, variably hairy with axillary hair tufts totally absent to very conspicuous. *Male catkins* up to 3 terminal and one lateral; to 18–25 × 4.5 mm in winter, brown, expanding in spring. *Fruiting catkin* erect to pendent, ovoid to oblong, 15–35 × 8–15 mm; peduncle to 6 mm. *Scales* 5–10 mm, lateral lobes 1/3 to 1/2 length of mid-lobe. *Nutlet* 2.5–3 × 1.8–2 mm with wing about half as wide as nutlet.

DISTRIBUTION. East Asia from Kamtchatka, Sakhalin, Japan and Korea west through the Russian Far East and north China. The typical form with less hairy buds and twigs probably more or less confined to Japan and Korea.

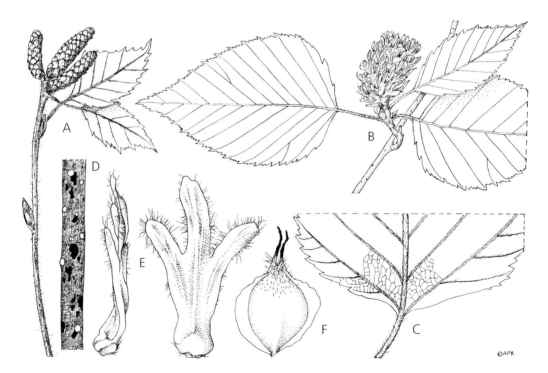

Fig. 207. *Betula ermanii* var. *lanata*. **A** twig with buds and immature male catkins × 1; **B** fruiting shoot with mature fruiting catkin × 1; **C** underside of leaf base × 2; **D** bark of twig × 2; **E** female catkin scale, back and side views × 7; **F** seed × 10. **A** from Leningrad sheet det. *Vassiljev*, 27 Aug. 1966; **B–F** from *Kuzeneve & Prochorov* 4552. DRAWN BY ANDREW BROWN.

HABITAT. Forming pure forests but also a component of coniferous and mixed broad-leaved forests, often up to the treeline.

CHROMOSOME NUMBER. Tetraploid: 2n=56.

Var. **lanata** Regel, *Nouv. Mém. Soc. Imp. Naturalistes Moscou* 13 (2): 122 (1861). Type: (northeastern Asia) prope Ajan, *H. Tiling* (holotype LE).

Betula ircutensis Sukaczev, *Trudy. Bot. Muz. Imp. Akad. Nauk* 8: 226 (1911). Type: ?? Podgolechnaya river between the Lena and Kirenga rivers (LE)

DESCRIPTION. Var. *lanata* differs from var. *ermanii* in having the bud scales and young shoots tomentose or even woolly. The leaves are triangular-ovate, shortly acute, with usually 7–9 lateral veins.

DISTRIBUTION. Russia: from the eastern shores of Lake Baikal, eastward to the Pacific coast except Korea. N China.

HABITAT. Montane woods.

Subgenus BETULA section BETULA

Bark usually peeling in small strips; *leaves* usually with fewer than 8 pairs of veins; *seed* wings usually broader than the nutlet.

31. BETULA PENDULA SILVER BIRCH

INTRODUCTION

In *The Circumpolar Plants* (1971), the Swedish botanist Eric Hultén states that

> The white birches are probably the most complicated of all circumpolar complexes ..... it is natural that the possibility to keep distinct species apart in such a mixture, hardly exists. The more species are described, the more the taxonomical difficulties increase. They have today reached a point, where no two taxonomists agree.... to treat the different *Betula* types of the complex as species would be comparable to treating different types of dogs as species.'

Hultén was familiar with the birches of his native Scandinavia and had botanised extensively in Alaska and Kamtchatka, so he had an appreciation of these birches over a very wide area and experienced, at first hand, the difficulties of attempting to name them.

Other authors have dealt with *Betula pendula* agg. in restricted geographical areas, e.g. in North America (Furlow 1990; Alam & Grant 1972; Brittain & Grant 1967b, 1972; Dugle 1966), in NE Asia and China (Jansson 1962), and several species which they have treated as distinct are not recognised here (Ashburner 1993; Huxley 1992; Kuzeneva 1936).

Japanese birches brought back to Europe as seed and herbarium specimens were clearly different from European *Betula pendula*, so they were named *B. japonica* in 1830, and other collections from the far eastern Asiatic mainland were described as distinct species in the early 1900s. The discontinuity between European and Japanese silver birches is bridged by intermediate-looking trees across vast areas of Siberian birch forest, and, when their total variation is examined, it is not possible to consider *B. pendula* to be a distinct species from *B. japonica*; in this we agree with Skvortsov (Li & Skvortsov 1999).

Fig. 208. Stabilised block scree habitat of *Betula pendula*, Les Etages, Massif des Écrins, French Alps. June 1984 (KBA).

Fig. 209. *Betula pendula.* NW Mongolia in conifer forest with understorey of *Vaccinium* species and *B. humilis* (MF).

Fig. 210. *Betula pendula* NW Mongolia, grove of white-stemmed trees up to 9 m with rich steppe groundflora in foreground; including *Geranium pratense*, *Dianthus versicolor*, *Lilium martagon*, *Potentilla fruticosa*, *Thalictrum*, *Aster*, *Aconitum*, *Iris* (MF).

There is no question that trees from the extremes of the distribution in Europe (*Betula pendula*), north-east Asia and Japan and western North America (*B. platyphylla*, *B. mandshurica*, *B. neolaskana*), and SW China and Tibet (*B. szechuanica*) do look different and can be distinguished from one another. However, when the whole continuous geographical range of diploid silver birches is considered, all those named as species intergrade in geographically intervening areas. This is not unexpected, as the distribution of these trees appears to be more or less continuous from western Europe to Japan and western North America. However, the differences between the geographic extremes in Europe and Japan make it useful to have names by which to refer to them, so here the eastern silver birches are kept as subspecies *mandshurica*.

Similarly the silver birch of the eastern Himalaya and south-west China named *Betula japonica* var. *szechuanica* C. K. Schneid. is here retained as subspecies *szechuanica*.

In an attempt to arrive at a usable classification of these taxa, we have examined living and herbarium specimens from throughout the distribution of the complex, and especially from the area in east central Siberia and Mongolia (Figs 12, 13, 210–213) where the distributions of named taxa overlap. The availability of living trees has allowed us to study characters not easily seen in dried specimens, and to compare trees of different provenance at all stages of growth.

All three subspecies of *Betula pendula* are distinguished from other birches by their combination of white bark, more or less triangular or deltoid leaves and translucent, sticky warts on the young twigs.

Fig. 211 (left). *Betula pendula* NW Mongolia (MF).

Fig. 212 (right). *Betula pendula* NW Mongolia. Trunk of mature tree (MF).

Nearly all have a weak venation pattern with relatively few pairs of veins, the leaves have long stalks, and the ripe fruiting catkins break up readily, with the scales of the fruiting catkins having recurved lateral lobes longer than the central lobe and the wings of the seed being wider than the nutlet. They are also typical pioneers of unstable habitats, forming the first crop of trees on newly exposed soil, before slower-growing and longer-lived trees such as conifers begin to mature.

To sum up: we have concluded that the diploid *Betula pendula* is best divided into three subspecies, subsp. *pendula* in Europe and eastwards to central Asia, subsp. *mandschurica* in eastern Asia and western North America and subsp. *szechuanica* in western China from Qinghai and Gansu to Yunnan and southeast Tibet (Xizang). *B. populifolia*, from eastern North America is isolated geographically from *B. pendula* and is recognised here as a distinct species.

Fig. 213. *Betula pendula* NW Mongolia. Grove of white-stemmed trees with bark removed for birch box making (MF).

31A. BETULA PENDULA subsp. PENDULA SILVER BIRCH

And hark, the noise of a near waterfall!
I pass forth into light — I find myself
Beneath a weeping birch (most beautiful
Of forest trees, the lady of the woods)
Hard by the brink of a tall, weedy rock
That overflows the cataract.

The Picture or The Lover's Resolution (Coleridge 1802)

ECOLOGY AND STATUS IN THE WILD

The European silver birch, *Betula pendula* subsp. *pendula* is recognised by its combination of deltoid, double-toothed leaves and white bark, which splits into dark corky ridges and cubes at the base of the trunk (Figs 212, 215); no other birch has branchlets which droop so elegantly, or even hang vertically (Fig. 216).

It is an inescapable element of heathland in northern Europe, ubiquitous in its beauty and taken for granted on the light and poor Tertiary sands of Bagshot Heath in Britain for instance, in Touraine in France, the Veluw in Holland, and Luneburg Heath in Germany. It also colonises limestone, including chalk soils (Atkinson 1992), although it is less common on these than on more acid soils, perhaps because of competition from the wider range of woody species which colonise soils of a high pH.

Valley populations in Scotland, particularly in the east, and in Norway, are magnificent. Fine trees of great beauty and grace are abundant across the north European plain, in south Sweden and across the Baltic States. Seedlings invade gaps in the woods, or indeed any open ground. Young trees crowd abandoned industrial sites, quarries, and old railway yards.

Although many people love the silver birch for its grace and beauty, foresters in Britain and some other countries where conifers are the main timber trees are not so pleased by its presence. It may be a good nurse, but it whips around in the wind and damages other trees. Ecologists condemn it as a destroyer of heathland as it shades out the heather (*Calluna vulgaris*) and replaces the lowland heath by birch forest, although the lowland heaths are a man-made habitat created from woodland by cutting, burning, and grazing. The sandy soils of the heaths are wide open for birch invasion, especially after a fire. In theory the birch should in turn be shaded out, in classic succession, by pine and oak, but this seldom happens; pines and oaks are usually kept at bay by human disturbance, grazing, burning or felling — or fail to arrive if no seed source is available — and birch continues to dominate. Where birch replaces heathland it increases soil fertility so that it may be difficult to re-establish heathland on the site (Mitchell *et al.* 1997, 2007).

TAXONOMY, RECOGNITION AND RELATIONSHIPS

Although the white birches of Sweden were well known to Linnaeus, in *Species Plantarum* (1753) he did not distinguish between the two species, the weeping birch *Betula pendula* Roth. and the downy birch *B. pubescens* Ehrh. The reasons for rejecting his name, *Betula alba* L. are set out by Rafael Govaerts, in *Taxon* 45: 697–698 (1996). Hybrids are reported between *B. pendula* and *B. pubescens* and they are discussed on p. 307, under *B. pubescens*.

The name *Betula verrucosa* Ehrh. has long been used, in Europe especially, for the weeping birch, and is most apt, derived from the Latin *verruca* = a wart, in reference to the rough glands on the most vigorous shoots and youngest twigs which characterise almost all populations of this species; the older name *pendula*, however, is correct, although only the European populations have the characteristic weeping branches.

The thin flexible twigs of subsp. *pendula*, hanging down vertically especially on older trees, are unique among birches. This shows as a fine and beautiful tracery in winter, and in summer the foliage moves easily, with each individual leaf, long stalked, fluttering readily in the slightest wind. Some other species, *e.g. Betula nigra*, may weep to some extent.

The leaves of *Betula pendula* subsp. *pendula* are easy to recognise, with their combination of triangular or deltoid shape, acuminate apex and double toothing. Their bluish green and high gloss could only be confused with that of the Chinese subsp. *szechuanica*, but on that tree the leaves are usually coarser, more leathery and usually with only a single order of toothing. Except perhaps in some Iberian populations, the leaves of subsp. *pendula* lack the elegant, really long points of *B. populifolia*, and they nowhere approach the dull yellow-green, twisted coarseness and leathery quality seen in subsp. *mandshurica*.

Fig. 214 (above). *Betula pendula*. Siberia, Novosibirsk. One of the whitest trunked provenances of the species in cultivation (PW).

Fig. 215 (right). *Betula pendula* subsp. *pendula*. Sweden, Molon, with beginnings of corky fissures at trunk base (PW).

Bark character is not always a very good diagnostic feature in birches, but it is in this group, as only subsp. *pendula* appears to develop the dark corky ridges and cubes at the base of the trunk. We have seen some populations however, both in the north and south of Europe, in which the bark is smooth and white to the base, although this may be because the trees were of relatively small diameter. These have been distinguished as var. *lapponica* and var. *aetnensis*, although at least the latter does develop corky fissures after about thirty years. The absence of such corkiness is a good character to distinguish typical subsp. *mandshurica* and subsp. *szechuanica*. The disappearance of fissured bark in Siberia is one obvious character change where populations of subsp. *pendula* merge into subsp. *mandshurica*. In the south of Europe, many of the populations of *Betula pendula* still weep, some exaggeratedly, such as the beautiful trees in the valleys of the Massif des Écrins in France and the Gran Paradiso in Italy, in which the smooth whiteness of the bark extends into the crowns of the trees. The finest of these alpine birches are the somewhat twisted individuals in the highest valleys near their upper limit, when seen in brilliant young leaf against still melting snow in May or even as late as June (Fig. 208).

Trees at high elevations in Corsica are very similar, as apparently are those in Macedonia and Bulgaria; the renowned populations on Mt Etna in Sicily are the most southerly birches in Europe, magnificently white against the black of new lava flows. On Etna the lava becomes green in only a few years with colonising vegetation including *Genista aethnensis*, but, before that, its blackness seems to enhance the birches in their brilliance in spring and early summer, or in the gold of autumn, so much that local people come especially to see 'le betulle' at these times. These trees appear to form the tree line.

Farthest south are the white birches of the Sierra de Guaderrama in central Spain and El Rif mountains of Morocco, the latter described as *Betula fontqueri*, based on a minor difference in the sweep of the lateral lobes of the bract scales. Birches are known to have survived the glacial maxima in southern mountain areas such as Spain, Italy and the Balkan peninsula (Carrion 2002; Huntley & Birks 1983; Willis *et al*. 2000) where they

Fig. 216. *Betula pendula* subsp. *pendula* in Ness Gardens covered in hoar frost (HMcA).

Fig. 217. *Betula pendula* subsp. *pendula* in Ness Gardens covered in hoar frost (HMcA).

remain in isolated populations. Recent molecular evidence suggests that these southern populations did not contribute to the re-colonisation of northern Europe, which came from a more or less continuous belt in central Europe not far from the ice front (Palmé & Vendramin 2002; Palmé *et al.* 2003; Maliouchenko *et al.* 2007). In spite of their long isolation from one another, probably during both glacials and interglacials, neither *Betula* '*aetnensis* ' on Mt Etna, nor *B.* '*fontqueri*' in Morocco, nor any of the other populations seem to have evolved sufficiently to deserve taxonomic recognition.

CULTIVATION AND CULTIVARS

Such is the beauty and grace of *Betula pendula* subsp. *pendula* that it has always been a favourite in the landscape and frequently planted in gardens, despite the difficulty in growing anything beneath the trees whose shallow roots rob the surface soil of moisture. Most individuals are rewardingly graceful, but some are more so than others, and when raising a batch of trees intended for the garden, it is always a good idea to harvest seed from the best. *B. pendula* is reported to be among the most ozone tolerant woody plants of central Europe (OECD 2003), and so thrives in urban situations. It grows in any well-drained soil, and is relatively fast-growing, reaching four metres or more after ten years. One negative character is that its fallen twigs can be untidy, especially after a winter storm.

Several cultivars have been selected from unusual or aberrant variants, highly dissected, purple or yellow leaves for instance, and these are covered in detail in the chapter on birch cultivars (p. 376). These may be selections from the wild, or occur quite by chance in a row of seedlings in the

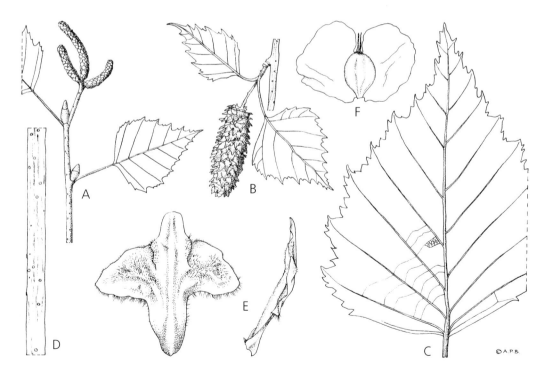

Fig. 218. *Betula pendula* subsp. *pendula*. **A** twig with buds and immature male catkins × 1; **B** fruiting shoot with mature fruiting catkin × 1; **C** underside of leaf × 2; **D** bark of twig × 2; **E** female catkin scale, back and side views (note hairs on inner face) × 8; **F** seed × 8. **A, C, D** from *J. Fraser* s.n. (Arbrook Common, Surrey, UK, 4 Sept. 1910); **B, E, F** from *Turrill* s.n. (Bladon Heath, Oxfordshire, UK, 18 June 1944). DRAWN BY ANDREW BROWN.

nursery. The most obvious feature to exaggerate in subsp. *pendula* is the weeping habit, and this has been achieved spectacularly in the selection of 'Tristis', possibly synonymous with 'Elegans' according to Bean, on which thin swaying branchlets hang like curtains. This clone exhibits the fine beautiful winter tracery of subsp. *pendula* even more than usual and its trunk is smooth to the base. Populations in valleys in the Maritime Alps are much like this and perhaps 'Tristis' originated in that region or somewhere in the south, as one suspects by its selection in Bonamy's nursery in Toulouse (Bean 1970).

There is remarkable concentration of cut-leaved and reduced-leaved forms of subsp. *pendula* in Scandinavia where they occur naturally as aberrant individuals, and many have been named and propagated; 'Dalecarlica' is a common example; a few unusual forms have appeared in Scotland too. These cut-leaved varieties may be homozygous for a recessive gene, as seedlings of one of the best known, the cultivar 'Crispa', revert to the type form. It is very interesting that the same phenomenon is seen in Scandinavia in *Alnus incana* and to a lesser extent *A. glutinosa*.

In contrast 'Rigida' has only the shallowest teeth. On 'Birkalensis' the teeth are single and coarse; and again in contrast, 'Ostrogothica' has shallow lobes and very small teeth. Many of these form trees of normal stature, and the cut leaves add to their elegance. Reduced forms such as 'Arbuscula', small-leaved and finely-toothed, may be natural hybrids with the dwarf birch *Betula nana*. It is interesting that these exaggerated leaf-forms have not been reported from populations of the other subspecies and appear to be concentrated in Norway, Sweden and Finland.

Unusual variants may be propagated by grafting or by cuttings of semi-ripe wood taken in July.

Betula pendula Roth, *Tent. Fl. Germ.* 1: 405 (1788), subsp. **pendula**

B. verrucosa Ehrh., *Beitr. Naturk.* 5: 161 (1790).

B. aetnensis Raf., *Giorn. Bot. Ital.* 1: 17 (1844).

B. gummifera Bertol., *Fl. Ital.* 10: 229 (1855).

B. alba f. *dalecarlica* Regel in A. P. de Candolle, *Prodr.* 16 (2): 164 (1868).

B. coriacea Pamp., *Nuov. Giorn. Bot. Ital.* n.s. xxii: 274 (1915).

B. fontqueri Rothm., *Bol. Soc. Brot.* sér. 2, 14: 149 (1940). Type: Habitat in montibus Atlantis riphaei Imperii Maroccani, Badú, 1600 m, It. mar., 1927, *Font Quer* 128 (BC).

B. verrucosa var. *lapponica* Lindq., *Svensk. Bot. Tidskr.* 41: 62 (1947).

B. platyphylloides V. N. Vassil., *Spisok. Rast. Gerb. Fl. S.S.S.R. Bot. Inst. Vsesoyuzn. Akad. Nauk.* 16: 81 (1966).

B. ferganensis V. N. Vassil., *Novosti Sist. Vyssh. Rast.* 7: 108 (1970, publ. 1971).

B. pendula var. *oycowiensis* (Besser) Dippel, *Handb. Laubholzk.* 2: 167 (1891) from SE Poland.

B. szaferi Jent.-Szaf. ex Staszk., *Acta Soc. Bot. Poloniae* 55 (3): 364 (1986).

DESCRIPTION. *Trees* to 25 m, but more usually 10–15 m; branches upright but outer branchlets on older trees becoming thin, drooping & flexible, forming a fine hanging haze of twigs in winter. *Trunks* generally cylindrical, not fluted, with inverted-V shaped black markings at branch junctions (Figs 7, 211). *Roots* shallow (except perhaps in some southern populations) and spreading. *Bark* white, smooth, scarcely peeling, and then only in thin shreds, most often dark, corky and fissured into rectangular corky bosses progressively from the base with age, sometimes to some way up the trunk, the fissures somewhat shallow or quite deep (Fig. 215); white and smooth and often less fissured in north Scandinavia (var. *lapponica*), the Alps, where bark white high into the branches; lenticels thin horizontal markings on trunks and main branches. *First year shoots* to 2.5 mm diam., often densely covered with pale resinous glands, especially on vigorous shoots, glabrescent except at seedling stage; weak shoots as thin as 1 mm, smooth with glands sparse or almost absent, commonly deep brown, but colour often masked by thin silver-grey peeling epidermis. Lenticels on twigs not conspicuous; slightly ovoid and as small as 0.25 mm. *Buds* ovoid, hairless. *Petioles* 20–30 mm. *Leaves* usually deltoid-triangular, flat, to 50–75 × 45–60 mm, acute to acuminate, mostly truncate to cuneate at base, with 5–8 pairs of lateral veins, often not arising opposite each other on the midrib; pale green and shiny below (abaxial), blue-green and shiny above (adaxial), with numerous resin dots on the upper surface when seen in reflected light, almost invariably double-toothed; petiole:blade-length ratio 1:3 to 1:4 *Male catkins* 30–60 mm, curving slightly, apparently clustered in 3s terminally, but in reality a terminal pair with a third immediately below (the usual pattern in this group). *Female catkins* thin & becoming pendulous. *Fruiting catkins* to 35 × 10 mm, pendulous, peduncle 10–20 mm, firm in summer but fragile in autumn with scales & fruits soon falling. *Scales* 5–6 mm, terminal lobe shorter than recurved lateral lobes (Fig. 34). *Nutlet* c. 2 × 1 mm, ovoid with membranous wings about twice as wide as nutlet (Fig. 34).

DISTRIBUTION. North Africa (Morocco), Europe from Spain, Ireland and Scotland east to Iraq, Iran (Elburz Mts), Uzbekistan, Kazakstan (S Tian Shan) and Lake Baikal.

HABITAT. Sandy and stony soils, especially acid, less frequently on limestone, often seeding after fire or disturbance of the soil.

CHROMOSOME NUMBER. Diploid: 2n=28.

Fig. 219. Habitat of *Betula pendula* subsp. *mandshurica*, South Korea (KBA).

31B. BETULA PENDULA subsp. MANDSHURICA

Japanese or Manchurian white birch, Alaskan white birch

HISTORY

Japanese silver birches brought back to Europe as seed and herbarium specimens were rather different from European *Betula pendula* and were named *B. japonica* by Siebold in 1830; unfortunately this name was illegitimate, because Thunberg had earlier used the name *Betula japonica* for the Japanese alder, *Alnus japonica* (Thunb.) Steud., and the Japanese silver birch has often been called *B. platyphylla* var. *japonica* (Miq.) Hara. Trees of Japanese provenance grew well in Europe, but few living or herbarium specimens were available to western botanists from the intervening areas of Siberia where silver birches are often dominant in the forests. This was partly because of inaccessibility and political factors after the Russian Revolution, but also because plants from the continental climate of central Siberia do not grow well in western Europe.

Betula pendula subsp. *mandshurica* was described by Edward Regel in his account of *Betula* and *Alnus*, published in 1865. Some other collections from far eastern Asia were described as distinct species in the early 1900s: *Betula grandifolia* was described from the Lena-Kolyma area in 1905, *B. platyphylla* from Dahuria, east of Lake Baikal in 1911, *B. ajanensis* from the Aldoma River basin in 1921, and *B. cajanderi* from the Lena river in 1929. The exploration of Alaska led to the description of *B. alaskana* by Sargent in 1901, but he realised that this name had already been used, so in 1922, he redescribed it as *B. neoalaskana*.

ECOLOGY AND STATUS IN THE WILD

In Manchuria and northeast China birches are abundant and very white-stemmed in the cold, dry continental climate and may be seen invading open habitats on light soils, equivalent to the heaths of western Europe. The trans-Siberian railway passes through at least one such area between Chita and Khabarovsk. Birches are very abundant in the country to the north of the Manchurian frontier, forming a simple mix with *Larix*. They are also common in the north-eastern provinces of China where, according to Wang (1961), they clothe the lower slopes of the mountains such as the Hsingan Range, where sturdy, wind-blasted montane *Betula pendula* subsp. *mandshurica* may be seen at higher elevations and up to the tree line, and also further west in the mountains of Jehol and Chahar in inner Mongolia, and Shanxi, frequently mixed with the Eurasian aspen *Populus tremula*. Further to the west, towards the drier grassland regions, birch and aspen groves are only found on the northern slopes of the isolated mountain ranges.

Throughout eastern Asia *Betula pendula* subsp. *mandshurica* is the common low altitude white birch. It does not usually ascend into the mountains in eastern Asia as does *B. ermanii*, but it is the tree line birch in western North America. It is most readily distinguished from the other subspecies by the presence of conspicuous hair tufts in the vein axils on the underside of the leaf, but intergrades with subsp. *pendula* in the Lake Baikal region and with subsp. *szechuanica* in northern China. The large, coarse leaves with a single order of toothing and stiff twigs make trees of this subspecies recognisable even from a distance.

TAXONOMY, RECOGNITION AND RELATIONSHIPS

The silver birches of eastern Asia and western North America were studied in detail by Jansson (1962) working with herbarium specimens and living material in the Botanic Garden at Göteborg. He recognised six distinct species in the area, but commented on the existence of intermediates between all the species he defined. The maps in Hultén (1971), which are taken from Jansson (*l.c.*) for east Asia and west North America, show that the distribution of this species group is more or less continuous in a belt around the Northern Hemisphere, with gaps only where there are oceans, and in central Ontario in Canada, where there is a gap between *Betula pendula* subsp. *mandshurica* in the extreme west of the province and *B. populifolia* in the east. There may also be a gap in central China between the circumboreal populations to the north and *B. pendula* subsp. *szechuanica* of Xizang (Tibet), Yunnan, Sichuan and Gansu in the south, but this seems unlikely as discussed below.

Jansson considered all populations in the vast area from the Atlantic seaboard of western Europe and North Africa as far east as the region around Lake Baikal and north west China, to be *Betula pendula* var. *pendula*. Interestingly he treated populations to the east and north-east of Lake Baikal as *B. pendula* var. *platyphylla*, but then he divided those to the east and south of Lake Baikal into six different taxa. He recognised *B. kamtschatica* in the Kamtchatka peninsula and western North America, and *B. resinifera* in western North America, to the southeast he recognised *B. mandshurica*, and to the south and west *B. szechuanica* and *B. rockii*.

Within the distribution given for *Betula mandshurica*, he plotted intermediates between the more northerly var. *platyphylla* (in Outer Mongolia) and the more southerly *B. mandshurica* (in northeastern China), there being no clear discontinuity between the distribution of these two species. Again,

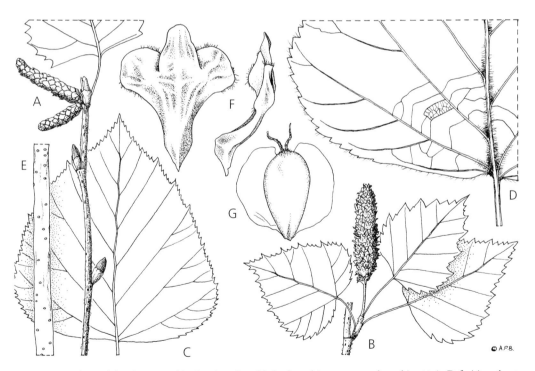

Fig. 220. *Betula pendula* subsp. *mandshurica*. **A** twig with buds and immature male catkins × 1; **B** fruiting shoot with mature fruiting catkin × 1; **C** outline of typical stem leaf × 1; **D** underside of leaf base × 2; **E** bark of twig × 2; **F** female catkin scale, back and side views × 10; **G** seed × 10. **A** from Kamchatka, no collector s.n. (collected as *B. ermanii* but det. *B. pendula* subsp. *mandshurica* by H. McAllister); **B, F, G** from *Hultén* 2612 (Swedish Expedition to Kamchatka, Shadutka Volcano, 2 Aug. 1921); **C–E** from *Kalinin* 216 (Blagoveshchensk-on-Amur, 3 June 1965). DRAWN BY ANDREW BROWN.

referring to var. *platyphylla*, he states: 'The distribution southward to China and Manchuria is difficult to follow on account of the forms intermediate between this type and *Betula mandshurica*.' This suggests to us that these two taxa are not really worth separating at species level, if at all, as the northern var. *platyphylla* intergrades southwards into what he is calling *B. mandshurica*.

Most interestingly, moving further east, despite his predilection to recognise separate species, Jansson treated the populations on either side of the Bering Strait as subspecies of *Betula kamtschatica*, accepting that the same species occurs on both sides of the strait, with the typical variety of *B. kamtschatica* on the Asian side and var. *kenaica* on the Alaskan side.

He stated that *Betula kamtschatica* is distinct from *B. mandshurica* and var. *platyphylla*, but then 'however, intermediate forms occur between these and nearly related birches.' He then gives a list of intermediates between *B. kamtschatica* var. *kenaica* and *B. resinifera* in western North America, listing almost half as many intermediates as 'pure' var. *kenaica*. Janssen's descriptions and comments suggest gradual (clinal) variation east of Lake Baikal south eastwards to Korea and Japan, and north eastwards towards and through Kamtchatka and across the Bering Strait into western North America. However, Jansson did not make chromosome counts and his specimens of var. *kenaica* illustrated have rather broad nutlets, so there must be a question as to whether these are in fact part of this diploid aggregate or polyploids (see discussion under *B. papyrifera* in which the complex variation in Alaska is discussed).

The overall situation is well summed up in Jansson's final comments (on p. 152): 'Where the peripheric regions of related species touch or merge into each other, there appears a characteristic

Fig. 221. Habitat of *Betula pendula* subsp. *mandshurica* and *B. dahurica*, South Korea (KBA).

zone of intermediate forms,' and again, on the same page: 'The multitude of intermediate forms is obvious.' Given these comments it is surprising that he still wished to maintain six distinct species and one subspecies in East Asia and western North America, while at the same time regarding all the extensive populations to the west of Lake Baikal as a single subspecies. Thus his data could even be used to support the amalgamation of all populations into a single variable species!

We have studied intermediate populations around Lake Baikal. In populations around Listvyanka on the south-west shores where the Angara River leaves the lake, well developed leaves regularly have a few, often only 1–3, conspicuous hairs in the vein axils. Interestingly, these trees are also intermediate in leaf toothing between the usually single toothed east Asian and double toothed west Siberian and European populations (see below). Thus, we could find no sharp break between the more glabrous trees to the west of Lake Baikal and the more hairy trees to the east, instead there seems to be a gradual transition from one to the other.

We can find no significant differences between the trees on either side of the Bering Straits, so we also refer those in western North America to *Betula pendula* subsp. *mandshurica*. Thus the distribution of *B. pendula* subsp. *mandshurica* is a vast arc stretching from east of Lake Baikal through eastern Siberia, China and Japan to western Ontario in Canada.

In the *Flora of China*, Li & Skvortsov (1999) recognised two species, *Betula pendula* and *B. platyphylla*, where *Betula pendula* represents the species in northern Xinjiang west of the Altai, and *B. platyphylla* the birches from Mongolia east of the Altai and into western and north-eastern China. Skvortsov himself, however, disagreed with the editorial decision that *B. platyphylla* be recognised as distinct from *B. pendula*, and considered the two synonymous.

Betula austrosichotensis V. N. Vassil. & V. I. Baranov, *Novosti Sist. Vyssh. Rast.* 21: 60 (1984). probably belongs here.

RECOGNITION

Subspecies *mandshurica* can be distinguished from the other subspecies by the presence of at least a few hairs in the vein axils on the leaf underside, by the leaf margin having usually only a single order of toothing, and by the often larger leaves lacking the blue-green colouration so typical of subsp. *pendula* and subsp. *szechuanica*, and by the absence of corky breaks in the white bark near the base of the trunk.

CULTIVATION

Subspecies *mandshurica* is not very commonly cultivated and only trees of well-documented provenance are likely to be correctly identified. Being generally coarser in foliage than subsp. *pendula*, subsp. *mandshurica* is probably a less attractive tree, except perhaps for autumn colour.

The early leafing of so many wild collections when planted in the milder parts of western Europe is very welcome for the brilliant translucency of the young leaves when other trees are still bare, and shoots and leaves are only damaged by the severest air-frosts. Collections from Japan and the Kurile Islands are growing well at Ness and Stone Lane Gardens.

Betula pendula subsp. mandshurica (Regel) Ashburner & McAll. comb. nov.

Betula alba subsp. *mandshurica* Regel, *Bull. Soc. Imp. Naturalistes Moscou* 38 (2): 399 (1865). Type: an der Olga-Bai, in der südöstlichen Mandshurei, *Schmidt* (holotype n.v.).

Betula alba var. *resinifera* Regel, *Bull. Soc. Imp. Naturalistes Moscou* 38 (2): 398 (1865). Type: bei Udskoi in Ostsiberien, *Middendorf* (holotype n.s.).

Betula alba var. *tauschii* Regel, *Bull. Soc. Imp. Naturalistes Moscou* 38 (2): 399 (1865). Type: Wächst in Siberien, im Amur und Ussuri-Gebiet, in der mandschurei, sowie in Japan auf insel Jesso, wo solche *C. Maximowicz* sammelte (holotype n.v.).

Betula alba var. *kamtschatica* Regel, *Bull. Soc. Imp. Naturalistes Moscou* 38 (2): 400 (1865). Type: In Kamschatka von *Rieder* und *Stewart* gesammelt (holotype n.v.).

Betula japonica Siebold, *Verh. Batav. Genootcsh. Kunsten* 12: 25 (1830) non Thunb. (1799).

Betula alba var. *humilis* Regel in A. P. de Candolle, *Prodr.* 16 (2): 166 (1868).

Betula alaskana Sarg., *Bot. Gaz.* 31: 236 (1901) non Lesquereux (1883).

Betula grandifolia Litv., *Trudy Bot. Muz. Imp. Akad. Nauk* 2: 98 (1905).

Betula platyphylla Sukaczev, *Trudy Bot. Muz. Imp. Akad. Nauk* 8: 220 (1911). Type: Hab. in Sibiria orientali (prov. Transbaicalia, prov. Jakutsk, part boreal. prov. Amur).

Betula mandschurica (Regel) Nakai, *Bot. Mag. (Tokyo)* 29: 42 (1915).

Betula ajanensis Kom., *Not. Syst. Herb. Horti Bot. Petrop.* 2: 33–34, 130 (1921). Type: Crescit in districtu Ochotensi, ad ripas fl. Aldoma, non proceel a portu Ajan, in silvis montanis, 4 Aug. 1912, *Th. V. Sokolov* 1544.

Betula neoalaskana Sarg., *J. Arnold Arbor.* 3: 206 (1922).

B. cajanderi Sukaczev, *Acta Forest. Fenn.* 34 (13): 3 (1929). Type: described from the Lena river. Type in Leningrad.

Betula papyrifera Marshall subsp. *neoalaskana* (Sarg.) A. E. Murray, *Kalmia* 13: 4 (1982).

Betula papyrifera Marshall subsp. *humilis* (Regel) A. E. Murray, *Kalmia* 12: 18 (1982).

?Betula austrosichotensis V. N. Vassil. & V. I. Baranov, *Novosti Sist. Vyssh. Rast.* 21: 60 (1984).

DESCRIPTION. *Tree* to 30 m. *Shoots and twigs* thick and therefore not flexible, often with densely crowded peltate glands which are clear brown initially; but not becoming so obviously white-warty as in the other subspecies. *Buds* larger than any in the other subspecies, 4 × 2 mm. *Leaves* coarse-looking, very wavy and undulate on young vigorous shoots, often cordate at the base, and most often a dull, light green, in contrast to the deep blue-green of the other subspecies, usually with a single order of toothing, often persistently silky and puberulent hairy on both surfaces and with conspicuous hair tufts in vein axils on lower (abaxial) surface. Petiole:leaf-blade ratio from 1:4 to 1:6. Male catkins terminal only, paired or in threes, slightly curved. Female flowers thin, vertical. *Fruiting catkins* cylindric, not as thin as in subsp. *szechuanica*. *Scales* to 4 mm with lateral lobes spreading. *Nutlets* elliptic, c. 2 × 1 mm with each wing wider than nutlet.

DISTRIBUTION. Siberia east of Lake Baikal, eastwards to NE China, Japan and Alaska south to British Columbia and east to Ontario.

HABITAT. Hills and flood plains on stony soil.

31C. BETULA PENDULA subsp. SZECHUANICA CHINESE WHITE BIRCH

HISTORY

This subspecies was first described as a variety of *Betula japonica* in 1917, later transferred to a variety of *B. platyphylla*, and later raised to specific rank as *B. szechuanica*. It is here named *Betula pendula* subsp. *szechuanica*. This is a well-known birch in many arboreta and botanic gardens and forms a stocky, somewhat untidy-growing tree with deep blue-green leathery leaves and extraordinarily pure white bark, which sheds copious white dust on the hands. This must be the birch with the whitest bark of all, truly chalk-white in maturity, but with suggestions of buff or pale orange when younger.

ECOLOGY AND STATUS IN THE WILD

Betula pendula subsp. *szechuanica* appears to be confined to south-western China and south-east Tibet, and was frequently observed and collected from Tibet, north-west Yunnan and Sichuan in the early years of the 20th century by Wilson, Rock and Forrest; more recently by several expeditions from RBG Kew and RBG Edinburgh; and individuals such as Roy Lancaster, Maurice Foster, Martyn Rix, Keith Rushforth and Charles Howick have also introduced seed. Ecologically it seems to be a filler in gaps in the mixed secondary forest of those regions (Fig. 18), but it also forms subalpine montane forest in the drier inner ranges in south-east Tibet, but apparently not on the moister main Himalayan range. It is frequent in some valleys to judge by photographs taken by Charles Howick in south-east Tibet, and it would presumably invade any suitable open area.

TAXONOMY AND RELATIONSHIPS

The studies of Jansson (1962), discussed above under subsp. *mandshurica*, included specimens from western China. The distribution shown on his map suggests a discontinuity between the northern populations discussed above under subsp. *mandshurica* and those in the mountains of western and southwestern China, which he assigns to *Betula szechuanica*. However, within the northern edge of the distribution given for *B. szechuanica* he raises Rehder's *B. japonica* var. *rockii* from the Qinghai Hu (Kokonor) region of north-western China to a distinct species, *B. rockii*. This is despite describing it

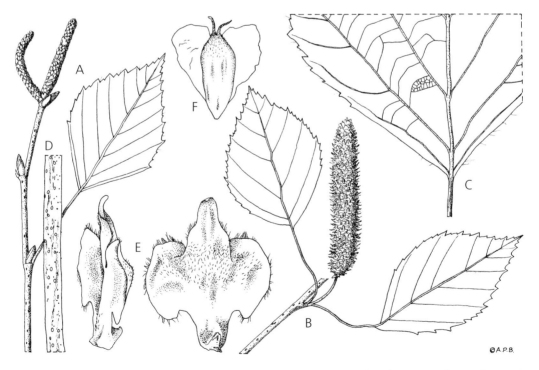

Fig. 222. *Betula pendula* subsp. *szechuanica*. **A** twig with buds and immature male catkins × 1; **B** fruiting shoot with mature fruiting catkin × 1; **C** underside of leaf base × 2; **D** bark of twig × 2; **E** female catkin scale, back and side views × 10; **F** seed × 10. All from fresh material cult. RBG Kew *SICH* 2113, collected 26 Aug. 2010. DRAWN BY ANDREW BROWN.

as resembling a hybrid between trees from further south in Tibet and west China (*B. szechuanica*) and those from further north and east (his var. *platyphylla*). The discontinuity shown in his maps in the distribution of the silver birches in China, with *B. szechuanica* in the provinces of Kansu and Sichuan to the west and south, no diploid silver birches in Shanxi and Shaanxi, and then *B. mandshurica* in the provinces to the east, is likely to be a discontinuity of the collections available to Janssen rather than a real discontinuity in distribution. Li & Skvortsov (1999) include Shanxi and Shaanxi in the distribution given for *B. platyphylla*, but it is not clear whether specimens from these provinces are closer to subsp. *szechuanica* or to subsp. *mandshurica*.

Intermediates occur between the northern subsp. *mandshurica*, in which axillary tufts of hairs are common, and the generally glabrous subsp. *szechuanica* in southern China. *Rock* 14718 from south-west Kansu province and *Purdom* 371 from Shensi (Shaanxi) province in north China have some axillary hairs and are geographically intermediate between the Mongolian populations and the Tibetan and south west Chinese populations of Yunnan and Sichuan. Much more surprising was to find some axillary hairs on *Forrest* 22520 from the Lijiang in NW Yunnan, well within the core area of typical *Betula szechuanica*. This finding blurs one of the possible distinguishing characteristics for *B. szechuanica* and links these south-western Chinese populations with those further north.

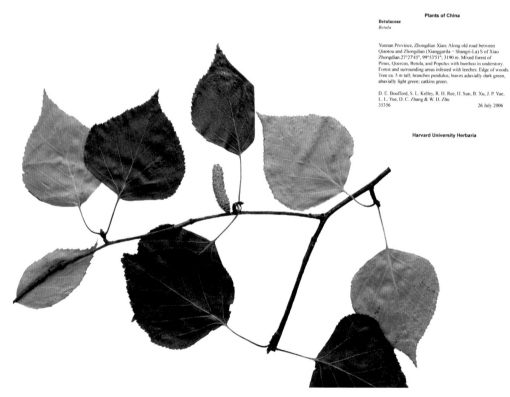

Fig. 223. *Betula pendula* subsp. *szechuanica* with typical leaf form with single toothing. China, Yunnan, Zhongdian Xian; along old road between Qiaotou and Zongdian S of Xiao Zongdian, 27°27'45"N 99°53'51"E; 3190 m. *D. E. Boufford, S. L. Kelly, R. H. Ree, H. Sun, B. Xu, J. P.Yue, L. L. Yue, D. C. Zhang & W. D. Zhu 35356,* 26 July 2006 (HUH) (PW).

RECOGNITION

The very white, dusty bark, together with the stocky spreading habit in older trees, and the large, leathery, dark, blue-green leaves, rather sparse on long shoots with long internodes, and the usually single-toothed leaf-margin (Figs 223, 225) and usual absence of axillary hair tufts on the leaf underside, identify this subspecies. The scales of the male catkins are sometimes humped, so the dormant catkins of some trees feel more or less 'knobbly'.

CULTIVATION

Recent introductions of this subspecies have grown well and are now widely distributed in gardens and arboreta. There seem to be no problems with early leafing or failure to mature the summer's growth. Occasionally there may be die-back due to damage from a late frost. There is a curious lack of autumn colour, certainly in cultivated trees in England, where the leaves remain persistently green then shrivel, turn brown and drop off; but quite possibly this may not be so in other climates, and is certainly not the case in the wild (Fig. 18).

No cultivars have been selected, but there has obviously been a tendency to clone Wilson and Forrest's original collections such as *W.* 4088 and *W.* 983. Perhaps most of the original seedlings disappeared or died, and increasing reliance was thereafter placed on vegetative propagation from those few that were kept and allowed to become specimens.

Fig. 224. Peeling trunk of young *Betula pendula* subsp. *szechuanica* (MF).

Betula pendula subsp. **szechuanica** (C. K. Schneid.) Ashburner & McAll. **comb. et stat. nov.**

B. japonica var. *szechuanica* C. K. Schneid., *Ill. Handb. Laubholzk.* 3: 44 (1917); *Plantae Wilsonianae* 3: 454 (1917). Type: western Sichuan: west of Tachienlu, Chetoshan, 2800–3400 m, woodlands, Sept. 1908, *Wilson* 983 (holotype HUH, isotype ???).

B. platyphylla var. *szechuanica* (C. K. Schneid.) Rehder, *J. Arnold Arbor.* 20: 411 (1939).

B. szechuanica (C. K. Schneid.) C.-A. Jansson, *Acta Horti Gothob.* 25:113 (1962).

B. japonica Siebold var. *rockii* Rehder, *J. Arnold Arbor.* 9: 25 (1928). Type: Eastern Tibet (now Qinghai), Kokonor Region, Koko Gorge, 11,000–11,500 m, Oct. 1925, *Rock* 13283 (holotype HUH, isotype, E).

B. rockii (Rehder) C.-A. Jansson, *Acta Horti Gothob.* 25: 120 (1962).

Plants of China

Betulaceae
Betula

Yunnan Province, Zhongdian Xian: Along old road between
Qiaotou and Zhongdian (Xianggarila = Shangri-La) S of Xiao
Zhongdian.27°27'45", 99°53'51"; 3190 m. Mixed forest of
Pinus, Quercus, Betula, and Populus with bamboo in understory.
Forest and surrounding areas infested with leeches. Edge of forest.
Tree ca. 5 m tall..

D. E. Boufford, S. L. Kelley, R. H. Ree, H. Sun, B. Xu, J. P. Yue,
L. L. Yue, D. C. Zhang & W. D. Zhu
35355   26 July 2006

Harvard University Herbaria

Fig. 225. *Betula pendula* subsp. *szechuanica* with leaves as double toothed as any subsp. *pendula*. China, Yunnan, Zhongdian Xian; along old road between Qiaotou and Zongdian S of Xiao Zongdian, 27°27'45"N 99°53'51"E; 3190 m. *D. E. Boufford, S. L. Kelly, R. H. Ree, H. Sun, B. Xu, J. P. Yue, L. L. Yue, D. C. Zhang & W. D. Zhu 35355*, 26 July 2006 (HUH) (PW).

DESCRIPTION. *Trees* up to 10 m with rather spreading, wayward habit. In cultivation older trees quite stocky and the winter silhouette often somewhat tangled, although in the wild at least young trees make neat pyramidal specimens. *Bark* brown or brownish-yellow when young, but this layer peeling off in quite thick sheets to reveal an eventual pure white with swollen cinnamon-coloured lenticels and shedding copious white dust (Fig. 224); older trunks with characteristic thick white flakes, again dusty. *Twigs* with quite large spot-like lenticels, and one-year shoots and twigs with rather long internodes, moderately to densely 'warty' with small resinous glands, but sometimes almost smooth. *Buds* narrower than those of subsp. *mandshurica* and quite pointed. Because of long internodes the foliage appears sparse. *Leaves*, dark, shiny blue-green, never subcordate, but usually truncate or cuneate at the base, often single-toothed with coarse teeth, rather leathery; lateral veins 4–7, crooked and wandering and proud of the upper leaf-surface; petiole:leaf-blade ratio 1:4 to 1:5. *Male catkins* thin and often curved, to c. 32 × 3 mm, largely terminal but often borne singly and sometimes axillary further down the twigs. *Fruiting catkins* rather thin, usually less than 4 mm wide; scales small. *Nutlets* quite rounded, c. 1 × 1 mm, the wings c. 1 mm wide.

DISTRIBUTION. Western China in SE Xizang, N Yunnan, Sichuan, Gansu, Qinghai.

HABITAT. Gaps in the mixed secondary forest; montane forest.

32. BETULA POPULIFOLIA   WIRE BIRCH, GRAY BIRCH

HISTORY

The common name WIRE BIRCH is apt for *Betula populifolia* as the twigs are thin and wiry, although the ultimate branches do not hang down as do those of *B. pendula*. Small thin fruiting catkins persist into winter when they enhance the twig tracery. They are remarkably hard and difficult to crumble and in this are quite different from the ripe catkins of other white barked birches of section *Betula*. The very long-pointed leaf is most elegant and distinctive and is unique in the genus, although the very longest leaves on *B. pendula* subsp. *pendula* (Fig. 218) can approach this sharp-pointed elegance. Another distinguishing feature is the usually solitary male catkin, projecting like a thin peg at the end of the twigs.

Marshall's name was published in *Arbustrum Americanum: the American Grove, or, an alphabetical catalogue of forest trees and shrubs, natives of the American United States*, printed in Philadelphia in 1785. Humphry Marshall (1722–1801) was a cousin of the pioneer collector John Bartram, and supplied seeds of American trees and shrubs to France as well as England (Raphael 1989).

ECOLOGY AND STATUS IN THE WILD

Betula populifolia is only found in eastern North America from Quebec and Maine to New York State, and westwards to Illinois. It has also been reported from Madison Co., Virginia in the Shenandoah National Park, and this is probably its southern limit. It is very frequently seen growing with *B. papyrifera* and *B. cordifolia*, and stabilised hybrid populations with the latter are so frequent that they have been named *B. ×caerulea* (Brittain & Grant 1972).

No geographical varieties appear to have been recognised and scarcely any aberrant forms, which is surprising as the leaves are so acuminate that exaggerations or reductions of this character would be expected to occur in the wild or be selected in cultivation. Only one appears to be mentioned in the literature, and that is forma *incisifolia*, recorded in two locations, one at Auburndale, Massachusetts and the other on Mt Penn, Pennsylvania (Fernald 1945). This is in contrast to the numerous named cultivars of *Betula pendula* and to the array of cut-leaved or small-leaved forms, which have appeared in *B. pendula* over the years in Scandinavia.

RECOGNITION

Betula populifolia is easily distinguished from all other birches by its very finely tapered leaf apices, usually solitary male catkins, and the small and persistent fruiting catkins with the scales more densely hairy on the adaxial surface than in those of *B. pendula* (Catling & Spicer 1988). In early spring, trees of *B. populifolia* are particularly beautiful in the freshness of their translucent, delicate, elegantly-pointed leaves. Although the bark is greyish, lustreless and does not peel, it is very attractive on older trunks, and is never fissured from the base upwards as it is in *B. pendula* subsp. *pendula*. As in all the species in the *B. pendula* aggregate, there are frequently inverted triangular 'fans' or 'eyebrows' of black on the trunks beneath each branch junction, which persist after the branch has fallen off; in *B. populifolia* these 'fans' are particularly well-marked (Fig. 31).

TAXONOMY AND RELATIONSHIPS

Betula populifolia is morphologically distinct from *B. pendula* and is the only species of the group whose populations are certainly geographically isolated from all others of the *pendula* complex. Chemical evidence supports the separation of *B. populifolia* from the rest of the complex, the compound platyphylloside being absent in this species, while it is present in all other members of the *B. pendula* complex which have been tested (Santamour & Lundgren 1996).

CULTIVATION

There is no doubt that *Betula populifolia* is well worth planting in the garden or using in landscaping, where, as with *B. pendula*, it is better planted in groups. In climates where late frosts are a hazard, such as in much of lowland maritime Britain, collections from Vermont with its continental-type climate, can suffer die-back of whole branches. Collections from coastal New England should grow better in western Europe.

Betula populifolia Marshall, *Arbust. Amer.*: 19 (1785).

Betula alba subsp. *populifolia* (Marshall) Regel, *Bull. Soc. Imp. Naturalistes Moscou* 38 (2): 399 (1865).

DESCRIPTION. *Tree* to 10 m, eventually, with more or less upright habit, occasionally suckering from roots (Fig. 31). *Trunks* often several, somewhat twisted and rarely straight. *Bark* rather grey-white or ashen-white, close and not peeling, nor even flaking. *Twigs* thin, but not flexible and pendulous; warts never profuse. *Buds* rounded, never more than 3 mm. *Leaves* distinctively acuminate, double-toothed but not deeply so, 30–100 × 30–80 mm, with 5–18 pairs of veins, glabrous or soon glabrescent. *Male catkin* in winter thin, and often solitary on twig, always terminal (paired or in 3s in the closely-allied *B. ×caerulea*). *Fruiting catkins* with strong peduncle c. 10 mm, ± erect to pendulous, cylindrical, firm, persistent, 10–25(–30) × c. 5 mm, [8–10 mm wide, fide Furlow (1997)]. *Nutlets* about 2 × 1 mm, wing broader than nutlet. *Scales* about 4–4.5 × 3.5 mm, the forward-pointing central lobes quite conspicuous in the intact fruiting catkin, lateral lobes spreading to recurved.

DISTRIBUTION. Eastern North America from eastern Ontario, Quebec, Nova Scotia south to South Carolina and west to Illinois.

HABITAT. On dry sandy or rocky soils but also in swamps and waste places and a common coloniser after fire or any kind of disturbance. Said to be short-lived, rarely living longer than 50 years (Hosie 1963) so may be the shortest lived of all birches.

CHROMOSOME NUMBER. 2n=28.

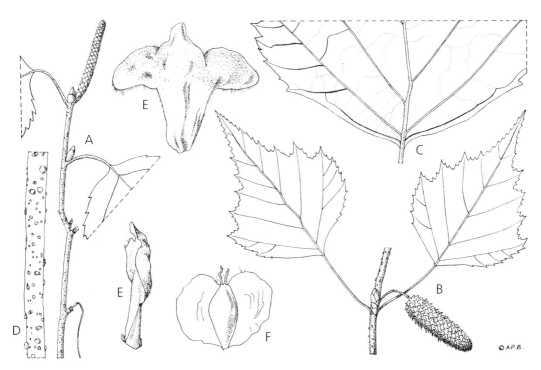

Fig. 226. *Betula populifolia.* **A** twig with buds and immature male catkin × 1; **B** fruiting shoot with mature fruiting catkin × 1; **C** underside of leaf base × 2; **D** bark of twig × 2; **E** female catkin scale, back (upper) and side (lower) views × 10; **F** seed × 10. **A, B, D–F** from *Howick & Warner* WH31 (Quabbin Reservoir, Mass. USA, 8 Oct. 1985); **C** from *Beattie* s.n. (Merrimack R., Lowell, Mass., USA, 30 July 1927). DRAWN BY ANDREW BROWN.

33. BETULA OCCIDENTALIS RIVER BIRCH, WATER BIRCH

HISTORY

Betula occidentalis was first described in his flora of North America by William Hooker from specimens collected by Dr Scouler, David Douglas and Thomas Drummond in the mountains of western Canada. Thomas Drummond was assistant to the second Land Arctic Expedition under Sir John Franklin in 1825–1827.

This species forms a shrub or small tree, reaching 10 m, and once seen the peculiar sea-green of its young leaves is unmistakable. These are also very gummy and glutinous, as are the young shoots. The dark bark is another sure indicator in the wild, but this does not develop properly in cultivation in the British Isles as the shrubs rarely develop to any size.

The name *Betula fontinalis* Sarg. was widely used for this species during the 20th century, but Hooker's earlier name is now considered correct (Furlow 1997).

ECOLOGY AND STATUS IN THE WILD

Betula occidentalis occurs as tall upright-growing, brown or brownish-black barked bushes, for example in the foothill country of southern Alberta east of the Rocky Mountains (Fig. 228), marking out watercourses and damp patches here and there in an otherwise low-rainfall and fascinatingly low-profile landscape in the rain shadow of the mountains. It also occurs commonly along rivers further westwards into the mountains in higher-rainfall areas, but is less conspicuous there amongst the richer vegetation except when in its golden-orange autumn colour. The few patches seen by the author (KBA) in eastern British Columbia and Alberta are only a minute part of its enormous range from Alaska and the Northern Territories of Canada southwards in the mountains to Arizona. Although mostly confined to the Cordillera, populations extend north-eastwards into Manitoba and even north-western Ontario. Its distribution and range seem to follow *Alnus tenuifolia* fairly closely, although this does not extend as far east (Hosie 1963; Furlow 1997).

Towards the northern end of its distribution, and especially in Alaska, hybridisation of *Betula occidentalis* with other birches has been recorded. There is no doubt that *B. occidentalis* hybridises with *B. pendula* subsp. *mandshurica* where they meet, for example along the Elbow River in Calgary, Alberta (KBA pers. obs.). Hybrids also occur with the shrub birch *B. glandulosa* to give the stout-twigged shrub *B. ×eastwoodiae*. According to Dugle (1966) this hybrid is found from Alberta northwards, but a population near Babb in Montana is evidence of its presence further south.

Hultén (1968), quoted in *Flora of North America* (Furlow 1997), suggested that what is referred to as *Betula occidentalis* in Alaska is in fact an extensive hybrid swarm between *B. pendula* subsp. *mandshurica* (as *B. resinifera*) and *B. glandulosa*. As all these species are diploid, knowledge of chromosome numbers will not resolve the question, but careful field observations combined with molecular work should do so. On the other hand, it is quite possible that 'true' *B. occidentalis* might also occur and be hybridising with the supposed hybrid swarm.

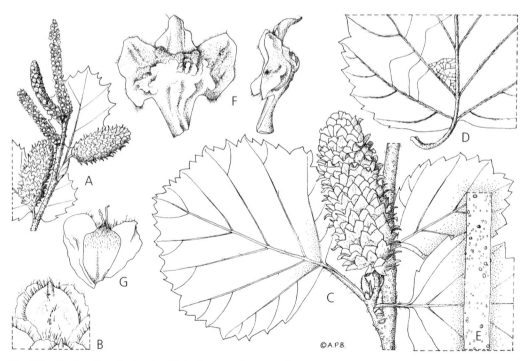

Fig. 227. *Betula occidentalis.* **A** twig with bud, mature fruiting catkins and mature male catkins × 1; **B** male catkin scale × 10; **C** fruiting shoot with mature fruiting catkin × 2; **D** underside of leaf base × 2; **E** bark of twig × 2; **F** female catkin scale, back and side views × 7; **G** seed × 7. All from *Howick & Warner* WH313 (Onion Valley, Inyo National Forest, California, USA, 24 Sept. 1986). DRAWN BY ANDREW BROWN.

Fig. 228. Shrubby *Betula occidentalis* on lake margin. Banff, Alberta (KBA).

CULTIVATION

It is a great disappointment that such an attractive birch fails to grow well in most sites in the British Isles and also in the Arnold Arboretum in Boston, Massachussetts. A good exception is in the Cambridge Botanic Garden where the climate of eastern England is more continental. The polished brown of the shining stems can almost rival those of *Prunus serrula*, but are often a duller black–brown, and their autumn colour can be most attractive. However, in practice the all too common appearance is of an untidy tangle of dead twigs, with fresh growth every summer which inevitably dies back during the following winter.

Where *Betula occidentalis* proves difficult to grow it is better to plant *B.* ×*utahensis*, its hybrid with *B. papyrifera*, especially the form from the Lochsa Valley, Idaho, which has most of the attractive features of *B. occidentalis* but can be paler in bark. The hybrid is always more vigorous, with larger leaves which are often sticky and frequently very shiny. Plants also sprout vigorous shoots from the base in the same way as the *B. occidentalis* parent.

Betula occidentalis Hook., *Fl. Bor.-Amer*. 2: 155 (1838). Type: Straits of De Fouca, *Dr Scouler*, near springs on the west side of the Rocky Mountains, *Douglas*; and on the east side, from the mountains to Edmonton House, *Drummond*.
Betula fontinalis Sarg., *Bot. Gaz*. 31: 239 (1901).

DESCRIPTION. *Shrubs* with several trunks, to 10 m. *Trunks* to 34 cm diam., dark reddish brown with conspicuous paler lenticels, not peeling. *Twigs* slender, densely resinous glandular, sparsely pubescent. *Young shoots* initially green, very glandular with translucent glands quickly turning brown, very sticky, especially towards apex. *Buds* about 6 mm, brown, acute, resinous and sticky. *Leaf blades* with petioles c. 10–13 mm, ovate to rhombic, base cuneate to truncate, with 2–6 pairs of veins, 40–50(–75) × 25–45 mm, sharply and coarsely toothed, often double toothed with 3–5 minor teeth between major; youngest leaves light green, thin, shiny and sticky, especially above, with many transparent glands. Petiole:blade ratio c. 1:3 to 1:5. *Fruiting catkins* ± erect to nearly pendulous, cylindrical-oblong, 20–30(–39) × 8–15 mm. *Scales* 3–6 × 4.5–5 mm with central lobe longer and narrower than the ascending rhombic lateral lobes. *Seed* c. 1.75 × 4 mm; nutlet c. 1 mm wide, wing broader than nutlet.

DISTRIBUTION. Rocky Mountains from California and New Mexico to Alaska (see above), and eastwards to Western Ontario.

HABITAT. Moist soils, often along stream banks, by springs or along lake shores; 100–3000 m.

CHROMOSOME NUMBER. 2n=28.

33A. BETULA ×CAERULEA BLUE BIRCH

HISTORY

The BLUE BIRCH, *Betula ×caerulea*, was originally named as a species by Blanchard in 1904. In the same year Blanchard also named *B. caerulea-grandis*, here treated as a synonym of *B. ×caerulea*. It has been shown that both are stabilised hybrids between *B. populifolia* (p. 297) and *B. cordifolia* (p. 333). The surface of the leaf is liberally dotted with glands — hence the name blue birch.

ECOLOGY AND STATUS IN THE WILD

Betula ×caerulea occurs in eastern North America from Massachusetts northward to Maine and New Brunswick and from New York State to Nova Scotia. *B. ×caerulea* is very beautiful and its brilliant white bark has pronounced 'eyebrows' which are as conspicuous as those on the trunks of *B. populifolia*. Stands of *B. ×caerulea* may be seen in cut-over and burned openings in secondary forest in the area where both the parent species grow.

TAXONOMY, RECOGNITION AND RELATIONSHIPS

In a series of thorough papers Brittain & Grant (1967a, 1969, 1972), Guerriero *et al.* (1970), Koshy *et al.* (1972), Grant & Thompson (1975), and De Hond & Campbell (1989) demonstrated that *Betula ×caerulea* is of hybrid origin between *B. populifolia* and *B. cordifolia*.

The differences between *Betula populifolia* and *B. ×caerulea* can be subtle: trees of *B. ×caerulea* have an upright habit, with a coarser silhouette than *B. populifolia* owing to thicker twigs, which are rough with many glandular warts. Bark on twigs and branches is pale brown to brown, tight, on trunks often pinkish, white or very white. Lenticels on the bark are thin to somewhat swollen, not so closely superimposed as in *B. populifolia*, and sometimes cinnamon-coloured, as in *B. pendula* subsp. *szechuanica*. The leaves are often much like those of *B. populifolia*, but longer and larger, and often subcordate at the base. The male catkins are quite long, up to 3.5 cm, not much curved and usually paired; the female flowers thin and erect, developing into fruiting catkins which are shortly cylindrical to somewhat ovate as in *B. populifolia*, but larger, equally persistent but not so hard and all hanging vertically.

Many authors, e.g. Britton & Brown (1913) have commented on the similarity between *Betula* ×*caerulea* and *B. pendula*, both subsp. *pendula* which is sometimes naturalised in North America, and subsp. *szechuanica*.

CULTIVATION

In cultivation in the British Isles, individuals of wild origin have been planted in various arboreta, but none are yet of any great size. It seems that the large tree called *Betula* ×*caerulea* at Hergest Croft cited in Bean (1970) and now dead, was in fact *B. pendula* subsp. *mandshurica*.

Betula ×caerulea Blanch., *Betula* 1: 1 (1904).
Betula caerulea-grandis Blanch., *Betula* 1 (2): 1=2 (1904).

DESCRIPTION. *Tree* up to c. 10 m, with upright habit, sometimes quite twisted; *bark* on twigs and branches pale brown to brown, tight; on trunks often reddish, pinkish, white or very white. *Lenticels* on bark thin to somewhat swollen, not so closely superimposed as in *B. populifolia*, and sometimes cinnamon as in subsp. *szechuanica*. *Twigs* rough with many glandular warts, or sometimes rather smooth even when young and strong, owing to a lack of glands. *Buds* to 6 mm. *Leaves* similar to those of *B. populifolia*, acuminate, but longer and larger, to 120 mm long, rather straight-sided, and with 5–7 pairs of lateral veins, liberally dotted with glands, not deeply double-toothed, truncate, cuneate or often subcordate at the base. *Male catkins* to 3.5 cm, slightly curved and usually paired, though occasionally in 3s or single. *Female catkins* thin and erect at flowering, at fruiting short-cylindrical to somewhat ovate as in *B. populifolia*, but larger, equally persistent and all hanging vertically. *Scales* to 5 × 5 mm, with body about equal to middle lobe, lateral lobes spreading, recurved. *Nutlet* elliptical, scarcely 1 mm, with wing as slightly wider than the nutlet.

DISTRIBUTION. North-eastern North America, from Newfoundland, and Quebec, to Manitoba, Maine and Vermont.

HABITAT. In secondary forest, usually with the parents but sometimes forming populations of the stabilised hybrid.

CHROMOSOME NUMBER. 2n=28.

34. BETULA CELTIBERICA

HISTORY, ECOLOGY AND STATUS IN THE WILD

Betula celtiberica was recognised as a distinct species by Werner Rothmaler and João de Carvalho de Vasconcellos in 1940, in a paper on the systematics of European birches. It was described from many of the mountain ranges of northwest Spain and eastern Portugal. Here we are only recognising the tetraploid birch populations of the Cantabrian Mountains as *B. celtiberica*. These are found in beautiful montane forests, and were the goal of the author's (KBA) first visit to European birch populations in 1977. The trees, bent by soil-creep and possibly also by winter snow, cling to very steep mountainsides, up to the treeline, above the upper limit of beech. Other populations of birch in isolated ranges in central Spain may belong to this species, but are thought to be more likely to be relict populations of *B. pendula* and perhaps also of *B. pubescens* (Peinado & Moreno 1989).

Morphologically, *Betula celtiberica* resembles *B. pendula* rather than *B. pubescens*, although it is mentioned under the latter by Max Walters in *Flora Europaea* (Tutin *et al.* 1964), probably on the basis of its chromosome number. It is perhaps a stabilised population derived from *B. pubescens* heavily introgressed with *B. pendula*. In addition to having twice as many chromosomes, *B. celtiberica* differs from European *B. pendula* subsp. *pendula* in the presence of axillary tufts of hair in the vein axils on the underside of the leaf, the greater hairiness of young shoots, and the less markedly double-toothed leaves. From *B. pubescens* it differs in the shape of the fruiting catkin scale, the lateral lobes usually being more or less recurved as in *B. pendula*, the shoots and twigs being rough with warts and more or less hairless, and the mature leaves being often rhomboid rather than ovoid and (apart from the hair tufts) more or less glabrous.

It might be questioned whether this taxon should be recognised as a distinct species. Its geographical isolation and the fact that it can be distinguished morphologically suggest that it is worth recognising. In *Flora Iberica*, Peinado & Moreno (1990) did not distinguish it from *B. pubescens*. However, it is equally out of place in *B. pendula*. In fact, as can be seen from the key, it is probably most similar to another 'species' of stabilised hybrid origin, the east North American *B. ×caerulea*.

CULTIVATION

In cultivation in the British Isles the male catkins develop earlier relative to the leaves than in related species so that the tree is showy in catkin (Fig. 230). It is one of the last birches to lose its leaves, which remain green well into November. It could therefore be expected to grow well in climates with a long, frost-free growing season where there is adequate summer rainfall.

The bark is variable in colour, but in some individuals is very white, so selecting individuals with the most attractive bark could yield useful amenity trees.

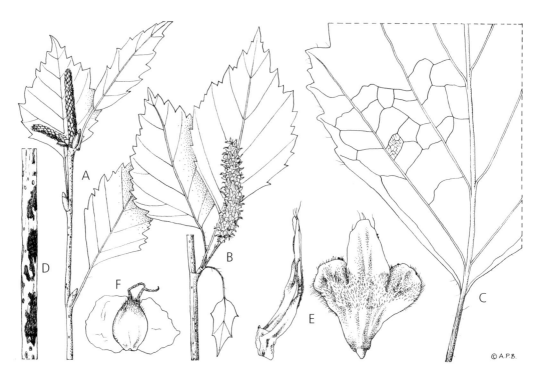

Fig. 229. *Betula celtiberica.* **A** twig with buds and immature male catkins × 1; **B** fruiting shoot with mature fruiting catkin × 1; **C** underside of leaf base × 2; **D** bark of twig × 2; **E** female catkin scale, side and back views (resin glands and hairs on inner surface) × 10; **F** seed × 10. All from fresh material collected from Stone Lane Gardens (cult. *K. Ashburner*, 19 Aug. 2010). DRAWN BY ANDREW BROWN.

Fig. 230. *Betula celtiberica* with male catkins before emergence of leaves (PW).

Betula celtiberica Rothm. & Vasc., *Bol. Soc. Brot.* ser. 2, 14: 147 (1940). Holotype (selected here): Cantabria: Puerto de Leitariegos, 1600 m, *Rothmaler* (COIMBRA).

Betula pubescens subsp. *celtiberica* (Rothm. & Vasc.) Rivas Mart., *Trab. Dept. Bot. Fisiol. Veg. Madrid* 3: 78 (1971).

DESCRIPTION. *Tree* to 10 m. Bark white or yellowish to the base. *Bud* sticky, *Shoot* initially puberulent, soon glabrescent, glandular. *Leaves* rhombic-ovate to broadly ovate, cuneate to ± truncate at base, single to somewhat double toothed, soon glabrescent except for axillary hair tufts in lower vein axils, these often absent on long-shoot leaves; petiole to 23 mm, leaf blade to 86 × 70 mm but usually smaller. *Fruiting catkin* with peduncle c. 10 mm, c. 28 × 7 mm. *Scale* c. 4–4.5 × 3.25 mm. *Seed* c. 2.25 × 3 mm with nutlet 2.25 × 1 mm; wings broader than nutlet.

DISTRIBUTION. Cordillera Cantabrica, north-west Spain.

HABITAT. On steep slopes, by streams and lakes in the mountains.

CHROMOSOME NUMBER. 2n=56.

35. BETULA PUBESCENS DOWNY BIRCH, WHITE BIRCH

HISTORY

Downy birch, *Betula pubescens*, is the most widespread birch in Europe and western Asia. The species was first described by Linnaeus in *Species Plantarum* (1753) as *Betula alba*, but since Linnaeus did not distinguish between the two different Swedish white birches, his name has generally been rejected in favour of the less ambiguous *B. pubescens* Ehrh. and *B. pendula* Roth (Govaerts 1996; Tuley 1973). *B. pubescens* is now known from Newfoundland, Greenland and Iceland, south to Spain and east to the Caucasus and Lake Baikal; as might be expected, it is very variable and numerous local varieties have been named as separate species.

From what is known of the history of trees with a similar distribution, it is likely that *Betula pubescens* survived cold periods during the Pleistocene in refugia not very far from the ice fronts (Willis & van Andel 2004; Hewitt 2004; Abbott & Brochmann 2003; Palmé *et al.* 2004; Svenning *et al.* 2008). Isolated populations would be expected to differentiate to some extent during the dry, cold glacial periods, but to merge and hybridise as they expanded to form more or less continuous populations during the warmer, moister interglacials. The most southerly populations have probably remained genetically isolated since at least the last (Eemian) interglacial, as has been suggested for *B. pendula* (Palmé *et al.* 2004), but have not, with the exception of *B. celtiberica*, diverged sufficiently to be recognised as new species (Maliouchenko *et al.* 2007).

The more or less isolated southern populations of *Betula pubescens* in the Pyrenees, Balkans, Caucasus, northeast Turkey and the Altai Mountains indicate where the most southerly surviving populations might have been (Willis *et al.* 2000). Thus, considerable variation probably survived within *B. pubescens* through the Full Glacial, to be added to by gene flow from *B. pendula*, mainly in the south, and from *B. nana* in the north.

There is increasing evidence for the survival of woody species much further north than was once thought (Stewart & Lister 2001). *Pinus sylvestris* and other warmth requiring trees might have survived in a refugium in Atlantic northwestern Europe off Ireland (Sinclair *et al.* 1998; Rowe *et al.* 2006; Teacher *et al.* 2009), and perhaps even off Norway (Stewart & Dalén 2008). Unlike the pine, the more cold-tolerant *Betula pubescens* is capable of sprouting from stumps, so is likely to have been the most northerly surviving tree, as it still is in western Europe. These North Atlantic per-glacial survivors could be the origin of the thin twigged small-leaved birches of the north and west of the British Isles, which are here called var. *fragrans*. They differ from the Scandinavian and Icelandic *B. pubescens* in lacking many of the characters (such as stiff hairy twigs and more or less erect fruiting catkins), which suggest introgression from *B. nana*. The variation found in Scandinavia is very great and might suggest derivation from several local per-glacial surviving populations (Provan & Bennett 2008; Binney *et al.* 2008), but the possibility of this is much disputed (Birks *et al.* 2006 and references therein).

TAXONOMY, RECOGNITION AND RELATIONSHIPS

Comparison of *Betula pubescens* and *Betula pendula*

A consistent character to separate the two birches, *Betula pubescens* and *B. pendula*, is that, in almost all *B. pubescens* the young shoots and leaves are densely velvety with very short, fine puberulent hairs and usually with few resinous glands. *B. pendula* subsp. *pendula*, the only subspecies sympatric with *B. pubescens*, rarely shows this and its twigs are usually soon totally glabrous and usually with conspicuous resinous warty glands. Even where introgression from *B. pendula* has occurred, as suggested by somewhat double-toothed leaves, young shoots and leaves of *B. pubescens* are velvety although the hairs may fall later in the season leaving glabrous twigs. The shoots of the small leaved *B. pubescens* var. *fragrans* in western and highland Scotland are also often soon glabrescent, but the small oval leaves with only a single order of toothing, and fruiting catkin scales with forward pointing lateral lobes are very different from those of any *B. pendula*.

The commonest leaf-shape in *Betula pubescens* is broadly ovate, perhaps fancifully spade-shaped, as seen in a pack of cards, with a single order of shallow teeth, although leaves of more southerly origin are often more or less double toothed. This oval leaf contrasts with the triangular to rhombic shape with double toothing typical of *B. pendula* subsp. *pendula*. Glandular warts on twigs are often scarce and a clear brown where present. After introgression from *B. pendula* there may be a thin scattering of blue-white warts. The species most similar to *B. pubescens* is probably *B. papyrifera*, but this can usually be distinguished by its larger (> 5 cm), less glossy leaves with more numerous (7–9) pairs of veins, and by the often spreading or reflexed lateral lobes on the scales of the fruiting catkins (erect in *B. pubescens*), and wider seed wings.

GENETIC ORIGIN AND INTROGRESSION

All individuals of *Betula pubescens* so far examined are tetraploid and hybridisation appears to take the form of introgression from other species into *B. pubescens*, through unreduced gametes from the

Fig. 231. Mistletoe on birch, Poland March 1995 (MBM).

diploids *B. pendula* and *B. nana*, pollen from partially fertile triploid hybrids, and possibly fertilisation of unreduced triploid gametes by haploid gametes from the diploids, producing tetraploids infertile with *B. pubescens*. Molecular work supports this conclusion, the genetic and morphological variation in *B. pubescens* being explicable by gene flow from these two species (Vaarama & Valanne 1973; Nokes 1979; Atkinson 1992; Palmé *et al.* 2004; Maliouchenko *et al.* 2007; Thórsson *et al.* 2007).

The only other species which might contribute to the genome of *Betula pubescens* in its core area is diploid *B. humilis*, which certainly does hybridise with *B. pubescens*, but little work seems to have been done on this beyond the statement that obvious hybrids exist. Only at the extreme limits of the distribution of *B. pubescens* might other species be involved: in Labrador, *B. pumila*; in the Lake Baikal region in east central Asia, *B. ermanii*; and in the Altai and Pamir region in central Asia *B. fruticosa*, *B. microphylla* and *B. tianshanica*. All of these are tetraploid like *B. pubescens*, so hybridisation with them would be easy and the offspring are likely to be highly fertile. Perhaps due to lack of study in these relatively remote areas, hybrids with these species have hardly ever been reported in the literature, and would probably be difficult to detect.

Throughout its distribution from Europe eastwards to Lake Baikal, *Betula pubescens* has much the same distribution as *B. pendula*, although extending further north and not as far south (Jalas & Suominen 1976), and often growing in colder and wetter habitats.

Introgression from *Betula pendula* is so frequent that it can sometimes be difficult to tell the two species apart in the wild, or for that matter in cultivation. The distinctions between *B. pubescens* and *B. pendula* are discussed above. Introgression proceeds not only when populations of the two species overlap, as when *B. pubescens* grows adjacent to *B. pendula*, but even when they are apart, for birch pollen is carried over great distances.

Fig. 232. *Betula pubescens.* Les Pontins, St. Imier, Switzerland, wet ground in conifer forest (IR).

Figs 233 (left) and **234** (right). *Betula pubescens*. Les Pontins, St. Imier, Switzerland (IR).

In the past *Betula pubescens* and *B. pendula* probably grew in different habitats because of their different ecological preferences, *B. pubescens* in wetter, colder sites and *B. pendula* in drier, warmer habitats. Today habitat disturbance allows both species to colonise open habitats where there is little or no competition. These two birches, with some willows (especially *Salix caprea*) and *Buddleja davidii*, are the main species in the British Isles whose seeds can reach derelict industrial or moorland sites.

GEOGRAPHY, ECOLOGY AND STATUS IN THE WILD

Betula pubescens occurs from Newfoundland, Iceland, the British Isles and Spain eastwards across all but the very south of Europe to the Lake Baikal region in Siberia, and from the northern limit of tree growth throughout this area to as far south as 40° in Europe; the southernmost localities are in Turkey, the Caucasus and the Altai Mountains. The most northerly shrubby form occurs in Greenland, in Newfoundland and probably Labrador in North America, and probably elsewhere. Some of what has been named *B. borealis* (see *B. pumila*) and some of what has been named *B. minor* (see Chapter 2) in North America is probably this form. Other records from North America are introductions from Europe.

Betula pubescens is absent from eastern Asia and may reach its eastern limit on the Barguzin Ridge on the north-east shore of Lake Baikal; east of this it is replaced by *B. ermanii* var. *lanata*.

Betula celtiberica Rothm. & Vasc. (see p. 303) is a group of isolated populations in the Cantabrian Mountains in Northern Spain, tetraploid like *B. pubescens*, but morphologically close to *B. pendula* and possibly much influenced by introgression from *B. pendula*. After long isolation it is now sufficiently distinct to be recognised as a separate species.

REGIONAL VARIANTS OF BETULA PUBESCENS

Although there is great variation in habit and stature in the trees associated with different habitats, there are no clear boundaries between the different forms. The most useful taxonomic treatment is to give names to the most distinct, but only at the varietal level. Following Väre (2001), we therefore use var. *pubescens* for the tall central European and southern British and Scandinavian populations; var. *pumila* for northern populations, which are probably derived from var. *pubescens* through adaptive selection and introgression from *B. nana*; var. *fragrans* for aromatic, small-leaved trees from northwest Britain; and var. *litwinowii* for Anatolian and Caucasian populations which are probably largely the result of introgression from *B. pendula*.

CENTRAL, SOUTHERN AND WESTERN EUROPE (*Betula pubescens* var. *pubescens*)

Betula pubescens var. *pubescens* is most widely distributed and abundant in forests in mid-latitude Scandinavia and northern Russia where succession to climax *Pinus* or *Picea* has been constantly frustrated by human influence such as logging or fire. Trees are white-stemmed with bark that sometimes peels quite spectacularly. The dense populations in wet habitats further south in Europe are also traditionally referred to this variety. This low latitude, low altitude *B. pubescens* which grows in fens and other wet places, for example in north Germany and Poland and at similar latitudes in Russia, is tall and flexible and often similar to *B. pendula* in habit (Fig. 231).

In southern Europe the populations are disjunct and confined to mountains and hills such as the Pyrenees, the Massif Central, the Jura, and the Alps and the Carpathians. Populations in the Carpathians were named *Betula carpatica* or var. *glabrata*, but do not differ significantly from var. *pubescens*.

Fig. 235. *Betula pubescens* forest, Storlien, Sweden (KBA).

Fig. 236. Populations of *Betula pubescens* in birch Arboretum, Stone Lane Gardens, Devon, England (KBA).

In the Jura these birches grow in the peat which accumulates in shallow valleys known as 'combes' formed from the splitting of the anticlines of the Jurassic limestones, (rather like the cracking of the upper surface of a Swiss roll), and there is a particularly fine population at Les Pontins in the Swiss Jura where the bark of the trees is very white (Figs 232–234).

Betula murithii was described from Mauvoisin, at the head of the Val de Bagnes in Valais, and also from the valley of the Aar — both in Switzerland. Its leaves appear a little broader than those from populations in the Massif Centrale and the Jura, and they approach the shape seen in var. *pumila*, but the toothing is double rather than single. We consider these populations to fall within var. *pubescens*. The trees are very white barked and slightly twisted and form attractive groves at the head of these beautiful valleys.

Betula pubescens is seen in the Vosges, and in the Ardennes on the wet and flat tableland of the Hoch Venn and Haute Fagnes. Further east into central Europe such montane trees are found mostly in isolated patches of fen or on rocky outcrops, especially when interrupting thick spruce forest, as in the highlands of the Bayerischer Wald, Thuringer Wald and Harz mountains in Germany, as well as in the highlands of the Czech Rebublic, the Tatras and Carpathians of Slovakia and Poland. Such scattered, white barked, rugged trees are often splendidly gnarled and twisted.

Betula pubescens appears to be rare in the Alps proper, apart from the small populations in the Valais mentioned above. It occurs scattered through the mountains in the Transylvanian Alps in Romania and in the Balkans, but is apparently absent from Bulgaria (Jalas & Suominen 1976).

Fig. 237. *Betula pubescens.* Fastigiate '*callosa*' from above Narvik, N Norway (KBA).

These montane populations of *Betula pubescens* in southern Europe and Asia have been treated in the past under several names — often at specific level — such as *B. odorata*, *B. glutinosa* and *B. rhombifolia*, to name three. The differential characters used such as leaf and bud shape appear to fall within those of *B. pubescens*. However, their winter buds are commonly sticky and resinous, a feature which only develops towards the end of summer. Barks are more consistently whiter than those of var. *pumila*, and indeed sometimes very white and, as with trees of var. *pumila*, these montane southern populations are more twisted than lowland individuals.

In Atlantic western Europe, populations in Brittany and those occurring on the fringes of Dartmoor in south-west England are sometimes known as the brown birch, with perhaps the most velvety-hairy shoots, twigs and leaves of any *Betula pubescens*. These trees can be very tall. The young leaves are strongly crimped and wrinkled by more deeply impressed veins, both secondary and tertiary, than is usual in *B. pubescens*, and this persists in older leaves. Young twigs are a warm crimson-brown and stand out in the winter sun. There may be as many as eight or even nine pairs of lateral veins as on some trees in Brittany, and twigs are not only velvety, but more or less rough with brown resinous warts. The leaf margins are crenate rather than serrate, and barks are whitish-brown or brown, with trunks of older trees deeply fluted. Quite possibly these are remnants of once extensive montane populations which would have reached the tree line on Dartmoor and the moors of Brittany; but are now much reduced by centuries of grazing. Although recognisable in the wild,

Fig. 238. Almost impenetrable scrub dominated by *Betula pubescens* var. *pumila* with *Salix lanata* (left foreground), *Sorbus aucuparia* (taller trees), and *Salix phylicifolia* (yellow-green in sunshine in centre distance). Trostansfjordur, west Iceland (HMcA).

Fig. 239. *Betula pubescens* var. *pumila* and *Betula nana* colonising 5,000 year old lava flow, Husafell, W Iceland (HMcA).

there are no definite characters by which these can be distinguished from more eastern var. *pubescens*. However, we (KBA) speculate that it is among the brown barked birches of Brittany, Devon and Cornwall that the elusive diploid *pubescens* may one day be found if such exists.

Scotland (*Betula pubescens* var. *fragrans*)

The mountain birches of the otherwise bare hills of the Scottish Highlands, were referred to subsp. *odorata* (Bechst.) E. F. Warb. in Clapham, Tutin and Warburg's *Flora of the British Isles* (1952), but they are not the same as Bechstein's *Betula odorata*. They are smaller-leaved and with much thinner shoots and twigs than the European mountain trees or the northern trees and bushes here treated as var. *pumila*. The thin twigs on mature trees droop slightly and bear sticky buds and fragrant young leaves. This scent is carried on the wind, and is particularly strong after an early summer shower (Figs 248, 251–254). We here describe this variety as var. *fragrans* Ashburner & McAll.

Caucasus and Turkey (*B. pubescens* var. *litwinowii*)

In his account of *Betula* in *Flora of Turkey* (Davis 1972), K. Browicz recognised four species: *Betula medwediewii* (see p. 189), *B. pendula* (see p. 277), *B. litwinowii* and *B. recurvata*. *B. litwinowii*, from the Caucasus and the mountains of north-eastern Turkey as far west as Gümüşane, Tunceli and Erzinçan, is difficult to accept as anything other than a variety of *B. pubescens*. With their double-toothed leaves, pubescent at first on the veins of the underside, these trees are hardly distinguishable from those of the Swiss Jura or the Pyrenees. *B. recurvata*, described from Sarikamiş differs mainly in its larger leaves and absence of resin glands, and the recently described *B. browicziana* also falls within *B. pubescens* var. *litwinowii*.

Fennoscandia and the Kola Peninsula (*B. pubescens* var. *pubescens* and var. *pumila*)

Betula pubescens of the mountains and the arctic and subarctic is wind-blasted, slow-growing, rugged and often twisted, perhaps partly due to the effects of the weight of winter snow, partly as a result of gene flow from *B. nana*. *B. pubescens* is strongly associated with the wild, rugged habitats — with remoteness and exposure, where it adds an extra and evocative element and enhances these landscapes, whether they be the barren lava flows of Iceland (Figs 239, 246), or the vast expanses of *Betula* forest in Fennoscandia which extend northwards to the treeless tundra (Fig. 235).

The appearance of trees in montane and subarctic populations in Scandinavia is strikingly distinct. They frequently show a great increase in the number of short, rigid spur shoots on the branches in place of flexible long shoots. The trees appear more rigid and are certainly slower growing, and they can be splendidly gnarled and twisted, or, in contrast, almost perfectly straight.

As is suggested by Väre (2001), and exemplified in detail by Kenworthy *et al.* (1972) and Thórsson *et al.* (2007) for variation in northern *Betula pubescens*, similar variants of *B. pubescens* have arisen independently several times following gene flow from *B. nana*; indeed, named variants such as '*tortuosa*' (described from the Altai in central Asia, but often applied to Scandinavian trees) are likely to be similar looking trees of similar hybrid origin.

In a rigorous account of the history of the classification of Nordic *Betula pubescens*, Väre (2001), describes some of the previous ridiculous, and presumably quite unworkable, classifications. No less than six birch species were listed as native in a single Norwegian county: Peter Benum's *The Flora of Troms fylke. A floristic and phytogeographical survey of the vascular flora of Tromsø fylke in Northern Norway* (1958), quoted by Vaarama & Valanne (1973), gives four species, *B. tortuosa*, *B. odorata*, *B. callosa*, and *B. coriacea*, within what we are here treating as *B. pubescens*. It is unlikely that four wind-pollinated species with the same chromosome number could remain distinct in such a small area.

The fine subarctic and arctic forests of Fennoscandia and the Kola Peninsula represent the most distinguishable group of populations within *Betula pubescens*, and were referred to as *B. alba* var. *pumila* by Linnaeus in *Flora Suecica* in 1745, and, more recently, to subsp. *tortuosa* or subsp. *czerapanovii*. These populations are extensive and dominate the vegetation over large areas. To ascend the ski lift at Abisko in the mountain country in north Sweden in midsummer is to look down on a sea of birch and see a vast carpet of translucent green clothing the lower slopes of the mountains down to the shores of Lake Torneträsk. In late September, the peak of autumn at these latitudes, this becomes a spectacular blanket of gold.

Fig. 240. HMcA collecting seed from a shrub of *Betula pubescens* var. *pumila*, Husafell, W Iceland (PP).

Within these forests, the barks are magnificent in their variety. They are mostly white, but there are reds and even near-blacks as well. Textures range from the rough and flaky to the polished and smooth, the prominent lenticels often gape. Enjoyment of such barks in cultivation is frustrated by the poor performance of seedlings at lower latitudes in more temperate climates. They are impossibly precocious in budburst, especially when young, and nearly all are doomed to succumb to spring frosts as young seedlings; great patience is needed to nurse any that survive as they are slow growing and easily smothered by weeds.

Väre (2001) quotes Hylander (1996) as stating that the change from var. *pubescens* to var. *pumila* is completely clinal, *i.e.* gradual, and yet it can in fact appear abrupt where the lowland and montane populations are contiguous as they are in Scandinavia. The boundary is easily detected wherever there is a sudden gain or loss of altitude throughout these northern latitudes, whether in Scandinavia or further east into Russia and Siberia.

The sudden change with increasing altitude is noticed even when on foot, but especially so when travelling comparatively swiftly by car or train. One example from Scandinavia which provides a good opportunity for observing this quick transition is the journey from Storlien on the Swedish-Norwegian border westwards towards Åre. The increasing frequency of the more lax-growing var. *pubescens* is very noticeable as you descend. Far to the north of this, a walk from the town of Gällivare to the slopes of Dundret reveals unmistakeable change with a surprisingly sharp altitudinal boundary. After passing through abundant, flexible and vigorous low-level var. *pubescens*, with young seedlings crowded by the roadside, there is quite a sudden increase in the stiffness of the trees, a reduction in numbers of seedlings, and a prolific suckering from the bases of the tree trunks, all marking the transition to var. *pumila*.

Fig. 241. Almost impenetrable scrub of *Betula pubescens* var. *pumila*, Trostansfjordur, W Iceland (HMcA).

Fig. 242. *Betula pubescens* var. *pumila*, Canada, Newfoundland, Daniel's Harbour: cultivated at Ness Gardens. Fruiting catkins and hairy twig (PW).

There are other variants of note within *Betula pubescens*. Trees of *B. pubescens* in some northern Scandinavian forests, as around Narvik in north-western Norway, are of moderate size but have strikingly narrow crowns so that they look almost fastigiate, perhaps an adaptation to heavy snowfall (see below). After such emphasis on the wayward shapes and contorted nature of trees in the mountain forests, it seems odd to find these neatly upright, sometimes near-fastigiate, trees growing on these mountain slopes (Fig. 237). According to Lindquist (1945), these fastigiate trees grow in other places in the Scandinavian montane forests and he considered that they, especially the ones with upright fruiting catkins, represent a different species, *B. callosa*, scattered here and there in the forests of var. *pumila*.

However, *Betula callosa* cannot be accepted as a species or even subspecies, so many are the intermediates with other forms of *B. pubescens*. The sharp single toothing of its leaves is similar to that in var. *pumila* and distinct from the double or even crenate toothing in var. *pubescens*. Possibly, therefore, it is an ancient and major component of the var. *pumila* populations with little evidence of introgression from *B. nana*. A study of the genetic basis of the distinctness of *B. callosa* might show whether it is derived from a relict population which survived at least the last glaciation in isolation and perhaps further north than other populations (Stewart & Lister 2001; Palmé & Vendramin 2002).

Betula coriacea Gunnarsson is the name sometimes given to the very distinct type of tree seen everywhere in cultivation in towns such as Boden, Bästuträsk, and Gällivare in northern Sweden. Populations, or perhaps only isolated individuals, presumably exist in the wild. Its only rival in stature

Fig. 243. *Betula pubescens* var. *pumila*, Daniel's harbour, Newfoundland, Canada, female catkins, 26 April 2006 (PW).

at these latitudes is the tall var. *lapponica* of *B. pendula*, also frequently planted. The bark of *coriacea* is very white and scored with deep, dark, gash-like lenticels. The leaf in its broadness and short-acute apex resembles that of var. *pumila*, but is generally far larger, correlating with the much greater vigour and size of the trees.

Betula concinna Gunnarsson comes from the south of Sweden and appears from herbarium specimens to be very similar to *B. coriacea*. Both the species that Gunnarsson named remain enigmatic, similar to one another, and distinct from other forms of *B. pubescens*.

RUSSIA AND NORTHERN ASIA (*B. pubescens* var. *pubescens* and var. *pumila*)

Betula tortuosa described by Ledebour from the Altai, *B. borysthenica* Klokov from central Ukraine and central Russia, *B. kusmischeffii* (Regel) Sukaczev from the White Sea coast, *B. rezniczenkoana* (Litv.) Schischk and probably also *B. kelleriana* Sukaczev, from west Siberia, *B. ircutensis* Sukaczev and *B. baicalensis* Sukaczev, from near Lake Baikal in east central Siberia, are all best treated as synonyms of *B. pubescens*, although we have not seen the type specimens which are in St Petersburg and a final decision must await their study.

Betula tortuosa from the Altai Mountains is likely to be only distantly related to the birches of Scandinavia. Orlova (quoted in Väre 2001) concluded that subsp. *tortuosa* in the Altai was a naturally occurring hybrid between *B. microphylla* and *B. rotundifolia* (= *B. glandulosa*), geographically very far away from, and differing from the montane/subarctic *B. pubescens* of Fennoscandia, for which he accepted the name subsp. *czerepanovii* (= var. *pumila*), a very probable interpretation.

Another complication is the possible penetration of *Betula pubescens* into south-eastern Asia as far as Tibet and Nepal. It is even listed in Wang's *Forests of China* (Wang 1961) as occurring on the mountains in Sinkiang and Mongolia.

In eastern Siberia the appearance of *Betula ermanii* var. *lanata* in the mountains to the east of Lake Balkal produces further intermediates. Several collections obtained from the Barguzin ridge on the north-east shores of the lake showed clear evidence of introgression. At the seedling stage one batch was clearly *B. ermanii* on the basis of hairiness and leaf toothing, but these died during their first winter. Others were intermediate in appearance between these and *B. pubescens* and these survived for a few years but eventually died. Only those which could be considered more or less 'pure' *B. pubescens* have survived in the long term and grown to maturity, but even they suffer in some years from late spring frosts.

CENTRAL ASIA (*B. pubescens* var. *pubescens* and var. *pumila*)

Southwest of the Altai Mountains in Central Asia, in the Tian Shan and Pamirs, in Kirghizstan, Tadjikistan, Turkmenistan, the situation is uncertain. Many of these isolated birch populations have been described as separate species: they are discussed and listed below under *Betula tianshanica*. Some may be forms of *B. pubescens*, but most probably represent isolated populations of hybrid origin involving *B. pubescens*, *B. fruticosa*, *B. microphylla*, or even *B. utilis* subsp. *occidentalis* (Eastwood *et al.* 2009).

Fig. 244 (left). *Betula pubescens* var. *pumila* with small leaves and erect fruiting catkin, Skutustadhir near Myvatn, Iceland (PW).

Fig. 245 (right). *Betula pubescens* var. *pumila*, Skutustadhir near Myvatn, Iceland, with undergrowth of *Equisetum pratense* with *Archangelica officinalis* and *Alchemilla* sp. (PW).

To sum up: it seems that *Betula pubescens* occurs as far south and east as the Altai mountains in central Asia. Further south and west in the Tian Shan mountains its place is taken by the *B. tianshanica-microphylla* aggregate, which may itself be of *B. pubescens* × *B. fruticosa* hybrid origin, and this may meet and intergrade further south and east in the Pamir and Hindu Kush with *B. utilis* of the Himalaya.

GREENLAND AND NORTH AMERICA (*B. pubescens* var. *pumila*)

In North America the tree form of *Betula pubescens* only occurs as an introduction in cultivation and occasionally an escape (Furlow 1997). In the wild its ecological niche near the tree-line is occupied by *B. cordifolia* in the east, and *B. pendula* subsp. *mandshurica* (syn. *B. neoalaskana*) in the west and possibly occasionally *B. papyrifera* elsewhere.

Betula pubescens var. *pumila* (subsp. *tortuosa* (Ledeb.) Nyman of *Flora of North America*, Furlow 1997) is reported from protected inland valleys in south-west Greenland where it grows to at least 6.5 m, but with many populations being much lower growing and containing obvious hybrids with *B. nana*, often confirmed as triploids (Sulkinoja 1990). From pollen analysis, *B. pubescens* is calculated to have arrived from Iceland around 3,500 years ago (Fredskild 1991). He concluded that, like *B. nana*, it did not survive in Iceland or Greenland during the last glacial period and recolonised in the post glacial, presumably from Iceland, which in turn is proposed to have been recolonised from Scandinavia. However, this scenario is not universally accepted (Rundgren & Ingólfsson 1999).

As discussed in Chapter 2 on Hybridisation, dwarf shrubby forms referable to var. *pumila* have also been found at Daniel's Harbour in Newfoundland, perhaps somewhat introgressed with *Betula pumila*, and similar shrubs may occur elsewhere in Newfoundland, Labrador and Quebec.

Fig. 246. *Betula pubescens* var. *pumila* colonising lava, Dimmuborgir, Iceland (PW).

Fig. 247. *Betula pubescens* var. *pumila* growing along the line of an active (i.e moving) ridge, Ludentarborgir, Iceland (PW).

CULTIVATION AND PROPAGATION

Betula pubescens thrives in moist habitats in gardens or arboreta, with a tolerance of water-logging almost matching that of *Alnus*. As with all woody plants, the usual difficulties arise in attempting to grow plants which originate in a continental climate, in a maritime climate with cool summers and usually mild winters. In such climates budburst is precocious and may be stimulated by mild temperatures as early as January; any survivors of late frosts develop a poor habit. In the case of var. *pumila* this precociousness seems to vary a little, so that it is often possible to select the small proportion of survivors with later bud-break.

Soft cuttings of *Betula pubescens*, taken in May or early June, root relatively easily under mist; and scions may easily be grafted on to *B. pendula* or *B. pubescens* stock from January until March. Propagation is most easy by seed, so that parents with the desired characters may be used. Where whiteness of stem is required, seedlings from continental populations, particularly those of the Swiss Jura, are most rewarding.

Betula pubescens Ehrh., *Beitr. Naturk.* 6: 98 (1791). Type: Europa. *Ehrh.* arb. n. 67.

DESCRIPTION. *Trees* or *shrubby trees* to 20 m, reduced to less than 1 m in extreme habitats, particularly in northern tundra and on mountains. *Habit* either straight and upright with conical crown, or more spreading and twisted with flatter crown. *Ultimate shoots* and *twigs* often tending to be somewhat pendent except in some mountain shrubby forms. Main trunks often fluted, and with burrs either on

Fig. 248. *Betula pubescens* var. *fragrans*, Sutherland, Scotland, UK (KBA).

trunk or at base, or cylindrical and unfluted. *Twigs* grey to reddish-crimson in winter, 1–3 mm diam. *Lenticels* on twigs small, dot-like and most often inconspicuous. *Buds* ovoid, most often pointed, but rounded at apex in montane taxa, from 4–10 × 1.5–3 mm, scales brown where exposed to sunlight, otherwise green, either dry and not fragrant, or sticky and sweetly so, with clear resin. *Young shoots* more or less densely covered with very short (c. 0.3 mm) erect puberulent hairs, often also with longer hairs, these hairs persistent through the summer to glabrescent, variably glandular with clear brown resinous glands. *Petiole* to about 13–32 mm. *Leaf blades* flat, 20–105 × 15–70 mm, ovate to rhombic, occasionally deltoid, shortly pointed, rounded or cuneate at base, less often shallow-cordate; bluish-green above and paler below with resin-dots, margins sharply serrate, sometimes doubly so with forward-pointing teeth, or with single serration; surfaces always shiny, glabrous on older shoots but with dense pubescence and velvety texture on young shoots. Midrib most often slightly twisted towards apex. Lateral veins 5–7 pairs. Petiole:leaf blade ratio from 1:2.5 to 1:6.5. *Male catkins* straight or slightly curved, paired terminally, sometimes with an axillary third close below, from 7.0–25.0 × 2.5–3.5 mm. *Fruiting catkins* with peduncle (5)10–15 mm, pendent, or upright in var. *pumila*, cylindrical, 10–57 × 6–7 mm; *scales* 3–4.5 × 3–5 mm, with a few hairs on both upper and lower surfaces, margin ciliate; lobes forward pointing, occasionally ± spreading. *Seed* 2.25 × 3.25 mm; style 0.7 mm; wing c. 1 mm, wider than nutlet, translucent, one cell thick.

DISTRIBUTION. Newfoundland and probably adjacent mainland Canada, S Greenland, Iceland, the British Isles and Spain eastwards to the Lake Baikal region in Siberia, and from the northern limit of tree growth to as far south as 40° in Europe and east to Turkey and the Caucasus and the Altai Mountains in Central Asia.

Var. **pubescens**

Betula odorata Bechst., *Roem. Arch. Bot.* 2: 73 (1799).

Betula carpatica Waldst. & Kit. ex Willd., *Sp. Pl.*, ed. 4 [Willdenow] 4 (1): 464 (1805).

Betula pubescens var. *glabrata* Wahlenb., *Fl. Carpat. Princ.*: 306 (1814).

Betula glutinosa Wallr., *Sched. Crit.* 497 (1822).

Betula rhombifolia Tausch, *Flora* 21 (2): 752 (1838).

Betula callosa Notø, *Tromso Mus. Aarsh.* 23: 158 (1901); Lindquist, *Svensk Bot. Tidskr.* 39: 183 (1945), descr. emend.

Betula murithii Gaudin ex Regel, *Prodr.* (DC.) 16 (2.2): 168 (1868).

Betula coriacea Gunnarsson in Lindm., *Svensk Fanerogamfl.*: 201 (1918) non Pamp., *Nuovo Giorn. Bot. Ital.*, n.s. 22: 274 (1915).

Betula concinna Gunnarsson in Lindm., *Svensk Fanerogamfl.*: 202 (1918).

Betula borysthenica Klokov, *Bot. Zhurn.* (*Kiev*) 3, No. 1–2: 18 (1947).

DESCRIPTION. *Tree* usually with single trunk to 20 m, bark white to very white. *Twigs* often remaining puberulent into second year. *Buds* very sticky or not. *Young shoots* densely velvety puberulent hairy with or without longer hairs. *Petiole* 15–32 mm. *Leaves* ovate to rhombic, cuneate to rounded at base, initially pubescent above and below, becoming glabrescent except in vein axils of preformed leaves, to 56–105 × 38–70 mm; vein pairs c. 6. *Fruiting catkins* pendulous.

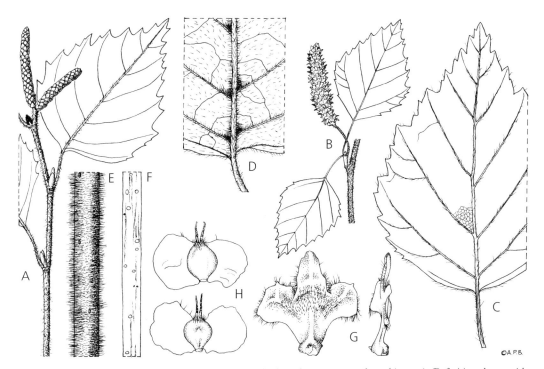

Fig. 249. *Betula pubescens* var. *pubescens*. **A** twig with buds and immature male catkins × 1; **B** fruiting shoot with mature fruiting catkin × 1; **C** underside of leaf × 2; **D** underside of leaf base of a very hairy individual × 2; **E** young twig showing dense indumentum × 4; **F** bark of twig × 2; **G** female catkin scale, back and side views × 8; **H** two seeds showing varying proportions of seed body and wing × 8. **A, D, F** from *Fraser* s.n. (Arbrook Common, UK, 4 Sept. 1910); **B, C, E, G, H** from *Fraser* s.n. (Sheen Common, Surrey, UK, 20 June 1930). DRAWN BY ANDREW BROWN.

DISTRIBUTION. As the species, except for Greenland, North America and Arctic Fennoscandia.

HABITAT. In wet woods, fens, wet mountainsides, gaps in conifer forest, usually of *Picea*, *Pinus*, or *Larix*.

Var. **pumila** (L.) Govaerts, *World Checkl. Seed Pl.* 2 (1): 10 (1996).

Betula pumila L., *Mant. Pl.*: 124 (1767).

Betula tortuosa Ledeb., *Fl. Ross.* (Ledeb.) 3 (2, 10): 652 (1850).

Betula pubescens var. *tortuosa* (Ledeb.) Koehne, *Deutsch. Dendr.*: 109 (1893).

Betula borealis Spach, *Ann. Sci. Nat., Bot. sér.* 2, 15: 196 (1841), *sensu* Fernald (1950).

Betula czerepanovïi N. I. Orlova, *Vestn. Leningrad. Univ., Biol.* 1978 (3): 60 (1978).

Betula pubescens var. *czerepanovii* (N. I. Orlova) Hämet-Ahti, *Memoranda Soc. Fauna Fl. Fenn.* 65: 9 (1989).

DESCRIPTION. *Small tree* to *dwarf shrub* with stiff branchlets. *Twigs* thicker than in the type to as thin as in *B. nana*. *Young shoots* densely hairy. *Leaves* usually smaller than in type; vein pairs 6 or fewer. *Fruiting catkins* often erect.

DISTRIBUTION. Newfoundland (and adjacent mainland North America?), Greenland, Iceland, Arctic Fennoscandia, N Russia.

HABITAT. In sheltered rocky valleys in the arctic, and on mountains further south.

Probably derived repeatedly from var. *pubescens* through adaptive selection of northern populations and introgression from *Betula nana*. Variation appears to be continuous between var. *pubescens* and tetraploid individuals indistinguishable from *B. nana* (Thórsson *et al.* 2001, 2007).

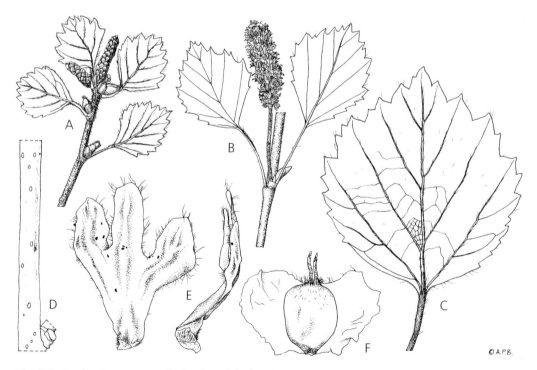

Fig. 250. *Betula pubescens* var. *pumila*. **A** twig with buds and immature male catkins × 1; **B** fruiting shoot with mature fruiting catkin × 1; **C** underside of leaf × 2; **D** bark of twig × 2; **E** female catkin scale, back and side views (resin glands present) × 10; **F** seed × 10. All from Norway, Nordland, SE of Narvik, head of Beisfjord, (cult. Ness). DRAWN BY ANDREW BROWN.

Fig. 251. *Betula pubescens* var. *fragrans*. Old tree near Fort Augustus, Invernessshire, Scotland (EMR).

Fig. 252. *Betula pubescens* var. *fragrans*. Old tree near Fort Augustus, Invernessshire, Scotland (EMR).

Var. **litwinowii** (Doluch.) Ashburner & McAll., **comb. et stat. nov.**

Betula litwinowii Doluch., *Zametki Sist. Geogr. Rast.* 7: 14 (1939). Type: in montibus totius Caucasi (Talysch excepto), atque in montibus Armeniae Turcicae, Kurdistaniae et Asiae minoris, 1700–2000 m.

Betula litwinowii Doluch. var. *recurvata* I. V. Vassil., *Bot. Zhurn.* (*Moscow & Leningrad*) 36 (6): 617 f. 7 (1951). Type: Kars opp. Sary-kamysch (Sarikamiş), in horto, 17 April 1914, *Litvinov*.

Betula recurvata (I. V. Vassil.) V. N. Vassil., *Sched. Herb. Fl. URSS* 16: 82 (1966).

Betula lazistanica Browicz, *Arbor. Kórnickie* 20: 42 & 44 (1975) nom. nud.

Betula browicziana Güner, *Doğa Bilim Derg, A, 2 Biyol.* 9 (2): 270, f.1 & map 1 (1985). Type: Turkey Rize: Çamlihemşin, Hisarcik Köyü–Ortasirt Yalyasi arasi, c. 1800 m, 20 Aug. 1981, *A. Güner* 4247 (holotype HUB, isotype KOR).

Fig. 253. *Betula pubescns* var. *fragrans* with understorey of bracken, Sutherland, Scotland, UK (KBA).

Fig. 254 (opposite). *Betula pubescens* **A–F** var. *pubescens*, **G–N** var. *fragrans*. **A** twig with buds and immature male catkins × 1; **B** fruiting shoot with mature fruiting catkin × 1; **C** underside of leaf base × 2; **D** bark of twig × 2; **E** female catkin scale, back and side views × 7; **F** seed × 7; **G** twig with buds and immature male catkins × 1; **H** fruiting shoot with mature fruiting catkin × 2; **J** underside of leaf × 2; **K** young twig showing glands × 4; **L** bark of twig × 2; **M** female catkin scale, back and side views × 7; **N** seed × 7. A from *Hippe* s.n. (Böhmen, Schönwald, 17 June 1877); **B, D–F** from *Schuhwerk* 92/165 (Estergebirge, Garmisch-Partenkirchen, Bavaria, 24 June 1992); **C** from *W. Kit?* 1321 (Blankenberg am Harze); **G–N** from *Brennan* s.n. (Creag na Faolinn, Loch Eribol, Sutherland, 28 Aug. 1968). DRAWN BY ANDREW BROWN.

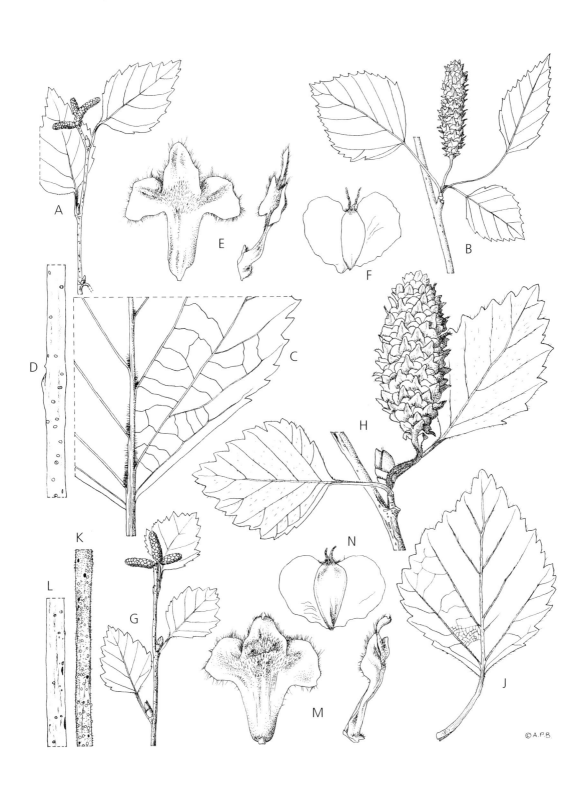

©A.P.B.

DESCRIPTION. Shrubby *tree*. *Young shoots* densely puberulent with a few long hairs, few glands. *Petiole* c. 10 mm, thinly puberulent and thinly long hairy. *Leaf blade* thinly puberulent and thinly long hairy on veins above and below, puberulence disappearing with age or never present, axillary hair tufts present; to c. 70 × 40 mm, broadly elliptical-rhombic, base cuneate, 6(–7) pairs veins. *Fruiting catkins* pendulous, with peduncle c. 10 mm, c. 25 × 10 mm; bract lateral scales erect to spreading to almost reflexed.

DISTRIBUTION. Caucasus, Georgia, E Turkey.

HABITAT. Shaded rocks, *Rhododendron* scrub, 1300–2000 m.

Var. **fragrans** Ashburner & McAll., **var. nov.** Type: Scotland, Argyll and Bute, Isle of Bute northern shore, mouth of Laggan burn, *H. McAllister*, 27 July 2000 (holotype LIV; isotypes, BM, E, HUH, K).

B. odorata subsp. *odorata sensu* E. F. Warb. in Clapham, Tutin & Warburg, *Flora of the British Isles*: 729 (1952), non Bechst.

DESCRIPTION. Small *trees* to often no more than about 5 m with twisted trunks, bark greyish-white. *Twig* often puberulent, persistent even in second year. *Bud* ovoid. *Young shoot* puberulent with a few ciliate hairs on youngest internodes, glandular warty. *Petiole* c. 13 mm. *Leaves* very small, to c. 33 × 30 mm, broadly ovate, cordate at base (not on long shoots). *Fruiting catkins* pendulous to 24–33 × 5–8 mm. *Scales* with lateral lobes longer than terminal. *Nutlet* broadly elliptical to more or less circular, to 2 × 1.5 mm with wing up to as wide as nutlet.

DISTRIBUTION. Scotland, Wales (Lake Bala), N England (Northumberland).

HABITAT. Steep hillsides and rocky streambanks in the mountains, and throughout the Scottish Highlands at all altitudes.

Probable hybrid between varieties:

Betula kusmisscheffii (Regel) Sukaczev, *Izv. Imp. Akad. Nauk*, ser. 6, 8: 233 (1914). (var. *pumila* × var. *pubescens*).

36. BETULA MICROPHYLLA

Betula microphylla was described from the Altai mountains. In the wild it forms a small tree, often with several stems. Its small, few-veined leaves and very white-warty twigs greatly resemble those of *B. humilis*, which may have contributed one of the genomes to this tetraploid.

Neither *Betula microphylla* nor *B. tianschanica* have any unique features not present in *B. pubescens* or *B. fruticosa* which suggests that the former two species may be of hybrid origin. In the Altai Mountains and Xinjiang and neighbouring Kazakstan and Mongolia, *B. microphylla* seems to intergrade with the shrubby *B. fruticosa* of eastern Asia. It shows characteristics of both *B. pubescens* and *B. fruticosa*, having hairy shoots, the taller habit of *B. pubescens*, the whitish warts and fat, often more or less erect fruiting catkins of *B. fruticosa*, and intermediate leaf shapes and sizes. It might be the result of introgression from *B. pubescens* or *B. pendula* into *B. fruticosa* or an allopolyploid of *B. humilis* × *B. pendula* agg. origin, or a mixture of all of these. Moving southward from the Altai Mountains through the Tian Shan there seems to be no clear distinction between *B. microphylla* and *B. tianschanica*. Further south still *B. tianschanica* appears to intergrade with *B. utilis* subsp. *occidentalis* in the south east/Pamir.

Fig. 255. *Betula microphylla* in cultivation at Ness showing winter death of buds on lower twigs (PW).

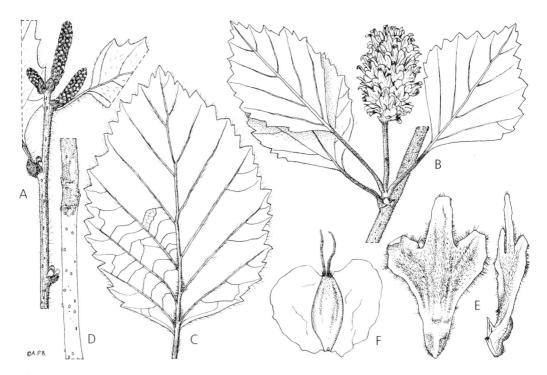

Fig. 256. *Betula microphylla*. **A** twig with buds and immature male catkins (catkins possibly not fully developed) × 2; **B** fruiting shoot with mature fruiting catkin × 2; **C** underside of leaf × 2; **D** bark of twig (upper part has juvenile epidermis with fine hairs, lower part has lost this) × 2; **E** female catkin scale, back and side views × 8; **F** seed × 8. **A, C, D** from *Timochina & Egarova* 934 (Tuva ASSR, 16 July 1974); **B, E, F** from *Krasnoborov & Hanminchyn* 898 (Kara-Xem, Tuva, 1500 m, 12 July 1974). DRAWN BY ANDREW BROWN.

Numerous isolated birch populations in the mountains of central Asia have been described as separate species: they are discussed below under *Betula tianshanica*.

Betula halophila was described from saline areas in Balibagi of the lower reaches of the Kelang River, Altay County, Xinjiang, China.

CULTIVATION

Betula microphylla is very rare in cultivation. As it originates in an extreme continental climate, it has not grown well in the British Isles. It forms a straggly small tree with several stems and usually suffers severe winter dieback of twigs, especially in the lower part of the canopy; only the most vigorous shoots exposed to the strongest sunlight mature enough to survive most winters (Fig. 255). It is illustrated in Kuzeneva (1936).

Betula microphylla Bunge, *Mém. Acad. Imp. Sci. St-Petersbourg divers Savans.* 2: 606 (1835). Type: described from the Altai (holotype LE).

B. fruticosa Pallas var. *cuneifolia* Regel, *Monogr.* 35 (1861).

B. halophila Ching in P. C. Li, *Acta Phytotax. Sin.* 17 (1): 88 (1979). Type: Sinkiang, A-le-tai Hsien, Ba-er-ba-gei, near salty swamp, at the bottom of moist valley, alt. 1500 m, 11 Sept. 1956 *R. C. Ching* 3151.

DESCRIPTION. A multi-stemmed small *tree*, to 6 m with greyish white bark. *Shoot* brown with colourless glands, densely hairy; *Twig* to 2.5 mm diam., hairs as shoot; white glands quite dense, bun-shaped, and smaller flat reddish-brown glands; lenticels small, oval, white. *Bud* to 3 × 1.75 mm, small, narrower than twig, grey-brown, scale margins ciliate, otherwise glabrous. Petiole c. 10 mm with only a few long hairs. *Leaves* triangular to rhombic-ovate, truncate to cuneate at base, with 4–6 pairs of veins, singly or doubly serrate, ± glabrous with a few long hairs on the main veins on both leaf surfaces and leaf axils on underside [although described as densely pubescent in *Flora of China* (Li & Skvortsov 1999)]. *Fruiting catkin* erect, 1–2.5 × 7–8 mm; peduncle 5–10 mm; bracts 5–6 mm, lateral lobes erect or slightly spreading. *Nutlet* c. 2.5 × 1.5 mm with wing about half as wide as nutlet.

DISTRIBUTION. Kazakstan, Mongolia, China (Altai Shan, Hami Xian).

HABITAT. Broad leaved forests and by streams in desert-steppe valleys.

CHROMOSOME NUMBER. Tetraploid: 2n=56 (McAllister unpubl.)

37. BETULA TIANSCHANICA

HISTORY AND NOMENCLATURE

Betula tianschanica was described from specimens collected by Baron von der Osten-Sacken near Lake Song Kol in Kyrgystan.

The distributions of *B. tianschanica* and *B. utilis* subsp. *occidentalis* may overlap in the Hindu Kush in north-west Pakistan, and hybrids may occur; these may have contributed to the small leaves with few veins found in some populations at the western end of the distribution of *B. utilis*, such as those from north-west Pakistan. Most are sturdy, wide-branching trees with white or pinkish, peeling bark.

CULTIVATION

This central Asian species is very rare in cultivation and somewhat similar in appearance and behaviour to *Betula microphylla*, although also with some resemblance to *B. pendula*. In cultivation in Britain it forms a single trunked tree but, due to late frost damage and winter dieback of twigs, does not usually make an attractive specimen.

Betula tianschanica Rupr., *Mem. Acad. Imp. Sci. St.-Petersbourg*, ser. 7, 4: 72 (1867). Type: Prov. Semireczje, montes Tian-Schanices, in angustiis Mol-da-asa ad lac. Son-Kul, ubi in rivulos cum salicibus crescentem invenit, 15 July 1867, *Baron F. V. Osten-Sacken* (LE). For synonyms see below.

DESCRIPTION. *Tree* to 12 m with whitish to yellow-brown peeling bark. *Twig* densely hairy, resinous, glandular. *Petiole* 5–7 mm. *Leaf* blade triangular to ovate-rhombic, 2–7 × 1–6 cm, glandular when young, doubly serrate, with 4–7 pairs of veins. *Fruiting catkin* pendulous on 5–8 mm peduncle, 10–40 × 5–7 mm; bracts 5–8 mm, lateral lobes spreading to recurved, slightly wider and shorter than mid-lobe; nutlet c. 2.5 mm with wing as wide as or wider than nutlet.

DISTRIBUTION. Uzbekistan, Kazakstan, Kyrgyzstan, Tajikistan, Mongolia, China (Xinjiang, in the Tian Shan).

HABITAT. Grassy slopes, margins of *Picea schrenkiana* forest.

CHROMOSOME NUMBER. Tetraploid: 2n=56.

TETRAPLOID BIRCHES IN CENTRAL ASIA

In the dry mountains of central Asia tree birches occur as small isolated populations which have been given many names; many of them are described as very rare (Eastwood *et al.* 2009), and therefore of conservation significance. Chromosome counts on several collections from the area have all been tetraploid, so these birches probably consist of a complex of isolated relict populations.

Pakhomov (2006) discusses the Pleistocene vegetation history of the now largely desert region east of the Caspian Sea between the Ural and Altai Mountains in the north and the Elburz/Caucasus and Pamir-Alai mountains in the south. He provides fossil evidence for the presence of boreal forest with *Betula*, *Picea*, *Sphagnum* and *Selaginella selaginoides*. This indicates that the climate was moister during cold periods and the northern forest extended south to join up with the forests of the Caucasus and the Pamir-Alai.

The following species were accepted in *Flora of the USSR* (Kuzeneva 1936), but most were included under *Betula tianschanica* by Skvortsov in *Conspectus Florae Asiae Mediae* (Vvedensky 1972); some of them may be synonymous with, or intermediate between, *B. tianschanica*, *B. microphylla*, or *B. utilis* subsp. *occidentalis*. They await further study.

Betula alba var. *schugnanica* B. Fedtsch., *Trudy Bot. Muz. Imp. Akad. Nauk* 1: 163 (1902).

Betula kirghisorum Sawicz., *Monit. Jard. Bot. Tiflis,* 25: 11 (1912).

Betula saposhnikovii Sukaczev, *Izv. Imp. Akad. Nauk.* ser. 6, 8: 235 (1914).

Betula korshinskyi Litv., *Trav. Mus. Bot. Acad. Sci. Petersb.* 12: 89 (1914). Type: Prov. Fergana, distr. Andishan, montes Ferganenses in declivibus versus trajectum Kug-art, 5–6,500 ft alt., 4 Aug. 1895, *S. Korshinsky* 4907.

Betula turkestanica Litv., *Trav. Mus. Bot. Acad. Sci. Petersb.* 12: 90 (1914).

Fig. 257 (left). *Betula* 'tianshanica', W Tian Shan.
Fig. 258 (above). Leaves of *Betula* 'tianshanica', W Tian Shan.

Betula pamirica Litv., *Trav. Mus. Bot. Acad. Sci. Petersb.* 12: 91 (1914). Type: Pamir, ad fl. Kisil-ssu pr. Irch, 19 Aug. 1897, *S. Korshinsky* 5530.

Betula alajica Litv., *Trav. Mus. Bot. Acad. Sci. Petersb.* 12: 92 (1914). Type: from Pamir, therefore probably *B. utilis* subsp. *occidentalis*.

Betula procurva Litv., *Trav. Mus. Bot. Acad. Sci. Petersb.* 12: 93 (1914).

Betula schugnanica Litv., *Trav. Mus. Bot. Acad. Sci. Petersb.* 12: 93 (1914). Type: Prov. Schugnan, in valle Gunt pr. P. Rivak, ad rivulum, 29 July 1901, *Fedczenko*.

Betula rezniczenkoana (Litv.) Schischk. in Krylov, *Fl. Zap. Sib.* IV: 793 (1930). Probably *B. microphylla*.

Betula jarmolenkoana Golosk., *Vestn. Akad. Nauk Kazakhsk. S.S.R.* 2 (131): 92 (1956).

Betula bucharica V. N. Vassil., *Bot. Zhurn. (Moscow & Leningrad)* 48: 905 (1963). Type: Schugnan, In valle fl. Garm-Czaschma, Kischlak (pag.) Anderob, 15 Aug. 1913, *N. Tuturin* 157 (LE).

Betula hissarica V. N. Vassil., *Bot. Zhurn. (Moscow & Leningrad)* 48: 903 (1963). Type: Jugum Hissaricum, declivitas australis, ripa fl. Varsob., supra pag. Kobuty, 28 Aug. 1944, *A. Pojarkova* (LE).

Betula seravshanica V. N. Vassil., *Bot. Zhurn. (Moscow & Leningrad)* 48: 903(1963). Type: Systema fl. Seravschan, lacus Iscander-kul, 16–23 July 1878, *V. Russov* (LE).

Betula tadzhikistanica V. N. Vassil., *Bot. Zhurn. (Moscow & Leningrad)* 48: 903 (1963). Type: Tadzhikistan Orientalis pag. Agacz-Kurgan, 1932, *N. Gorbunov & L. Lanina* (LE).

Betula pseudoalajica V. N. Vassil, *Bot. Zhurn. (Moscow & Leningrad)* 48: 904 (1963). Type: Darvaz, Faux a trajecto Kamczirak ad Kischlak (pag.) Czil-dara, 27 June 1911, *A. K. Golbek* 203 (LE).

Betula saposhnikovii subsp. *jarmolenkoana* (Golosk.) Ovcz., Czukav. & Shibkova, *Fl. Tadzhikskoi SSR* 3: 140 (1968).

Betula dolicholepis Ovcz., Czukav. & Shibkova, *Fl. Tadzhikskoi SSR* 3: 126, 655 (1968). Type: probably *B. utilis* subsp. *occidentalis*.

Betula heptopotamica V. N. Vassil., *Novosti Sist. Vyssh. Rast.* 7: 115 (1970 publ. 1971). Type: Regio Heptopotamica, distr. Dzharkent, vallis in montibus Bas-Ogly-tau, 30 July 1915, *V. Rezniczenko* 188 (LE). This is probably best treated as a *B. pubescens* introgressed with *B. fruticosa*.

B. ferganensis V. N. Vassil., *Novosti Sist. Vyssh. Rast.* 7: 108 (1970, publ. 1971). Type: Jugum Ferganense, lacus Sary-Czilek, silva Piceeta in declivitate ad lacum, 23 Aug. 1950, *A. Pojarkova* 636 (LE). This is probably *B. pendula* subsp. *pendula*.

Betula pyrolifolia V. N. Vassil., *Novosti Sist. Vyssh. Rast.* 7: 111 (1970 publ. 1971). Type: Asia Media, jugum Seravschanicum, declivitas australis, fl. Iskander-darja, «tugai» prope Narvat, 21 May 1914, *B. Dubianskyi* (LE). Probably *B. utilis* subsp. *occidentalis*.

38. BETULA CORDIFOLIA   MOUNTAIN WHITE BIRCH, HEARTLEAF BIRCH, SPECKLED BIRCH

HISTORY AND NOMENCLATURE

Betula cordifolia was described by Regel in 1861. Eduard von Regel (1815–1892) was at this time director of the Imperial Botanical Gardens in St Petersburg, and one of the foremost botanists of the day. *Betula* was one of the genera he studied in detail, and he published an account of the genus in 1861, with additions in 1866.

Naturalists, foresters, fieldworkers and all interested and connected to the outdoors in eastern Canada and USA recognise the speckled birch as something quite apart from ordinary white birch — i.e. *Betula papyrifera*. The name is apt for the twigs are spotted with comparatively large lenticels, in contrast to those of *B. papyrifera* in which the lenticels vary considerably in size. These same naturalists and foresters see *B. cordifolia* as quite distinct in other respects too. Apart from their fall colour, individual trees are not difficult to recognise in their somewhat fastigiate, erect, rather stiff growth and large rounded buds.

ECOLOGY AND STATUS IN THE WILD

The distribution of *Betula cordifolia* is not fully known due to confusion with *B. papyrifera*. It is abundant in Newfoundland and Labrador and Atlantic Canada generally, and occurs apart from *B. papyrifera*, generally growing at higher altitudes or further north. In Maritime Canada it grows as scattered trees on the hills — showing up well in the fall as an orange-coloured patch in the ubiquitous and dark green *Picea mariana* — for instance along the dirt road between Goose Bay and the interior, or again along the iron ore railway from Labrador City to Sept-Iles on the St Lawrence. It is abundant in the Lawrentides too, and even forms patches of climax forest, as at the foot of the Gros Morne in western Newfoundland (see below).

Although the main concentration of the species appears to be in Labrador, Nova Scotia and Newfoundland, the big surprise is that there are relict populations of stately trees far to the south in the Appalachians, on Black Mountain and neighbouring Mount Mitchell in North Carolina. These stand further south than any existing records of *Betula papyrifera*, and immediately recall the disjunct distribution of *Alnus viridis* subsp. *crispa* which is common in the northern latitudes and has also left a remarkable relict population far to the south on Roan Mountain on the border of North Carolina and Tennessee. It is tempting to speculate that *B. cordifolia* might once have formed the montane

Fig. 259. *Betula cordifolia* on sea cliff in conifer forest on Forillon National Park, Gaspé Peninsula, Quebec, Canada (HMcA).

forest of the Appalachians, from Maritime Canada to Georgia, growing with *A. crispa*. These southern relics are imposing, their bark peels freely and they stand, apparently barren, showing all the signs of dwindling — for replacement seedlings appear to be absent.

Today *Betula cordifolia* is scarce in the Appalachians south of New England. In Vermont it is scattered sporadically through a dominant scrub of rather low *B. alleghaniensis*. It extends westwards into Ontario and even into Minnesota and may be seen in the hills to the north east of Lake Superior and again in the hills to the south of Thunder Bay.

In the far north in Labrador and Quebec, details of its distribution appear to be unrecorded, and yet may be deduced from the population on the screes of the Gros Morne in north-western Newfoundland. The trail to the summit of the mountain leads through a mix of *Abies balsamifera* and sometimes rather battered older trees of *Betula cordifolia*, then climbs through low thickets of it, and towards the summit through curious, horizontally-growing, smaller-leaved birches (probably *B. cordifolia × B. glandulosa*) among the rocks, before eventually meeting scrub of *B. glandulosa*. These wide spreading or prostrate shrubs appear to be the primary hybrid between the two species, which has been named *B. ×minor*. It seems likely that the same sequence would occur northwards towards the tundra, and this theory is corroborated by the birches on the barrens above Labrador City where, after an approach through much reduced *B. cordifolia*, carpets of *B. glandulosa* are punctured by numerous verticals of *B. ×minor*.

TAXONOMY, RECOGNITION AND RELATIONSHIPS

Until recently *Betula cordifolia* has been regarded as a mere variety, or even a northerly or montane form, of *B. papyrifera*. Even Furlow (1997) in the *Flora of North America* questions whether it deserves specific status. However, the studies of Brittain & Grant (1965b, 1967b) and Grant & Thompson (1975) confirmed the diploid status of *B. cordifolia,* defined its morphological distinctness, and suggest that it is likely to be one of the constituent genomes of hexaploid *B. papyrifera*. Reports of tetraploids in *B. cordifolia* are likely to be due to the great difficulty in distinguishing between *B. cordifolia* and the *B. cordifolia × B. papyrifera* hybrid.

The leaves of *Betula cordifolia*, particularly those on the spur shoots, are somewhat reminiscent of those of a mulberry, or even *Davidia*. *B. cordifolia* may also be recognised by the uniformly large lenticels on twigs (a wide range of sizes in *B. papyrifera*), the big buds, stout twigs and cordate leaves. Other characters are the oblong, parallel-sided central lobe of the fruiting catkin scale, (more or less triangular in *B. papyrifera*), the forward-pointing lateral lobes (very variable in attitude in *B. papyrifera*), and especially the young female catkin being pendent from emergence from the bud (initially erect in *B. papyrifera*). Less definitive is the larger vein number with somewhat impressed veins, Furlow (1997) giving 9–12 as against up to 9 in *B. papyrifera,* but this varies greatly with vigour and environment. The cordate leaf is also found in some individuals of *B. papyrifera*.

CULTIVATION

Betula cordifolia grows well in the British Isles but can be killed back by late spring frosts and is reported to be sensitive to drought. Growth is not so rapid as in *B. papyrifera*, but the trees are sturdier in appearance and of neat habit. It has been planted so seldom and is rarely recognised as distinct from *B. papyrifera* that quite probably its attractive qualities are not yet appreciated. These include sturdiness and upright habit with an ability to withstand exposure. To add to this is the peeling pink to white bark and the autumn colour. It is noticeable that, in cultivation, individuals of *B. cordifolia* come into fruit at a younger age than those of *B. papyrifera* from a similar provenance.

Betula cordifolia Regel, *Nouv. Mém. Soc. Imp. Naturalistes Moscou* 13 (2): 86 (1861).
Betula papyrifera var. *cordifolia* (Regel) Fernald, *Rhodora* 3: 173 (1901).

DESCRIPTION. *Tree* of erect habit to 20 m, typically somewhat fastigiate with narrow crown, shrubby in exposed habitats. *Bark* mostly red brown, but often white. *Twigs* dark red-brown, relatively thick and stiff. *Shoots* variably silky and puberulent hairy with very small and brown glands, clear on very youngest shoots, scarcely visible even under the lens, becoming opaque white later. Lenticels numerous becoming elliptic or rounded, some up to 1 mm across. *Buds* to 10 mm long, acute, ovate, projecting outwards rather than pressed close to the stem. *Leaves* 6–14(20) × 4–9 cm, becoming matt, ovate, usually cordate; with 6–12 pairs of lateral veins, numerous glands on veins on both upper and lower surfaces; marginal toothing shallow, single or more rarely double; petioles 4 mm wide at base; Petiole:blade ratio c. 1:5. *Male catkin* short and stubby to 22 × 6 mm. *Female flowering catkin* pendent from emergence from bud. *Fruiting catkin* pendent, 25–55 × 6–10 mm; scales to c. 9.5 × 6 mm with central lobe oblong with more or less parallel sides, to c. 4 mm, lateral lobes

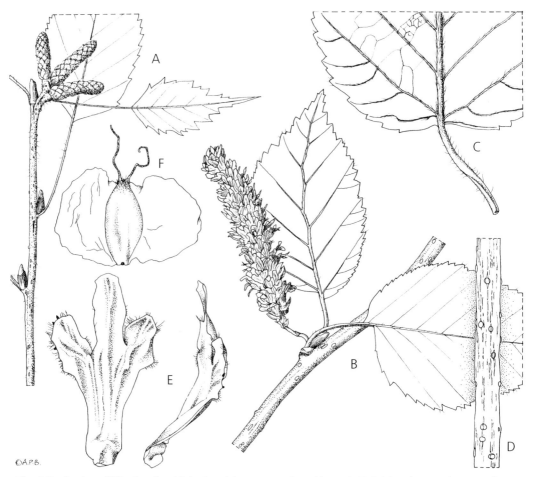

Fig. 260. *Betula cordifolia*. **A** twig with buds and immature male catkins × 1; **B** fruiting shoot with mature fruiting catkin × 1; **C** underside of leaf base × 2; **D** bark of twig × 2; **E** female catkin scale, back and side views × 7; **F** seed × 7. All from Gros Morne, Newfoundland, cult. Ness. DRAWN BY ANDREW BROWN.

forward pointing, 1.5 – 2.5 mm. *Seeds* to c. 2.5 × 6 mm with nutlet to c. 2.5 × 1.75 mm and wing c. 2.0 mm each side.

DISTRIBUTION. Newfoundland, Quebec and Ontario, to Minnesota and Wisconsin, and east to Pennsylvania, scattered south in the Appelachians to West Virginia and North Carolina.

HABITAT. Rocky slopes, often with *Betula alleghaniensis* from Maine southwards above 800–2000 m, or in *Picea mariana* forest in the north.

CHROMOSOME NUMBER. Diploid: 2n=28.

39. BETULA PAPYRIFERA PAPER BIRCH, CANOE BIRCH, OR WHITE BIRCH

HISTORY AND NOMENCLATURE

Betula papyrifera was described by Humphrey Marshall in 1785 in *Arbustrum Americanum*, *The American Grove*, published in Philadelphia in 1785. This was an alphabetical catalogue of trees and shrubs native to North America and is one of the earliest botanical books published in America. Marshall was a cousin of the Bartrams, and had his own botanic garden in Chester Co., Pennsylvania. This white-stemmed paper birch is a familiar tree throughout the northern parts of North America.

ECOLOGY AND STATUS IN THE WILD

Betula papyrifera ranges across North America from the Atlantic to the Pacific coast. Some see it is as the American equivalent of *B. pubescens* in Eurasia, but the parallel cannot be pushed far, if at all, because its ecology in North America seems to resemble that of the Eurasian *B. pendula* in being mainly a lowland tree. In north-west North America *B. pendula* subsp. *mandshurica* appears to be the tree birch which extends to the tree-line (Brittain & Grant 1968b), while in the east of the continent *B. cordifolia* occupies this ecological niche; both these are diploids whereas *B. papyrifera* is hexaploid. Human disturbance of the North American coniferous forest is more recent and has historically been less than in Eurasia, so that *B. papyrifera* does not so often form extensive stands as do *B. pubescens* and *B. pendula*.

Like the Eurasian tetraploid *Betula pubescens*, *B. papyrifera* is exceptionally variable. The mature bark is usually white, but may be brown. Leaves are usually ovate and can be large, or they can be quite small. Shoots and twigs may be densely hairy with scarcely any glands and soft to the touch, or they may be quite glabrous and rough with dense resinous warts.

Fig. 261. *Betula papyrifera.* Glacier National Park, Montana (KBA).

Fig. 262. Conifer forest with *Betula papyrifera* confined to disturbed ground along roads. From Mt Xalibu, Parc Nationale de la Gaspésie, Quebec, Canada (HMcA).

The most abundant and well known form is the big stately tree with very white bark, seen so frequently in the wild on the east of the continent, for example in Vermont and New Hampshire, and commonly planted in botanic gardens and in towns and neighbourhoods from Ontario eastward to the Atlantic. Gaps in deciduous forest and in coniferous forest at higher latitudes are their natural habitat, and here their occurrence is patchy and the trees themselves are slender, drawn up by the shade of the climax forest (Figs 29, 261, 262, 263).

This variant grows very commonly indeed in eastern and in middle Canada and in New England. In the eastern USA its southern limit approximates to the Maryland/Pennsylvania border, which can be confirmed by driving northwards through the Shenandoah valley in West Virginia where there is little or no white birch, and meeting the first *Betula papyrifera* north of the very narrow neck of Maryland at Hancock.

Further west, southern Manitoba, Saskatchewan and Alberta appear empty of birches although abundantly endowed with poplars such as *Populus balsamifera* and suckering clumps of *P. tremuloides*. Beyond these, through the Rockies and into southwestern British Columbia, birches are quite scarce and fail to stand out prominently, perhaps because of their frequently brown bark, but there are remarkable trees growing in the lower basin of the Fraser River, both in British Columbia and in Washington, and huge older trees are found on the coastal plain between the mountains and Vancouver. The naming of these splendid orange-brown barked birches is uncertain, but they were named *Betula commutata* by Regel, and *B. papyrifera* var. *commutata* (Regel) Fernald. They possess one persistent and potentially useful physiological property — the ability to regenerate powerfully from even the most closely-levelled stump. All seedlings originating from the Vancouver area in 1981 sprouted vigorously from stumps when felled at Stone Farm in subsequent years. In contrast, there has been no sign of regeneration from stumps of the white barked *B. papyrifera* from further east.

Brown barked *Betula papyrifera* individuals are, however, not confined to the West. There are many in Atlantic Canada, in Nova Scotia and Newfoundland, scattered sporadically yet forming small populations here and there. Many trees are pink barked, as for example in the hills north of the east end of Lake Superior. A var. *commutata* seedling with lingering orange-yellow foliage has been selected at Stone Farm and named 'Vancouver', and another, from Belle Vue in western Newfoundland with the same attributes is named 'St George'.

Betula ×utahensis, which can form stable tetraploid populations, is one of the most distinct hybrids, probably between brown or even black barked *B. occidentalis* and *B. papyrifera*. It effectively behaves as a species in the absence of the parents.

A single birch tree high on a volcano in the island of Flores in the Azores in the middle of the Atlantic (Rumsey, pers. comm., specimen in BM!) is probably *Betula papyrifera*, suggesting natural long distance dispersal. Otherwise *B. papyrifera* is confined to North America.

HYBRIDISATION

Betula papyrifera is hexaploid, 2n=84, but counts of 2n=56 (tetraploids) and 70 (pentaploids) are recorded, e.g. by Furlow (1997) following Johnsson (1945) and Brittain & Grant (1965a, b; 1966; 1967b; 1968a, b). Tetraploids are here interpreted as hybrids with the diploids *B. cordifolia* and *B. populifolia,* in the east (Grant & Thompson 1975) and *B. pendula* subsp. *mandschurica,* and *B. occidentalis* in the west, and the pentaploids as the backcrosses of these with *B. papyrifera*. Both groups are very

Fig. 263. *Betula papyrifera* (orange yellow), and *Populus grandientata* (yellow-green by road) from Mt Olivine near the Gite de Mont Albert, Parc Nationale de la Gaspésie, Quebec, Canada (HMcA).

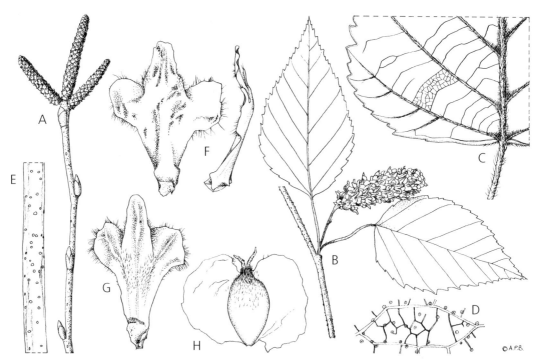

Fig. 264. *Betula papyrifera.* **A** twig with buds and immature male catkins × 1; **B** fruiting shoot with mature fruiting catkin × 1; **C** underside of leaf base × 2; **D** glands on leaf underside × 10; **E** bark of twig × 2; **F** female catkin scale (with relatively few hairs on back), back and side views × 7; **G** female catkin scale (with much hair on the back) × 7; **H** seed × 7. **A** from *Bourgeau* s.n. (Carlton, Canada, 21 Jan. 1858); **B, G** from *Scoggan* 2550 (Norway House, N end of Lake Winnipeg, 24 June 1948); **C–E** from *Muenscher* 18807 (Connecticut Hill, Ithaca, NY, USA, 25 June 1925); **F, H** from *McCalla* 2386 (Devil's Head Lake, Banff, Alberta, 4 Aug. 1899). DRAWN BY ANDREW BROWN.

difficult to distinguish morphologically from 'pure' *B. papyrifera,* hence the records of the tetraploid and pentaploid chromosome counts for *B. papyrifera.* The most distinct hybrid is *Betula ×utahensis* (*B. papyrifera × B. occidentalis*), and this can form stable tetraploid populations behaving as a species in the absence of both parents.

The possibility of gene flow from these diploids into *Betula papyrifera* has been proposed, but has not been demonstrated experimentally. It is not immediately obvious how gametes from tetraploids and pentaploids could regularly cause interbreeding with a hexaploid unless unreduced pentaploid gametes (n=70), fertilised by a diploid (n=14), generate a hexaploid which would be likely to be interfertile with *B. papyrifera.*

Betula papyrifera also hybridises with *B. glandulosa* to produce hybrids which would probably be referred to *B. ×minor* (correctly *B. ×ungavensis*) as in tetrapoid seedlings raised from an isolated plant of *B. glandulosa* from Mt Olivine, Gaspé, Quebec growing in the presence of *B. papyrifera.*

CULTIVATION

Betula papyrifera is grown for its bark, most often very white, but sometimes brown or pink and quite dark — perhaps with a 'bloom' of betulin in some populations. It makes splendid specimens in the arboretum or the larger garden, and is often planted as a street tree in urban areas of Ontario, Quebec and the northeastern states of the USA and also in the UK.

Trees grow rapidly, probably more so than any other birch, with quickly extending root systems. First year seedlings, at least in cultivation, can reach in excess of two metres and be well furnished with side branches. The autumn colours are always very fine and of a deep orange-yellow, so this species is splendid for the garden as long as there is plenty of space. Some of the selected and named clones are mentioned in the cultivar section below.

Named varieties such as var. *pensilis* Fernald would be worth propagating. This, according to Fernald (1945) 'is very striking, not only as a weeping birch but on account of the mostly acute-based leaves and the very abundant fruiting aments....often clustered on the spurs in fascicles of 2–4'. The locations he gives are scattered in eastern Canada and USA, so probably this should be seen as merely a form, probably comparable to the *oycoviensis* form of *B. pendula*, which also has numerous clustered fruiting catkins.

A major deterrent to growing this species in a park or arboretum or in neighbourhoods in North America is its susceptibility to the bronze birch borer (Santamour 1999), although the insect appears to present no problems in wild stands and does not yet occur in Eurasia.

Betula papyrifera Marshall, *Arbust. Amer.*: 19 (1785).
Betula papyrifera var. *commutata* (Regel) Fernald, *Rhodora* 47: 312 (1945).
Betula alba var. *commutata* Regel, *Bull. Soc. Imp. Naturalistes Moscou* 38: 401 (1865). Type: Wächst in Nord-america, Sumass Prairie (*Lyall*), Topsfield Mass. (*Asa Gray*), Oregon (Lyall).
Betula subcordata Rydb. ex E. J. Butler, *Bull. Torrey Bot. Club* 36: 436 (1909). Type: Hatwai Creek, Nez Perce Co., Idaho, *J. H. Sandberg 33*.

DESCRIPTION. *Tree* to 30 m, often with a single trunk. *Bark* on twigs and young branches dark reddish brown, on trunks very white to dark brown with yellow, orange or purple colouration also present, often with a white bloom. When trees become very white this only develops when trunks reach a much larger diameter than in other species. *Twig* glabrescent. *Bud* sticky. *Shoot* initially coarsely hairy and usually with dense puberulence of short erect hairs under a variable density of long silky hairs which are ± persistent, with resinous glands present at varying densities. *Leaf* ovate with base very variable, cuneate to cordate; with up to 9 pairs of veins; to 50–90(120) × 40–70 mm; margins often regularly toothed, sometimes double toothed, especially on long shoot leaves; initially coarsely silky hairy and puberulent along veins above and below, persistently puberulent on at least petiole upper (adaxial) surface near junction with blade, with persistent hair tufts in vein axils extending along midrib (Fig. 55). *Male catkins* often in groups of 3. *Female catkin* initially erect, soon pendent. *Fruiting catkin* with peduncle c. 15 mm; 25–50 × 6–13 mm (when dry and open), scales non-glandular. *Scale* very variable, c. 5.5 × 5.5 mm, central lobe triangular to rhombic, lateral very variable in orientation, forward pointing to recurved. *Seed* c. 2.5 × 5.5 mm, nutlet c. 2.5 × 1.5 mm with wing wider, c. 2 mm each side, usually extending beyond styles.
DISTRIBUTION. North America from coast to coast from Alaska and Oregon east to Newfoundland and south to Colorado, Indiana, and Virginia.
HABITAT. Forests, rocky woods and swamps.
CHROMOSOME NUMBER. 2n=usually 84, (sometimes reported to be 56 or 70, but these probably represent hybrids with *B. occidentalis*, *B. populifolia*, *B. pendula* subsp. *mandschurica*, *B. cordifolia* and backcrosses).

Betula kenaica W. H. Evans, *Bot. Gaz*. 27: 481 (1899). Types: Alaska, Kenai peninsula in the vicinity of Cook inlet, Sunrise at the head of Turnagain Arm, 1897, *W. H. Evans* 492; Kussiloff, on the W side of the peninsula, 1898, *W. H. Evans* 664.

Westwards into Alaska the closely related *Betula kenaica* is recorded for the Kenai peninsula and adjacent areas and is quoted as hybridising with *B. papyrifera* (Hultén 1968). In *Flora of North America*, Furlow (1997) recognised *B. kenaica* as a distinct species and, presumably quoting Woodworth (1931), reports *B. kenaica* to be pentaploid (2n=70), the count having been made on a tree of that name in the Arnold Arboretum. As 2n=70 is pentaploid, and so unlikely to be the chromosome number of a species, we suggest it arose through hybridisation and backcrossing involving a hexaploid (*B. papyrifera*) and a diploid (in this case probably *B. pendula* subsp. *mandshurica* or *B. occidentalis*), but there is no evidence given of reliability or otherwise of the supposed wild source of the tree apart from its name.

Betula kenaica is said to differ from *B. papyrifera* 'primarily in its smaller stature and in its smaller, blunter-tipped, more coarsely and regularly serrate leaves'; Hultén adds that in *B. kenaica* the nutlet wings are narrower than the nutlet, as against broader than the nutlet in *B. papyrifera* — hardly good characters to distinguish *B. kenaica* considering how variable *B. papyrifera* is. Dugle (1966), who had not seen the type specimen, interprets later collections of *B. kenaica* as being partly *B. papyrifera*, partly *B. pendula* subsp. *mandshurica*, and partly hybrids, an interpretation we suspect may be correct.

Brittain & Grant report an isolated hexaploid at Bonanza Creek, Yukon (presumably near Dawson City) which Hultén identified as *Betula kenaica*. This is far from the Kenai peninsula of Alaska and nearer the general area of distribution of *B. papyrifera* and so should probably be referred to that species.

Herbarium specimens, illustrations, and descriptions of *Betula kenaica*, together with the reported chromosome numbers of 2n=70 and 2n=84, suggest that it is best considered a westward extension of *B. papyrifera* into southern Alaska, perhaps affected by introgression from *B. pendula* subsp. *mandshurica*.

Furlow's map in the *Flora of North America* (1997) shows *Betula papyrifera* occurring north to about 56°N in Canada, and in Alaska only in the southern panhandle. However, Brittain & Grant (1968b) report hexaploid trees similar to *B. papyrifera*, and suggest them to be hybrids. These differ from the typical form only in their rather more resinous, warty twigs, and have been seen in several stations further north — to at least 68°N near Aklavik, almost on the shores of the Arctic Beaufort Sea. The fact that seedlings from the Aklavik tree proved to be hexaploid (2n=84) suggests that this population should perhaps be regarded as a variant of *B. papyrifera*, perhaps affected by introgression from *B. pendula* subsp. *mandshurica*.

Subgenus BETULA section APTEROCARYON

Section *Apterocaryon* Spach, *Ann. Sci. Nat., Bot.* 15: 195 (1844).
  Syn. Sect. *Humiles* W. D. J. Koch, *Synopsis Fl. German.*, ed. 2: 761 (1844).
    Sect. *Nanae* Regel in DC., *Prodr.* 16, 2: 162 (1868).
      Subsect. *Nanae* (Regel) Winkl.

Shrubs; *twigs* densely puberulent, variably long-hairy; *leaves* small (< 5 cm) with crenate or acute teeth and fewer than 6 (7) pairs of veins; stipules more or less persistent (except *Betula glandulosa*); *fruiting catkin* scale lobes more or less equal in length, no more than twice as long as broad (or lateral lobes absent in *B. michauxii*); *seed* wings usually narrower than nutlet.

INTRODUCTION

The shrubby dwarf tundra birches which are described here include four distinct diploids, *Betula nana*, *B. michauxii*, *B. glandulosa*, and *B. humilis*, and two or three tetraploids, the North American *B. pumila* and East Asian *B. fruticosa*, with another of probable hybrid origin, *B. gmelinii* (which is circumscribed to include *B. apoiensis*). All have small leaves with up to six pairs of veins, and, except for *B. glandulosa*, persistent stipules. The fruiting catkins are usually upright, with similar short scale lobes no more than twice as long as broad and more or less equal in length; the mostly small seeds have wings narrower than, or equal to, the nutlet. *B. michauxii* differs in having no lateral lobes on the scale and no distinct wing to the seed.

These dwarf birches are usually referred to the section *Humiles* W. D. J. Koch, although Furlow (1990), used sect. *Nanae* (Regel) Winkl. Unfortunately, sect. *Apterocaryon* has priority and so is the valid name for this group at the level of section. The name was coined to distinguish *Betula michauxii* with its apparently wingless nutlet from all other species, which were placed in section *Pterocaryon* (Skvortsov 2002).

Typical species of this section have male catkins with only brown dry parts of the scales visible. This gives the impression that the male catkins are enclosed within buds over winter (Furlow 1990), but in fact they can be clearly distinguished from non-flowering buds by being larger and more elongated with many more scales visible (Figs 51, 271, 282, 288). Dissection reveals that the lower, greenish, concealed part of the brown scales is attached to male flowers. In all other birches the male catkin scales are different from the bud scales, often, as in the white barked birches, being glossy, sometimes resinous and very tightly packed together (Figs 61, 63, 68, 279).

The taller growing tetraploid *Betula microphylla* of the Altai mountains and central Asia bears some resemblance to *B. humilis* and *B. fruticosa* and may intergrade with the latter, but is placed here in sect. *Betula* (see p. 328).

Fig. 265. Habitat of *Betula nana* (foreground) with hillside of *Salix phylicifolia* in yellow autumn colour, *Vaccinium uliginosum* (red) and *B. pubescens* var. *pumila*. Trostansfjordur, NW Iceland (HMcA).

In North America, hybrids between the diploid *Betula glandulosa* and two other diploids, *B. cordifolia* in the east and *B. pendula* subsp. *mandshurica* in the west, are reported (Furlow 1997). These have caused much confusion. In the *Flora of North America* Furlow (1997) describes *B. minor* of north-eastern North America as a problematic 'species'. Its suggested parents are *B. glandulosa* and *B. cordifolia*, for example in the Adirondaks where these two species occur together. However, tetraploid counts are also reported for *B. minor* and were found in seedlings from an isolated plant of *B. glandulosa* on Mt Xalibu, Gaspé, Quebec, where *B. papyrifera* was the only other birch present.

Chromosome number will not help to elucidate whether the Alaskan dwarf birches are largely *Betula occidentalis* or a *B. glandulosa* × *B. pendula* subsp. *mandshurica* hybrid complex, as proposed by Hultén (1968), as all three species are diploid (Furlow 1997),

40. BETULA NANA DWARF BIRCH

HISTORY AND NOMENCLATURE

Betula nana, DWARF BIRCH, was described by Linnaeus from Lapland, Sweden and northern Russia. Linnaeus visited Lapland in 1732, his first major field trip, when he was twenty-five. He covered 3000 miles, and developed great admiration for the Lapps and their way of life. His studies resulted in *Flora Lapponica* (1739), and his trip is described by Wilfrid Blunt in his life of Linnaeus (1971).

ECOLOGY AND STATUS IN THE WILD

Betula nana is often a major component of arctic tundra vegetation, extending to the shores of the arctic sea from Greenland through northern Europe to west Siberia, but it also occurs quite further south in mountain ranges such as the Alps and Carpathians in Europe. It may be the longest-lived birch; a stem with 147 growth rings has been recorded from south-east Greenland, and its potential to layer means that individual clones could be very long lived (De Groot *et al.* 1997).

Fig. 266. *Betula nana* with *Empetrum* spp., *Vaccinium myrtillus* and *V. vitis-idaea* above Narvik, Norway (KBA).

Fig. 267 (top left). *Betula nana* with *Vaccinium uliginosum* and *Carex* spp., Trostansfjordur, NW Iceland (HMcA.)

Fig. 268 (top right). *Betula nana* showing typical branching habit and characteristic uniform small leaves and erect fruiting catkins (Iceland) (PW).

Fig. 269 (above left). *Betula nana* showing characteristic uniform small leaves (Iceland) (PW).

Fig. 270 (above right). Patch of *Betula nana* near Ilulissat, W Greenland (MW & PW).

TAXONOMY, RECOGNITION AND RELATIONSHIPS

Betula nana is a low shrub up to 1 m tall and a most distinct and easily identified species. It is immediately recognisable, even in the fossil state, by its small, more or less circular leaves with relatively few rounded teeth, and hairy twigs with few glands (Figs 269, 271). Only the related N American species *B. michauxii*, *B. pumila*, and *B. glandulosa* (Figs 52, 283, 290), and hybrids of *B. nana* (Fig. 41), have somewhat similar leaves.

CULTIVATION

These small shrubs are not commonly grown in gardens as they often develop a straggly habit, and do not have particularly showy catkins. Being largely arctic or high mountain plants of immature or peaty soils, they do not thrive on clay. However, if grown in full sun in a cool climate they make neat, attractive small shrubs and could be potential subjects for bonsai or a stone trough. They are best grown on humus-rich soils, unless you are gardening in cold moist climates as in Newfoundland, Iceland or arctic Norway where they are likely to thrive under normal garden conditions.

Betula nana is easily propagated by seed but, as it is largely self-incompatible, more than one clone needs to be grown for seed production. Cuttings, which root easily, produce larger plants more rapidly.

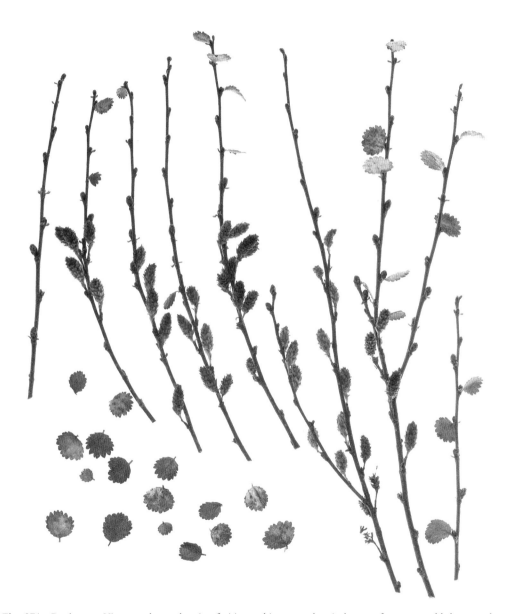

Fig. 271. *Betula nana*. Vigorous shoots showing fruiting catkins towards apical parts of two-year-old shoots and, on vigorous current year's growth increments, male catkins towards base (looking like elongated buds) and buds towards apex which will give rise to female catkins and vegetative shoots. The persistent stipules are conspicuous (PW).

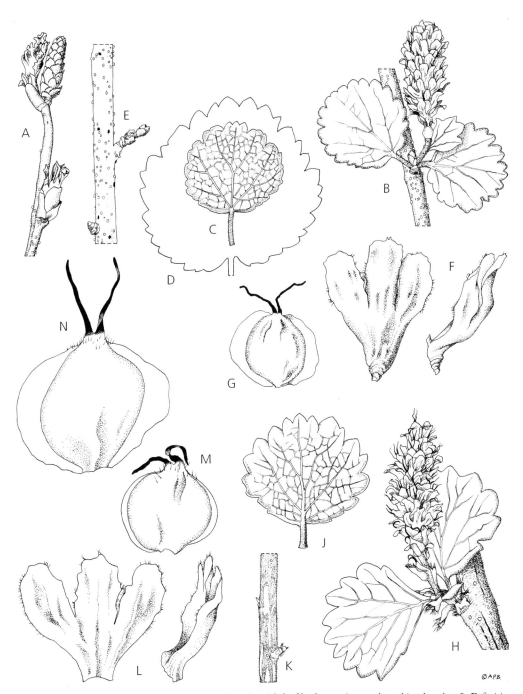

Fig. 272. A–G *Betula glandulosa*; **H–M** *B. nana*. **A** twig with leaf buds opening, male catkin closed × 2; **B** fruiting shoot with ripe fruiting catkin × 2; **C** leaf from **B**, underside × 2; **D** silhouette of larger leaf from non-fruiting branch × 2; **E** bark of twig × 2; **F** female catkin scale back and side views × 10; **G** seed × 10; **H** fruiting shoot with ripe fruiting catkin × 4; **J** leaf from **H** × 4; **K** bark of twig × 2; **L** female catkin scale, back and side views × 20; **M** seed from **K** × 20; **N** larger seed × 20. **A** from *Mason* (Cascade Creek, Norton Sound, Alaska, June 1931); **B–G** *McDonald & Farjon* 79 (Kamchatka, SE of Nalychevo, 53°07'N 159°09'E, 28 July 2003). **H–L** from *F. S. Chapman* 290 (Lake Fjord, Greenland, July 1933); **M** from *Køie* 4217 (Cape Stosch, Greenland, Aug. 1932). DRAWN BY ANDREW BROWN.

Fig. 273 (left). *Betula nana* showing hairy, non-warty twig and fruiting catkins with *Vaccinium microphyllum* (diploid *V. uliginosum*) with green and red autumn colouring leaves, fruiting *Empetrum hermaphroditum*, and *Salix herbacea* near Ilulissat, W Greenland (MW & PW).

Fig. 274 (above right). *Betula nana* showing hairy, non-warty, current year shoot (centre of picture) and fruiting catkins near Ilulissat, W Greenland (MW & PW).

Betula nana L., *Sp. Pl.* 2: 983 (1753). Type: Herb. Linn. 1109.9.

B. tundrarum Perfilj., *Bot. Zhurn.* (*Moscow & Leningrad*) 48: 1139 (1963). Type: not found.

DESCRIPTION. *Shrub* with spreading habit up to 1 m in height. *Bark* grey to dark brown, not peeling. *Twig* densely hairy with few if any glands visible. *Bud* to c. 2.5 × 1.75 mm, brown, ± spherical, less than twig diam. *Petiole* to c. 3 mm. *Leaf* ovate to obovate to orbicular or kidney-shaped, often broader than long, variably hairy when young, glabrescent, leaf blade to c. 10–15 mm with usually 3 pairs of veins, margin crenate with up to 10 (–11) rounded teeth on each side. *Male and female catkins* small, erect, males lateral below females on previous year's long shoots and terminal on short shoots on older wood. *Male catkins* c. 4 × 2.25 mm with rounded brown scales (similar to bud scales), ciliate at margins, not expanding much at anthesis. *Fruiting catkins* erect, more or less cylindrical, to 10 (–12) × c. 4 mm. *Bracts* c. 4 mm with three almost equal, forward-pointing, parallel-sided lobes. *Seed* less than 2 mm, wing to half as wide as nutlet; style c. 0.3 mm.

DISTRIBUTION. Greenland, Iceland, Scandinavia, Scotland, the Baltic area, western Siberia east to the Yenisey-Lena watershed and in lower latitude mountain ranges in the Alps and Carpathians in Europe.

HABITAT. Tundra and subarctic and mountain moorland.

CHROMOSOME NUMBER. 2n=28.

41. BETULA GLANDULOSA RESIN BIRCH

HISTORY AND NOMENCLATURE

The North American *Betula glandulosa* was described by André Michaux in his *Flora boreali-americana* in 1803 from specimens collected southeast of Hudson Bay. After extensive study of herbarium specimens we have decided to combine *B. nana* subsp. *exilis* and *B. glandulosa*. This disagrees with the treatment in *Flora of North America* (Furlow 1997), but follows Moss & Packer (1983), Porsild &

Cody (1980) and Fries in Hultén & Fries (1986), all authors with extensive experience of the flora of arctic America. We interpret the material traditionally referred to *B. nana* subsp. *exilis* as dwarf, mainly more northern, variants of *B. glandulosa*.

ECOLOGY AND STATUS IN THE WILD

According to the above interpretation, *Betula glandulosa* grows from SW Greenland westwards through the whole of northern North America and across the Bering Straights into Siberia as far west as the Lena-Yenisey watershed and the Altai Mountains. It is an invader of gaps in northern coniferous forest (Fig. 278), growing on the whole in dry, exposed places, such as rocky barrens to the north (Fig. 277), and mountain tops (Figs 280, 281) in the south of its range.

TAXONOMY, RECOGNITION AND RELATIONSHIPS

Betula glandulosa can be recognised by its very resinous warty shoots with very fine (to 0.05 mm long) puberulence, and small, gland dotted, reniform to obovate, crenately toothed leaves. It differs significantly from *B. nana* and its close relatives by its often solitary terminal male catkins (Hjelmqvist 1948) and non-persistent stipules; in these characters it is similar to species in other sections and somewhat resembles a dwarf *B. occidentalis* although most specimens of that species lack the puberulence (Figs 52, 53). *B. pumila* is said to intergrade with *B. glandulosa* 'creating a confusing complex of intermediate forms' (Furlow 1997; De Groot *et al*. 1997). This relationship should be easy to resolve as *B. glandulosa* has always been reported to be diploid and *B. pumila* (including what Dugle names *B. glandulifera*) tetraploid. Although Dugle (1966) concludes that many of the populations in the Rocky Mountains are of hybrid origin between these two species (*B.* ×*sargentii*), this seems rather unlikely given the difference in chromosome number, and most of the Rocky Mountain specimens studied by the present authors fall within the range of *B. glandulosa*.

Fig. 275. *Betula glandulosa*. Altai Mountains (JG).

Fig. 276. *Betula glandulosa*. Altai Mountains (JG).

Fig. 277. Habitat of *Betula glandulosa* above treeline, Smoky Mountain, Labrador City, Canada (KBA).

Fig. 278. Shrubs of *Betula glandulosa* on wet ground in clearing in conifer forest, Banff National Park, Alberta (KBA).

CULTIVATION

In cultivation at Ness all provenances of *Betula glandulosa* grew poorly, usually losing their leaves in June and producing a second flush in July. Low latitude origins might thrive better in cultivation in temperate climates and certainly develop brilliant autumn colouration in the wild (Figs 275, 276). A collection from Goose Bay, Labrador, initially considered to be *B. glandulosa,* has now been realised to the hybrid with *B. cordifolia, B. ×minor.* From its wild origin it is clearly tolerant of exposure and the leaves develop deep red and orange autumn colour, so it might be an attractive garden shrub for exposed situations as long as grown in full sun in soil which is not too dry.

B. glandulosa Michx., *Fl. Bor.-Amer.* 2:180 (1803). Type: 'circa lacus, a sinu Hudsonis ad Mistassins' (holotype).
Betula rotundifolia Spach, *Ann. Sci. Nat., Bot.* sér. 2. 15: 194 (1841).
Betula glandulosa var. *sibirica* S. F. Blake, *Rhodora* 17: 87 (1915).

DESCRIPTION. *Shrub* with prostrate to spreading to erect habit up to 3 m in height, but often under 1 m in exposed situations. *Twig* persistently finely (0.05 mm) puberulent-hairy, often initially with a few longer silky hairs, resinous-warty with reddish-brown glands covered with translucent whitish resin. *Bud* brown, blunt, oblong-ovoid, wider than twig, c. 3 × 1.5 mm. *Petiole* to 5 mm. *Leaf* obovate to orbicular, glabrescent, leaf blade to 30 × 25 mm with 2–6 pairs of veins and (5)9–13(16) crenate

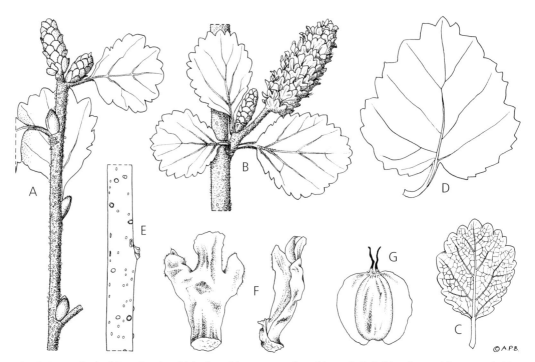

Fig. 279. *Betula glandulosa.* **A** twig with buds and immature male catkins × 2; **B** fruiting shoot with mature fruiting catkin and immature male catkin × 2; **C** underside of leaf × 2; **D** leaf × 2; **E** bark of twig × 2; **F** female catkin scale, back and side views × 10; **G** seed × 10. **A** from *Moir* 421 (Consolation Lake, Alberta, 28 Aug. 1938); **B, C, E–G** from *Kirkwood* 2440 (Donabe Ranch, Flatbed National Forest, Montana, 6 Sept. 1925); **D** from *Looff & Looff* 515 (Olga Bay, Alaska (Kodiak Island), Aug. 1938). DRAWN BY ANDREW BROWN.

Fig. 280. Habitat of *Betula glandulosa* on bare gravelly soil with dwarf, wind-blasted *Picea mariana*. Xalibu, Parc Nationale de la Gaspésie, Quebec, Canada (HMcA).

Fig. 281. *Betula glandulosa* (foreground with fruiting catkins), *B. cordifolia* (right centre with yellow-green leaves) with dwarfed, wind-blasted *Picea mariana*. Xalibu, Parc Nationale de la Gaspésie, Quebec, Canada (HMcA).

teeth on each side, initially abaxially densely glandular but glands falling, base cuneate to truncate. *Male catkins* terminal, about 9 × 3.5 mm. *Fruiting catkins* with pedicel to 9 mm, to 15 × 5 mm, erect; *seed* < 2 mm.

The sometimes very similar *Betula ×minor* can be distinguished by: *Bud* to 6 × 3 mm. *Petiole* to 11 mm. *Leaf* obovate to orbicular, glabrescent, leaf blade to 40 × 31 mm with 2–6 pairs of veins and 9–16 shortly acute teeth on each side. *Male catkins* terminal, to 15 × 5 mm, erect; seed < 2m. *Fruiting catkins* with pedicel to 9 mm, to 25 × 12 mm, becoming pendent at maturity.

DISTRIBUTION. Eastern Siberia west to Lena/Yenisey watershed and Altai Mts, North America from Alaska eastwards to Labrador and SW Greenland and south in the mountains to California, Colorado and New York.

HABITAT. Arctic and alpine tundra, whether wet or dry, exposed summits, peat bogs and streamsides (Fig. 17).

CHROMOSOME NUMBER. Diploid: 2n=28.

42. BETULA MICHAUXII Newfoundland dwarf birch; Michaux's birch

HISTORY AND NOMENCLATURE

Betula michauxii was described by Edouard Spach (1801–1879) from specimens collected in Newfoundland. It is named after André Michaux, a French botanist who collected in North America, setting up two gardens there and sending many specimens and seeds to France.

ECOLOGY AND STATUS IN THE WILD

Betula michauxii is very similar in general appearance to *B. nana*, but upright and even more slender and delicate (Fig. 282). It is a species of wet bogs in eastern mainland Canada and Newfoundland, easily distinguished by the absence of lateral lobes to the scales of the female and fruiting catkin, only one nutlet (instead of the usual three) per fruiting catkin scale, and nutlet with a swollen wing such that the wing and nutlet are not distinguishable.

TAXONOMY, RECOGNITION AND RELATIONSHIPS

Betula michauxii is easily distinguished by these characters and so it is treated in this book as a distinct species, although we realise that the differences could be due to one or a very few genes, as is the case in *B. oycoviensis* (Ch. 1), *B. uber* (McAllister 2005a) or the fern *Athyrium*

Fig. 282 (right). *Betula michauxii* Canada, Newfoundland, Come By Chance, Bog. Characteristic structure of vigorous shoots of this species (and of *B. nana* and *B. pumila*). Current year's growth with persistent stipules, distal buds which will produce vegetative and female catkin bearing shoots, and below these the elongated male catkins. The previous year's shoot bears fruiting catkins distally with, below these, the persistent stipules and bud scales on the remnants of the lateral male catkins. This produces the characteristic appearance of the twigs of these three species (PW).

flexile (McHaffie *et al.* 2001). Although its distribution overlaps with that of *B. glandulosa* (Furlow 1997), it remains distinct and no hybrids have been reported, perhaps because *B. michauxii* grows in wetter habitats, but more observations are required in the area in which both species occur.

CULTIVATION

Being upright in habit, *Betula michauxii* makes a neater plant than most forms of *B. nana*. It has small leaves which develop good autumn colour. Although apparently a peat bog plant in the wild, it grows reasonably well in ordinary garden soil as long as this is not allowed to dry out. Surprisingly, it even grows quite well on wet calcareous clay although there the leaves do become rather chlorotic late in the season.

Betula michauxii Spach, *Ann. Sci. Nat., Bot.*, sér. 2, 15: 195. 1841.
Betula terra-novae Fernald, *Rhodora* 47: 326 (1945).

DESCRIPTION. *Shrub* with upright habit up to 1 m in height, although given as to 0.5 m in wild. *Bark* grey to dark brown, not peeling. *Twig* puberulent and with longer hairs, glandular but hardly resinous, with stipules and bud scales often persisting for more than one year (Figs 282, 283). *Petiole* to c. 2 mm. *Leaf* broadly obovate, base cuneate, variably hairy when young, glabrescent, leaf blade

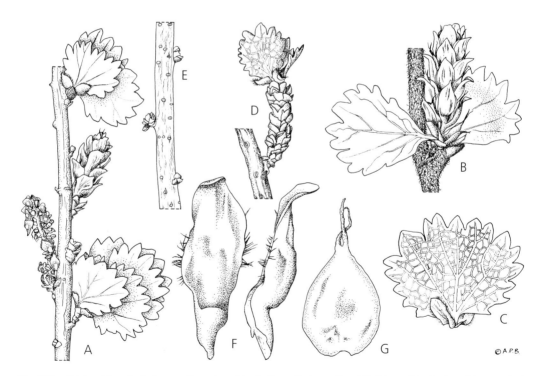

Fig. 283. *Betula michauxii*. **A** shoot with male catkin (anthers discharged) and fruiting catkin (male catkins are sited on mid-stem nodes, one per node, sometimes in a series of 5 or 6 nodes) × 3; **B** fruiting shoot with mature fruiting catkin × 4; **C** underside of leaf (with persistent bud scales) × 4; **D** old short shoot × 4; **E** bark of twig × 2; **F** female catkin scale, back and side views × 15; **G** seed ×15. **A** from *Rouleau* 730 (Gaff Topsail, Newfoundland, July 1950); **B**, **C**, **E–G** from *Fernald & Wiegend* 3272 (Goose Pond, Newfoundland, 10 July 1910); **D** from *Moir* 57 (Middle Ridge, Newfoundland, 27 July 1937). DRAWN BY ANDREW BROWN.

to 0.5–1 × 0.5–1.2 cm with 2–3 pairs of veins, margin crenate with rounded teeth. *Bud* more or less spherical, c. 3 × 1.5 mm, less than twig diameter. *Male and female catkins* small, erect, male lateral below female on previous year's long shoots, rarely terminal (Fig. 282). *Male catkins* to 5 (–6) × 2 (–2.5) mm with rounded brown ciliate-margined scales. *Fruiting catkins* erect, broadly elliptic, to 11 × 4 mm, peduncle very short. *Bracts* elliptic, ovate, unlobed, each subtending single nutlet. *Nutlet* less than 2 mm, wing seeming inflated and not distinct from nutlet; style c. 0.5 mm.

DISTRIBUTION. Eastern Canada: St. Pierre and Miquelon, Newfoundland, Nova Scotia, Quebec.

HABITAT. Wet peaty situations, *Sphagnum* bogs.

CHROMOSOME NUMBER. 2n=28.

43. BETULA HUMILIS

HISTORY AND NOMENCLATURE

Betula humilis was described from Bavaria, which is the western limit of its very wide distribution from Germany to eastern Siberia and Korea, by Franz von Paula von Schrank (1747–1835) who was born in Bavaria. He became a Jesuit, and was professor at various universities before becoming, in 1809, the director of the newly-formed Botanic Gardens in Munich. He was a prolific author, both of his travels studying the Natural History of southern Germany, and of Floras, notably *Baiersche Flora* (1789), and the wonderfully illustrated *Flora Monacensis* (1811–1820).

ECOLOGY AND STATUS IN THE WILD

Betula humilis is rare in western Europe, in fact only reported from a few stations in Germany south of Munich and further north in eastern Germany. It seems equally scarce further east in Poland (Figs

Fig. 284. *Betula humilis* in swamp in *B. pubesens* forest, Slésin, Torun, Poland, Oct. 1984 (KBA).

Fig. 285 (left). *Betula humilis* winter twigs with shoot on right showing typical arrangement on vigorous shoots with buds which will give rise to vegetative and female catkin bearing shoots towards tip of shoot and laterally borne male catkins below. Other less vigorous shoots bear lateral and terminal male catkins (PW).

Fig. 286 (right). Kenneth Ashburner collecting seed of *Betula humilis*, Slésin, Torun, Poland, Oct. 1984 (Pru Barnes).

284, 286), where it seems to be absent from formerly recorded sites (KBA pers. obs.). *Atlas Florae Europaeae* (Jalas & Suominen 1976) records many stations in European Russia.

It seems to have declined steadily over the years, or even centuries, with the advance of cultivation and drainage, perhaps combined with competition from *Salix* species which respond more positively to eutrophication. Although a species of wet ground in the wild, *Betula humilis* grows successfully in even quite dry garden conditions in normal soil. Its restriction to wet soils in the wild may therefore be because of low competitive ability in more favourable conditions. Recent collections by Maurice Foster (*MF* 325) (Figs 14, 15) confirm the presence of *B. humilis* as far east as northern Mongolia and there is an herbarium specimen at Kew (K!) (as *B. extremiorientalis*) from just to the north east of Kamchatka which make plausible Li & Skvortsov's (1999) reports of *B. humilis* (as *B. fruticosa* which Skvortsov considered a probable synonym) occurring throughout Asia to north-east Siberia.

TAXONOMY, RECOGNITION AND RELATIONSHIPS

The numerous translucent glands on the shoots which become white with age and persist on second year and older twigs are very characteristic of *Betula humilis*, although they are also seen in the tetraploid *B. fruticosa* and in *B. microphylla*, probably indicating the presence of the *B. humilis* genome in their parentage.

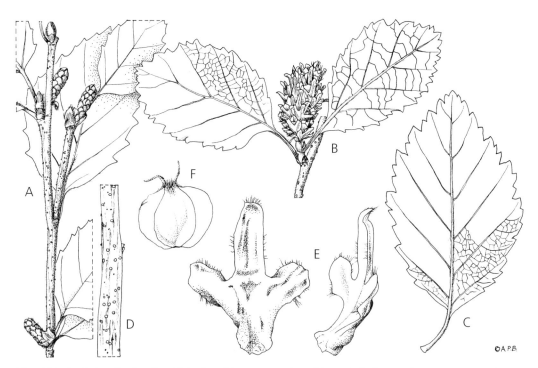

Fig. 287. *Betula humilis.* **A** twig with buds and immature male catkins × 2; **B** fruiting shoot with mature fruiting catkin × 2; **C** underside of leaf × 2; **D** bark of twig × 2; **E** female catkin scale, back and side views × 12; **F** seed × 12. **A** from *F. Kraviec* (Poznan, 6 Oct. 1929); **B, D** from *Kraenzle* 1429a (Upper Bavaria, Haspelmoor between Munich & Augsburg, 21 June 1911); **C** from *T. Mgdalski* (Bitohorszca pr. Leopolin, Poland, 8 Aug. 1929); **E, F** from Herb. *Churchillianum Proprium* (ex Königsberg, Pomerania, det. Ashburner Nov. 2011). DRAWN BY ANDREW BROWN.

CULTIVATION

Even though *Betula humilis* is a reasonably tall shrub, the thinness of its twigs, the minuscule buds and the elegance of its very small leaves match the neatness of *B. nana*. The leaves are characteristically coarsely toothed in proportion to their size. In cultivation, although not vigorous, it is very persistent, even in competition with grass and other shrubs, and seems to be more shade tolerant than most birches, although, of course, it will not make an attractive plant except in good light.

Betula humilis Schrank, *Baier. Fl.* 1: 420 (1789).

Betula sibirica Lodd. ex Regel, *Prodr.* (DC.) 16 (2.2): 173 (1868).

?*Betula extremiorientalis* Kuzen. & V. N. Vassil., *Vestn. Dal'nevost. Fil. Akad. Nauk S.S.S.R.* 21: 161 (1936).

Betula fruticosa sensu Skvortsov.

DESCRIPTION. *Shrub* to 2(–3) m. *Twigs* thin, less than 2 mm, even at base of current year's annual growth increment, twigs rough with many opaque white glands and a dense covering of short (0.05 mm) puberulent hairs. *Buds* very small, rounded, scarcely to 2.0 × 1.0 mm, scales brown to near black, fringed with very short hairs. *Petiole* 5–8 mm, petiole:blade ratios 1:9–1:13. *Leaves* ovate to ovate-elliptic, 10–40 × 6–30 mm, acute, cuneate or rounded at base with 4–5 pairs of veins, when young glossy green above, paler below but still quite shiny, and gland-dotted above, slightly hairy on veins above and below, glabrescent, margins with coarse acute teeth which are strikingly large in proportion to leaf size, 1 or 2 minor teeth between major and sinuses as deep as 5 mm, so quite markedly double-toothed but usually only on long-shoot leaves which are more ovate triangular in shape and with shorter petioles than pre-formed leaves on spur shoots and base of long shoots. *Male catkins* terminal and lateral, 5–8 × 2 mm. *Female catkins* lateral and usually above (distal to) males on long shoots (Fig. 285). *Fruiting catkins* erect to 18 × 6 mm on peduncles of up to 3 mm. *Bracts* 2.75–4 mm, ciliate, terminal lobe oblong, 1–1.25 mm; lateral lobes 1–1.5 mm. *Nutlet* to c. 2.25 × 1.25 mm with wing ¼ to ½ as wide as nutlet; style c. 0.75 mm.

DISTRIBUTION. Europe (Germany, Austria, Romania, Poland, Lithuania, Latvia, Estonia, Russia), W Siberia, Kazakstan, to N Mongolia to the Pacific coastal regions of Far Eastern Russia and Korea.

HABITAT. Wet ground in forests, and the edges of lakes.

CHROMOSOME NUMBER. 2n=28.

44. BETULA PUMILA BOG BIRCH

HISTORY AND NOMENCLATURE

Betula pumila was described by Linnaeus in *Mantissa plantarum*: 124 (1767). In America septentrionale.

ECOLOGY AND STATUS IN THE WILD

Our studies of herbarium specimens and of the few wild collections we have in cultivation suggest that typical hairy, relatively eglandular *Betula pumila* is more or less confined to the north eastern corner of North America, perhaps as relatively isolated populations. Populations of less hairy, more glandular shrubs occur sporadically in wet habitats across North America to Vancouver Island (Dugle 1966). It is the ecological equivalent of *B. humilis* and *fruticosa* of Eurasia, occurring in bogs and on the fringes of lakes. All these shrub birches may be modest in themselves, but their appeal lies in their association with the freshness of the unspoilt, unpolluted outdoors in these sparsely inhabited northern latitudes.

In the wild, fringes or patches of *Betula pumila* are common bordering creeks, as west of Thunder Bay in Ontario, and in northern Minnesota. But it is not ubiquitous, as many such habitats are occupied by *Salix* species. *B. glandulifera* (Dugle 1966) is not recognised as distinguishable.

TAXONOMY, RECOGNITION AND RELATIONSHIPS

The often densely hairy and usually stiffer twigs, shorter male catkins (Fig. 51), and male and female catkin arrangement (Figs 65, 288, 289) distinguish this species from *Betula glandulosa*.

It can be defined by the presence of persistent coarse crisped hairs (0.5 mm and longer, as against 0.05 mm with few longer silky hairs in *Betula glandulosa*) on the shoots, coarser and sharper leaf teeth, and, on vigorous shoots, the catkin arrangement typical of *B. nana*, with lateral males below females. In typical eastern *B. pumila* most leaves are soon eglandular on the underside, although some, especially in the west, have persistently glandular leaves. Using these characters with herbarium specimens, it seems possible to distinguish between the coarser, hairier, largely more southerly *B. pumila* of wet habitats, and the more delicate less hairy *B. glandulosa* of usually drier habitats.

Betula pumila appears to be the only regularly tetraploid birch species in North America.

CULTIVATION

Perhaps *Betula pumila* is not a shrub for the garden, in western Europe anyway, because of its sparse open straggly habit. The shrubs may be upright, tall and brown barked, and thus could be confused

Fig. 288. *Betula pumila* Canada, Newfoundland, Woody Point. Three shoots to left showing two years of growth increments with larger buds containing female catkins towards apex (terminal bud will produce a vegetative long shoot) and male catkins on spur shoots on terminal part of previous year's growth. Twigs to the right with shorter current year's growth increments have terminal male catkins and fruiting catkins towards terminal part of previous year's growth increment (PW).

with *B. humilis*, but generally are more vigorous, have much stouter twigs, bigger buds and more or less obovate leaves. It is never as neat and elegant as *B. humilis* can be. However, on wet soils it may form quite a neat bush if pruned; the crimson tipped female flowers (Fig. 65) can be conspicuous in the spring and it produces attractive autumn colour.

Like many plants of wet, open habitats it is very intolerant of shade, only growing well in full sun. It has grown happily in north Devon in well-drained poor soil, forming a shrub 2 m high in ten years, flowering and fruiting freely.

Betula pumila L., *Mant. Pl.* 124 (1767). Type: Canada.

?*Betula pumila* var. *glandulifera* Regel, *Bull. Soc. Imp. Naturalistes Moscou* 38 (2): 410 (1865).

?*Betula glandulifera* (Regel) B. T. Butler, *Bull. Torrey. Bot. Club* 36: 424, Fig. 2 (1909).

?*B. obovata* B. T. Butler, *Bull. Torrey. Bot. Club* 36: 427, Fig. 5 (1909). Type: valley of the Jocko river, Ravalli, Montana, *B. T. Butler* 317 (NY). No Latin description.

?*B. hallii* Howell, *Fl. N.W. Amer.* 1: 614 (1902). Type: Lake Labish, Marion Co., Oregon: also Mr Gorman from Fort Selkirk, Yukon Territory appear to be of this species. Described as the *B. glandulosa* of Oregon. No Latin description.

Betula exilis Sukaczev, *Trav. Mus. Bot. Acad. Petersb.* 8: 213 (1911). Type: described from the surroundings of Eniseisk (holotype LE).

B. nana subsp. *exilis* Hultén, *Fl. Alaska Yukon* 4: 579 (1944).

DESCRIPTION. *Shrubs* to 4 m, often rather straggling or spreading. *Twigs* stout, to 4 mm across at base, mid-brown, slightly shining, typically densely hairy with longer (to 0.5 mm) hairs than *Betula glandulosa* (0.05 mm) present at least in bud axils (*B. glandulosa* may have occasional longer silky hairs on young shoots) and sometimes eglandular in the east. *Buds* ovoid, rounded, c. 4 × 2 mm, narrower than twig, scales brownish-black, fringed with hairs. *Leaves* on long-shoots larger than preformed spur-shoot and basal long-shoot leaves, elliptic to obovate to sometimes almost orbicular (especially when stressed) with cuneate bases, 18–58(–70) × 13–40(–50) mm, with 2–6 pairs of veins often branching before reaching the margin and kinked where the tertiaries join, teeth often ± triangular and less rounded than in *B. glandulosa*. *Male catkins* to 3 terminal on short extension growths of the previous year (Fig. 51), 2.5–5 × 2.5–3.5 mm, vigorous long shoots often with numerous, clustered, lateral males towards base of current year's extension growth below buds which will give rise to female catkins (Figs 65, 289) or towards tip of previous year's growth (Fig. 288), erect to somewhat pendulous at anthesis (Fig. 65). *Fruiting catkins* to 20 (–30) × 10 mm, often smaller. *Scales* 3–4.5 × 3.5–5 mm. *Seed* 1.6–2.5 × 2.5–3.2 mm, with wing narrower than the nutlet.

Fig. 289. *Betula pumila* Canada, Ontario, Shebandowan, Rossmere Creek. The four shoots to the right are of the current year's growth showing the male catkins on the lower part of the shoots with buds on the more apical parts which will give rise to female catkins or vegetative shoots (PW).

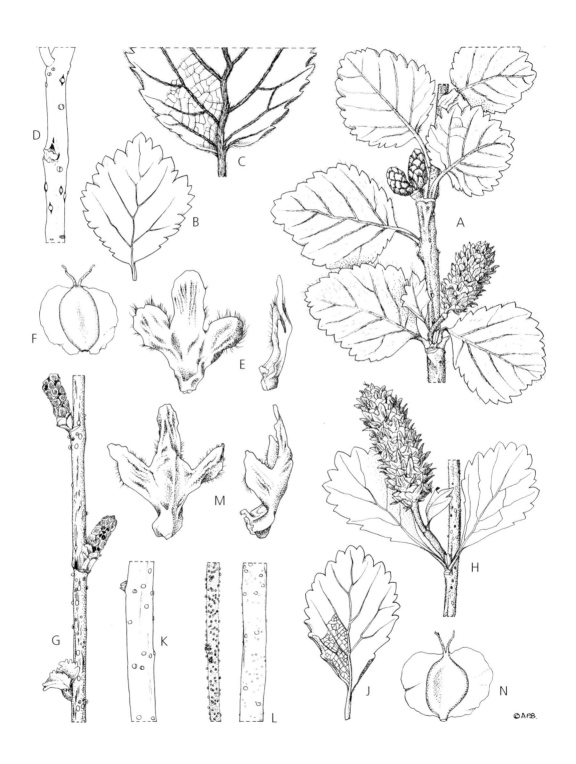

DISTRIBUTION. Newfoundland, Eastern Canada and north-eastern USA with scattered occurrences westward to the Pacific coast. Probably much confused with *Betula glandulosa*.

HABITAT. Wet ground, whether acid bog or calcareous fen or streamside.

CHROMOSOME NUMBER. 2n=56.

45. BETULA FRUTICOSA Japanese bog birch

HISTORY AND NOMENCLATURE

Under *Betula fruticosa* and *B. gmelinii* we are treating a difficult complex of presumably tetraploid shrubby birches of eastern and central northern Asia. The type specimen of *B. fruticosa* (BM!) was collected in the Lake Baikal region and, from its larger fruiting catkin and seed, is clearly tetraploid and so synonymous with *B. ovalifolia*, not *B. humilis* as suggested in *Flora of China* (Li & Skvortsov 1999).

Betula fruticosa was described by Pallas in 1776 in his account of his travels across Russia, at the same time as he described *B. dahurica* q.v. It thus has priority over *B. ovalifolia*, described from the Shantar islands off the coast of Siberia in the sea of Okhotsk in 1857.

ECOLOGY AND STATUS IN THE WILD

The name *Betula fruticosa* is here used for slender shrubs of wetter habitats (Fig. 292), mainly with elliptical to rhombic to obovate leaves, while those of drier habitats with more sharply toothed triangular to rhombic leaves are considered under *B. gmelinii* (Figs 39, 294, 296). Recent collections by Maurice Foster from northern Mongolia to the southwest of Lake Baikal are somewhat intermediate morphologically as well as geographically between *B. fruticosa* and *B. microphylla*.

TAXONOMY, RECOGNITION AND RELATIONSHIPS

The taller growing tetraploid *Betula microphylla* (see p. 328) of the Altai mountains and central Asia is very similar to *B. fruticosa*, and may be the result of introgression from *B. pubescens* into *B. fruticosa*. *B. microphylla* bears a close resemblance to both *B. humilis* and *B. fruticosa*, especially in the white wartiness of its thin twigs, erect fruiting catkins, and the overall appearance of its leaves.

Betula baicalensis Sukaczev, *Izv. Imp. Akad. Nauk* 8: 233 (1914), described from the shores of Lake Baikal, is possibly a hybrid between *B. fruticosa* and *B. pendula*.

Betula tatewakiana was described in 1959 from bogs in Hokkaido, but is generally considered a minor variant of *B. fruticosa*.

Round leaved tetraploid shrubs in Eastern Asia usually identified as *Betula middendorffii* or *B. divaricata* probably belong here, but are very similar to the tetraploid *B. glandulosa* from Goose Bay in Labrador.

Fig. 290 (opposite). *Betula pumila* **A–F** 'subsp. *pumila*', **G–M** 'subsp. *glandulifera*'. **A** fruiting shoot with ripe fruiting catkin and immature male catkins × 2; **B** leaf from young shoot of another plant, adaxial surface, hairs omitted × 1; **C** underside of leaf base × 2; **D** bark of twig × 2; **E** female catkin scale back and side views × 10; **F** seed × 10; **G** twig with foliage bud and male catkins just starting to open × 2; **H** fruiting shoot with ripe fruiting catkin × 2; **J** underside of leaf × 2; **K** bark of twig, specimen with few glands × 2; **L** bark of very young and older twig, specimen with abundant glands × 2; **M** female catkin scale, back and side views × 10; **N** seed × 10. **A, D–F** from *Fernald & Long* 28070 (Burnt Cape, Newfoundland, 5 Aug. 1925); **B** from *Moir* 53 (Middle Ridge, Newfoundland, 26 July 1937); **C** from *J. Faroles* (New Brunswick, ex Herb. Blake); **G** from *Baldwin* 5648 (Ford Lake, Ontario, 4 June 1954); **H–K, M, N** from *Baldwin & Breitung* 2834 (La Sarre, Québec, 27 June 1952); **K** from *Scoggan* 5307 (Oxford Lake, Manitoba, 2 July 1949). DRAWN BY ANDREW BROWN.

CULTIVATION

Betula fruticosa is rare in cultivation, with the collection of Japanese origin, formerly referred to as *B. tatewakiana*, probably being the commonest provenance in cultivation (Fig. 292). It is a neat upright shrub with attractive autumn colours of muted orange and yellow. Most collections from continental Asia do not thrive in the British Isles due to damage by spring frosts, although one from 'Olga' (presumably Olga south of Magadan on the Russian Far Eastern coast), obtained through Tallin Botanic Garden, has grown well at Ness and forms an attractive upright shrub with whitish bark. In the wild in Japan it reaches 1.5 m (Satake *et al.* 1989), to 2 m in eastern Asia (Li & Skvortsov 1999), but at least 3 m in cultivation.

Betula fruticosa Pall., *Reise Russ. Reich.* 3: 758 & tab. Kk (1776). Type: Abundat in paludibus saxosis, inque alpinis frigidis Sibiriae orientalis, praesertim circa Baikalem, *Rhododendro daurico* ubique conterranea et semper sibi similis (BM!).
Betula divaricata Ledeb., *Denkschr. Königl.-Baier. Bot. Ges. Regensburg* 3: 59 (1841).
Betula ovalifolia Rupr., *Bull. Cl. Phys.-Math. Acad. Imp. Sci. Saint-Pétersbourg* 15: 378 (1857). Type: described from the Shantar Islands (LE).
Betula reticulata Rupr., *Bull. Cl. Phys.-Math. Acad. Imp. Sci. Saint-Pétersbourg* 15: 378 (1857).
Betula fruticosa var. *ruprechtiana* Trautv. in Maxim., *Prim. Fl. Amur.*: 254 (1859). Type: Am untern Amur: häufig auf nassem Wiesen in der Niederung bei Kitsi, 29 Juni 1856 (frut. submat.) — Wahrscheinlich gehören hierher auch einige bei Myllki auf sampfigen Wiesen den 16 Mai 1855 gesammelte Exemplare (flor.).

Fig. 291 (left). Spreading, semi-weeping *Betula fruticosa/microphylla* multi-stemmed up to 5 m with fruiting catkins up to 2 cm, NW Mongolia (MF).

Fig. 292 (right). *Betula fruticosa* (*B. tatewakiana*) fruiting catkins, male catkins and leaves; cult. Ness Gardens (Sarabetsu Bog, Hokkaido, Japan) (PW).

Fig. 293. *Betula fruticosa.* **A** twig with buds and immature male catkins × 1; **B** fruiting short shoot with mature fruiting catkin × 1; **C** underside of leaf base (distribution of resin glands indicated) × 2; **D** bark of twig × 2; **E** female catkin scale, back and side views × 8; **F** seed × 8. **A** from *Furuse* 9629 (Kami-sarabetsu, Takachi, Hokkaido, 1 Sept. 1975); **B, C, E, F** from *Furuse* 9631 (Kami-sarabetsu, Takachi, Hokkaido, 1 Sept. 1975); **D** from *Furuse* 9630 (Kami-sarabetsu, Takachi, Hokkaido, 1 Sept. 1975). DRAWN BY ANDREW BROWN.

Betula humilis var. *ovalifolia* (Rupr.) Regel, *Nouv. Mém. Soc. Imp. Naturalistes Moscou* 13: 109 (1861).
Betula humilis var. *reticulata* (Rupr.) Regel, *Nouv. Mém. Soc. Imp. Naturalistes Moscou* 13: 109 (1861).
Betula humilis var. *rupechtii* Regel, *Nouv. Mém. Soc. Imp. Naturalistes Moscou* 13: 109 (1861).
Betula tatewakiana M. Ohki & S. Watan., *J. Jap. Bot.* 34: 329 (1959). Type: Hokkaido, Prov. Tokachi, Sarabetsu moor, 18 Aug. 1958, *M. Ohki* (SAPA).
Betula yoshimurae Miyabe & Tatew., *Trans. Sapporo Nat. Hist. Soc.* 15: 129 (1938). Type: S. Saghalien, Distr. Shikka, the upper Rukutama, 14 July 1937, *B. Yoshimura & M. Hara*.
Betula middendorffii Trautv. & C. A. Mey. in Middend., *Reise Sibir.*, 1 (2), *Fl. Ochot.*

DESCRIPTION. *Shrub* with several stems, to 2 (–3 m). *Twig* to 1.5 mm diam., puberulent, densely covered with white glands which become darker with ageing on second year twigs. *Bud* about 4 × 2 mm, brown, scale margins ciliate, otherwise glabrous. *Shoots* hairy and resinous-glandular when young, the red and colourless glands becoming white on ageing. *Petiole* 3–7 mm. *Leaves* elliptic to obovate, to 30–35 × 20–40 mm, cuneate at base, with 5–6 (7) pairs of veins, often untoothed in lower third; teeth ± equal in size. *Male catkin* about 9 × 3 mm in winter, expanding to c. 34 mm in spring; scales 1.5 × 1.5 mm; stamens red (Japan). *Fruiting catkin* erect, 15–30 × 7–12 mm on short peduncle of 2–6 mm. *Bracts* 5–6 mm, terminal lobe c. 2 mm, laterals c. 1 mm. *Nutlet* 2.25–3 × 1.25–1.5 mm, wing ½ nutlet width; style c. 0.75 mm.

DISTRIBUTION. East Asian Pacific coastal regions from Hokkaido in Japan and Russian Far East south to Korea, and westwards to Inner (Nei) Mongolia.

HABITAT. Wet ground in peat bogs, marshes and streamsides.

CHROMOSOME NUMBER. 2n=56.

46. BETULA GMELINII (including *Betula apoiensis*)

HISTORY AND NOMENCLATURE

Betula gmelinii was described in 1835 by Bunge from collections made by Johann Georg Gmelin in 1737 in eastern Siberia while he was a scientist on Bering's second Kamschatka expedition (1731–1742). It has often been treated as a variety or synonym of *B. fruticosa* (e.g. by Govaerts & Frodin 1998), but is said to differ primarily in having a wider wing to its nutlet (1–2 times as wide as against about half as wide in *B. fruticosa*, Li & Skvortsov 1999).

The name *Betula apoiensis* was given by Nakai to an isolated and very distinctive population of shrubby birches growing on the summit of Mt Apoi in Hokkaido in Japan. Nagamitsu *et al.* (2006a, b) have now shown that *B. apoiensis* is the result of genetic introgression into the Japanese bog birch, *B. fruticosa*, from the Japanese mountain tree birch, *B. ermanii*. As *B. ermanii* is the treeline birch of eastern Asia, it has ample opportunity to hybridise with the shrubs of *B. fruticosa* above and beyond the treeline. Both species are tetraploid, so there is even less impediment to gene flow than in Europe where there are ploidy level differences between the treeline tree birch, tetraploid *B. pubescens*, and the tundra shrub species, diploid *B. nana*.

Fig. 294. *Betula gmelinii* on sand dunes near Khuzhir, Olkhon Island in Lake Baikal, *Pinus sylvestris* in background (HMcA).

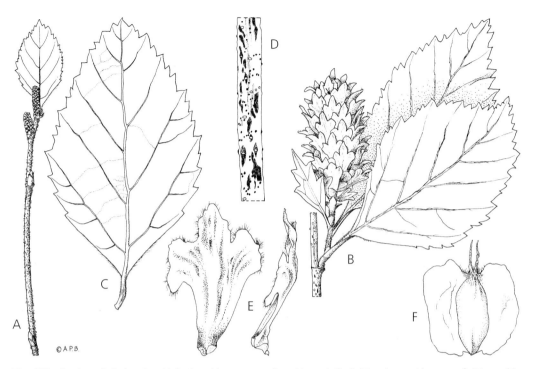

Fig. 295. *Betula gmelinii*. **A** twig with buds and immature male catkins × 1; **B** fruiting shoot with mature fruiting catkin × 2; **C** underside of leaf × 2; **D** bark of twig × 2; **E** female catkin scale, back and side views × 7; **F** seed × 10. **A, C–F** from *Svortskov* (Nizhni Tsassatshei, Transbaikalika, 10 Aug. 1988); **B** from *Smirnov* 4235 (E Siberia, 6 July 1911). DRAWN BY ANDREW BROWN.

From our limited experience, *Betula apoiensis* appears to be very similar to *B. gmelinii* in habit and morphology and it seems likely that *B. gmelinii* may be also of hybrid origin with the same parents; therefore we have decided to sink *B. apoiensis* under the earlier name *B. gmelinii*.

ECOLOGY AND STATUS IN THE WILD

Betula gmelinii is recorded as growing in dry habitats, sandy hills and deserts (Fig. 294) (Li & Skvortsov 1999). The illustration here shows heavily grazed shrubs on the sand dunes on Olkhon Island in Lake Baikal; unfortunately they bore no fruiting catkins. Their leaves are sharply toothed and similar to those of *B. dahurica*, *B. 'apoiensis'* (Fig. 39), and herbarium specimens labelled *B. gmelinii*. Further collection and study of these shrubby east Asian birches of dry habitats is clearly required. In the wild, *B. apoiensis* forms prostrate scrub on serpentine towards the exposed summit of Mt Apoi in southern Hokkaido, Japan. Whereas *B. ermanii* and *B. fruticosa* appear to require abundant moisture even when well-established, *B. apoiensis* is drought tolerant and very persistent in cultivation, surviving competition from weeds to a remarkable degree.

CULTIVATION

Betula apoiensis was first introduced to cultivation in the British Isles in 1984 when Professor Ohba of Tokyo sent seed to the authors. The numerous seedlings raised were widely distributed. They were variable in time of bud break and only those individuals with later bud break survived recurrent shoot death caused by late spring frost. One particularly attractive clone of *Betula apoiensis* has been

Fig. 296. *Betula gmelinii* ('*apoiensis*') at Ness showing typical habit in cultivation (PW).

released into the trade as 'Mount Apoi'. In cultivation plants from seed from Mount Apoi form twiggy low shrubs (Fig. 296) up to about one metre in height with most of the long shoots growing more or less horizontally and ending in clusters of male catkins. After flowering these die back and growth is continued by lateral shoots. The very dark and rather thick twigs are striking and make the numerous yellow male catkins conspicuous; in contrast, the stamens of Japanese collections of *B. fruticosa* are reddish-pink.

Betula gmelinii Bunge, *Verz. Altai Pfl.*: 113 (1835). *Mem. Acad. Imp. Sci. St.-Petersbourg divers Savans* 2: 607 (1835). Type: Dahuria (LE).

Betula apoiensis Nakai, *Tenn.-kinenb.-chosa-hokoku Shokub.* 12: 33, 46 (1930); Hara, *J. Jap. Bot.* 10: 227 (1934). Type: Yezo, prov. Hidaka, in locis glareosis jugi montis Apoi, Aug. 1928, *T. Nakai* (SAPA).

DESCRIPTION. *Shrubs* with one or more stems to 1 m, main stems becoming horizontal. *Twig* robust, rough, to 3.5 mm diam., dark brown and densely (up to 50%) covered with dark purple-brown glands on one-year old twigs, glands becoming white on two-year–old twigs. Lenticels white, circular, about 0.5 mm diam. *Bud* not sticky, large. *Shoots* thinly covered with long silky hairs, with a layer of short, patent, velvety puberulent hairs; young shoots densely covered with clear glands interspersed with a few red glands, the clear glands becoming dark purple brown. *Leaves* with long silky hairs and a layer of short, patent, velvety puberulent hairs beneath on the petiole, but only on the veins on the leaf upper surface; leaf midrib with red colleters and clear glands on the upper surface; petiole (2) 4–10 mm, blade very variable in size with basal long shoot (preformed in bud) leaves usually by far the largest, 1.2–6 (9) × 1–4.5 (7) cm (the higher figures from shrubs in cultivation), broadly elliptic-ovate to ovate-orbicular, acute, usually rounded at base, and serrate acute teeth. *Male catkins* up to 2 terminal with numerous (to 12 or more) laterals on distal nodes of twig, to 23 × 4 mm, expanding to 55 × 7 mm; scales peltate, 1.5 × 1.5 mm, ± circular-shield-shaped, dark brown and conspicuous among the yellow stamens, margins finely ciliate. *Female catkins* red, to 15 × 4.5 mm, often clustered towards ends of branches, developing from the buds behind (proximal to) the male catkins, each bud giving rise to a short shoot bearing up to 4 leaves and either a single catkin or 2 catkins. (As the leaves on these female catkin bearing short shoots do not develop axillary buds, the terminal parts of many branches which bore male and female catkins die following fruit maturation). *Fruiting catkin* upright, to 33 × 11 mm, peduncle to 5 mm. *Bract* to 4–5 mm, terminal lobe c. 2 mm, laterals c. 1.5 mm. *Nutlet* c. 2 × 1 mm with wing c. ½ nutlet width.

DISTRIBUTION. Russia (Baikal, E Siberia), Mongolia, Korea, Japan (Mt Apoi, Hokkaido), China (N Heilongjiang, N Liaoning, Nei Mongol).

HABITAT. Sandy hills, deserts (Li & Skvortsov 1999); rocky exposed places on serpentine in Japan (*Betula apoiensis*).

CHROMOSOME NUMBER. 2n=56 (*Betula apoiensis*).

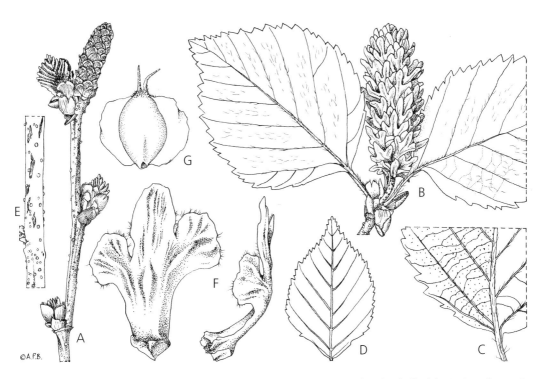

Fig. 297. *Betula apoiensis* (synonym of *B. gmelinii*). **A** winter twig with immature male catkin, leaf buds beginning to burst × 2; **B** fruiting shoot with mature fruiting catkin × 2; **C** underside of leaf × 2; **D** variant leaf shape (hairs and petiole omitted) × 1; **E** bark of twig × 2; **F** female catkin scale, back and side views × 10; **G** seed × 10. **A** from *Furuse 8600* (Mt Apoi, 25 May 1995); **B, C, E, F, G** from *Furuse 29557* (W ridge, Mt Apoi, 29 July 1955); **D** from *Furuse 6122* (Mt Apoi, 3 July 1974). DRAWN BY ANDREW BROWN.

Fig. 298. Type specimen of *Betula ×dosmannii* (*B. maximowicziana* × *B. ermanii*) from the only known tree of this hybrid which is growing in the Arnold Arboretum (Fig. 43). The leaves are very similar to those of *B. maximowicziana* but smaller and less cordate at the base. The fruiting catkins more resemble those of *B. ermanii*, *B. utilis* and *B. papyrifera*.

8. BIRCH CULTIVARS

Compiled by Paul Bartlett, Garden Manager, Stone Lane Gardens, Devon.

Although birches are easily grown from seed collected in the wild, the individual trees are relatively short-lived and specimens grown from original seed collected in the wild in the late 1800s and early 1900s have often died. Any offspring, especially of the white barked birches, propagated by seed, are likely to be hybrids because most are self-incompatible and many gardens grow single individuals of each collection or species (Green 1966; Clarke 1970; Santamour & Lundgren 1996).

Even grafted specimens may be incorrectly named, as the similarity of the long-shoot leaves of different species may make the death of the scion and growth of the rootstock difficult to detect. Therefore, labels on trees in arboreta should be treated with extreme caution, even when a tree has supposedly been grown from wild collected seed.

The species names can be used with cultivar names, to designate selected clones, and a list is given below. However there are many, often attractive, clones in cultivation which are more or less unidentifiable as species and may be of hybrid origin. These are named as cultivars and described below in the section on hybrids.

Further research into the species cultivars (particularly the older ones) will no doubt identify more as hybrids.

Cultivar names are in bold.

Betula chinensis

'Rhinegold'
Named by Robert Vernon of Bluebell nursery. Grown from seed given by Tim Whiteley. Superb golden bark. Robert's parent stock plant has since died.

Betula cordifolia

'Clarenville'
From a tree at Stone Lane Gardens, Devon. Collected and named by Kenneth Ashburner from Clarenville, Newfoundland, Canada.

Pink-brown bark roughly peeling. Good autumn colour.

Betula costata

This species was often confused with *Betula ermanii* in the past. All the known *B. costata* cultivars have now been proven to be *B. ermanii* and there are no cultivars of true *B. costata* that we know of.

Betula dahurica

'Maurice Foster'

Selected by Maurice Foster from a tree at Stone Lane Gardens, Devon. Named by Kenneth Ashburner. Seed collected from a forest plot of wild Hokkaido provenance in Yamabe, Furano, Hokkaido, Japan.

Quite vigorous and very shaggy barked, with less susceptibility to late frosts than most *Betula dahurica* of continental provenance. Bark ultimately pale but red when young, with a showy red-brown colour to the inner surface of the bark flakes typical of hexaploid populations (Hokkaido, Honshu and Iturup).

'Stone Farm'

Taken from a tree at Stone Lane Gardens, Devon. From the same group of seedlings as 'Maurice Foster'. Named by Kevin Croucher of Thornhayes Nursery in 1997.

The very dark bark ex-foliates in curly cinnamon-like flakes to reveal a pinky layer beneath.

Betula ermanii

'Blush'

Named by Piet de Jong. The standard *Betula ermanii* of the Dutch trade which had long been wrongly named as *B. costata*. Similar to 'Grayswood Hill'. Origin unknown.

A graceful large birch with creamy white bark with horizontal lenticels. Rich yellow leaves in autumn.

'Fincham Cream'

From the late Maurice Mason's arboretum at Talbot Manor, Fincham. Originally sold as *Betula costata* but now recognised as being *B. ermanii*. A batch of seedlings were grown at Brentry / Hillier arboretum and five of these trees were sent by Harold Hillier to Maurice Mason as *B. costata*. The best one was selected by Brian Humphrey and called 'Fincham Cream'.

Very attractive pale cream smooth bark and superb gold autumn colour.

'Grayswood Hill'

From Grayswood Hill, Haslemere, Surrey. Home of the Pilkington family (of glass manufacturing fame). Origin unknown.

Good, creamy orange bark with attractive lenticels. Strong grower. Reliable rich golden autumn leaf colour.

'Hakkoda Orange'

From a tree at Stone Lane Gardens, Devon grown from seed collected by Kenneth Ashburner from Mt Hakkoda, Japan.

A beautiful form with creamy orange bark and reliable rich golden autumn leaf colour.

'Holland'

New name for a plant grown in the Netherlands as *Betula ermanii*, although it may be a hybrid.

'Kwanak Weeping'

Originated in Ness Gardens, Wirral as a spreading shrub from Kwanak Botanic Garden, Korea in the 1970s. Initially given the cultivar name of 'Pendula' but changed to comply with the rules of nomenclature.

Propagated by Maurice Foster as a standard to create a beautiful small weeping tree with a gently spreading crown of cascading twigs, ideal for a small garden specimen. Cream coloured peeling bark, strong yellow autumn tints.

'Mt Apoi' — see *Betula fruticosa* var. *apoiensis* in Hybrids at end.

'Mt Zao Purple'
From a tree at Stone Lane Gardens, Devon grown from seed collected by Kenneth Ashburner from Mt Zao, North Honshu, Japan. The names 'Mt Zao' and 'Zao Purple' were created about the same time, so it is difficult to decide which is correct. We now call it 'Mt Zao Purple', in the hope that this will tie the two names together. 'Zoo purple' is simply a mis-spelling that has been perpetuated.

The bark has a base colour of purple overlain with creamy orange, combined with very prominent horizontal bands of lenticels, creating an almost striped appearance. Reliable rich golden autumn leaf colour.

'Pendula' — see 'Kwanak Weeping'

'Polar Bear' — see Hybrids at end

'Saitoana' — see *Betula ermanii* var. *saitoana*

'Zao Purple' — see **'Mt Zao Purple'**

'Zoo Purple' — misspelling: see **'Mt Zao Purple'**

Betula ermanii var. *saitoana*

Collected in Korea by Mark Fillan. Sometimes given the cultivar name 'Saitoana' but correctly called *Betula ermanii* var. *saitoana* (coll. no. *MSF* 865). This is obviously not a true cultivar, but is sometimes incorrectly treated as such by the nursery trade.

A true dwarf genotype that will make a bushy, white-stemmed little tree. As a variety, it is slow growing and after 20 years varies from a nondescript twiggy 2.5 m shrub to a small single stemmed tree to 3.5 m with a typical cream *Betula ermanii* stem, rather like a miniature version of the species.

Betula medwediewii

At Westonbirt there are a couple of *Betula medwediewii* that were planted in 1938 by Mr Mitchell. The Westonbirt notes show that these trees are grafted, presumably on *B. pendula* rootstocks. The Westonbirt trees are now quite large (huge compared with other *B. medwediewii* I have seen). So it seems that the *B. pendula* rootstock is having an effect on their growth (not surprising) and making them much larger trees than their seed-grown counterparts. For this reason, grafted *B. medwediewii* being sold by nurseries should not be described as small shrubs as this is misleading to the public. Only plants grown from seed or cuttings are going to retain the characteristic of low growth.

'Gold Bark'
A selection from Alphons van der Bom Nursery, Netherlands, about 1965; origin unknown. Supposed to be more tree-like, but all the photos I have seen appear almost identical to normal seed-grown *Betula medwediewii*.

Silvery-brown metallic stems with a slight golden cast. Multi-stemmed shrub.

'Winkworth Form'

A selection by Kevin Croucher of Thornhayes Nursery, Devon from trees in Stone Lane Gardens, Devon, acquired from Winkworth Arboretum. Origin unknown.

A form with a slightly paler bark, though not very different from seed-grown trees.

Betula nana

'Glengarry'

A Scottish selection. Origin unknown.

Betula nigra

A word of warning about *Betula nigra* cultivars. *B. nigra* is one of the most heat tolerant birches, coming as it does from areas with hot summers. Many of the cultivars below state that they have a dwarf or compact shrubby nature. However, it has been observed that in more temperate climates, these cultivars will grow larger.

'Black Star'

Introduced by Minier nursery, France in 2006.

Superb satin cinnamon bark standing out in shreds. Graceful and airy with plenty of flexible branches.

'Cinnamon Bark'

Introduced by Forrest Keeling Nursery, Elsberry, Missouri, USA.

Exfoliating bark, lighter-coloured than most native trees.

'City Slicker'®

Raised by Carl Whitcomb of Stillwater, Oklahoma, USA.

Dark leaved and a whiter than usual trunk.

Dura-Heat = 'BNMTF'

Raised by Moon Nursery, Loganville, Georgia, USA.

A more heat and drought tolerant form of this species. Densely pyramidal habit with glossy green leaves that fade to clear butter yellow in fall. Winter bark is pinkish-orange with prominent exfoliation. Bark exfoliates early.

Fox Valley® = 'Little King'

Discovered by Jim King, Oswego, Illinois. There is a mature specimen at RHS Wisley.

This cultivar features orange to brownish bark which exfoliates at an early age to reveal a somewhat lighter shaded brownish inner bark. Compact growing with many branches, almost shrub-like.

'Graceful Arms'

Found in Wisconsin by Darrell Kromm and Tom Dilatush.

Another weeper, more open and larger leaved than 'Summer Cascade', and later to produce exfoliating bark.

'Heritage' = 'Cully'

From a tree in St Louis, Missouri, USA. Origin of parent tree unknown. Cultivated by Earl Cully in 1979.

A paler form with vivid gold autumn colour and creamy shaggy bark. Selected for its notable resistance to birch borer and leaf spot. It has been noted that this cultivar does not do well in the colder east side of the UK and certainly many *Betula nigra* of different provenance struggle with the late frosts of the UK.

Northern Tribute® = 'Dickinson'
Introduced by North Dakota State University. In 1989, a seed-propagated tree growing in Dickinson, North Dakota was tissue-culture propagated. This 40-year old tree was planted in 1966 and is the largest tree of this species observed in the upper Northern Plains. It has performed very well under rather compacted, dry and alkaline soil conditions in USDA zone 3. Many seed sources of river birch suffer winter injury in zone 3 and also may die from iron chlorosis in alkaline pH soils.

The bark on this cultivar is ivory coloured with striking coppery-bronze exfoliating bark which contrasts to add landscape interest. As trees reach 20 to 25 years of age, the shaggy bark becomes more uniformly tannish-brown. Foliage quality has been good with yellow autumn colour.

'Peter Collinson'
This selection was collected in Canada by French plantsman Aurélien Hemono and is propagated by Junkers Nursery, Somerset. Named in honour of the man responsible for introducing *Betula nigra* to the nursery trade. *B. nigra* is not native to Canada, so although the parent tree was found in a plantation in Quebec, the trees must have been imported seedlings from the USA.

Darker young bark and more compact than 'Heritage'.

'Shiloh Splash'
An introduction by Shiloh Nursery, Harmony, North Carolina, USA.

Variegated, shrubby form, white margined green leaves, flushed pink when young, very attractive but reverts to green quite easily.

'Summer Cascade'
A weeping form picked in 1996 from a batch of seedlings planted in 1992 in Shiloh Nursery, Harmony, North Carolina, USA. Origin of seedlings unknown.

'Suwanee'
Introduced by R. L. Byrnes, Trail Ridge nursery, Keystone Heights, Jacksonville, Florida, USA in 1985.

With larger and shinier leaves than is typical for the species, and with salmon-white exfoliating bark.

Tecumseh Compact = 'Studetec'
Introduced by Studebaker Nurseries in New Carlisle, Ohio, USA.

A compact form with graceful, pendulous branches and a low yet wide and spreading habit. The leaves are oval, glossy green and the bark tends to be pink when young before turning brown and ridged on older trees.

'Wakehurst'
From a vigorous seedling (now dead) at Wakehurst showing good bark qualities (Acc. no. 1983 8312). Named by Tony Schilling in 1990. Seed was from Kenneth Ashburner, in which case it would have been sent from a US seed firm, so the provenance will never be known. Propagated by Chris Lane.

A good form with more pink in the bark than most and showing early bark colour, although the coloured bark fades as it matures.

Betula papyrifera

'Belle Vue'

From a tree at Stone Lane Gardens, Devon. Collected and named by Kenneth Ashburner from Belle Vue, Newfoundland, Canada.

Pink-brown slightly rough peeling bark. Autumn leaf colour is a rich yellow persisting on the trees for longer than most *Betula papyrifera*.

'Chickadee'

From a tree collected 25 miles northwest of Whitecourt, Alberta, Canada. Propagated by Beaverlodge nursery. The parent tree is at the University of Guelph arboretum, Guelph, Ontario, Canada.

This form is slow-growing with a dense, pyramidal habit. Possibly a hybrid between *Betula papyrifera* and *B. pendula* subsp. *mandshurica*.

'Grandis'

Named by H. J. Grootendorst. Not used much outside the Netherlands.

Large tree with large, heart-shaped leaves; leaves deeply and doubly serrate.

'Peter Mills'

Originated from seedlings from the Hemlock Valley ski resort near Mission to the east.

'Prairie Dream'

Selected at North Dakota State University, USA.

It has dark green leaves, golden yellow in autumn and snow white bark.

Renaissance Compact® = **'Cenci'**

White barked tree forms a tight pyramid, 30 × 15 feet with semi-glossy rich green leaves. Said to have good birch borer resistance.

Renaissance Oasis® = **'Oenci'**

Broadly pyramidal tree that is said to have good birch borer resistance. Dark green leaves. Rich mahogany-reddish bark matures to white.

Renaissance Reflection® = **'Renci'**

From a tree growing in a trial plot at Wooster, Ohio in 1986. It is an open-pollinated cross of two *Betula papyrifera*, origin unknown. Selected by Thomas Pinney of Sturgeon Bay, Wisconsin, USA and propagated by the Evergreen Nursery company of the same location.

A white barked, fast growing form that is said to have good birch borer resistance.

Renaissance Upright® = **'Uenci'**

White barked tree forming a narrow pyramid with a strong central leader. Said to have good birch borer resistance.

'Saint George'

From a tree at Stone Lane Gardens, Devon. Collected by Kenneth Ashburner from St Johns, Newfoundland, Canada. Selected by Kevin Croucher of Thornhayes nursery, Devon.

It is red barked, peeling to white when young, ultimately white. Excellent form with good persistent golden yellow leaves in autumn.

'Snowy'

From a tree at Michigan state University, USA. White Barked.

'Vancouver'

From a tree at Stone Lane Gardens, Devon, collected by Kenneth Ashburner from Abbotsford, Vancouver, Canada. to the south-west of the city. Selected by Kevin Croucher of Thornhayes Nursery, Devon.

It is very vigorous, has rich red brown stems and rich gold autumn colour in November.

Betula pendula subsp. *mandshurica*

Dakota Pinnacle = 'Fargo'

Developed and introduced by North Dakota State University, USA. It was selected from an open-pollinated population of *Betula pendula* subsp. *mandshurica* sown in 1986.

Narrow pyramidal habit. Dark green leaves, bark creamy white. Good for heavy and drought prone soils.

'Whitespire' or 'Whitespire Senior' — see *Betula populifolia*

Betula pendula subsp. *pendula*

'Bangor'

This was named and propagated by Chris Sanders whilst at Bridgemere nursery and released in small numbers from around 1997 onwards. It originated in a Forestry Commission seed bed at Bangor in North Wales and cuttings were given to Chris Sanders by the forester who had spotted it and transplanted it to his private garden in Bangor.

It is a cut-leaved form similar to 'Dalecarlica/Laciniata' but much more vigorous than either.

'Bircalensis'

Found and cultivated in Finland.

With deeply incised triangular leaves.

'Boeugh's Variety'

Thought to be the same as Tristis.

'Crimson Frost'

Betula pendula subsp. *szechuanica* × *B. pendula* 'Purpurea'

Named by Tom Pinney Jr in 1978. Propagated by Evergreen Nursery, Sturgeon Bay, Wisconsin, USA.

Bark of tan and white. Bright red new leaves become dark red later.

'Crispa'

Form with deeply cut leaves. Similar to 'Dalecarlica' and 'Laciniata'.

'Dalecarlica'

The original tree grew at Lilla Ornäs, in the Swedish province Dalarna (Dalecarlia), but was killed in a storm in 1887. Photographs of the tree, of the leaves and branches survive (Hylander 1957). Vegetative propagations had been made, however, and the cultivar was imported into the Netherlands in 1932. H. J. Grootendorst noted that most of the trees produced in the Netherlands were exported to Scandinavia, and very few planted specimens could be found in the Netherlands. He also was able to distinguish 'Dalecarlica' from 'Laciniata' on the basis of vegetative buds during the winter; rounded in 'Dalecarlica' and pointed in 'Laciniata'.

Tall slender tree. Cut leaves and drooping branchlets. Strong autumn colour.

'Dark Prince'
Purple-leaved, darker than 'Purpurea'.

'Elegans'
Originated at Bonamy's Nursery, Toulouse, France, about 1866.

Has erect leader and branches hanging almost perpendicular; may be the same as 'Tristis'.

'Fastigiata'
First distributed by Simon-Louis Freres, France before 1870.

A columnar form of the native silver birch.

'Golden Beauty'
Similar to 'Golden Cloud' below, but does not scorch as badly. Origin unknown.

'Golden Cloud' correctly 'Schneverdinger Goldbirke'
Raised by G. Horstmann of Schneverdingen, Germany in 1975.

A true golden leaved form. Apparently scorches badly in the UK.

'Gracilis'
Apparently originated in Moscow about 1888, and was grown by Späth Nursery, Germany as *Betula alba laciniata gracilis pendula* about 1930.

Small tree with various ascending branches (no main trunk), but with smaller twigs pendulous, hanging together in bundles; leaves more deeply lobed than 'Laciniata'.

'Laciniata'
A tree commonly grown in Sweden but thought to be a clone of German origin.

A handsome cut-leaved form. Usually wrongly sold as 'Dalecarlica'.

'Long Trunk' — see Hybrids at end.

'Obelisk'
Found in the wild in northern France by P. L. M. van der Bom. Brought into cultivation in 1956 by Royal Nurseries Alphons van der Bom, Oudenbosch, Netherlands.

Extremely fastigiate form with stiffer branches than 'Fastigiata', bark very white.

'Purple Rain'
A selection made at Monrovia Nursery Company, Azusa, California, from open-pollinated seedlings of *Betula pendula* 'Purpurea' in 1987.

Slightly pendulous branches, new foliage displays lustrous, vivid purple colouring that is retained through the season.

'Purple Splendor'

Tree named and propagated by Darrold D. Belcher, Gresham, Oregon, USA. (Plant Patent No. 2107, November 28, 1961) Found as a red leaved seedling in nursery that contained other green-leaved *Betula pendula*.

Tree with white bark, purple leaves, and drooping leader and branches.

'Purpurea'

Created in about 1870 in France. Origin unknown.

A most ornamental slow growing birch with purple leaves.

'Silver Cascade'

Graceful, weeping branches.

'Silver Grace'

This was selected by Bernard Ticknor of Fullers Mill, near Bury St. Edmunds and first propagated by Ivan Dickings at Notcutts Nursery in 1980/81. It was originally thought to be a hybrid between *Betula pendula* and *B. utilis* or subsp. *jacquemontii* but is almost certainly just a good white stemmed form of *B. pendula*.

Graceful, semi-weeping form with larger, toothed leaves taking on butter-yellow autumn colour. Silver-white bark develops in time.

'Spider Alley' (syn. Silver Trestles)

Sourced and selected by Matt Lohan in Ireland.

It is a semi pendulous form with the branches twisting and curling in the manner of the contorted hazel.

'Swiss Glory'

A compact and upright form of *Betula pendula*. May be the same as 'Fastigiata'. Originated in Switzerland as 'Zwitser's Glorie'.

'Tristis'

Appeared in the Netherlands in 1867. Origin unknown.

Tall graceful tree with slender pendulous branches.

'Trost Dwarf'

Registered with the U.S. National Arboretum in 1984 by Richard Bush. A truly dwarf birch discovered in 1976 by Dieter Trost at Southern Oregon Nurseries, Medford, Oregon.

Leaves deeply dissected; a slow growing cut-leaf form.

'Viscosa' and 'Dentata Viscosa'

Appeared about 1912 in France. Origin unknown.

A small bushy tree with coarsely toothed leaves; young growths sticky.

'Wades Golden'

Named and propagated by Andrew Norfield from a New Zealand tree (Obviously imported from northern hemisphere, not native).

Leaves emerge yellow in the spring later greening out, followed by an amazing clear bright yellow in the autumn. Stems turn orange/yellow in winter resembling a dogwood.

'White Edge' and 'Silver Edge'

Variagated leaf, with a white edge all around the leaf. Can have a tendency to revert to green.

'Youngii'

Appeared in England about 1873. Origin unknown.

A mushroom headed, small weeping tree, used to be grafted high up on a standard but now propagated by tissue culture which seems to give it a higher stem.

Betula pendula subsp. *szechuanica*

'Liuba White'

From a tree collected by Roy Lancaster at 3500 m in 1981 from Pa La Ho (Pa La river) above Liuba, SW of Kangding, Sichuan, China. Selected by Kevin Croucher of Thornhayes Nursery, Devon.

'Moonlight'

I can find no information about the origins of this cultivar name. It may be an invalid name.

'Pendula' — see 'Zhongdian Weeping' below.

'Zhongdian Weeping'

Named by Maurice Foster and Chris Brickell from a tree raised from seed collected at Zhongdian in NW Yunnan at 3350 m (collection no. *CLD* 0126). Initially given the cultivar name of 'Pendula' but changed to comply with the rules of nomenclature.

A weeping, less vigorous form, with an open spreading mushroom shaped crown and slender twigs weeping to the ground.

Betula populifolia

'Whitespire Senior' and 'Whitespire'

Originally thought to be from a *Betula pendula* subsp. *mandshurica* (syn. *B. platyphylla* var. *japonica*) tree at Madison arboretum, University of Wisconsin, USA. Collected by John Creech (coll. no. 235128) Shibuyu Onsen, Yatsugatake Mts, Honshu, Japan, 1956.

The 'Whitespire' Story: It was selected and introduced, in 1983, by Edward R. Hasselkus, University of Wisconsin-Madison and named *Betula platyphylla* var. *japonica* 'Whitespire', Whitespire Asian white birch. Later it was realised that there had been an error in record keeping and that 'Whitespire' was a gray birch (*Betula populifolia*), not a selection of Asian white birch, and it did not originate from the seeds collected by John L. Creech in Japan in 1951 as had been reported.

Around 1993 this selection was renamed *Betula populifolia* 'Whitespire'. Unfortunately some nurseries still offer it as 'Whitespire' Asian white birch even when they have adopted the correct scientific name. The cultivar name 'Whitespire' was applied to a vegetatively propagated clone of the original Wisconsin tree, which had not succumbed to bronze birch borer whereas many of the surrounding birch trees at this site had. As of 2006, the original 'Whitespire' tree, now 49 years old, is still free of bronze birch borer (Ed Hasselkus, pers. comm.). Unfortunately, the name 'Whitespire' has also been applied to seed propagated material, that is to trees not genetically identical to the original tree. To emphasise this important difference, the cultivar name 'Whitespire Senior' is now used to designate clonal material from the original tree and 'Whitespire Junior' is used for trees propagated from seed of 'Whitespire Senior'.

Betula pubescens

'Armenian Gold'
This cultivar has yellow leaves.

'Arnold Brembo' — see *Betula pubescens* subsp. *pumila*

'Combe de Lechaud'
There is known to be a tree at RHS Wisley that appeared in 2000 from their propagation department but we have no more information.

'Crenata Nana'
A dwarf, round bush growing at the rate of 5–7 cm annually. Given its name and habit, it could possibly be a hybrid of *Betula nana*.

'Incisa'
Found in Sweden in 1914.
   Deeply cut leaves.

'Silver Fountain'
I cannot find any information about this cultivar.

'*Tortuosa*' — see *Betula pubescens* subsp. *pumila*

'Variegata'
A tree of unknown origin described by C. Schneider from a specimen at the Moscow Botanic Garden.
   White-variegated-striped leaves that turn more green during the growing season, dense crown with somewhat twisted branches, and whiter bark than normal *Betula pubescens*.

'Yellow Wings'
This cultivar has yellow leaves.

Betula pubescens subsp. *pumila*

'Arnold Brembo'
A selection by Kenneth Ashburner from a tree at Stone Lane Gardens, Devon; most likely the Norwegian Namsskogan provenance of *Betula pubescens* subsp. *pumila* — see 'Tortuosa'. Named after a retired Norwegian army officer that Kenneth met and stayed with in Fauske, Norway. There is a fine specimen of this in Cambridge University Botanic Garden.
   It has a perfect pyramidal shape — just as if it had been clipped, and a very respectable white trunk. It leafs very early, usually by the end of January when its bright green young foliage is very conspicuous and apparently undamaged by frost. Slow growing.

'Tortuosa'
Name given to slow growing, short branched tree from Norway. Not a cultivar name but really *Betula pubescens* subsp. *pumila*.

Betula raddeana

'Hugh McAllister'

From a tree at Ness grown from seed collected at the edge of grazed meadows in Karachai-Cherkessk autonomous region, nr Karachayevsk, Northern Caucasus, Russia, Altitude 1600 m. Collected by an American expedition with seed being distributed by the Cary Arboretum.

Betula utilis subsp. *albosinensis*

It seems that almost all of the early selections of this species come from *Wilson* 4106 from "Western Szech'uan, west and near Wen-ch'uan at 2600 m–3600 m". Wilson's note after the entries for the collections says, "It is a tree from 20 to 26 m tall with a trunk 2 to 3.5 m in girth. The bark is bright orange to orange red, peeling off in very thin sheets, each successive sheet being covered with a white glaucous bloom,". It sounds as if the seed came from a group of trees rather than one individual.

I can find no evidence that any of the earlier Wilson collections were ever distributed or survived in Britain. We know that Wilson collections from the 1910 expedition for the Arnold Arboretum came to Caerhays and Werrington where J. C. Williams raised many of his plants and from seed he sent to Stanage Park in Radnorshire, Hergest Croft in Herefordshire and maybe the RBG Edinburgh. Named cultivars from this source include 'Bowling Green', 'China Ruby', 'Chinese Garden' and 'Kenneth Ashburner'. Many other named plants including B. 'Hergest 'seem to be hybrids of the original collection whilst B. 'Fascination' is either a hybrid or a selection of B. *utilis*. The only other early collection surviving in cultivation appears to be B. *utilis* subsp. *albosinensis Purdom* 752

More recent introductions include 'Red Panda' a collection by Steven Spongberg in Hubei in 1980 and 'China Rose', 'Kansu' and 'Pink Champagne' from Gansu."

'Bowling Green'

From the tree (now dead) which formerly grew beside the Bowling Green at Werrington Park, Cornwall. Collected by Wilson (*W* 4106) from Sichuan, China.

It has a rich chestnut bark, with similarly coloured buds and catkins which are freely produced. Rich outer bark peels to reveal a paler honey-coloured under layer.

'China Rose'

From a tree at Stone Lane Gardens, Devon. Sent out as seed by Dr Pan of the Chinese Academy of Forestry. Collected at Kaolan, near Lanchow, Gansu, China. 36°N, 103°E.

This tree has more solidly red bark than most, with very little blooming.

'China Ruby'

Named by Brian Humphrey in 1994 from a 3-stemmed tree growing at Cambridge Botanic Garden (Accession no. 197800202 A, planted 1978/79) The Cambridge tree came from Hilliers nursery as *Betula albosinensis septentrionalis*. The Hilliers tree was apparently grafted from a mature tree at Hilliers West Hill nursery in Winchester, which was lost when the site was built over. Harold Hillier said the West Hill tree came from Werrington Park, Cornwall (pers. comm. Roy Lancaster). Werrington Park had several of Wilson's introductions and this points to the cultivar 'China Ruby' being a clone of W4106 (see 'Chinese Garden' and 'Bowling Green'). Thanks to Lawrence Banks (Hergest Croft) for collating this information. This name was initially given in error by Stone Lane Gardens to the 'China Rose' cultivar. This error has now been rectified.

'Chinese Garden'

From a tree in the Chinese Garden (originally the American Garden) at Werrington Park, Cornwall. Collected by Wilson (*W* 4106) from Sichuan, China.

 The rich pinkish bark is even darker than that of 'Bowling Green'.

'Conyngham'

Grown from seed received by Kenneth Ashburner from a contact in China. Thought by Kenneth to be a cross of *Betula utilis* subsp. *albosinensis* and some other species, but the mature trees I have seen appear to be pure subsp. *albosinensis*. The trees were never planted at Stone Lane Gardens, but a batch of the original wild-origin trees were sold to Charles Lenox-Conyngham (pronounced Cunningham) in Kent and so Kenneth named the tree after him, as Kenneth did not think they were pure *B. utilis* subsp. *albosinensis*. A specimen was planted at Bicton College by Dick Fulcher probably between 1985–1987. Kevin Croucher of Thornhayes nursery took grafts of the Bicton tree from 1989 until the early 1990s.

 It has a stem that changes colour through the year from cream to pink and semi pendant branches with prolific large catkins.

'Fascination' — see Hybrids at end.

'Joseph Rock'

A selection of unknown origin. Material given to Karan Junker of Junker's nursery by Kenneth Lorentson.

 It is particularly narrow and upright in habit, forming a column. The bark is a good purple.

'Kansu'

From Gansu province, China. Probably sent out as seed by Dr Pan of the Chinese Academy of Forestry. Collected at Kaolan, near Lanchow, Gansu, China. 36°N, 103°E.

 A good strong growing form with multi-coloured copper and pink bark.

'Kenneth Ashburner'

From a group of *controlled-pollination* seedlings of *Betula utilis* subsp. *albosinensis* 'Chinese Garden' at Stone Lane Gardens, Devon. These trees are now dead. Named by Kevin Croucher of Thornhayes Nursery, Devon.

 It is a richer, deeper colour and more vigorous than most forms of this variety.

'Ness'

From a seedling at Stone Lane Gardens, Devon. Probably from seed sent out by Dr Pan of the Chinese Academy of Forestry. Collected at Kaolan, near Lanchow, Gansu, China. 36°N, 103°E. Selected by Kevin Croucher of Thornhayes Nursery, Devon. Initially known as Clone F from 1995, but then changed to 'Ness' in 1998.

'Pink Champagne'

From a tree at Stone Lane Gardens, Devon. Sent out as seed by Dr Pan of the Chinese Academy of Forestry. Collected at Kaolan, near Lanchow, Gansu, China. 36°N, 103°E.

 This particular tree has soft pale pink bark, caused by a persistent bloom of white betulin over the trunk.

'Purdom'

Collected by William Purdom, plant hunter in the early 1900s. Thought to be from the *Purdom* 752 collection, now rare, which was collected from the Tow (Tao) River in western Gansu.

A beautiful stem of smoky mauves, buffs and satiny pinks.

'Red Panda'

From a tree at Wakehurst Place (acc. Number 140- 87 01144) raised from seed collected by Steve Sponberg at 2500 m from Shennongji a county, Hubei, China as part of the Sino-American Botanical Expedition 1980 (1980 *SABE* 31). Named 'Red Panda' by Jim Gardiner (RHS Wisley) in 2008, due to the red pandas living in that locality. There is also a good specimen in Howard's Field at RHS Wisley.

'Rhinegold'

Despite being shown in the *RHS Plantfinder*, I can find no record of any nursery or garden having this cultivar.

'Ridgebourne' — see *Betula* 'Haywood' in Hybrids below.

'Sable'

Sold for a short while by Perryhill Nursery and reported as being from a tree at Wakehurst Place.

Origin and propagator unknown. Dark barked.

Betula utilis subsp. *jacquemontii*

'Doorenbos'

Introduced by the famous Dutch plantsman, Albert Doorenbos. From a batch of subsp. *jacquemontii* seed received around 1933, origin unknown, of which the 'Doorenbos' tree stood out as distinctly different. Possibly a hybrid, as male catkins not knobbly. Probably *Betula utilis* subsp. *jacquemontii* × *B. ermanii* as some fruiting catkins are erect. The leaf is intermediate and the bark has the characteristics of both species. Trade name of 'Snowqueen' given to it by Nick Dunn of Frank Matthews Ltd, but 'Doorenbos' is the older, correct cultivar name.

A beautiful form producing a lovely white stem.

'Fastigiate'

An upright, columnar selection giving a pillar of the beautiful white bark.

'Grayswood Ghost'

From a tree (now dead) at Grayswood Hill, Haslemere, Surrey. Home of the Pilkington family. Origin unknown. Cultivar obtained by Tony Schilling just before the parent tree died. Named by Tony Schilling and Kenneth Ashburner in 1985. A tree from the original grafted material exists at Wakehurst Place (054-71-05258). Possibly a hybrid because the current season's shoots are too hairy and smooth, the leaf is too ovate, there are too many pairs of leaf veins and the veins are too deeply impressed.

A strong vigorous birch with brilliant white bark.

'Gregory Birch'

From a tree at Werrington Park, Cornwall. Named by the owner Robert Williams after his head gardener R. M. Gregory (died 1954). Frank Knight told Mr Williams that the trees were raised from seed collected by R. M. Gregory somewhere in North West India (probably north of Simla, Dalhousie, Kasauli or Dagshai) during the Great War (1914–1919). Gregory served with 6th

Devonshire regiment and Mr Williams discovered that they were in that part of India towards the end of the war.

Good, white bark. Strong growing tree.

'Inverleith'

From a tree at Royal Botanic Garden Edinburgh which has since died. There is a grafted clone of this at Wakehurst Place (pers. comm. Tony Schilling). Possibly a hybrid because the current season's shoots are too hairy and smooth, the leaf is too ovate, there are too many pairs of leaf veins and the veins are too deeply impressed. It may be derived from seed sent by Hooker from Kew around 1881. May be the same as **'Silver Shadow'** although there is some debate about this. Named by Kenneth Ashburner and Tony Schilling in 1985.

The **'Silver Shadow'**/**'Inverleith'** debate: Many plantsmen are of the opinion that these are one and the same. However, some strongly disagree. People who have compared material from the original trees and subsequent clones of these trees are generally of the opinion that they both differ from typical subsp. *jacquemontii* in their hairy twigs. The original 'Silver Shadow' tree came to Hillier's nursery from Edinburgh and this also points to a connection. However, when comparing named trees in nurseries today there is some disagreement over the similarities, and some examples appear to be quite different. It is possible that one or both of these names has been wrongly applied to recent stock and this has resulted in trees being wrongly identified. Sadly the two original trees are no longer alive, so we cannot simply compare the two, and it is likely that this issue will never be resolved to anyone's satisfaction.

'Jermyns'

From a tree at Hillier's nursery, Hampshire. Original trees arrived via Belgium in the 1950–60's (from a Dutch nursery) from the same source as 'Doorenbos'. Looked different to 'Silver Shadow' so it was named by Sir Harold Hillier and his head gardener Jack Bryce. It is possibly a hybrid as there are too many leaf veins and the veins are too deeply impressed.

Outstanding for the pure white bark on its trunk and branches. Produces long catkins in spring. Good vigour.

'Kashmir White'

From seed collected from Lashpathri, above Sonnamarg, Kashmir in 1978 (*L128A*) by Roy Lancaster. Originally thought to be *Betula utilis* subsp. *occidentalis*, but unlikely as, like trees of Kyelang provenance, it is too far east and its characteristics are more like *jacquemontii*. Named by Roy Lancaster. Propagated by Brian Humphrey and then by a nurseryman in Oregon, USA. It is susceptible to late frosts.

'Knightshayes'

From a tree at Knightshayes, Devon, origin unknown. Selected by Kevin Croucher of Thornhayes Nursery, Devon.

A form with dazzling white stems and an unusual semi-pendulous habit of growth.

'Kyelang'

From a tree at Stone Farm. Sticky winter buds and quite twiggy growth. Not as clean a white as many of the *jacquemontii* selections. Collected by Kenneth Ashburner from Kyelang, Lahaul, NW India. Originally thought to be *Betula utilis* subsp. *occidentalis*, but unlikely as it is too far east and its characteristics are more like *jacquemontii*.

'Macbeth'

White bark. Larger than normal leaves with deep veins. It is highly probable that this is a misnaming of 'McBeath', as they are pronounced the same.

'McBeath'

From a tree grown from seed collected by Ron McBeath in Manali, Rhotang Pass, Himachal Pradesh, India at a height of 3000 m on an open mountain slope (*McB.* 1737). In the wild the trees grew to 10 m tall.

The original tree of the cultivar is thought to be one growing in Dundee Botanic Garden (Acc. no. 1985-0606).

'Moonbeam'

From a tree at Wakehurst Place. Sometimes named as *Betula utilis* subsp. *occidentalis*, but unlikely because *occidentalis* is usually a poorer grower in the UK with less attractive bark and leaves like *pendula*.

This has a good white trunk but is a smaller tree than other forms of *jacquemontii*.

'Pendula'

This may be the cultivar now known as 'Long Trunk'. See hybrids at the end. An old specimen exists at Miniers arboretum in Angiers, France.

'Pradhan'

I cannot find any information about this cultivar, except that Keshab C. Pradhan did collect *Betula utilis* seed at 10–12,000 feet in the Yumthang Valley, north Sikkim in the late 1980s and also from the Singalila Ridge between Sikkim and Nepal. The seed was probably widely distributed, so it may well have found its way to a nursery, either in Europe or the USA.

'Ramdana River'

From a tree at Stone Lane Gardens, Devon. Collected by Kenneth Ashburner from Ramdana River, Garwhal, Uttar Pradesh, India in 1991.

White bark, even on young trees, and glossy green foliage. Apparently it can suffer from late frost damage, though the trees in Stone Lane Gardens do not.

'Sauwala White'

Named by Tony Schilling and Kenneth Ashburner, reportedly from a wild collection by Stainton, Sykes & Williams (*SSW* 4382) on the 1954 RHS/British Museum expedition. From the valley of the Sauwala Khola, south of Dhaulagiri, central Nepal. The original tree was donated by Windsor Great Park to Wakehurst Place (acc. no 046-76-0308) but the cultivar was propagated from a tree (now dead) at RHS Wisley. There is divided opinion about the identity of this cultivar. No wild collected specimen from the area is remotely similar, so it may be a hybrid or a form of some other species.

'Silver Shadow'

From a tree at Hilliers West Hill nursery, Hampshire. Named by John Hillier. A tree from the original grafted material exists at RHS Wisley (W920277A). Like 'Inverleith' it has distinctly hairy shoots (unlike most *jacquemontii*), and the original certainly came from Edinburgh so may be the same thing. However, there is some debate about this.

Striking white bark, and large drooping dark leaves.

'Snow Leopard'

'Snow Leopard' is a clone from a lovely group of trees at Stone Lane Gardens, which were grown from seed collected by Oleg Polunin in the wild near Sonamarg in the Indian state of Jammu & Kashmir. This is a clone of true *Betula utilis* subsp. *jacquemontii,* whereas most *jacquemontii* cultivars are probable hybrids.

Lovely pure white bark, smooth and peeling dramatically from the trunk.

Snowqueen — see **'Doorenbos'**

'Somerford'

I can find no reference to this cultivar anywhere, except that there is a tree at Valley Gardens, part of Windsor Great Park.

'Thyangboche Monastery'

Thyangboche Monastery is in the Khumbu region of Nepal, near Mt Everest. There are no details available for this cultivar, but it is possibly from one of the trees collected by Tony Schilling just above Thyangboche (see *Betula utilis* subsp. *utilis* 'Buddha'). However, his are darker barked trees, not *jacquemontii.*

'Trinity College'

From a tree at Trinity College, Dublin grown from seed sent by Hooker from Kew around 1881, possibly the same batch of seed that was sent to Edinburgh (see 'Inverleith').

A good selection with gleaming, almost translucent white trunk from an early age and a narrow habit. Retains its leaves well into the autumn.

'Werrington'

This plant is known in the Dutch trade and appears to be a white barked *jacquemontii.* To judge by its name, it is possible that the original trees came from Werrington Park in Cornwall. It could be the same as 'Gregory Birch' which is known to have been propagated from a tree of northwest Indian origin in the Chinese Garden at Werrington. However, up till the 1980s there was also a white barked tree growing beside the *albosinensis* Wilson 4106 by the Bowling Green at Werrington.

'White Satin'™

Selection by Dr Hasselkus, Wisconsin, USA.

White bark, good yellow autumn colour.

Betula utilis subsp. *occidentalis*

We do not know of any cultivars having been produced; 'Kashmir White' and 'Kyelang' are both subsp. *jacquemontii.* Certainly it is a tree that has proved to be difficult to grow in Britain. Stone Lane Gardens are currently experimenting with *Betula utilis* subsp. *occidentalis* grafted onto *B. pendula* rootstocks, which appear to be giving stronger, hardier growth. So a cultivar may be available in the next few years.

Betula utilis subsp. *utilis*

'Bhutan Sienna'

From a tree at Stone Lane Gardens, Devon. Collected by Sinclair & Long in 1984 from West

Bhutan. Originally named 'Nepalese Red' because it was wrongly thought to have come from East Nepal. This name is now obviously redundant and has been replaced by the name 'Bhutan Sienna'.

This *utilis* has an attractive dark red bark with a brown tint.

'Buckland'

From the Garden House at Buckland Monachorum, Devon. It appears that there is a tree at the Garden House that was grown from seed given by Kenneth Ashburner to Vic Powlowski. This was selected for growing on as a cultivar by Vic and Keith Wiley and propagated for a while by Endsleigh Nursery, Devon.

The tree has pink bark.

'Buddha'

Selected by Chris Sanders from one of three trees growing at Ness Gardens, Wirral from seed collected upstream of Tengboche (Thyangboche) Monastery, Khumbu region, Nepal on 24th October 1976 by Tony Schilling at about 3800 m (*SCH* 2168). Given the name of 'Buddha' by Tony Schilling. Introduced by Bridgemere Nursery in 1994. There are also trees from this collection number growing at Castle Howard.

Attractive underlying pinkish-brown bark thinly washed with white and contrasting coppery exfoliating bark strips. Heavily-veined, large, glossy leaves.

'China Bronze'

From a tree in Stone Lane Gardens, Devon. Collected by T. T. Yu in Yunnan, China. (Coll. no. *YU* 10561).

Dark metallic bark with a somewhat weeping habit.

'Dark-Ness'

Named by Chris Lane & Chris Sanders from a plant growing at Ness Gardens, Wirral in 2000. The Ness tree was collected in Thimphu District, below Shodu, Bhutan at 3500 m by Sinclair & Long (*S&L* 5540) in 1984. Propagated by Chris Lane.

Nice dark shiny bark with conspicuous white lenticels and good upright habit.

'Edinburgh'

Named *Betula* 'Edinburgh' in the trade and thought to be a cross of *B. utilis* and *B. utilis* subsp. *albosinensis* var. *septentrionalis*, grown at Royal Botanic Garden Edinburgh. The name appeared in a 1986 *Dendroflora* article by Dr P. C. De Jong. Propagated by Brian Humphrey, who actually doubts it is a cross with *albosinensis* and thinks it may be a hybrid with *B. ermanii*.

Medium-size, fastigiate tree with ovate, glossy green leaves. Good whitish bark.

'Fascination' — see Hybrids at end.

'Forest Blush'

From *Forrest* 19505 from Yunnan, China. This is often mis-spelt as 'Forrest Blush' (and I for one am guilty of this), an understandable mistake but one we aim to correct here. Named by Brian Humphrey in 1994. Thought to come from a tree at Royal Botanic Garden Edinburgh. Also named as 'Marble Stem' by Kevin Croucher of Thornhayes nursery in 1998, as Kevin at the time was not aware of the previous name. However, 'Forest Blush' is the earlier and therefore correct name.

A fine white bark with a hint of pink. Smooth bark with prominent lenticels. Attractive elegant leaves.

'Himalayan Pink' — see 'Nepalese Orange'

'Khumbu'
Selected by Chris Sanders from a tree growing at Ness Gardens, Wirral from Tony Schilling's *SCH* 2168 collection (see 'Buddha' above). Similar to 'Buddha', but from a different seedling with somewhat darker bark. Named by Chris Lane & Chris Sanders and propagated by Chris Lane.

'Marble Stem' — see **'Forest Blush'**

'Moonbeam' — see *Betula utilis* subsp. *jacquemontii*

'Mt Luoji'
From a tree at Stone Lane Gardens, Devon. Collected by Lord Howick from Mt Luoji, Sichuan, (Coll. no. *H & M* 1480).
Stunning dark brown bark with a hint of red together with patches of white bloom. A very rich bark colouring.

'Nepalese Orange'
From a tree at Stone Lane Gardens, Devon. Collected by Beer, Lancaster & Morris. Seed collected on steep wooded slope at 12,500ft above Topke Gola, E Nepal on October 19th 1971 (coll. no. *BL&M* 100). Selected by Kenneth Ashburner. From 1998 it was also offered in the trade under the name of 'Himalayan Pink' by Kevin Croucher of Thornhayes nursery. However, it is not sold under this name any more.
This form has a rich orange bark, with very noticeable horizontal bands of lenticels and soft blooming. A very attractive and unusual *utilis*, but not suitable for very frosty or exposed areas.

'Nepalese Red' — see **'Bhutan Sienna'**

'Park Wood'
From a tree at Hergest Croft, Herefordshire (Cat. 1123). Origin unknown, although possibly from the Williams family. Misnamed *Betula utilis* var. *prattii* for many years, these chocolate barked birches (ranging from E Nepal to Sichuan, China) are now being treated as just dark barked *B. utilis* subsp. *utilis*.
A lovely birch much admired for its smooth dark chocolate bark and handsome shiny leaves.

'Ramdana River' — see *Betula utilis* subsp. *jacquemontii*

'Schilling'
This cultivar name is incorrect. It is actually the cultivar 'Buddha'.

'Sichuan Red'
From a tree at Stone Lane Gardens, Devon, grown from material donated by Lord Howick. Collected by Simmons, Erskine, Howick, & McNamara on 28th September 1991 at 3440 m on the west bank of Jiulong River N of Jiulong towards Jichou Pass, Sichuan, China. (Coll. no. *SICH* 667).
A lovely dark brown bark, overlain with orange and red. Very little bloom.

'Wakehurst Place Chocolate'
A dark barked *utilis*, the parent tree in Bethlehem Wood, Wakehurst Place. The Wakehurst trees were donated as grafts by Keith Rushforth from Westonbirt's Forrest collection (*Forrest* 15381, planted 1924, now dead) from Wa-Di-I Shan, c. 27°40'N.

Plain chocolate bark which forms early.

'Werrington' — see *Betula utilis* subsp. *jacquemontii*

HYBRIDS

The majority of named hybrids are the result of chance cross-pollination in gardens or, more rarely, in the wild, although a few are the result of deliberate cross-pollination.

It is worth noting that a significant number of the older, established cultivars have dubious parentage which could easily have been the result of chance cross-pollination in botanical gardens and nurseries. Recent cultivars have generally come from trees of known wild-origin, where it is easier to establish their provenance. I am sure that further research into the characteristics of the cultivars will result in this Hybrid list becoming longer.

'Charlotte'
Named and propagated by Chris Grey-Wilson from seed in an American garden and named after Chris's daughter.

Vigorous tree with creamy-white/apricot bark and good autumn colour.

'Crimson Frost' — see *Betula pendula* subsp. *pendula*

'Dick Banks'
A sister seedling of 'Hergest' which is white stemmed and has more characters of *Betula ermanii*.

'Edinburgh' — see *Betula utilis* subsp. *utilis*

'Fascination'
Originally sent from England to the Netherlands wrongly labelled as *Betula caerulea-grandis*. One particular seedling was noticed by Herman Oterdoom and was considered worthy of cultivation. It was named by Herman Oterdoom and Dick van Gelderen in the late 70s to early 80s and grown by Dick van Gelderen at his Planten Tuin Esveld nursery, at Boskop, Netherlands. It has been labelled as both *utilis* and *albosinensis* by nurseries. Photos of the parent tree in Esvelt suggest it is more closely associated to the *B. utilis* aggregate, but other photos and limited study of some other trees hint at a cross with *B. ermanii*. Clearly more research needs to be done to determine its pedigree. In the meantime, I am putting it with the hybrids, as we cannot categorically state that it is a species.

It has deep orange satin peeling bark turning a pale salmon–white, revealing layers of colours. Great vigour and good autumn colour.

'Fetisowii'
A slow growing excellent hybrid with creamy white bark.

'Haywood'

Another sister seedling of *Betula* 'Hergest', which is darker stemmed. This has also been called *B. albosinensis* 'Ridgebourne' but should in future be identified by the earlier name of *Betula* 'Haywood'. The mis-naming as 'Ridgebourne' came about after Chris Lane selected it from a tree in the garden of one Margaret Owen. It has since been discovered by Lawrence Banks that Margaret Owen's tree is in fact a tree of the hybrid *Betula* 'Haywood', given to her by his father, Dick Banks.

A good vigorous form with dark coppery brown bark from an early age.

'Hergest'

From a fine mature tree at Hergest Croft, Herefordshire. The parent tree is one of a group of chance hybrids of *Betula ermanii* and *B. utilis* subsp. *albosinensis* that happened at Hergest Croft. It exhibits more qualities of subsp. *albosinensis* than of *B. ermanii*. The *albosinensis* parent was very probably a sister seedling of the W4106 collection ('Bowling Green', 'Chinese Garden'), as Hergest Croft received many plants from the Williams family c. 1910.

The bark is bronze when in shade and paler when grown in full sun.

'Kerscott Charm'

Named by Jessica Duncan and propagated by Paul Bartlett of Stone Lane Gardens in 2014. Hybrid tree produced from open pollinated seed of *Betula ermanii* given to Jessica Duncan by Kenneth Ashburner. The parent tree grows at Kerscott in North Devon. A hybrid of *Betula ermanii* and another birch species, probably *B. pendula*.

Beautiful cream bark with a hint of peach, peeling in large scrolls. Prominent and attractively long lenticels swirl around the trunk. Very open structure, good autumn colour and slightly weeping branchlets.

'Long Trunk'

This is commonly shown as either *Betula utilis* or *B. pendula* in the nursery trade. Having looked at immature plants in our nursery, I wonder if it is a cross between the two species. It certainly appears to be a hybrid of some kind. Origin unknown.

'Mt Apoi'

Named by Danish workers at a *Betula* conference, or possibly by Kenneth Lorentzon. This genotype is thought to be a natural hybrid of *Betula ermanii* and *B. fruticosa* and the correct species name is *B. gmelinii* (= *apoiensis*).

From seed collected by Prof. Hideaki Ohba from Mt Apoi, Hokkaido, Japan. Not a true *Betula ermanii* cultivar, but sometimes treated as one by the trade.

A small shrubby tree with creamy white bark.

'Polar Bear'

Whilst most people in the nursery trade think of this as *Betula ermanii*, our observations would suggest that it is in fact a hybrid. The teeth on the leaf margins are blunter than *B. ermanii* and the fruit is pendant rather than horizontal or upright. The scales of the fruit have side lobes that are not spreading, the central lobe is too short and the wings of the seed are too big.

A strong grower with a pure white heavy trunk and branch structure. Stunning in winter. Reliable rich golden autumn leaf colour.

'Royal Frost'

Betula populifolia 'Whitespire' × *Betula* 'Crimson Frost' .
    Persistent purple leaves and white bark.

'White Light'

Named and propagated by the late John Buckley, Birdhill, Co. Tipperary, Ireland. A cross
reputedly between *Betula costata* and *B. utilis* subsp *jacquemontii*, although it is possible that the *B. costata* was in fact *B. ermanii*.
    Interesting pale bark which peels attractively to reveal a pale brown underside when first exposed
to light.

Betula ×jackii

Seedlings raised from open pollinated *Betula ×jackii* [2n=42 (Woodworth 1929)] from the Arnold
Arboretum fall into two distinct types from which can be deduced the breeding behaviour of the
parent hybrid. a) Very slow growing highly branched hexaploids (2n=84), presumably the result
of self-fertilisation by unreduced gametes. b) Seedlings more 'normal' in appearance, larger,
more vigorous growing and less branched, pentaploids (2n=70), hexaploids (2n=84), and an
octoploid (2n=112), presumably the result of fertilisation of an unreduced female gamete (n=42) of
B. ×jackii by normal reduced pollen from a tetraploid (n=28), presumably *B. pubescens* or *B. utilis*
as these are the only tetraploids present, a hexaploid (n=42), probably *B. alleghaniensis* judging from
the appearance of the seedling, and a decaploid (n=70), presumably *B. globispica* as this is the only
decaploid present. Other explanations are possible (Anamthawat-Jonsson &Tomasson 1990), but the
above would seem to be the most probable scenario, suggesting only unreduced gametes function in
birch hybrids between distantly related species.

GLOSSARY

Agamospermous — asexual production of seed so that seedlings are genetically identical to parent

Allele — different state of a gene (e.g. A, B, and O are different states of a human blood group gene)

Allopolyploidy — the doubling of the chromosome number following hybridisation between two distinct species, usually creating a new species containing the full chromosome sets of both parental species

Angiosperms — flowering plants (as distinct from conifers and other Gymnosperms)

Apomixis — asexual reproduction

Arboretum (pl. **arboreta**) — collection of (usually many different) tree species

Asexual (of reproduction) — method of reproduction which does not involve sex, in which recombination of genes takes place. Offspring are therefore identical to the parent.

Autopolyploidy — doubling of the chromosome number within a species creating an organism with two sets of the parental chromosomes

Boreal — northern

Bract — reduced leaf, often scale-like as in inflorescence of alders; or derived from several reduced leaves as in the 'bracts' or scales of the fruiting catkins of birches

Broadleaved (trees) — angiosperm (flowering plant) trees which mostly have broad, net-veined leaves, in contrast to needle-leaved conifers

Cambium — layer of dividing cells between bark and wood which allow stem to grow in thickness. If exposed by removal of bark this layer will die so that 'ring barking' leads to the death of branches beyond the damage.

Camptodromous (of leaf venation) — main lateral veins not reaching margin but curving away from margin or dividing up into many branches as margin is approached

Carpel — unit of the female structure of the flower containing ovules which develop to seeds. The carpels together make up the ovary which develops to the fruit

Catkin — dense, often showy, spike of many small flowers

Chloroplast — cell organelle which contains the green pigment chlorophyll

Chromosome(s) — structure(s) within the nucleus of cells which contains the genetic material. Every cell of a higher organism (e.g. flowering plant) contains two sets of chromosomes, one from each parent.

Clade — group of organisms presumed to be derived from a single ancestral population because they share one or more characteristics in common which are thought to have been present in that ancestral population

Cladistics — system of classification based on deduced presumed ancestry, usually constructed using many characters and assuming fewest changes in character states

Classification — ordering so that a hierarchy of groupings reflects degree of similarity

Cline — gradual change in a character from one extreme to another along a geographical gradient

Clone — group of genetically identical organisms

Colleters — small peg-like structures, often reddish, found on adaxial (upper) surface of petiole and leaf midrib

Conifer — cone bearing tree (e.g. pine, spruce, fir, larch)

Cordate (of base of leaf blade) — heart shaped

Contact herbicide — weedkiller which kills only the tissues in contact with the herbicide, usually leaves and young stems. It is not moved within the plant.

Craspedodromous (of leaf venation) — main lateral veins running to margin and usually ending in the tips of the largest marginal teeth

Cuneate (of base of leaf blade) — tapering into leaf stalk (petiole), as distinct from truncate (straight across so that petiole makes a T-junction with leaf blade) or cordate (heart shaped)

Cytology — study of cells, in the sense used here the study of chromosome numbers

Cytotaxonomic — taxonomic study using information on chromosome number and structure

Diploid — the chromosome number of a higher plant (e.g. flowering plant) with two sets of chromosomes, one from each parent.

Ecology — the study of the inter-relationships among living things and their environment

Endemic — native (to a region, usually used in the sense of native only to that region) see Paleoendemic/Neoendemic

Epicormic (of shoot) — new shoot growing from main trunk or large branch, usually vigorous, very soft and juvenile with leaf shape and hairiness very different from mature shoots in the twiggy canopy

Epigenetic — inherited characteristics affecting gene expression but not involving changes in DNA

Facultative (apomict) — individual or taxon which can reproduce both sexually and apomictically

Fertilisation — fusion of male and female gametes (sex cells) to form zygote (fertilised egg) which will develop into seed (in plants)

Form — minor variant of a species, usually very distinct, but determined by one or very few genes (e.g. red hair in humans, cut-leaved and purple-leaved forms in birch)

Genus — group of related species

Glabrescent — becoming almost hairless

Glabrous — hairless

Glaucous — whitish to bluish colouration

Grafting — getting a twig of one plant to grow on the rootstock of another

Gynoecium — female part of flower, composed of carpels

Haploid — the chromosome number in a gamete (egg, sperm in pollen) and half the number in an adult

Herbarium — collection of dried plant specimens

Herbicide — weedkiller (see: contact, residual, translocated)

Hyaline — more or less translucent

Hybrid (interspecific) — organism resulting from cross between individuals of two different species

Hybridise — to produce offspring following pollination and fertilisation by an individual of another species

Inbreeding — self fertilisation or breeding among close relatives

Inbreeding depression — lack of vigour in offspring of fertilisation among close relatives resulting from recessive alleles present in homozygous condition

Indehiscent (of fruits) — which don't break up on ripening and therefore usually only slowly dispersing their seed

Inferior (of ovary or carpels) — sunk within the apex of stem such that flower parts appear to arise on top of ovary

Meiosis — cell divisions in which gametes (male and female sex cells, sperm and eggs) are produced, each gamete having half the chromsome number of the parent

Miocene — geological period extending from 23 to 5.3 million years ago. Following Oligocene and before Pliocene.

Molecular (of systematics) — information on evolutionary relationships obtained from knowledge of the structure or base sequence of DNA

Neoendemic — recently evolved species confined to an area because it has not yet spread beyond it

Nutlet — term used for the small, often winged, fruits (often called seeds) of birch

Oligocene — geological period extending from 34 to 23 million years ago. Following Eocene and before Miocene.

Ovary — female part of the flower which will develop into the fruit

Paleoendemic — relict endemic, usually of a species which has survived in an area for a very long time, having become extinct in surrounding areas (cf Neoendemic)

Phylogeography — study of the evolution and distribution of living organisms

Phytogeography — study of the distribution of plants and how these distributions have been achieved

Pleistocene — recent geological period of Ice Ages from 1.8 million years ago

Pliocene — geological period from 5.3 million until 1.8 million years ago and leading up to the Pleistocene Ice Ages. After Miocene

Pollination — transfer of pollen from a stamen to the stigma of a carpel. The pollen than produces a male gamete which usually fertilises the female gamete (egg) in the ovule so that a seed can be produced.

Polyphyletic (of a taxon) — containing more than one evolutionary line and therefore not a 'natural' unit of classification, being descended from more than one ancestral population

Provenance — wild collection locality. This term has been extended to refer to a group of seedlings introduced from a known locality.

Puberulent — with short hairs standing out at right angles to plant part like the pile on velvet when dense

Residual herbicide — weedkiller applied to soil which is taken up by roots and remains active in the soil for some time

Revolute (of leaf margins) — curled downwards at margins

Rhizomatous — with underground stems

Rhizome — underground stem. Plants with rhizomes spread by means of these to form patches with aerial stems emerging from the ground at some distance from one another.

Rootstock — stem with root onto which a scion is grafted

Scion — twig to be grafted onto rootstock

Self-compatible — plant which can produce seed by self-pollination and self-fertilisation

Self-incompatible — plant which cannot produce seed by self-pollination and self-fertilisation so that plants of two different genotypes (clones) are required for seed production

Shrub — multi-stemmed woody plant of smaller stature than most trees

Species — group of populations which are distinguishable from other groups of populations. The basic unit of classification and the second name in standard binomials.

Stamen — floral organ consisting of a stalk (filament) and pollen sac (anther) which bears the pollen

Standard (of a tree) — on a single, unbranched stem, usually of 1–2 m

Stigma — part of carpel at apex of style which receives the pollen

Stipule — paired, usually leafy or scale-like, structures on either side of the base of the leaf stalk where it is attached to the stem

Strobile — term for the fruiting 'catkin' of a birch which develops from the female catkin

Style — part of carpel which bears the stigma at its apex

Sucker — shoot arising from below ground at the base of the trunk, usually vigorous, very soft and juvenile with leaf shape and hairiness very different from mature shoots in the twiggy canopy

Syncarpic — of flowers whose carpels are fused to form a single structure, the ovary

Taxon — a unit of classification which has been given a name. A useful term to refer to a unit which may for instance be regarded by different authors as a species or subspecies, genus or subgenus.

Taxonomy — the science of classification

Tepal (of perianth) — single element of perianth, the term used for individual structure when there is no differentiation into a calyx of sepals and a corolla of petals

Tetraploid — chromosome number which is four times the base (x) number for the species group (usually genus)

Translocated herbicide — weedkiller (usually applied to leaves) which moves within the plant and usually kills the whole plant, including underground parts

Triploid — chromosome number which is three times the base (x) number for the species group (usually genus)

Truncate (of base of leaf blade) — straight across so that leaf stalk (petiole) makes a T-junction with leaf blade, compare Cuneate (tapering into petiole), and Cordate (heart-shaped)

Tundra — treeless vegetation north of the treeline (in the Northern Hemisphere)

Type specimen — herbarium specimen specified by the taxonomist who named a species and on which the name depends

REFERENCES AND BIBLIOGRAPHY

Abbe, E. C. (1935). Studies in the phylogeny of the Betulaceae. I. Floral and inflorescence anatomy and morphology. *Bot. Gaz.* **97**(1): 1–67.

—— (1938). Studies in the phylogeny of the Betulaceae. II. Extremes in the range of variation of floral and inflorescence morphology. *Bot. Gaz.* **99**(3): 413–469.

—— (1974). Flowers and inflorescences of the "Amentiferae". *Bot. Rev. (Lancaster)* **40** (2): 159–261.

Abbott, R. J. & Brochmann, C. (2003). History and evolution of the Arctic flora. *Molec. Ecol.* **12**(2): 299–313.

—— & Comes, H. P. (2003). Evolution in the Arctic: a phylogeographic analysis of the circumarctic plant, *Saxifraga oppositifolia* (Purple saxifrage). *New Phytol.* **161**: 211–224.

——, Smith, L. C., Milne, R. I. Crawford, R. M. M., Wolff, K. & Balfour, J. (2000). Molecular analysis of plant migration and refugia in the arctic. *Science* **289**: 1343–1346.

Abod, S. A. & Webster, A. D. (1990). Shoot and root pruning effects on the growth and water relations of young *Malus*, *Tilia* and *Betula* transplants. *J. Hort. Sci.* **65**(4): 451–459.

——, —— & Quinlan, J. D. (1991). Carbohydrates and their effects on the growth and establishment of *Tilia* and *Betula*: II. The early season movement of carbohydrates between shoots and roots. *J. Hort. Sci.* **66**(3): 345–355.

Aerts, R. (1995). The advantages of being evergreen. *Trends Ecol. Evol.* **10**(10): 402–407.

Afzal-Rafii, Z. & Dodd, R. S. (2007). Chloroplast DNA supports supports a hypothesis of glacial refugia over postglacial recolonisation in disjunct populations of black pine (*Pinus nigra*) in western Europe. *Molec. Ecol.* **16**: 723–736.

Ainouchi, M. L., Baumel, A., Salmon, A. & Yannik, G. (2003). Hybridisation, polyploidy and speciation in *Spartina* (Poaceae). *New Phytol.* **161**: 163–172.

Alam, M. T. & Grant, W. F. (1972). Interspecific hybridisation in birch (*Betula*). *Naturaliste Canad.* **99**: 33–40.

Albertson, F. W. & Weaver, J. E. (1945). Injury and death or recovery of trees in prairie climate. *Ecol. Monogr.* **15**: 393–443.

Alice, L. A. & Campbell, C. (1999). Phylogeny of *Rubus* (Rosaceae) based on nuclear ribosomal DNA internal transcribed spacer region sequences. *Amer. J. Bot.* **86**(1): 81–97.

Alsos, I. G., Engelskjøn, T., Gielly, L., Taberlet, P. & Brochmann, C. (2005). Impact of ice ages on circumpolar molecular diversity: insights from an ecological key species. *Molec. Ecol.* **14**: 2739–2753.

Álvarez, I. & Wendel, J. F. (2003). Ribosomal ITS sequences and plant phylogenetic inference. *Molec. Phylogenet. Evol.* **29**: 417–434.

Amos, W. & Balmford, A. (2001). When does conservation genetics matter? *Heredity* **87**: 257–265.

Anamthawat-Jonsson, K., Atipanumpai, L., Tigerstedt, P. M. A. & Tomasson, T. (1986). The Feulgen-Giemsa method for chromosomes of *Betula* species. *Hereditas* **104**: 321–322.

—— & Tomasson, T. (1990). Cytogenetics of hybrid introgression in Icelandic birch. *Hereditas* **112**: 65–70.

Anderson, E. & Abbe, E. C. (1934). A quantitative comparison of specific and generic differences in the Betulaceae. *J. Arnold Arbor.* **15**: 43–49.

Anon (1999). Volcano botany. *Harrisiana (The Newsletter of the Friends of the Harris Garden, Reading, UK)* **35**: 5–6.

Anon (1996). Returning chocolate *Cosmos* to Mexico. *The Garden* **111**(9): 529.

Appiah, A. A., Jennings, P. & Turner, J. A. (2004). *Phytophthora ramorum*: one pathogen and many diseases, and emerging threat to forest ecosystems and ornamental plant life. *Mycologist* **18**: 145–150.

Aradottir, A. L. & Arnalds, O. (2001). Ecosystem degradation and restoration of birch woodlands in Iceland. In: F. E. Wielgolaski (ed.), *Nordic Mountain Birch Ecosystems*, Ch. 24. Man and the Biosphere Series, **27**. UNESCO, Paris.

——, Thorsteinsson, I. & Sigurdson, S. (2001). Distribution and characteristics of birch woodlands in north Iceland. In: F. E. Wielgolaski (ed.), *Nordic Mountain Birch Ecosystems*, Ch. 6. Man and the Biosphere Series, **27**. UNESCO, Paris.

Arber, A. (1954). *The mind and the eye*. Cambridge University Press.

Archibald, B. & Farrell, B. D. (2003). Wheeler's dilemma. *Acta zoologica cracoviensia* **46** (suppl.–fossil insects): 17–23.

Ashburner, K. B. (1993). Birches in the wild, their habitats and ecology. In: D. Hunt, *Proceedings of the IDS Betula Symposium:* 19–28. International Dendrological Society.

—— & McAllister, H. A. (2005). *Betula lenta* forma *uber*. *Curtis's Bot. Mag.* **205**: Tab. 487.

—— & Schilling, T. (1985). *Betula utilis* and its varieties. *Plantsman* **7**(2): 116–125.

—— & Walters, S. M. (1989). *Betula*. In: J. C. M. Alexander, A. Brady, C. D. Brickell, J. Cullen, P. S. Green, V. H. Heywood, V. A. Matthews, N. K. B. Robson, S. M. Walters, P. F. Yeo & S. G. Knees (eds), *The European Garden Flora* **3**. pp. 49–55. Cambridge University Press.

Ashe, W. W. (1918). Notes on *Betula*. *Rhodora* **20**: 63–64.

Atkinson, M. D. (1992). Biological flora of the British Isles no. 175. *Betula pendula* Roth (*B. verrucosa* Ehrh.) and *B. pubescens* Ehrh. *J. Ecology* **80**: 837–870.

Austerlitz, F., Mariette, S., Machon, N., Gouyon, P-H. & Godelle, B. (2000). Effects of colonising processes on genetic diversity: differences between annual plants and tree species. *Genetics*. **154**: 1309–1321.

Ayers, D. R. & Strong, D. R. (2001). Origin and genetic diversity of *Spartina anglica*. *Amer. J. Bot.* **88**(10): 1863–1867.

Azuma, T., Kajita, T., Yokoyama, J. & Ohashi, H. (2000). Phylogenetic relationships of *Salix* (Salicaceae) based on *RBCL* sequence data. *Amer. J. Bot.* **87**(1): 67–75.

Bachmann, K. (2000). Molecules, morphology and maps: New directions in evolutionary genetics. *Pl. Spec. Biol.* **15**: 197–210.

Bacles, C. F. E., Lowe, A. J. & Enos, R. A. (2004). Genetic effects of chronic habitat fragmentation on tree species: the case of *Sorbus aucuparia* in a deforested Scottish landscape. *Molec. Ecol.* **13**: 1–12.

—— & —— (1962). Further notes on *Veronica filiformis*. *Watsonia* **4**(4): 384–397.

Bannister, P. (1964). The water relations of certain heath plants with reference to their ecological amplitude. III. Experimental studies: General Conclusions. *J. Ecol.* **52**(3): 499–509.

Bartholomew, B., Boufford, D. E., Chang, A. L., Cheng, Z., Dudley, T. R., He, S. A., Jin, Y. X., Li, Q. Y., Luteyn, J. L., Spongberg, S. A., Sun, S. C., Tang, Y. C., Wan, J. X. & Ying, T. S. (1983). The 1980 Sino American Botanical Expedition to Western Hubei Province, Peoples Republic of China. *J. Arnold Arbor.* **64**: 1–103.

Barclay, A. M. & Crawford, R. H. M. (1982). Winter desiccating stress and resting bud viability in relation to high altitude survival. *Flora* **172**: 21–34.

Bard, E. Hamelin, B. & Delanghe-Sabatier, D. (2010). Deglacial meltwater pulse 1B and Younger Dryas sea levels revisited with boreholes at Tahiti. *Science* **327**: 1235–1237.

Barnes, B. V. & Dancik, B. P. (1985). Characteristics and origin of a new birch species, *Betula murrayana*, from southeastern Michigan. *Canad. J. Bot.* **63**(2): 223–226.

——, —— & Sharik, T. L. (1974). Natural hybridization of yellow birch and paper birch. *Forest Sci.* **20**(3): 215–221.

Bartholomew, B., Boufford, D. E. & Spongberg, S. A. (1983). *Metasequoia glyptostroboides* its present status in Central China. *J. Arnold Arbor.* **64**: 105–128.

Basinger, J. F. (1984). Seed cones of *Metasequoia milleri* from the Middle Eocene of southern British Columbia. *Canad. J. Bot.* **62**: 281–289.

—— (1991). The fossil forests of the Buchanan Lake formation (Early Tertiary), Axel Heiberg Island, Canadian Arctic Archipelago: Preliminary floristics and paleoclimate. In: R. L. Christie & N. J. McMillan (eds), Tertiary fossil forests of the Geodetic Hills, Axel Heiberg Isleand, Arctic Archipelago. *Bull. Geol. Surv. Canada* **403**: 39–65.

Bateman, R. (2007) Making the biological case for joined-up thinking in government. *Research Fortnight* 7th March, 2007: 16.

Bean, W. J. (1970). *Trees and shrubs hardy in the British Isles*. 1. 8th ed. John Murray, London.

—— (1988). *Trees and shrubs hardy in the British Isles*. 8th ed. (supplement). John Murray, London.

Bell, F. G. (1969). The occurrence of southern, steppe and halophyte elements in Weischelian (Last Glacial) floras from southern Britain. *New Phytol.* **68**: 913–922.

—— (1970a). Fossil of an American sedge, *Dulichium arundinaceum* (L.) Britt., in Britain. *Nature* **227**: 629–630.

—— (1970b). Late Pleistocene flora from Earith, Huntingdonshire. *Philos. Trans.*, *Ser. B* **258** (No. 826): 347–378.

Bell, N. (ed.) (1994). *The ecological effects of increased aerial deposition of nitrogen*. British Ecological Society, Ecological Issues No. 5. Field Studies Council. Shrewsbury.

Bennett, M. D. (2004). Perspectives on polyploidy in plants — ancient and neo. *Biol. J. Linn. Soc.* **82**: 411–423.

Bennike, O. & Bocher, J. (1990). Forest–tundra neighbouring the North Pole: plant and insect remains from the Plio-Pleistocene Kap København formation, north Greenland. *Arctic* **43**(4): 331–338.

Bergmann, F. & Gillet, E. M. (1997) Phylogenetic relationships among *Pinus* species (Pinaceae) inferred from different numbers of 6PGDH loci. *Pl. Syst. Evol.* **208**(1): 25 –38.

Bergman, J., Hammarlund, D., Hannon, G. Barnekow, L. & Wohlfarth, B. (2005). Deglacial vegetation succession and Holocene tree-line dynamics in the Scandes Mountains, west-central Sweden: stratigraphic data compared to megafossil evidence. *Rev. Paleobot. Palynol.* **134**: 129–151.

Bertolani-Marchetti, D. (1985). Pollen paleoclimatology in the mediterranean since Messinian time. In: D. J. Stanley & F.-C. Wezel (eds), *Geological evolution of the Mediterranean basin* Ch. 24: 525–543. Springer-Verlag, New York.

Bevington, J. (1986). Geographic differences in the seed germination of paper birch (*Betula papyrifera*) *Amer. J. Bot.* **73**(4): 564–573.

Bhagwat, S. A. & Willis, K. J. (2008). Species persistence in northerly glacial refugia of Europe: matter of chance or biogeographical traits. *J. Biogeogr.* **35**: 464–482.

Bialobrzeska, M. (1955). Morphological and biological characteristics of *Betula oycoviensis* Besser, *B. verrucosa* Ehrh, and their hybrid. (in Polish). *Rocz. Dendr.* **10**: 165–189.

Binney, H. A., Willis, K. J., Edwards, M. E., Bhagwat, S. A., Anderson, P. M., Andreeve, A. A., Blaauw, M., Damblon, F., Haesaerts, P., Kienast, F., Kremenetski, K. V., Krivonogov, S. K., Lozhkin, A. V., MacDonald, G. M., Novenko, E. L., Oksanen, P., Sapelko, T. V., Valiranta, M. & Vazhenina, L. (2008). The distribution of late-Quaternary woody taxa in northern Eurasia: evidence fom a new macrofossil database. *Quatern. Sci. Rev.* **28**: 2445–2464.

Birks, H. H. & Birks, H. J. B. (2000). Future Uses of Pollen Analysis Must Include Plant Macrofossils. *J. Biogeogr.* **27**(1): 31–35.

—— Larsen, E. & Birks H. J. B. (2006). On the presence of late-glacial trees in western Norway and the Scandes: a further comment. *J. Biogeogr.* **33**: 376.

Bjune, A. E. (2005). Holocene vegetation history and tree-line changes on a north-south transect crossing major gradients in southern Norway — evidence fom pollen and plant macrofossils in lake sediments. *Rev. Paleobot. Palynol.* **135**: 249–275.

Black, M. & Wareing, P. F. (1954). Photoperiodic control of germination in seed of birch (*Betula pubescens* Ehrh.) *Nature* **174**: 705–706.

—— & —— (1955). Growth studies in woody species. VII. Photoperiodic control of germination in *Betula pubescens* Ehrh. *Physiol. Pl.* (Copenhagen) **8**: 300–316.

Blackmore, S., Steinmann, J. A. J., Hoen, P. P. & Punt, W. (2003). The Northwest European Pollen Flora, 65. Betulaceae and Corylaceae. *Rev. Paleobot. Palynol.* **123**: 71–98.

Blunt, W. (1971). *The compleat naturalist: a life of Linnaeus*. Collins, London.

—— & Stearn, W. T. (1994). *The Art of Botanical Illustration*. Antique Collectors Club, Woodbridge, Suffolk, in association with the Royal Botanic Gardens, Kew.

Blyakharchuk, T. A., Wright, H. E., Borodavko, P. S., van der Knaap, W. O. & Ammann, B. (2004). Late Glacial and Holocene vegetational changees on the Ulagan high-mountain plateau, Altai Mountains, southern Siberia. *Palaeogeogr., Palaeoclimat. Palaeoecol.* **209**: 250–279.

Bobbink, R., Hornung, M. & Roelofs, J. G. M. (1998). The effects of air-borne nitrogen pollutants on species in natural and semi-natural European vegetation. *J. Ecol.* **86**: 717–738.

Bogdanowicz, S. M., Schaefer, P. W. & Harrison, R. G. (2000). Mitochondrial DNA variation among world wide populations of Gypsy Moths, *Lymantria dispar*. *Molec. Phylogenet. Evol.* **15**: 487–495.

Bond, W. J. (1989). The tortoise and the hare: ecology of angiosperm dominance and gymnosperm persistence. *Biol. J. Linn. Soc.* **36**: 227–249.

—— & Midgley, J. J. (2001). Ecology of sprouting in woody plants: the persistence niche. *Trends Ecol. Evol.* **16**(1): 45–51.

Bopp, W. (1994). *Mapping the 'find spots' of natural source Betula specimens, to assist the management and systematic representation of a living botanical collection*. Kew Diploma, Course 30. Taxonomy Project.

Boratynska, K. (1982). Corologia i rejonizacja lesna brzozowatych (*Betulaceae*), cz. I. (Chorology and forest regionalization in family Betulaceae, Part 1). *Arbor. Kornickie* **27**: 31–100.

Boratynski, A. (1998). *Betula oycoviensis*. In: IUCN 2010. *IUCN Red List of Threatened Species. Version 2010.2*. <www.iucnredlist.org>. Downloaded on 22 August 2010.

Boulter, M. C., Benfield, J. N., Fisher, H. C., Gee, D. A. & Lhotak, M. (1996). The evolution and global migration of the Aceraceae. *Philos. Trans., Ser. B* **351**: 589–603.

Bourque, C. P. A., Cox, R. M., Allen, D. J., Arp, P. A. & Meng F. R. (2005). Spacial extent of winter thaw events in eastern North America: historical weather records in relation to yellow birch decline. *Global Change Biol.* **11**: 1477–1492.

Bousquet, J., Strauss, S. H. & Li, P. (1992). Complete congruence between morphological and rbcL-based molecular phylogenies in birches and related species (Betulaceae). *Molec. Biol. Evol.* **9**(6): 1076–1088.

Bouillé, M. & Bousquet, J. (2005). Trans-species shared polymorphisms at orthologous nuclear gene loci among distant species in the conifer *Picea* (Pinaceae): implications for the long-term maintenance of genetic diversity in trees. *Amer. J. Bot.* **92**: 63–73.

Boyd, W. E. & Dickson, J. H. (1986). Patterns in the geographical distribution of the early Flandrian *Corylus* rise in southwest Scotland. *New Phytol.* **102**: 615–623.

Bradshaw, A. D., Hunt, B. & Walmsley, T. (1995). Trees in the urban landscape. E. & F. N. Spon, London.

Brasier, C. M. (2001). Rapid evolution of introduced plant pathogens via interspecific hybridisation. *BioScience* **51**(2): 123–133.

——, Beales, P. A., Kirk, S. A., Denman, S. & Rose, J. (2005). *Phytophthora kernoviae* sp. nov., an invasive pathogen causing bleeding stem lesions on forest trees and foliar necrosis of ornamentals in the UK. *Mycol. Res.* **109**: 853–859.

——, Rose, J., Kirk, S. A. & Webber, J. F. (2002). Pathogenicity of *Phytophthora ramorum* isolates from North America and Europe to bark of European Fagaceae, American *Quercus rubra* and other forest trees. Sudden Oak Death Science Symposium, Monterey California. 12.2002. Paper Abstract: Pathogenicity and Resistance Session. http://danr.ucop.edu/ihrmp/sodsymp/paper/paper09.html

Braum, E. L. (1961). *The woody plants of Ohio*. Hafner, New York.

Bretagnolle, F. & Thompson, J. D. (1995). Gametes with somatic chromosome number: mechanism of their formation and role in the evolution of autopolyploid plants. *New Phytol.* **129**: 1–22.

Brickell, C. D. (ed.) (1994). *The RHS gardeners' encyclopedia of plants and flowers*. New ed. Dorling Kindersley, London.

—— (ed.) (1996). *The RHS A–Z encyclopaedia of garden plants*. Dorling Kindersley, London.

Brighton, C. A. (1978). Telocentric chromosomes in Corsican *Crocus* L. (Iridaceae). *Pl. Syst. Evol.* **129**: 299–314.

Brittain, W. H. & Grant, W. F. (1965a). Observations on Canadian birch (*Betula*) collections at the Morgan Arboretum. I. *B. papyrifera* in eastern Canada. *Canad. Field-Naturalist* **79**(3): 189–197.

—— & —— (1965b). Observations on Canadian birch (*Betula*) collections at the Morgan Arboretum. II. *B. papyrifera* var. *cordifolia*. *Canad. Field-Naturalist* **79**(4): 253–57.

—— & —— (1966). Observations on Canadian birch (*Betula*) collections at the Morgan Arboretum. III. *B. papyrifera* of British Columbia. *Canad. Field-Naturalist* **80**(3): 147–157.

—— & —— (1967a). Observations on Canadian birch (*Betula*) collections at the Morgan Arboretum. IV. *B. caerulea-grandis* and hybrids. *Canad. Field-Naturalist* **81**(2): 116–127.

—— & —— (1967b). Observations on Canadian birch (*Betula*) collections at the Morgan Arboretum. V. *B. papyrifera* and *B. cordifolia* from eastern Canada. *Canad. Field-Naturalist* **81**(4): 251–262.

—— & —— (1968a). Observations on Canadian birch (*Betula*) collections at the Morgan Arboretum. VI. *B. papyrifera* from the Rocky Mountains. *Canad. Field-Naturalist* **82**(1): 44–48.

—— & —— (1968b). Observations on Canadian birch (*Betula*) collections at the Morgan Arboretum. VII. *B. papyrifera* and *B. resinifera* from northwestern Canada. *Canad. Field-Naturalist* **82**(3): 185–202.

—— & —— (1969). Observations on Canadian birch (*Betula*) collections at the Morgan Arboretum. VIII. *Betula* from Grand Manan Island, New Brunswick. *Canad. Field-Naturalist* **83**(4): 161–183.

—— & —— (1972). Observations on the *Betula caerulea* complex. *Naturaliste Canad.* **98**: 49–58.

Britton, N. L. & Brown, A. (1913). *An illustrated flora of the northern United States and Canada* II. Dover Publications Inc., New York.

Brochmann, C., Brysting, A. K., Alsos, I. G., Borgen, L., Grundt, H. H., Scheen, A-C. & Elven, R. (2004). Polyploidy in arctic plants. *Biol. J. Linn. Soc.* **82**: 521–536.

——, Gabrielsen, T. M., Nordal, I., Landvik, J. I. & Elven, R. (2003). Glacial survival or *tabula rasa*. *Taxon* **52**: 417–450.

Browicz, K. (1972). *Betula* L. In: P. H. Davis (ed.), *Flora of Turkey and the East Aegean Islands* **4**: 6. Edinburgh University Press.

—— (1975). Distribution of species of the genus *Betula* L. in Turkey, Iran and Iraq. *Arbor. Kornickie* **20**: 37–46.

Brown, G. E. & Kirkham, T. (2004). *The pruning of trees, shrubs and conifers*. Timber Press, Portland, Oregon and Cambridge.

Brown, I. R. & Al-Dawoody, D. M. (1977). Cytotype diversity in a population of *Betula alba* L. *New Phytol.* **79**: 441–453.

—— & Williams, D. A. (1984). Cytology of the *Betula alba* L. complex. *Proc. Roy. Soc. Edinburgh* **85B**: 49–64.

Brubaker, L. B., Anderson, P. M., Edwards, M. E. & Lozhkin, A. V. (2005). Beringia as a glacial refugium for boreal trees and shrubs: new perspectives from mapped pollen data. *J. Biogeogr.* **32**: 833–848.

Buckland, S. M., Grime, J. P., Hodgson, J. G. & Thompson, K. (1997). A comparison of plant responses to the extreme drought of 1995 in northern England. *J. Ecol.* **85**(6): 875–882.

Buczacki, S. & Harris, K. (1981). *Collins guide to the pests, diseases and disorders of garden plants.* Collins, London.

Budantsev, L. Y. (1992). Early stages of formation and dispersal of the temperate flora in the boreal region. *Bot. Rev.* (*Lancaster*) **58**(1): 1–48.

Butler, B. T. (1909). The western American birches. *Bull. Torrey Bot. Club* **36**(8): 421–440.

Byers, D. L. & Waller, D. M. (1999). Do plant populations purge their genetic load? Effects of population size and mating history on inbreeding depression. *Annual Rev. Ecol. Syst.* **30**: 479–513.

Byun, A. S., Koop, B. & Reimchen, T. E. (1999). Coastal refugia and postglacial recolonisation routes: a reply to Demboski, Stone, and Cook. *Evolution* **53**(6): 2013–2015.

Cadot, P., Diaz. J/F., Proost, P., Van Damme, J., Engelborghs, Y., Stevens, E. A. M. & Ceuppens, J. L. (2000). Purification and characterisation of an 18-kd allergen of birch (*Betula verrucosa*) pollen: Identification as a cyclophilin. *J. Allergy Clinical Immunology* **104**(2): 286–291.

Cain, M. L., Milligan, B. G. & Strand, A. E. (2000). Long-distance seed dispersal in plant populations. *Amer. J. Bot.* **87**(9): 1217–1227.

Cannell, M. G. R. (1987). Photosynthesis, foliage development and productivity of Sitka spruce. *Proc. Roy. Soc. Edinburgh* **93B**: 61–73.

Cappaert, D., McCullough, D. G., Pland, T. M. & Siegert, N. W. (2005). Emerald ash borer in North America: a research and regulatory challenge. *Amer. Entomol.* **51**(3): 152–165.

Carcaillet, C. & Vernet, J-L. (2001). Comments on "The full-Glacial forests of central and southeastern Europe" by Willis *et al.*. *Quatern. Res.* **55**: 385–387.

Carrion, J. S. (2002). Patterns and processes of Late Quaternary environmental change in a montane region of southwestern Europe. *Quatern. Sci. Rev.* **21**: 2047–2066.

Catling, P. M. & Spicer, K. W. (1988). The separation of *Betula populifolia* and *Betula pendula* and their status in Ontario. *Canad. J. Forest Res.* **18**: 1017–1026.

Chaffey, N. (2002). Why is there so little research into the cell biology of the secondary vascular system of trees. *New Phytol.* **153**: 213–223.

Chen, Z-D, Manchester, S. R. & Sun, H-Y. (1999). Phylogeny and evolution of the Betulaceae as inferred from DNA sequences, morphology, and paleobotany. *Amer. J. Bot.* **86**(8): 1168–1181.

Clapham, A. R., Tutin, T. G. & Warburg, E. F. (1952). *Flora of the British Isles.* Cambridge University Press.

Clark, J. S., Beckage, B., Camill, P., Cleveland, B., HilleRisLambers, J., Lichter, J., McLachlan, J., Mohan, J. & Wyckoff, P. (1999). Interpreting recruitment limitation in forests. *Amer. J. Bot.* **86**(1): 1–16.

Clarke, G. P. (1998) Plants in peril, 24. Notes on lowland African violets (*Saintpaulia*) in the wild. *Curtis's Bot. Mag.* t. 332. **15**(1): 62–67.

Clausen, J. J. & Kozlowski, T. T. (1965). Heterophyllous shoots in *Betula. Nature* **205**: 1030–1031.

Clausen, K. E. (1963). Characteristics of a hybrid birch and its parent species. *Canad. J. Bot.* **41**: 441–458.

—— (1966). Studies of compatibility in *Betula*.In: Joint Proceedings, Second Genetics Workshop of Society of American Foresters and Seventh Lake States Forest Tree Improvement Conference. USDA Forest Service Research Paper NC 6. North Central Forest Experimental Station, St. Paul, MN.

—— (1970). Interspecific crossability in *Betula. Proceedings of the IUFRO section 22 working group on sexual reproduction of forest trees.* Varparanta, Finland. pp. 1–10.

Clegg, B. F., Tinner, W., Gavin, D. G. & Hu, F. S. (2005). Morphological differentiation of *Betula* (birch) pollen in northwest North America and its palaeoecological application. *The Holocene* **15**(2): 229–237.

Clennet, C. & Sanderson, H. (2002). *Betula papyrifera. Curtis's Bot. Mag.* **19**(1): 40–48. Tab. 436.

Coates, A. M. (1969). *The quest for plants.* Studio Vista Ltd., London.

Coates, D. J. (1992). Genetic consequences of a bottleneck and spatial genetic structure in the triggerplant *Stylidium coronoforme* (Stylidiaceae). *Heredity* **69**: 512–520.

Coder, K. D. (1999). *Tree selection for drought resistance.* University of Georgia, Daniel B. Warnell School of Forest Resources, Extension publication FOR99-088.

Collinson, M. E. (1990). Plant evolution and ecology during the Early Cainozoic diversification. *Advances Bot. Res.* **17**: 1–98.

—— (2000). Cenozoic evolution of modern plant communities and vegetation. In: S. J. Culver & P. F. Rawson (eds), *Biotic response to global change: The last 145 million years.* 223–243. Cambridge University Press, Cambridge.

Comai, L. (2000). Genetic and epigenetic interactions in allopolyploid plants. *Pl. Molec. Biol.* **43**: 387–399.

Cox, K., Storm, K. & Baker, I. (2001). *Frank Kingdon-Ward's Riddle of the Tsangpo Gorges.* Antique Collectors' Club, Woodbridge, Suffolk.

Crane, M. B. (1940a). Reproductive versatility in *Rubus* I. Morphology and inheritance. *J. Genetics* **40**: 109–118.

—— (1940b). The origin of new forms in *Rubus* II. The loganberry, *R. loganobaccus* Bailey. *J. Genetics* **40**: 129–140.

Crane, P. R. (1981). Betulaceous leaves and fruits from the British Upper Paleocene *Bot. J. Linn. Soc.* **83**(2): 103–136.

—— (2004). Documenting plant diversity: unfinished business. *Philos. Trans. Ser. B* **359**: 735–737.

—— & Carvel, W. N. (2007). The importance of history. *Curtis's Bot. Mag.* **24**(3): 196 –216.

—— & Stockey, R. A. (1987). *Betula* leaves and reproductive structures from the Middle Eocene of British Columbia. *Canad. J. Bot.* **65**: 2490–2500.

Cronquist, A. (1987). A botanical critique of cladism. *Bot. Rev. (Lancaster)* **53**(1): 1–52.

Culley, T. M., Weller, S. G. & Sakai, A. K. (2002). The evolution of wind pollination in angiosperms. *Trends Ecol. Evol.* **17**(8): 361–369.

Dancik, B. P. & Barnes, B. V. (1969). Dark barked birches of southern Michigan. *Michigan Bot.* **8**: 38–41.

—— & —— (1971). Variability in bark morphology of yellow birch in an even-aged stand. *Michigan Bot.* **10**: 34–38.

—— & —— (1972). Natural variation and hybridisation of yellow birch and bog birch in south east Michigan. *Silvae Genetica* **21**(1– 2): 1–64.

—— & —— (1974). Leaf diversity on yellow birch (*Betula alleghaniensis*). *Canad. J. Forest Res.* **5**(2): 149–159.

—— & —— (1975). Leaf variability in yellow birch (*Betula alleghaniensis*) in relation to environment. *Canad. J. Bot.* **52**(11): 2407–2414.

——, —— & Wagner, W. H. (1974). Aberrant pistillate catkins of *Betula alleghaniensis*. *Michigan Bot.* **13**: 177–179.

Davidson, W. D. (1981). A revision of *Rhododendron* a horticulturalists' view. *Rhododendrons Magnolias Camellias* 1981/2: 78–81.

Davis, P. H. (1972). Betulaceae. In: *Flora of Turkey* **7**: 688–693. Edinburgh.

—— & Heywood, V. H. (1963). *Principles of Angiosperm taxonomy.* Oliver and Boyd, Edinburgh and London.

Dawson, M. R., West, R. M., Langstron, W. & Hutchison, J. H. (1976). Paleogene terrestrial vertebrates: northernmost occurrence, Ellesmere island, Canada. *Science* **192**: 781–782.

De Groot, W. J., Thomas, P. A. & Wein, R. W. (1997). *Betula nana* L. and *Betula glandulosa* Michx. *J. Ecol.* **85**: 241–264.

DeHond, P. E. & Campbell, C. S. (1989). Multivariate analyses of hybridisation between *Betula cordifolia* and *B. populifolia* (Betulaceae). *Canad. J. Bot.* **67**: 2252–2260.

De Jong, P. C. (1993). An introduction to *Betula*: its morphology, evolution, classification and distribution, with a survey of recent work. In: D. Hunt, *Proceedings of the IDS Betula Symposium:* 7–18. International Dendrological Society.

—— (1996). The genetic diversity of trees in cultivation. In: D. Hunt (ed), *Temperate trees under threat*, pp. 179–184. International Dendrological Society Symposium, Morpeth.

Del Tredici, P. (2001). Sprouting in temperate trees: a morphological and ecological review. *Bot. Rev. (Lancaster)* **67**(2): 121–140.

Denk, T. & Dillhoff, R. M. (2005). *Ulmus* leaves and fruit from the early-middle Eocene of northwestern America: systematics and implications for character evolution within Ulmaceae. *Canad. J. Bot.* **83**(12): 1663–1681.

—— & Grimm, G. W. (2005). Phylogeny and biogeography of *Zelkova* (Ulmaceae *sensu stricto*) as inferred from leaf morphology, ITS sequence data and the fossil record. *Bot. J. Linn. Soc.* **147**(2): 129–157.

—— & —— (2009). The biogeographic history of beech trees. *Rev. Palaeobot. Palynol.* **158**(1–2): 83–100.

——, —— & Hemleben (2005). Patterns of molecular and morphological differentiation in *Fagus* (Fagaceae): phylogenetic implications. *Amer. J. Bot.* **92**: 1006–1016.

——, ——, Stogerer, K. & Hemleben, V. (2002). The evolutionary history of *Fagus* in western Eurasia: evidence from genes, morphology and the fossil record. *Pl. Syst. Evol.* **232:** 213–236.

——, Grímsson, F. & Kvacek, Z. (2005). The Miocene floras of Iceland. *Pl. Syst. Evol.* **232:** 213–236.

——, —— & Uhl, D. (2007). The Miocene floras of Iceland — a unique archive of terrestrial climate evolution in the northern North Atlantic. In: F. Grimsson, *The Miocene floras of Iceland, Origin and evolution of fossil floras from North-West and Western Iceland, 15 to 6 Ma*. University of Iceland. Reykjavik.

——, —— & Zetter, R. (2010). Episodic migration of oaks to Iceland. *Amer. J. Bot.* **92**: 1006–1016.

De Smet, K. (1993). Cheetahs teetered on the brink in the ice age. *New Scientist* **138** (Issue 1875): 16.

Dickson, J. H. (1984). Pleistocene history of *Betula* with special reference to the British Isles. *Proc. Roy. Soc. Edinburgh* **85** B: 1–11.

Diels, L. (1912a). Plantae chinenses Forrestianae. *Notes Roy. Bot. Gard. Edinburgh.* 1 (no. 31): 1–411.

—— (1912b). Plantae chinenses Forrestianae. New and imperfectly known species. *Notes Roy. Bot. Gard. Edinburgh* **5**: 161–308, (272).

Dilcher, D. L. (1974). Approaches to the identification of Angiosperm leaf remains. *Bot. Rev. (Lancaster)* **40**(1): 1–157.

Dmitrieva, A. A. (1960). *Opredelitelj rastenii Adzharii.* Tbilisi. [Identification key to the plants of Adzhariya]

Dodd, M. E., Silvertown, J. & Chase, M. W. (1999). Phylogenetic analysis of trait evolution and species diversity variation among angiospern families. *Evolution* **53**(3): 732–744.

Doluchanov, A. G. (1939). *Zametki Sist. Geogr. Rast.* **7**: 14.

Don, D. (1825). *Prodromus florae Nepalensis*. J. Gale, Londini.

Donoghue, M. J. & Smith, S. A. (2004). Patterns in the assembly of temperate forests around the Northern Hemisphere. *Philos. Trans. Ser. B* **359**: 1633–1644.

Dosmann, M. & Del Tredici, P. (2003). Plant introduction, distribution and survival: a case study of the 1980 Sino-American Botanical Expedition. *Bioscience* **53**(6): 588–597.

Dubatolov, V. V. & Kosterin, O. E. (2000). Nemoral species of Lepidoptera (Insecta) in Siberia: a novel view on their history and dating of range of disjunctions. *Entomol. Fenn.* **11**: 141–166.

Dubey, S., Zaitsev, M. Cosson, J-F., Abdukadier, A. & Vogel, P. (2006). Pliocene and Pleistocene diversification and multiple refugia in a Eurasian shrew (*Crocidura suaveolens* group). *Molec. Phylogenet. Evol.* **38**: 635–647.

Dufresne, F. & Herbert, P. D. N. (1994). Hybridization and origins of polyploidy. *Proc. Roy. Soc. London. Ser. B, Biol. Sci.* **258:** 141–146.

Dugle, J. R. (1966). A taxonomic study of western Canadian species in the genus *Betula. Canad. J. Bot.* **44**: 929–1007.

Dyer, A. F. (1963). The use of lacto-propioinic orcein in rapid squash methods for chromosome preparations. *Stain Technol.* **38**: 85–90.

Eaker, T. A., Ranney, T. G., Viloria, Z. J. & Mowrey, J. A. (2004). Variation in ploidy level among birch taxa. *Southern Nursery Association (SNA) Research Conference* **49**: 548–551.

Eastwood, A., Bytebier, B., Tye, H., Tye, A., Robertson., A. & Maunder, M. (1998). The conservation status of *Saintpaulia. Curtis's Bot. Mag.* **15**(1): 49–62.

Eastwood, E., Lazkov, G. & Newton, A. (2009). *The Red List of Trees of Central Asia*. Fauna & Flora International, Cambridge.

Eberle, J. J. (2006). Early Eocene Brontotheriidae (Perissodactyla) from the Eureka Sound Group. Ellesmere Island, Canadian High Arctic — implications for Brontothere origins and high latitude dispersal. *J. Vertebr. Paleontol.* **26**(2): 381–386.

Ehrhart, F. (1791). *Beiträge zur Naturkunde, und den damit verwandten Wissenschaften: besonders der Botanik, Chemie, Haus- und Landwirthschaft, Arzneigelahrtheit und Apothe-kerkunst* 6: 98. Schmidt, Hannover & Osnabrück.

Ehrich, D., Alsos, I. G. & Brochman, C. (2008). Where did the northern peatland species survive the dry glacials: cloudberry (*Rubus chamaemorus*) as an example. *J. Biogeogr.* **35**: 801–614.

Eidesen, P. B., Alsos, I. G., Popp, M., Stensrud, Ø., Suda, J. & Brochmann, C. (2007a). Nuclear vs. plastid data: complex Pleistocene history of a circumpolar key species. *Molec. Ecol.* **16**: 3902–3925.

——, Carlsen, T., Molau, U. & Brochmann, C. (2007b). Repeatedly out of Beringia: *Cassiope tetragona* embraces the arctic. *J. Biogeogr.* **34**: 1559–1574.

Eldrett, J. S., Harding, I. C., Wilson, P. A., Butler, E. & Roberts, A. P. (2007). Continental ice in Greenland during the Eocene and Oligocene. *Nature* **446**: 176–179.

Elias, S. A. (2000). Late Pleistocene climates of Beringia, based on analysis of fossil beetles. *Quatern. Res.* **53**: 229–235.

Elkington, T. T. (1968). Introgressive hybridisation between *Betula nana* L. and *B. pubescens* Ehrh. In north-west Iceland. *New Phytol.* **67**: 109–118.

Ellstrand, N. C. & Schierenbeck, K. A. (2000). Hybridization as a stimulus for the evolution of invasiveness in plants? *Proc. Natl. Acad. Sci. U.S.A.* **97**: 7043–7050.

Ellsworth, J. W. & McComb, B. C. (2003). Potential effects of passenger pigeon flocks on the structure and composition of presettlement forests of eastern North America. *Conserv. Biol.* **17**: 1548–1558.

Elton, C. S. (1958). *The ecology of invasions by animals and plants.* Methuen, London.

Eminagaoğlu, O., Kutbay, H. G., Bilgin, A. & Yalçin, E. (2006). Contribution to the phytosociology and conservation of tertiary relict species in northeastern Anatolia (Turkey). *Belg. J. Bot.* **139**(1): 124–30.

Erdogan, V. & Mehlenbacher, S. A. (2000). Phyogenetic relationships of *Corylus* species (Betulaceae) based on nuclear ribosomal DNA ITS region and chloroplast matK gene sequences. *Syst. Bot.* **25**(4): 727–737.

Eriksson, G. & Jonsson, A. (1986). A review of the genetics of *Betula*. *Scandinavian J. Forest Res.* **1**: 421–434.

Ermolli, E. R. & Di Pasquale, G. (2002). Vegetation dynamics of south-western Italy in the last 28kyr inferred from pollen analysis of a Tyrrhenian sea core. *Veg. Hist. Archeobot.* **11**: 211–219.

Esen, D., Zedaker, S. M., Kirwan, J. L. & Mou, P. (2004). Soil and site factors influencing purple flowered rhododendron (*Rhododendron ponticum* L.) and eastern beech forests (*Fagus sylvatica* Lipsky) in Turkey. *Forest Ecol. Managem.* **203**: 229–240.

Evert, R. F. (2006). *Esau's Plant Anatomy, Meristems, Cells, and Tissues of the Plant Body: their Structure, Function, and Development.* 3rd ed. John Wiley & Sons, Inc., New Jersey.

Eyde, R. & Ferguson, K. (1989). The little lost dogwood of Frank Kingdon Ward. *Kew Mag.* **6**(2): 74–83.

Falk, D. A. & Holsinger, K. E.(eds) (1991). *Genetics and conservation of rare plants.* Centre for Plant conservation. Oxford University Press.

Farjon, A. (1996). A world list of threatened conifers: How much do we know? In: D. Hunt (ed.), *Temperate trees under threat*, pp. 151–160. International Dendrological Society.

Farell, B. & Miter, C. (1990). Phylogenesis of insect/plant interactions: have *Phyllobritica* leaf beetles (Chrysomelidae) and the Lamiales diversified in parallel? *Evolution* **44**(6): 1389–1403.

Fang, J., Wang, Z. & Tang, Z. (2009). *Atlas of Woody Plants in China* 1. Higher Education Press, Beijing.

Fauquette, S. & Bertini, A. (2003). Quantification of the northern Italy Pliocene climate from pollen data: evidence for a very peculiar climate pattern. *Boreas* **32**: 361–369.

Favarger, C. (1978). Philosophie des comptages de chromosomes. *Taxon* **27**(5/6): 441–448.

Fedorov V. B. & Stenseth, N. C. (2002). Multiple glacial refugia in the North American Arctic: inference from phylogeography of the collared lemming (*Dicrostonyx groenlandicus*). *Proc. Roy. Soc. London Ser. B, Biol. Sci.* **269**: 2071–2077.

Feliner, G. N. & Rosselló, J. A. (2007). Better the devil you know? Guidelines for insightful utilisation of nrDNA ITS in species-level evolutionary studies in plants. *Molec. Phylogenet. Evol.* **44**: 911–919.

Ferguson, D. K. & Knobloch, E. (1998). A fresh look at the rich assemblage from the Pliocene sink-hole of Willershausen, Germany. *Rev. Palaeobot. Palynol.* **101**: 271–286.

Fernald, M. L. (1945). Some North American Corylaceae (Betulaceae). I. Notes on *Betula* in eastern North America. *Rhodora* **47**: 303–329.

—— (1950). *Gray's Manual of Botany* 8th ed. American Book Company, New York.

Fontaine, F. J. (1970). Het geslacht *Betula* (bijdrage tot een monografie). *Belmontiana* **6**: 99–180.

Forest, F. & Bruneau, A. (2000). Phylogenetic analysis, organisation, and molecular evolution of the nontranscribed spacer of 5s ribosomal RNA genes in *Corylus* (Betulaceae). *Int. J. Pl. Sci.* **161**(5): 793–806.

——, Savolainen, V., Chase, M., Lupia, R., Bruneau, A. & Crane, P. R. (2005). Teasing apart molecular-versus fossil-based estimates when dating phylogenetic trees: a case study in the birch family (Betulaceae). *Syst. Bot.* **30**(1): 118–133.

——, Drouin, J. N., Charest, R., Brouillet, L. & Bruneau, A. (2001). A morphological phylogenetic analysis of *Aesculus* L. and *Billia* Peyr. (Sapindaceae). *Canad. J. Bot.* **79**(2): 154–169.

Forrest, G. (1924). Plantae Chinenses Forrestianae. *Notes Roy. Bot. Gard. Edinburgh.* **14**: 75–393.

—— (1929a). *Plantae Chinenses Forrestianae. Notes Roy. Bot. Gard. Edinburgh.* **17**: 1–406.

—— (1929b). *Field Notes of Trees, Shrubs and Plants other than Rhododendrons collected in Western China by Mr. George Forrest 1917–19.* Royal Horticultural Society, London.

Frahm, J-P. (1997). Systematics of the Bryophytes. *Progr. Bot.* **58**: 455–469.

Franchet, A. R. (1889). *Plantae Delavayanae.* Paul Klincksieck, Paris.

—— (1899). Plantarum sinensium ecloge tertia. *J. Bot. (Morot)* **13**: 253–266.

—— & Savatier, P. A. L. (1879). *Enumeratio Plantarum in Japonia Sponte Crescentium.* **2**: 351. F. Savy, Paris.

Francis, J. E. (1988). A 50-million-year-old fossil forest from Strathcona Fjord, Ellesmere Island, Arctic

Canada: Evidence for a warm polar climate. *Arctic* **41**(4): 314–318.

Fredskild, B. (1991). The genus *Betula* in Greenland — Holocene history, present distribution and synecology. *Nordic J. Bot.* **11**(4): 393–412.

Frenzel, B. (1997). History of flora and vegetation during the Quaternary. *Progr. Bot.* **59**: 599–633.

Freund, H., Birks, H. H. & Birks, H. J. B. (2001). The identification of wingless *Betula* fruits in Weischelian sediments in the Gross Todtshorn borehole (Lower Saxony, Germany) — the occurrence of *Betula humilis*. *Veg. Hist. Archeobot.* **10**: 117–115.

Friis, E. M., Pedersen, K. R. & Crane, P. R. (2006). Cretaceous angiosperm flowers: Innovation and evolution in plant reproduction. *Paleogeogr. Paleoclimatol. Paleoecol.* **232**: 251–293.

——, —— & Schönenberger, J. (2003). *Endressianthus*, a new Normapolles producing plant genus of fagalean affinity from the Late Cretaceous of Portugal. *Int. J. Pl. Sci.* **154** (5 Suppl.): S201–S223.

Fu, L-K. & Jin, J-M. (eds) (1992). *China Plant Red Data Book*. Science Press, Beijing.

Furlow, J. J. (1990). The genera of Betulaceae in the southeastern United States. *J. Arnold Arbor.* **71**: 1–67.

—— (1997). Betulaceae. In: Flora of North America Editorial Committee (eds), *Flora of North America* **3**. Oxford University Press.

—— & Mitchell, R. S. (1990). Betulaceae through Cactaceae of New York State. In: R. S. Mitchell (ed.), *Contributions to a Flora of New York State* VIII. Bulletin No. 476. New York State Museum. New York.

Gale, A. S. (2000). The Cretaceous world. In: S. J. Culver & P. F. Rawson (eds), *Biotic response to global change: The last 145 million years*, pp. 4–19. Cambridge University Press. Cambridge.

Gardiner, A. S. (1968). The reputation of birch for soil improvement. *Forestry Commission, Res. Developm. Pap. No. 67.*

—— (1984). Taxonomy of infraspecific variation in *Betula pubescens* Ehrh. With particular reference to the Scottish Highlands. *Proc. Roy. Soc. Edinburgh* **85B**: 13–26.

Gardner, A. R. & Willis, K. J. (1999). Prehistoric farming and the postglacial expansion of beech and hornbeam: a comment on Küster. *The Holocene* **9**: 119–121.

Gardner, D. (1998). *Corydalis flexuosa. Curtis's Bot. Mag.* Plate 332. **15**(1): 12–19.

Gasson, P. E. & Cutler, D. F. (1990). Tree root plate morphology. *Arboric. J.* **14**(3): 193–264.

Geml, J., Laursen, A., O'Neil, K., Nusbaum, H. C. & Taylor, D. L. (2006). Beringian origins and cryptic speciation events in the fly agaric (*Amanita muscaria*). *Molec. Ecol.* **15**: 225–239.

——, Tulloss, R. E., Laursen, G. A., Sazanova, N. A. & Taylor, D. L. (2008). Evidence for strong inter- and intraconinental phylogeographic structure in *Amanita muscaria*, a wind-dispersed ectomycorrhizal basidiomycete. *Molec. Phylogenet. Evol.* **48**: 694–701.

Gernandt, D. S. & Liston, A. (1999). Internal transcribed spacer region evolution in *Larix* and *Pseudotsuga* (Pinaceae). *Amer. J. Bot.* **86**(5): 711–723.

Ghorbani, J., Das, P. M., Das, A. B., Hughes, J. M., McAllister, H. A., Pallai, S. K., Pakeman, R. J., Marrs, R. H. & Le Duc, M. G. (2003a). Effects of different bracken control and vegetation restoration treatments on the soil diaspore bank size and composition. *Aspects Appl. Biol.* **69**: 29–37.

——, ——, ——, ——, ——, ——, ——, —— & —— (2003b). Effects of restoration treatments on the diaspore bank under dense *Pteridium* stands in the UK. *Appl. Veg. Sci.* **6**: 189–198.

Gibby, M. & Walker, S. (1972). Further cytogenetic studies and a reappraisal of the diploid ancestry of the *Dryopteris carthusiana* complex. *Brit. Fern Gaz.* **11**(5): 315–324.

Gilbertson, P. & Bradshaw, A. D. (1990). The survival of newly planted trees in inner cities. *Arboric. J.* **14**: 287–309.

Godfray, H. C. J. & Knapp, S. (2004). Taxonomy for the twenty-first century. Introduction. *Philos. Trans., Ser. B* **359**: 559–569.

Goldblatt, P. (2007). The index to plant chromosome numbers — past and future. *Taxon* **56**(4): 984–986.

Gómez-Campo, C. (1985). *Plant conservation in the mediterranean area*. Dr W. Junk, Dordrecht.

Gómez-Pompa, A. (2004). The role of biodiversity scientists in a troubled world. *Biosci.* **54**(3): 217–225.

Goropashnaya, A. V., Fedorov, V. B., Seifert, B. & Pamilo, P. (2004). Limited phylogeographical structure across Eurasia in two red wood ant species *Formica pratensis* and *F. lugubris* (Hymenoptera, Formicidae). *Molec. Ecol.* **13**: 1849–1858.

Gosling, W. D. & Bunting, M. J. (2008). A role for palaeoecology in anticipating future change in mountain ranges? *Palaeogeogr. Palaeoclimatol. Palaeoecol.* **259**(1): 1–5.

Govaerts, R. (1996). Proposal to reject the name *Betula alba* (Betulaceae). *Taxon* **45**: 697–698.

—— & Frodin, D. G. (1998). *World Checklist and Bibliography of Fagales*. Royal Botanic Gardens, Kew.

Govier, R., Walter, K. S., Chamberlain, D., Gardner, M., Thomas, P., Alexander, C., Maxwell, H. S. & Watson, M. F. (2001). Royal Botanic Garden Edinburgh. Catalogue of Plants 2001. RBGE, Edinburgh.

Grace, J. (1987). Climatic tolerance and the distribution of plants. *New Phytol.* **106** (Suppl.): 113–130.

Gradstein, F. M., Ogg, J. G. & Smith, A. G. (2004). *A geological time scale* 2004. Cambridge University Press.

Graham, A. (1963). Systematic revision of the Sucker Creek and Trout Creek Miocene floras of southeastern Oregon. *Amer. J. Bot.* **50**: 921–936.

Granoszewski, W., Demske, D., Nita, M., Heumann, G. & Andreev, A. A. (2005). Vegetation and climate variability during the Last Interglacial evidenced in the pollen record from lake Baikal. *Global and Planetary Change* **46**: 187–198.

Grant, B. R. & Grant, P. R. (2003). What Darwin's finches can teach us about evolutionary origin and regulation of biodiversity. *Biosci.* **53**(10): 965–975.

Grant, W. F. (1976). The evolution of karyotype and polyploidy in arboreal plants. *Taxon* **25**(1): 71–84.

—— & Thompson, B. K. (1975). Observations on Canadian birches, *Betula cordifolia, B. neoalaskana, B. populifolia, B. papyrifera,* and *B. ×caerulea. Canad. J. Bot.* **53**(15): 1478–1490.

Greene, H. W. (2005). Organisms in nature as a central focus for biology. *Trends Ecol. Evol.* **20**(1): 23–27.

Green, P. S. (1966). Thoughts on authenticated and documented living collections. *Taxon* **15**(8): 289–291.

Greenwood, D. R. & Basinger, J. F. (1993). Stratigraphy and floristics of Eocene swamp forests from Axel Heiberg Island, Canadian Arctic Archipeligo. *Canad. J. Earth Sci.* **30**: 1914–1923.

—— & —— (1994). The paleocology of high-latitude Eocene swamp forests from Axel Heiberg Island, Canadian High Arctic. *Rev. Paleobot. Palynol.* **81**: 83–97.

Grey-Wilson, C. (1988). Yunnan, journey to the Jade Dragon Snow Mountains, 2. *Alpine Gard. Soc. Bull.* **56**(2): 115–132.

Grierson, A. J. C. & Long, D. G. (1983). *Flora of Bhutan* I. Royal Botanic Gardens, Edinburgh.

Grimaldi, D. A. & Engl, M. S. (2007). Why descriptive science still matters. *Biosci.* **57**(8): 646–647.

Grimanelli, D., Leblanc, O., Perotti, E. & Grossniklaus, U. (2001). Development genetics of gametophytic apomixis. *Trends Genet.* **17**(10): 597–604.

Grime, J. P., Hodgson, J. G. & Hunt, R. (1988). *Comparative plant ecology.* Unwin Hyman, London.

Grimsson, F. & Denk, T. (2005). *Fagus* from the Miocene of Iceland: systematics and bioeographical considerations. *Rev. Paleobot. Palynol.* **134**: 27–54.

——, —— & Simonarson, L. A. (2007). Middle Miocene floras of Iceland — the early colonisation of an Island. *Rev. Paleobot. Palynol.* **144**: 181–219.

——, —— & Zetter, R. (2008). Pollen, fruits, and leaves of *Tetracentron* (Trochodendraceae) from the Cainozoic of Iceland and western North America and their palaeobiogeographic implications. *Grana* **47**(1): 1–14.

Grubb, P. J. (1977). The maintenance of species-richness in plant communities: the importance of the regeneration niche. *Biol. Rev.* **52**: 107–145.

Gu, L., Hanson, P. J., Mac Post, W., Kaiser, D. P., Yang, B., Nemani, R., Pallardy, S. G. & Meyers, T. (2008). The 2007 Eastern US Spring Freeze: Increased Cold Damage in a Warming World? *BioSci.* **58**(3): 253–262.

Guerriero, A. G., Grant, W. F. & Brittain, W. H. (1970). Interspecific hybridisation between *Betula cordifolia* and *Betula populifolia* at Valcartier, Quebec. *Canad. J. Bot.* **48**(12): 2241–2247.

Haak, R. A. (1996). Will global warming alter paper birch susceptibility to bronze birch borer attack? In: W. J. Mattson, P. Niemila & M. Rossi (eds), *Dynamics of Forest Herbivory: Quest for Pattern and Principle,* pp. 234–247. United States Department of Agriculture Forest Service, General Technical Report NC-183.

Hacke, U. & Sauter J. J. (1996). Xylem dysfunction during winter and recovery of hydraulic conductivity in diffuse-porous and ring-porous trees *Oecologia* **105**(4): 1432–1939.

Hagman, M. (1971). On self and cross-incompatibility shown by *Betula verrucosa* Ehrh. and *Betula pubescens* Ehrh. *Commun. Inst. Forest Fenn.* **73**(6): 1–125.

Hamilton, W. D. (1989). Significance of root severance on performance of established trees. *Arboric. J.* **13**: 249–257.

Hardin, J. W. & Bell, J. M. (1986). Atlas of foliar surface features in woody plants, IX. Betulaceae of eastern United States. *Brittonia* **38**(2): 133–144.

Harvey, R. B. (1923). Relation of the Color of Bark to the Temperature of the Cambium in Winter. *Ecology* **4**(4): 391–394.

Hayek, E. W. H., Jordis, U., Moche, W. & Sauter, F. (1989). A bicentennial of betulin. *Phytochemistry* **28**(9): 2229–2242.

Hedberg, K. O. (1992). Taxonomic differentiation in *Saxifraga hirculus* L. (Saxifragaceae) — a circumpolar arctic-boreal species of central Asiatic origin. *Bot. J. Linn. Soc.* **109**: 377–393.

Hedberg, O. (1997). The genus *Koenigia* L. emend. Hedberg (Polygonaceae). *Bot. J. Linn. Soc.* **124**(4): 295–330.

Hedrick, P. W. & Kalinowski, S. T. (2000). Inbreeding depression in conservation biology. *Annual Rev. Ecol. Syst.* **31**: 139–162.

Herman, A. B. (2002). Late Early-Late Cretaceous floras of the North Pacific region: florogenesis and early angiosperm invasion. *Rev. Paleobot. Palynol.* **122**: 1–11.

Hermanutz, L. A., Innes, D. J. & Weis, I. M. (1989). Clonal structure of arctic dwarf birch (*Betula glandulosa*) at its northern limit. *Amer. J. Bot.* **76**(5): 755–761.

Hewitt, G. (2000). The genetic legacy of the Quaternary ice ages. *Nature* **405**: 907–913.

—— (2004). Genetic consequences of climatic oscillations in the Quaternary. *Philos. Trans. Ser. B* **359**: 183–195.

Hind, N. & Fay, M. F. (2003). *Cosmos atrosanguinea*. Plants in peril 26. *Curtis's Bot. Mag.* **20**(1): 31–48.

Hjelmqvist, H. (1948). Studies on the floral morphology and phylogeny of the Amentiferae. *Bot. Not. Suppl.* **2**: 1: 1–171.

Hoffmann, M. H. & Röser, M. (2009). Taxon recruitment of the Arctic flora: and analysis of phylogenies. *New Phytol.* **182**: 774–780.

Hosie, R. C. (1963). Native trees of Canada. *Bull. Canada Dept. Forest.* 61.

Hou, J-H., Mi, X-C. & Ma, K-P. (2006). Tree competition and species coexistence in a *Quercus-Betula* forest in the Dongling Mountains in northern China. *Acta Oecologia* **30**(2): 117–125.

Howard, B. H. & Harrison-Murray, R. S. (1995). Responses of dark-preconditioned and normal light-grown cuttings of *Syringa vulgaris* 'Madame Lemoine' to light and wetness gradients in the propagation environment. *J. Hort. Sci.* **79**(6): 989–1001.

—— & —— (1991a). Rooting potential in plum hardwood cuttings: I. Relationship with shoot diameter. *J. Hort. Sci.* **66**(6): 673–680.

—— & Ridout, M.S. (1991b). Rooting potential in plum hardwood cuttings: II. Relationships between shoot variables and rooting in cuttings from different sources. *J. Hort. Sci.* **66**(6): 681–687.

—— & —— (1992). A mechanism to explain increased rooting in leafy cuttings of *Syringa vulgaris* 'Madame Lemoine' following dark treatment of the stock plant. *J. Hort. Sci.* **67**(1): 103–114.

Hsu, Y. & Wang, C. J. (1983) A new species of *Betula* from Yunnan. *Acta Bot. Yunnan.* **5**(4): 381–382.

Hubbell, S. P. (2001) *The unified theory of biodiversity and biogeography*. Monographs in population biology 32. Princeton University Press.

Huberten, H. W., Andreev, A., Henriksen, M. Hjort, C. Houmark-Nielsen, M., Jakobsen, M., Kuzmina, S., Larsen, E., Lunkka, J. P., Lyså, A., Mangerud, J., Moller, P., Saarnisto, M., Schirrmeister, L. Sher., A. V., Siegert, C., Siegert, M. J. & Inge, J. I. J.(2004) The periclacial climate and environment in northern Eurasia during the Last Glaciation. *Quaternary Sci. Rev.* **23**(11–12): 1333–1357.

Hultén, E. (1927). Flora of Kamtchatka and the adjacent islands. *Kongl. Sveska Vetensk. Acad. Handl.*, Ser. 3, 5–8.

—— (1944). *Flora of Alaska and Yukon*. Stanford. Hakan Ohlssons Boktryekeri, Lund.

—— (1958). *The Ampi-Atlantic plants. Kongl. Sveska Vetensk. Acad. Handl.*, Ser. 4, Band 7. Nr. 1.

—— (1962). The circumpolar plants. 1, Vascular crypto-gams, conifers, monocotyledons. *Kongl. Svenska Vetensk. Acad. Handl.*, Ser. 4, **8**(5): 1–275.

—— (1968). *Flora of Alaska and neighbouring territories*. Stanford University Press.

—— (1971). *The Circumpolar plants* II. *Kongl. Sveska Vetensk. Acad. Handl.*, Ser. 4, Band 13. Nr. 1.

—— & Fries, M. (1986). *Atlas of North European vascular plants: north of the Tropic of Cancer*, vol. 1–3. Koeltz Scientific Books, Königstein.

Hummel, A. (1991a). Revision of the oldest specimens of *Betula prisca* Ettinghausen. *Acta Paleobot.* **31**(1–2): 63–71.

—— (1991b) The Pliocene leaf flora form Ruszów near Zary in Lower Silesia, south west Poland. Part II Betulaceae. *Acta Paleobot.* 31(1–2):153–199.

Hundertmark, K. J., Shields, G. F., Udina, I. G., Bowyer, R. T., Danilkin, A. A. & Schwartz, C. C. (2002). Mitochondrial phylogeography of Moose (*Alces alces*): late Pleistocene divergence and population expansion. *Molec. Phylogenet. Evol.* **22**(2): 375–387.

Huntley, B. (1990). European post-glacial forests: compositional changes in response to climatic change. *J. Veg. Sci.* **1**: 507–518.

—— & Birks, H. J. B. (1983). *An atlas of past and present pollen maps for Europe: 0–13,000 years ago*. Cambridge University Press.

Husseinov, Sh. A. (1971). New species of the genus *Betula* L. from Dagestan. *Novosti Sist. Vyssh. Rast.* **8**: 128.

Hutchinson, T. (1966). The occurrence of living and sub-fossil remains of *Betula nana* L. in Upper Teesdale. *New Phytol.* **65**: 351–357.

Huxley, A. (ed.) (1992). *The New Royal Horticultural Society Dictionary of Gardening*. Macmillan, London & Basingstoke.

Hylander, N. (1957). On cut-leaved and small-leaved forms of Scandinavian birches. *Svensk. Bot. Tidskr.* **51**: 417–436 (+ 28 plates).

—— (1957). Om falsk och äkta ornäsbjörk och om några andra avvikande björkformer. *Fören. Dendrol. Parkvård Årsb. Lustgården* 1956–1957: 31–84.

Iliashenko, V. Yu. & Iliashenko, E. I. (2000). Krasnaya kniga Rossii: pravovye akty [Red Data Book of Russia: legislative acts]. State committee of the Russian Federation for Environmental Protection. Moscow. [In Russian].

Insley, H. & Buckley, G. P. (1985). The influence of desiccation and root pruning on the survival and growth of broadleaved seedlings. *J. Hort. Sci.* **60**(3): 377–387.

Ives, J. W. (1977). Pollen separation of three North American birches. *Arctic and Alpine Research.* **9**(1): 73–80.

Jackson, D. B., Fuller R. J. & Campbell, S. T. (2004). Long-term population changes among breeding shorebirds in the Outer Hebrides, Scotland, in relation to introduced hedgehogs (*Erinaceus europaeus*) *Biol. Conserv.* **117**: 151–166.

Jalas, J. & Suominen, J. (eds) (1976). *Atlas Florae Europaeae: distribution of vascular plants in Europe.* **3**: Salicaceae to Balanophoraceae, pp. 54–57. Societas Biologica Fennica Vanamo, Helsinki.

Jansson, C. A. (1962). Some species and varieties of *Betula* ser, *Verrucosae* Suk. in East Asia and N.W. America. *Acta Hort. Gotoburgensis* **25**: 103–155.

Jaramillo-Correa, J. P., Beaulieu, J. & Bousquet, J. (2004). Variation in mitochondrial DNA reveals multiple distant glacial refugia in black spruce (*Picea mariana*) *Molec. Ecol.* **13**: 2735–2747.

Järvinen, P. (2004). *Nucleotide variation of birch (Betula L.) species: population structure and phylogenetic relationships.* PhD Dissertation in Biology, No 34, University of Joensuu.

——, Palme, A., Morales, L. O., Lannenpaa, M., Keinanen, M., Sopanen, T. & Lascoux, M. (2004). Phylogenetic relationships of *Betula* species (Betulaceae) based on nuclear *ADH* and chloroplast *matK* sequences. *Amer. J. Bot.* **91**: 1834–1845.

Jarvis, C. E. (2007). *Order out of chaos: Linnean plant names and their types.* Linnean Society of London in association with the Natural History Museum.

Jeandroz, S., Roy, A. & Bousquet, J. (1997). Phylogeny and phylogeography of the circumpolar genus *Fraxinus* (Oleaceae) based on internal transcribed spacer sequences of nuclear ribosomal DNA. *Molec. Phylogenet. Evol.* **7**(2): 241–251.

Jeffrey, C. (1982). Rhododendron and classification — a comment. *Rhododendrons with Magnolias and Camellias*, Yearbook 1982/3. Royal Horticultural Society, London.

Jentys-Szaferowa, J. (1967). Experimental invstigation on the taxonomy of *Betula oycoviensis* Besser *Rocz. Seke. Dendr.* **21**: 5–56.

Jiang, J., Yang, C.-P., Liu, G.-F., Wu. J.-H. & Li. T.-H. (2002). Genetic relations of interspecies for eight birch species. *J. Forest Res.* **13**(4): 281–284.

Johnson, A. G. (1954). *Betula lenta* var. *uber* Ashe. *Rhodora* **56**: 129–131.

Johnsson, H. (1944). Triploidy in *Betula alba* L. *Bot. Not.* 1944: 85–96.

—— (1945). Interspecific hybridization within the genus *Betula. Hereditas* **31**: 163–176.

—— (1949). Studies on birch species hybrids I. *Betula verrucosa* × *B. japonica*, *B. verrucosa* × *B. papyrifera* and *B. pubescens* × *B. papyrifera. Hereditas* **35**: 115–135.

—— (1974). The hybrid *Betula lutea*, sect *Costatae* × *Betula occidentalis*, sect Albae. *Silvae Geneticae* **23**: 1–3.

Jump, B. A. & Woodward, S. (1994). Histology of witches' brooms on *Betula pubescens. European J. Forest Pathol.* **24**(4): 229–237.

Kallio, P., Niemi, S., Sulkinoja, M. & Vallane, T. (1983). The Fennoscandian birch and its evolution in the marginal forest zone. *Collection Nordicana.* **47**: 101–110.

Karlsdóttir, L., Hallsdóttir, M., Thórsson, A. Th. & Anamthawat-Jónsson, K. (2008). Characteristics of pollen from natural triploid *Betula* hybrids. *Grana* **47**(1): 52–59.

Kawano, S. (1966). Biosystematic studies of the *Deschampsia cespitosa* complex with special reference to the karyology of Icelandic populations. *Bot. Mag. Tokyo* **79**: 292–307.

Kawase, M. (1961). Dormancy in *Betula* as a quantitative state. *Plant Physiol.* **36**(5): 643–649.

Keinänen, M., Julkunen-Tiitto, R., Rousi, M. & Tahvanainen, J. (1999). Taxonomic implications of phenolic variation in leaves of birch (*Betula* L.) species. *Biochem. Syst. Ecol.* **27**: 243–254.

Kenworthy, J. B., Aston, D. & Bucknall, S. A. (1972). A study of hybrids between *Betula pubescens* Ehrh. and *Betula nana* L. from Sutherland — an integrated approach. *Trans. Bot. Soc. Edinburgh.* **42**: 517–539.

Kienast, F., Schirrmeister, L., Siegert, C. & Tarasov, P. (2005). Paleobotanical evidence for warm summers in the East Siberian Arctic during the last cold stage. *Quaternary Res*. **63**: 283–300.

Kikuzawa, K. (1982). Leaf survival and evolution in Betulaceae. *Ann. Bot*. **50**: 345–353.

Kinkhead, E. (1976). The search for *Betula uber*. *The Explorer* (Cleveland, Ohio). **18**(2): 12–22.

Kling, G. J., Perkins, L. M. & Nobles, R. (1985). Rooting *Betula platyphylla* var. *szechuanica* (Rehd.) cuttings. *Plant Propagator*. **31**(1): 9–10.

Kolman-Adamska, A., Ziembinska-Tworzydlo, M. & Zastawniak, E. (2004). In situ pollen in some flowers and inflorescences in the Late Miocene flora of Sośnica (S.W. Poland) *Rev. Paleobot. Palynol*. **132**(3–4): 261–280.

Koshy, T. K., Grant, W. F. & Brittain, W. H. (1972). A numerical chemotaxonomy of the *Betula caerulea* complex. *Symposium. Biol. Hung*. **12**: 201–211.

Koski, V. & Rousi, M. (2005). A review of the promises and constraints of breeding silver birch (*Betula pendula* Roth) in Finland. *Forestry* **78**: 187–198.

Kozlowski, T. T. & Clausen, J. J. (1966). Shoot growth characteristics of heterophyllous woody plants. *Canad. J. Bot*. **44**: 827–843.

—— & Pallardy, S. G. (2002). Acclimation and adaptive responses of woody plants to environmental stresses. *Bot. Rev*. **68**: 270–334.

Kriebel, H. B. (1966). Genetic implications of arboretum seed exchange. *Taxon* **15**(3): 94–95.

Kring, D. A. (2007). The Chicxulub impact event and its environmental consequences at the Cretaceous-Tertiary boundary. *Paleogeogr., Paleoclimatol., Paleoecol*. **255**: 4–21.

Kuser, J. (1983). Inbreeding depression in *Metasequoia*. *J. Arnold Arbor*. **64**: 475–481.

Kuzeneva, O. I. (1936). *Betula*. In: *Flora of the USSR*. 269–305.

Kvist, L., Martens, J., Ahola, A. & Orell, M. (2001). Phylogeography of a Palearctic sedentary passerine, the willow tit (*Parus montanus*). *J. Evol. Biol*. **14**: 930–941.

——, ——, Higuchi, H., Nazarenko, A. A., Valchuk, O. P. & Orell, M. (2003). Evolution and genetic structure of the great tit (*Parus major*) complex. *Proc. Roy. Soc. London. Ser B*. **B 270**: 1447–1454.

Lahtinen, M., Lempa, K., Salaminen, J-P. & Pihlaja, K. (2006). HPLC analysis of leaf surface flavonoids for preiminary classification of birch species. *Phytochem. Anal*. **17**: 197–203.

Lancaster, R. (2008). *Travels in China: A Plantsman's Paradise*. Garden Art Press, Woodbridge, Suffolk.

Lande, R. & Schemske, D. W. (1985). The evolution of self fertilization and inbreeding depression. I. Genetic models. *Evolution* **39** (1): 24–40.

Landrum, L. R. (2003) What has happened to descriptive systematics? What would make it thrive? *Syst. Bot*. **26**(2): 438–442.

Lane, C. G. (1993). Propagation of the genus *Betula*. In: D. Hunt, *Proceedings of the IDS Betula Symposium*, pp. 51–60. International Dendrological Society.

Lapinjoki, S. P., Elo, H. A. & Taipale, H. T. (1991). Development and structue of resin glands on tissues of *Betula pendula* Roth. during growth. *New Phytol*. **117**: 219–223.

Larsen, E., Funder, S. & Thiede, J. (1999). Late Quaternary history of northern Russia and adjacent shelves — a synopsis. *Boreas* **28**: 6–11.

Larson, D. W., Mattes, U. & Kelly, P. E. (2000a). *Cliff Ecology*. Cambridge University Press.

——, ——, Gerrath, J. A., Larson, N. W. K., Gerrath, J. M., Nekola, J. C., Walker, G. L., Porembski, S., & Charlton, A. (2000b). Evidence for the widespread occurrence of ancient forests on cliffs. *J. Biogeogr*. **27**: 319–331.

Lascoux, M., Palmé, A. E., Chedadi, R. & Latta, R. G. (2003). Impact of Ice Ages on the genetic structure of trees and shrubs. *Philos. Trans. Roy. Soc. London* B **359**: 197–207.

Lendzian, K. L. (2006). Survival strategies of plants during secondary growth: barrier properties of Phellems and lenticels towards water, oxygen and carbon dioxide. *J. Exper. Bot*. **57**(11): 2535–2546.

Lepage, E. (1976). Les bouleaux arbustifs du Canada et de L'Alaska. *Naturaliste Canadienne* **103**(3): 215–233.

Lev-Yadun, S. & Holopainen, J. K. (2009). Why red-dominated autumn leaves in America and yellow-dominated autumn leaves in Northern Europe. *New Phytol*. **183**: 506–512.

Li, W. L., Berlyn, G. P., Mark. P & Ashton, S. (1996). Polyploids and their structural and physiological characteristics relative to water deficit in *Betula papyrifera* (Betulaceae). *Amer. J. Bot*. **83**(1): 15–20.

Li, J., Shoup, S. & Chen, Z. (2007). Phylogenetic relationships of diploid species of *Betula* (Betulaceae) inferred from DNA sequences of nuclear nitrate reductase. *Syst. Bot*. **32**(2): 357–365.

——, —— & —— (2005). Phylogenetics of *Betula* (Betulaceae) inferred from nuclear ribosomal DNA. *Botanical Society of America*. Poster Session 33–94. Systematics Section / ASPT (American Association of Plant Taxonomists) http://www.2005. botanyconference. org/engine/search/index.php? func+detail&aid=125 (accessed 9/12/2005)

Li, P-C. (1979). New taxa of Betulaceae from China. *Acta Phytotax. Sin.* **17**(1): 87–91.

—— (1983). *Flora Xizangica* 1. Science Press [in Chinese].

—— & Cheng, S-X. (1979). Betulaceae. In: K-Z. Kuang & P-C. Li (eds), *Flora Republicae Popularis Sinicae* **21**: 44–137. Science Press, Beijing [In Chinese].

—— & Skvortsov, A. K. (1999). *Betula*. In: Z-Y. Wu & P. H. Raven (eds), *Flora of China.* **4:** 303–313. Science Press, Beijing and Missouri Botanical Garden Press, St. Louis.

Li, W-H. (1997). *Molecular Evolution.* Sinauer Associates, Sunderland.

Lindquist, B. (1945). *Betula callosa* Notö, a neglected species in the scandinavian subalpine forests. *Svensk Botanisk Tidskrift.* Bd **39**, H. 2: 161–186.

—— (1947). *Svensk Botanisk Tidskrift.* Bd 41, H.1.

Linnaeus, C. (1739). *Flora Lapponica.* Apud S. Schouten, Amstelaedami.

—— (1745). *Flora Suecica.* Apud Conradum & Georg Wishoff, Lugduni Batavorum.

—— (1753). *Species Plantarum.* Stockholm.

—— (1767). *Mantissa plantarum.* Holmiae.

Longman, K. A. & Wareing, P. F. (1959). Early induction of flowering in birch seedlings. *Nature* **184**: 2037–2038.

—— (1984). Physiological studies in birch. *Proc. Roy. Soc. Edinburgh* **85B**: 97–113.

Lorenzetti, F., Delagrange, S., Bouffard, D. & Nolet, P. (2008). Establishment, survivorship, and growth of yellow birch seedlings after site preparation treatments in large gaps. *Forest Ecol. Managem.* **254**: 350–361.

Lorentzon, K. (1993). A brief analysis of birch cultivation in Sweden and a new cultivar of *Betula ermanii.* In: D. Hunt, *Proceedings of the IDS Betula Symposium.* 67–72. International Dendrological Society.

Lucas, G. & Synge, H. (1978). *The IUCN plant red data book. Betula uber* (Ashe) Fernald p. 83–84. IUCN, Morges.

McAllister, H. A. (1987). Conservation and taxonomy of *Santolina chamaecyparissus* agg. *The National Council for the Conservation of Plants and Gardens Newsletter* **10**: 7–10.

—— (1993). Cytology and the conservation of rare birches. In: D. Hunt, *Proceedings of the IDS Betula Symposium,* pp. 61–66. International Dendrological Society.

—— (1999). The importance of living collections for taxonomy. In: S. Andrews, A. C. Leslie & C. Alexander

(eds), *Taxonomy of cultivated plants. Third International Symposium,* pp. 3–10. Royal Botanic Gardens, Kew.

—— (2005a). *Betula insignis. Curtis's Bot. Mag.* **22** (4): 220–224, Tab. 540.

—— (2005b). *The genus Sorbus: mountain ash and other rowans.* Royal Botanic Gardens, Kew.

—— & Ashburner, K. (2004). *Betula lenta* forma *uber. Curtis's Bot. Mag.* **21** (1): 54–60, Tab. 487.

—— & —— (2007). *Betula megrelica. Curtis's Bot. Mag.* **24** (3): 174–179, Tab. 593.

—— & Rutherford, A. (1990) *Hedera helix* L. and *H. hibernica* (Kirch.) Bean (Araliaceae) in the British Isles. *Watsonia* **18** (1): 7–15.

MacAloney, H. J. (1968). The bronze birch borer. U.S. Department of Agriculture Forest Service, Forest Pest Leaflet 111. pp 1–5.

McElwain, J. C. & Punyasena, S. W. (2007). Mass extinction and the plant fossil record. *Trends Ecol. Evol.* **22**(10): 548–557.

McHaffie, H. S., Legg, C. J. & Ennos, R. A. (2001). A single gene with pleitropic effects accounts for the Scottish endemic taxon *Athyrium distentifolium* var. *flexile. New Phytol.* **152**: 491–500.

McLean, B. (1997). *A pioneering plantsman. A. K. Bulley and the great plant hunters.* The Stationery Office, London.

—— (2004). *George Forrest plant hunter.* Antique Collectors' Club, Woodbridge, Suffolk.

Mace, G. M. (2004). The role of taxonomy in species conservation. *Philos. Trans. Roy. Soc. London* B **359**: 711–719.

Mäkelä, E. (1998). The Holocene history of *Betula* at Lake Lilompolo, Inari Lapland, northeastern Finland. *The Holocene.* **8**(1): 55–67.

Maliouchenko, O., Palmé, A. E., Buonamici, A., Vendramin, G. G. & Lascoux, M. (2007). Comparative phylogeography and population structure of European *Betula* species, with particular focus on *B. pendula* and *B. pubescens. J. Biogeogr.* **34**: 1601–1610.

Manchester, S. R. (1994). Inflorescence bracts of fossil and extant *Tilia* on North America, Europe, and Asia: patterns of morphologic divergence and biogeographic history. *Amer. J. Bot.* **81**(9): 1176–1185.

—— (1999). Biogeographical relationships of North American Tertiary floras. *Ann. Missouri Bot. Gard.* **86**(2): 472–522.

—— & Tiffney, B. H. (2001). Integration of paleobotanical and neobotanical data in the assessment of phytogeographic history of Holarctic Angiosperm clades. *Int. J. Pl. Sci.* **162**(6): 19–27.

Marx, L. & Walters, M. B. (2008). Survival of tree seedlings on different species of decaying wood maintains tree distribution in Michigan hemlock-hardwood forests. *J. Ecol.* **96**: 505–513.

Mascheretti, S., Rogatcheva, M. B., Gunduz, I., Fredga, K. & Searle, J. B. (2003). How did pygmy shrews colonize Ireland? Clues from a phylogenetic analysis of mitochondrial cytochrome *b* sequences. *Proc. Roy. Soc. London. Ser B.* **270**: 1593–1599.

Mason-Gamer, R. J. (2008). Allohexaploidy, introgression, and the complex phylogenetic history of *Elymus repens* (Poaceae). *Molec. Phylogenet. Evol.* **47**: 598–611.

Matlack, G. R. (1989). Secondary dispersal of seed across snow in *Betula lenta*, a gap colonising tree species. *J. Ecol.* **77**: 853–869.

Mayol, M. & Roselló, J. A. (2001). Why nuclear ribosomal DNA spacers (ITS) tell different stories in *Quercus. Molec. Phylogenet. Evol.* **19**(2): 167–176.

Mazzeo, P. M. (1974). *Betula uber*, — What is it and where is it? *Castanea* **39**(3): 273–278.

Meffert, L. M. (1999). How speciation experiments relate to conservation biology. *Biosci.* **49**(9): 701–711.

Mehra, P. N. & Sareen, T. S. (1969). In: A. Löve, IOPB Chromosome number reports. XX11. *Taxon* **18**: 433–442.

—— & —— (1973). Cytology of west Himalayan Betulaceae and Salicaceae. *J. Arnold Arbor.* **54**: 412–418.

Melville, R. (1939). *Betula medwediewii. Curtis's Bot. Mag.* 162. Tab. 9569.

Merilä, J., Björklund, M. & Baker, A. J. (1996). The successful founder: genetics of introduced *Carduelis chloris* (greenfinch) populations in New Zealand. *Heredity* **77**: 410–422.

——, ——, & —— (1997). Historical demography and pesent day population structure of the greenfinch, *Carduelis chloris* — an analysis of mtDNA control-region sequences. *Heredity* **5**: 946–956.

Michaux, A. (1803). *Flora boreali-americana.* Apud fratres Levrault, Parisiis et Argentorati.

Michaux, F. A. (1819). *The North American sylva.* C. D'Hautel, Paris.

Michaux, J. R., Chevret, P., Filippucci, M-G. & Macholan, M. (2003). Phylogeny of the genus *Apodemus* with a special emphasis on the subgenus *Sylvaemus* using the nuclear IRBP gene and two mitochondrial markers: cytochrome *b* and 12S rRNA. *Molec. Phylogenet. Evol.* **19**(2): 167–176.

Miles, J. (1981). *Effect of birch on moorlands.* Institute of Terrestrial Ecology, Cambridge.

Miller-Rushing, A. J. & Primack, R. B. (2008). Effects of winter temperatures on two birch (*Betula*) species.

Tree Physiol. **28**: 659–664.

Milne, R. I. (2006). Northern Hemisphere Plant Disjunctions: A Window on Tertiary Land Bridges and Climate Change? *Ann. Bot.* **98**(3): 465–472.

Mitchell, R. J., Campbell, C. D., Chapman, S. J., Osler, G. H. R., Vanbergen, A. J., Ross, L. C., Cameron, C. M. & Cole, L. (2007). The cascading effects of birch on heather moorland: a test for the top-down control of an ecosystem engineer. *J. Ecol.* **95**: 540–550.

——, Marrs, R. H., LeDuc, M. G. & Auld, M. H. D. (1997). A study of succession on lowland heaths in Dorset, southern England: changes in vegetation and soil chemical properties. *J. Appl. Ecol.* **34**: 1426–1444.

Moore, J. M. & Wein, R. W. (1977). Viable seed populations by soil depth and potential site recolonisation after disturbance. *Canad. J. Bot.* **55**: 2408–2412.

Moran, V. C. & Southwood, T. R. E. (1982). The guild composition of arthropod communities in trees. *J. Animal Ecol.* **51**: 289–306.

Morley, R. J. (1998). Palynological evidence for Tertiary plant dispersals in SE Asian region in relation to plate tectonics and climate. In: R. Hall & J. D. Holloway (eds), *Biogeography and geological evolution of SE Asia.* Backhuys Publishers, Leiden.

Morrison, D. A. (2009). Why would phylogeneticists ignore computerised sequence alignment? *Syst. Biol.* **58**(1): 150–158.

Moss, E. H. & Packer, J. G. (1983). *Flora of Alberta.* University of Toronto Press.

Muir, G. & Schlötterer, C. (2006). Moving beyond single-locus studies to characterise hybridization between oaks (*Quercus* spp.). *Molec. Ecol.* **15**: 2301–2304.

Nagamitsu, T., Kawahara, T. & Hotta, M. (2004). Phenotypic variation and leaf fluctuating asymmetry in isolated populations of an endangered dwarf birch *Betula ovalifolia* in Hokkaido, Japan. *Pl. Species Biol.* **19**: 13–21.

——, —— & Kanazashi, A. (2006a). Endemic dwarf birch *Betula apoiensis* (Betulaceae) is a hybrid that originated from *B. ermanii* and *B. ovalifolia. Pl. Species Biol.* **21**: 19–29.

——, —— & —— (2006b). Pollen-limited production of viable seeds in an endemic dwarf birch, *Betula apoiensis*, and incomplete reproductive barriers to a sympatric congener, *B. ermanii. Biol. Conserv.* **129**: 91–99.

Nakai, T. (1915). *Flora Sylvatica Koreana Pars II.* Betulaceae. Government of Chosen.

Nakashizuka, T. (1989). Role of uprooting in composition and dynamics of an old-growth Forest in Japan. *Ecology* **70**(5): 1273–1278.

—— & Numata, M. (1982). Regeneration process of climax beech forests. I. Structure of a beech forest with the undergrowth of *Sasa. Jap. J. Ecol.* **32**: 57–67.

Nic Lughadha, E. (2004). Towards a working list of all known plant species. *Philos. Trans., Ser. B* **359**: 681–687.

Nielsen, D. G., Muilenburg, V. L. & Herms, D. A. (2011). Interspecific Variation in Resistance of Asian, European, and North American Birches (*Betula* spp.) to Bronze Birch Borer (Coleoptera: Buprestidae). *Environm. Entomol.* **40**(3): 648–653.

Nokes, D. C. B. (1979). *Biosystematic studies of Betula pendula Roth and B. pubescens Ehrh. in Great Britain.* PhD thesis. The Polytechnic, Wolverhampton.

Noltie, H. (1996). Snapshots of China. *The Garden* **121** (5): 274–277.

OECD (2003). Consensus document on the biology of European white birch (*Betula pendula* Roth). OECD Environment, Health and Safety Publications. *Series on Harmonisation of Regulatory oversight in Biotechnology* No. 28. Environment Directorate OECD, Paris.

Ødum, S. (2003). Choice of Conifer plant material for southwest Greenland. *Acta Hort.* **615**: 273–279.

Ogle, D. W. & Mazzeo, P. M. (1976). *Betula uber*, The Virginia round-leaf birch, rediscovered in south west Virginia. *Castanea* **41** (3): 248–256.

Ohdachi, S., Dokuchaev, N. E., Hasegawa, M. & Masuda, R. (2001). Intraspecific phylogeny and geographical variation of six species of northeast Asiatic *Sorex* shrews based on the mitochondrial cytochrome *b* sequences. *Molec. Ecol.* **10**: 2199–2213.

Ohwi, J. (1965). *Flora of Japan.* Smithsonian Institution, Washington.

Oldfield, S., Lusty, C. & Mackinven, A. (1998). *The World List of Threatened Trees.* World Conservation Press, Cambridge.

Olivares, C. A., Antón, M. C., Manzaneque, F. G. & Juaristi, C. M. (2004). Paleoenvironmental interpretation of the Neogene locality Caranceja (Reocín, Cantabria, N Spain) from comparative studies of wood, charcoal and pollen. *Rev. Paleobot. Palynol.* **132**: 133–157.

Oshida, T., Abramov, A., Yanagawa, H. & Masuda, R. (2005). Phylogeography of the Russian flying squirrel (*Pteromys volans*): implication of refugial theory in arboreal small mammal of Eurasia. *Molec. Ecol.* **14**: 1191–1196.

Osumi, K. (2005). Recriprocal distribution of two congeneric trees, *Betula platyphylla* var. *japonica* and *Betula maximowicziana*, in a landscape dominated by anthropogenic disturbances in north-eastern Japan. *J. Biogeogr.* **32**: 2057–2058.

—— & Sakurai, S. (1997). Seedling emergence of *Betula maximowicziana* following human disturbance and the role of buried viable seed. *Forest Ecol. Managem.* **93**: 235–243.

Oterdoom, H. J. (1994). Paleobotany and evolution of maples. In: D. M. van Gelderen, P. C. de Jong & H. J. Oterdoom, *Maples of the world.* Timber Press, Portland, Oregon.

Päckert, M., Martens, J., Eck, S., Nazerenko, A. A., Valchuk, O. P., Petri, B. & Veith, M. (2005). The great tit (*Parus major*) — a misclassified ring species. *Biol. J. Linn. Soc.* **86**: 153–174.

Painter, J. N., Siitonen, J. & Hanski, I. (2007). Phylogeographical patterns and genetic diversity in three species of Eurasian boreal forest beetles. *Biol. J. Linn. Soc.* **91**: 267–279.

Pakhomov, M. M. (2006). Glacial-Interglacial cycles in arid regions of northern Eurasia. *Quatern. Intern.* **152–3**: 70–77.

Palmé, A. (2003). *Evolutionary history and chloropoast DNA variation in three plant genera: Betula, Corylus, and Salix: the impact of post-glacial colonisation and hybridisation.* PhD thesis. Uppsala University.

—— & Vendramin, C. C. (2002). Chloroplast DNA variation, postglacial recolonisation and hybridization in hazel, *Corylus avallana. Molec. Ecol.* 11: 1769–1779.

——, Su, Q., Palsson, S. & Lascoux, M. (2004). Extensive sharing of chloroplast haplotypes among European birches indicates hybridization among *Betula pendula*, *B. pubescens*, and *B. nana. Molec. Ecol.* **13** (1): 167–178.

——, ——, Rautenberg, A., Mann, F. & Lascoux, M. (2003). Postglacial recolonisation and cpDNA variation of silver birch, *Betula pendula. Molec. Ecol.* **12** (1): 201–212.

Palmen, A. & Hametahti, L. (1992). *Alnus incana* f. *rubra* and *Betula pendula* f. *palmeri* (Betulaceae) in Finland. *Ann. Bot. Fenn.* **29** (4): 329–330.

Pautasso, M. (2009). Geographical genetics and the conservation of forest trees. *Pl. Ecol., Evol. Syst.* **11**: 157–189.

Pawlowska, L. (1980a). Flavonoids in the leaves of Polish species of *Betula* L. I. The flavonoids of *B. pendula* Roth. and *B. obscura* Kot. leaves. *Acta Soc. Bot. Poloniae* **49** (3): 281–296.

—— (1980b). Flavonoids in the leaves of Polish species of *Betula* L. II. The flavonoids of *B.* "nova" Roth. and *B. humilis* Schrk. leaves. *Acta Soc. Bot. Poloniae* **49** (3): 297–310.

—— (1980c). Flavonoids in the leaves of Polish species of *Betula* L. III. The flavonoids of *B. oycoviensis* Besser leaves. *Acta Soc. Bot. Poloniae* **49** (3): 311–320.

Peinado, M. & Moreno, G. (1989). The genus *Betula* (Betulaceae) in the Sistema Central (Spain). *Willdenowia* **18** (2): 343–359.

—— & —— (1990). *Betula.* In: S. Castroviejo *et al.* (eds), *Floral Iberica* **2**: 38–43. Real Jardin Botanico, Madrid.

——, —— & Velasco, A. (1983). Sur les Boulais lusoextremadurenses (*Galio broteriani – Betuleto parvibracteatae* S.). *Willdenowia* **13**: 349–360.

Pelham, J., Gardiner, A. S., Smith, R. I. & Last, F. T. (1988). Variation in *Betula pubescens* Ehrh. (Betulaceae) in Scotland: its nature and association with environmental factors. *Bot. J. Linn. Soc.* **96**: 217–234.

Perala, R. T. & Alm, A. A. (1990). Reproductive ecology of birch: a review. *Forest Ecol. Managem.* **32**: 1–38.

Perez-Moreno, J. & Read, D. J. (2001). Exploitation of pollen by mycorrhizal mycelial systems with special reference to nutrient recycling in boreal forests. *Proc. Roy. Soc. London, Ser. B, Biol. Sci.* **268** (No. 1474): 1329–1335.

Petit, R. J. & Excoffier, L. (2009). Gene flow and species delimitation. *Trends Ecol. Evol.* **24** (7): 386–393.

——, Bialozyt, R., Garnier-Géré, P. & Hampe, A. (2004). Ecology and genetics of tree invasions: from recent introductions to Quaternary migrations. *Forest Ecol. Managem.* **197**: 117–137.

—— & Hampe, A. (2006). Some evolutionary consequences of being a tree. *Annual Rev. Ecol. Evol. Syst.* **37**: 187–214.

Pickering, K. T. (2000). The Cenozoic world. In: S. J. Culver & P. F. Rawson (eds), *Biotic response to global change: The last 145 million years.* 20–34. Cambridge University Press, Cambridge.

Pleijel, F., Jondelius, U., Norlinder, E., Nugren, A., Oxelman, B., Schander, C., Sundberg, P. & Tholleson, M. (2008). Phylogenies without roots? A plea for the use of vouchers in molecular phylogenetic studies. *Molec. Phylogenet. Evol.* **48**: 369–371.

Pollard, A. M. & Heron, C. (1996). *Archeological Chemistry.* The Royal Society of Chemistry, Letchworth.

Porsild, A. E. & W. J. Cody (1980). *Vascular Plants of Continental Northwest Territories.* National Museums of Canada, Ottawa.

Post, E. (2003). Climate — vegetation dynamics in the fast lane. *Trends Ecol. Evol.* **18** (11): 551–553.

Prévosto, B., Coquillard, P. & Gueugnot, J. (1999). Growth models of silver birch (*Betula pendula* Roth.) on two volcanic mountains in the French Massif Central. *Vegetatio* **144** (2): 231–242.

Price, M. P. & Simpson, N. D. (1913). An account of the plants collected by Mr. M. P. Price on the Carruthers-Miller-Price expedition through north-west Mongolia and Chinese Dzungaria in 1910. In: M. P. Price, Observations on the vegetation of the Siberian-Mongolian frontier, the north-west Mongolian plateau, and Chinese Dzungaria, pp. 385–398. II Simpson, N.D. Enumeration of the plants. *Bot. J. Linn. Soc.* **41**: 399–456.

Provan, J. & Bennett, K. (2008). Phylogeographic insights into cryptic glacial refugia. *Molec. Ecol.* **14**: 793–803.

Qian, H. & Ricklefs, R. E. (2000). Large-scale processes and the Asian bias in species diversity of temperate plants. *Nature* **407**: 180–182.

—— & —— (2001). Diversity of temperate plants in east Asia. *Nature* **413**: 130.

Qiu, Y.-L., Parks, C. R. & Chase, M. W. (1995). Molecular divergence in the eastern Asia — eastern North America disjunct section *Rhytidiospermum* of *Magnolia* (Magnoliaceae). *Amer. J. Bot.* **82** (12): 1589–1598.

Raatikainenen, O. J., Taipale, H. T., Pelttari, A. & Lapinjoki, S. P. (1992). An electron microscope study of resin production and secretion by glands of seedlings of *Betula pendula* Roth. *New Phytol.* **122**: 537–543.

Ramsey, J. & Schemske, D. W. (1998). Pathways, mechanisms, and rates of polyploid formation in flowering plants. *Annual Rev. Ecol. Syst.* **29**: 467–501.

Ran, J-H., Wei, X-X. & Wang, X-Q. (2006). Molecular phylogeny and biogeography of *Picea* (Pinaceae): implications for phylogeographical studies using cytoplasmic haplotypess. *Molec. Phylogenet. Evol.* **41**: 405–419.

Raphael, S. (1989). *An Oak Spring sylva.* A selection of the rare books on fruit in the Oak Spring Garden Library. Upperville, Virginia.

Rausher, M. D. (2007). New Phytologist on plant evolution. *New Phytol.* **174**: 705–707.

Raven, P. H. (1972). Plant species disjunctions: a summary. *Ann. Missouri Bot. Gard.* **59**: 234–246.

—— (2004). Taxonomy: where are we now? *Philos. Trans., Ser. B* **359**: 571–583.

Raymond, O., Piola, F. & Sanlaville-Boisson, C. (2002). Inference of reticulation in outcrossing allopolyploid taxa: caveats, likleyhood and perspectives. *Trends Ecol. Evol.* **17**(1): 3–6.

Rehder, A. (1940). *Manual of Cultivated Trees and Shrubs*, pp. 373–381. Macmillan, New York.

Renner, S. S., Beenken, L., Grimm, G. W. Kocyan, A. & Ricklefs, R. E. (2007). The evolution of dioecy, heterodichogamy, and labile sex expression in *Acer*. *Evolution* **61** (11): 2701–2719.

Repenning, C.A. (1990). Of mice and ice in the Late Pliocene of North America. *Arctic* **43** (4): 314–323.

Richardson-Calfee, L. E. (2003). *Post-transplant root production, mortality, and periodicity of landscape-sized shade trees.* PhD Thesis; Virginia Polytechnic Institute and State University. Blacksburg, Virginia.

Rinne, P. & Kauppi, A. (1987). Induction of adventitious buds and sprouts on birch sedlings (*Betula pendula* Roth and *B. pubescens* Ehrh.) *Canad. J. Forest Res.* **17**: 545–555.

Rogers, D. L. (2000). Genotype diversity and clone size in old growth populations of coast redwood (*Sequoia sempervirens*). *Canad. J. Bot.* **78** (11): 1408–1419.

Ronquist, F. & Nylin, S. (1990). Process and pattern in the evolution of species associations. *Syst. Zool.* **39** (4): 323–344.

Roskam, J. C. (1977). Biosystematics of insects living in female birch catkins I. Gall midges of the genus *Semudobia* Keffer (Diptera, Cecidomyiidae). *Tijdschr. Entomol.* **120** (6): 153–197.

—— (1979). Biosystematics of insects living in female birch catkins II. Inquiline and predaceous gall midges belonging to various genera. *Netherlands J. Zool.* **29** (3): 283–351.

—— (1985). Evolutionary patterns in gall midge — host plant associations (Diptera, Cecidomyiidae). *Tijdschr. Entomol.* **128**: 193–213.

—— & Van Uffelen, G. A. (1981). Biosystematics of insects living in female birch catkins III. Plant-insect relation between white birches, *Betula* L. section *excelsae* (Koch) and gall midges of the genus *Semudobia* Keffer (Diptera, Cecidomyiidae). *Netherlands J. Zool.* **31**(3): 533–553.

Rowe, G., Harris, D. J. & Beebee, T. J. C. (2006). Lusitania revisited: A phylogeographic analysis of the natterjack toad *Bufo calamita* across its entire geographical range. *Molec. Phylogenet. Evol.* **39**: 335–346.

Rundgren, M. & Ingólfsson, Ó. (1999). Plant survival in Iceland during periods of glaciation? *J. Biogeogr.* **26** (2): 387–396.

Rushforth, K. (1999). *Trees of Britain and Europe.* HarperCollins, London.

—— (2003). Lectotypification of *Betula bomiensis* P.C. Li (*Betulaceae*). *Edinburgh J. Bot.* **60**(2): 175–179.

Safford, L. O., Bjorkbom, J. C. & Zasada, J. C. (1990). *Betula papyrifera* Marsh. paper birch. In: R. M. Burns & B. H. Honkala (technical coordinators), *Silvics of North America. Vol. 2. Hardwoods. Agric. Handb.* **654**: 158–171. U.S. Department of Agriculture, Forest Service, Washington, DC.

Santamour, F. S. (1999). Progress in the development of borer-resistant white-barked birches. *J. Arboriculture.* **25**(3): 151–162.

—— (2001). Differential feeding by adult Japanese beetles on foliage of birch (*Betula*) species and hybrids. *J. Arboriculture* **27**(1): 18–23.

—— & Lundgren, L. N. (1996). Distribution and inheritance of platyphylloside in *Betula*. *Biochem. Syst. Ecol.* **24**(2): 145–156.

Sargent, C. S. (1916). *Plantae Wilsonianae 2.* John Murray, London

Särkilahti, E. & Valanne, T. (1990). Induced polyploidy in *Betula*. *Silva Fenn.* **24**: 227–234.

Satake, Y., Hara, H., Watari, S. & Tominari, T. (eds) (1989). *Wild flowers of Japan, woody plants*: I. Heibonsha, Tokyo.

Sax, D. F. & Brown, J. H. (2000). The paradox of invasion. *Global Ecol. Biogeogr.* **9**: 363–371.

Schemske, D. W. & Lande, R. (1985). The evolution of self fertilization and inbreeding depression. II. Empirical observations. *Evolution* **39**(1): 41–52.

Schenk, M. F., Thienpont, C-N., Koopman, W. J. M., Gilissen, L. J. W. J. & Smulders, J. M. (2008). Phylogenetic relationships in *Betula pubescens* (Betulaceae). *Tree Genet. Genomics* **4**: 911–924.

Schilling, A. D. (1984). Birch in cultivation. *Proc. Roy. Soc. Edinburgh* **85B**: 197–202.

—— (1989). The *Betula* collection at Wakehurst Place. *The Kew Mazazine* **6**(2): 65–73.

—— (1993). Himalayan birches and their companion plants, with special reference to *Betula utilis*. In: D. R. Hunt (ed.), *Betula: proceedings of the IDS Betula Symposium, 2–4 October 1992,* pp. 29–40. International Dendrology Society, Morpeth.

Schneider, C. (1916). Betulaceae. In: C. S. Sargent (ed.), *Plantae Wilsonianae* **2**. 423–508. Cambridge University Press. Reprinted 1988 by Dioscorides Press, Portland, Oregon.

Schrank, F. von P. von (1789). *Baierische Flora.* Joh. Bapt. Strohl, München.

Schwaegerle, K. E. & Schaal, B. A. (1979). Genetic variability and founder effect in the pitcher plant *Sarracenia purpurea* L. *Evolution* **33**(4): 1210–1218.

Seberg, O. & Petersen, G. (1998). A critical review of concepts and methods used in classical genome analysis. *Bot. Rev.* **64**: 372–417.

Seifriz, W. (1931). Sketches of the vegetation of some southern provinces of Soviet Russia: II. Plant life along the Georgian Military Way, north Caucasus. *J. Ecol.* **19**(2): 372–382.

Semerikov, V. L. & Lascoux, M. (1999). Genetic relationships among Eurasian and American *Larix* species based on allozymes. *Heredity* **83**: 62–70.

——, Zhang, H., Sun, M. & Lascoux, M. (1999). Conflicting phylogenies of *Larix (Pinaceae)* based on cytoplasmic and nuclear DNA. *Molec. Phylogenet. Evol.* **27**: 173–184.

Setliff, E. C. (2002). The wound pathogen *Chondrostereum purpureum*, its history and incidence on trees in North America. *Austral. J. Bot.* **50**(5): 645–651.

Sharik, T. L. (1984). Minutes to the *Betula uber* protection management and research co-ordinating committee meeting, May 31st, 1984. *Forest Ecol. Managem.* **6**: 115–128.

—— & Barnes, B. V. (1971). Hybridization in *Betula alleghaniensis* Britt. and *B. lenta* L.: A comparative analysis of controlled crosses. *Forest Sci.* **17**(4): 415–424.

—— & —— (1979). Natural variation in morphology among diverse populations of yellow birch (*Betula alleghaniensis*) and sweet birch (*Betula lenta*). *Canad. J. Bot.* **57**(18): 1932–1939.

—— & Ford, R. H. (1984). Variation and taxonomy of *Betula uber*, *B. lenta*, and *B. alleghaniensis*. *Brittonia* **36**(3): 307–316.

Sinclair, W. T., Morman, J. D. & Ennos, R. A. (1998). Multiple origins for Scots pine (*Pinus sylvestris* L.) in Scotland: evidence from mitochondrial DNA variation. *Heredity* **80**: 233–240.

Skan, S. A. (1910). *Betula maximowicziana. Curtis's Bot. Mag.* 136. Tab. 8337.

Skovsted, A. (1929). Cytological investigations of the genus *Aesculus* L. *Hereditas* **12**: 64–70.

Skvortsov, A. K. (1968). Willows of Russia and adjacent countries: taxonomical and geographical revision. English translation by I. N. Kadis, (ed. A. G. Zinovjev) (1999). University of Joensuu, Finland.

—— (1997). Taxonomic notes on *Betula* — I. Section *Betulaster. Harvard Pap. Bot.* No 11: 65–70.

—— (1998). Taxonomic notes on *Betula* — II. Two Chinese birches described by A. Franchet in 1899 and misunderstood by subsequent authors. *Harvard Pap. Bot.* 3(1): 109–112.

—— (2002). A new system of the genus *Betula*. *Bull. Moscow Soc. Naturalists* **107**(5): 73–76.

Smart, J. & Simmonds, N. J. (1995). *Evolution of crop plants*. 2nd ed. Longman, Harlow, Essex.

Smiley, E. T. & Booth, D. C. (2000). Grown to die? *Amer. Nurseryman* **191**(6): 48–54.

Smith, G. F. (2006). Herbaria in the real world. *Taxon* **55**(3): 571–572.

Sokolov, S. Ya. (1951). *Derevya i kustarniky SSSR (The trees and shrubs of USSR)*. II, izd. AN SSSR. Moskva-Leningrad.

Solomon, D. S. & Kenlan, K. W. (1982). Discriminant analysis of interspecific hybridisation in *Betula*. *Silvae Genet.* **31**: 138–145.

Soltis, D. E., Soltis, P. S. & Tate, J. A. (2003). Advances in the study of polyploidy since *Plant speciation*. *New Phytol.* **161**: 173–191.

Sorensen, F. C. & Miles, R. S. (1982). Inbreeding depression in height, height growth, and survival of Douglas fir, ponderosa pine, and noble fir to 10 years of age. *Forest Sci.* **28**(2): 283–292.

Sperling, F. A. & Harrison, R. G. (1994). Mitochondrial DNA variation within and between species of the *Papilio machaon* group of swallowtail butterflies. *Evolution* **48**: 408–422.

Spicer, R. A. (1989). Plants at the Cretaceous-Tertiary boundary. *Philos. Trans. Ser. B* **325**: 291–305.

Srodon, A. (1968). Pollen specta form Spitzbergen. *Polish Spitzbergen Expeditions 1957–1960*. (Ed. K. Birkenmajer), Polish Academy of Sciences III. I.G.Y./I.G.C. Committee, Warszawa.

Stace, C. A. (1975). *Hybridisation and the flora of the British Isles*. Academic Press, London.

—— (1991). *New Flora of the British Isles*. Cambridge University Press.

Staszkiewicz, L. (1986). *Betula szaferi* — a new species of the genus *Betula* from Poland. *Acta Soc. Bot. Poloniae* **55**(3): 361–366.

Stephens, P. A. & Sutherland, W. J. (1999). Consequences of the Allee effect for behaviour, ecology and conservation. *Trends Ecol. Evol.* **14**(10): 401–405.

Stewart, J. R. & Dalén, L. (2008). Is the glacial refugium concept relevant for northern species? A comment on Pruet and Winker 2005. *Climate Change* **86**: 19–22.

—— & Lister, A. M. (2001). Cryptic northern refugia and the origins of the modern biota. *Trends Ecol. Evol.* **16**(11): 608–613.

——, ——, Barnes, I. & Dalen, L. (2010). Refugia revisited: individualistic responses of species in space and time. *Proc. Roy. Soc. London, Ser. B, Biol. Sci.* **277**: 661–671.

Stults, D. Z. & Axsmith, B. J. (2009). Betulaceae form the Pliocene and Pleistocene of southwest Alabama, southeastern United States. *Rev. Palaeobot. Palynol.* **155**(1–2): 25–31.

Suda, J., Krahulcová, A., Trávnicek, P. & Krahulek, F. (2006). Ploidy level versus DNA ploidy level: and appeal for consistent terminology. *Taxon* **55**(2): 447–450.

Sulkinoja, M. (1981) Lapin koivulajien muuntelusta ja risteytymisestuä (On the variation and hybridisation of *Betula* species in Lapland). *Lapin tutkimusseura Vuosikirja* **22**: 22–30.

—— (1990). Hybridisation, introgression and taxonomy of the mountain birch in S.W. Greenland compared with related results in Iceland and Finnish Lappland. *Meddel. Gronland, Biosci.* **33**: 21–29.

—— & Valanne, T. (1980). Polyembryony and abnormal germination in *Betula pubescens* ssp. *tortuosa*. *Rep. Kevo Subarctic Res. Sta.* **16**: 31–37.

Sunnerheim, K., Palo, R. T., Theander, O. & Knutsson, P-G. (1988). Chemical defense in birch. Platyphylloside: a phenol from *Betula pendula* inhibiting digestibility. *J. Chem. Ecol.* **14**(2): 549–560.

Suzuki, H., Sato, J. J., Tsuchiya, K., Luo, J., Zang, Y-P., Wang, Y-W & Jiang, X-L. (2003). Molecular phylogeny of wood mice (*Apodemus*, Muridae). *Biol. J. Linn. Soc.* **80**: 469–481.

Svenning, J-C., Normand, S. & Kageyama, M. (2008). Glacial refugia of temperate trees in Europe: insights from species distribution modelling. *J. Ecol.* **96**: 1117–1127.

—— & Skov, F. (2004). Limited filling of the potential range in European tree species. *Ecol. Lett.* **6**: 565–573.

—— & —— (2007). Ice age legacies in the geographical distribution of tree species richness in Europe. *Global Ecol. Biogeogr.* **16**: 234–245.

Szwabowicz, A. (1972). Karyological studies on *Betula oycoviensis* Besser and its progeny. *Acta Soc. Bot. Poloniae* **41**(2): 235–252.

Tabata, H. (1966). A contribution to the biology of Japanese birches. *Mem. Fac. Sci. Kyoto Univ., Ser. Biol.* **32**(3): 239–271.

—— (1970). Root habit of Japanese birches, *Mem. Fac. Sci. Kyoto Univ., Ser. Biol.* IV: 130–138.

Taberlet, P., Fumagalli, L., Wust-Saucy, A-G. & Cosson, J-F. (1998). Comparative phylogeography and postglacial colonisation routes in Europe. *Molec. Ecol.* **7**: 453–464.

Taggart, J. B., McNally, S. F. & Sharp, P. M. (1990). Genetic variability and differentiation among founder populations of the pitcher plant (*Sarracenia purpurea* L.) in Ireland. *Heredity* **64**: 177–183.

Tanai, T. (1972). Tertiary history of vegetation in Japan In: A. Graham (ed.), *Floristics and paleo-floristics of Asia and E.N. America*. Elsevier, Amsterdam, London, New York.

Tao, J. R. & Du, N. Q. (1987). Miocene flora from Markam county and fossil record of Betulaceae. *Acta Bot. Sin.* **29**(6): 649–655.

Tarasov, P. E., Webb, T., Andreev, A. A., Afaas'eva, N. B., Berezina, N. A., Bezusko, L. G., Blyakharchuk, T. A., Bolikovskaya, N. S., Heddadi, R., Chernavskaya, M. M., Chernova, G. M., Dorofeyuk, N. I., Dirksen, V. G., Elina, G. A., Flimonova, L. M., Glebov, F. Z., Guiot, J., Gunova, V. S., Harrison, S. P., Jolly, D., Khomutova, V. I., Kvavadze, E. V., Osipova, I. M., Panova, N. K., Prentice, C., Saarse, L. Sevastynov, D. V., Volkova, V. S. & Zernitskaya, V. P. (1998). Present day and mid-Holocene biomes reconstructed from pollen and plant macrofossil data from the former Soviet Union and Mongolia. *J. Biogeogr.* **25**: 1029–1053.

Taulavuori, K. M. J., Taulavuori, E. B., Skre, O., Nilsen, J., Igeland, B. & Laine, K. M. (2004). Dehardening of mountain birch (*Betula pubescens* ssp. *czerepanovii*) ecotypes at elevated winter temperatures. *New Phytol.* **162**: 427–436.

Teacher, A. G. F., Garner, T. W. J. & Nichols, R. A. (2009). European phylogeography of the common frog (*Rana temporaria*): routes of postglacial colonization into the British Isles, and evidence for an Irish glacial refugium. *Heredity* **102**: 490–496.

Thompson, J. D. & Lumaret, R. (1992). The evolutionary dynamics of polyploid plants: origins, establishment and persistence. *Trends Ecol. Evol.* **7**: 302–307.

Thomas, P. T. (1940a). Reproductive versatility in *Rubus* II. The chromosomes and development. *J. Genet.* **40**: 119–128.

—— (1940b). The origin of new forms in *Rubus* III. The chromosome constitution of *R. loganobaccus* Bailey, its parents and derivatives. *J. Genet.* **40**: 141–156.

Thórsson, AE. Th., Pálsson, S., Sigurgeirsson, A. & Anamthawat-Jonsson, K. (2007). Morphological variation among *Betula nana* (diploid), *B. pubescens* (tetraploid) and their triploid hybrids in Iceland. *Ann. Bot.* **99**(6): 1183–1193.

——, Salmela, E. & Anamthawat-Jónsson, K. (2001). Morphological, cytogenetic, and molecular evidence for introgressive hybridization in birch. *J. Heredity* **92**(5): 404–408.

Tiffney, B. H. (1985). Perspectives on the origin of the floristic similarity between eastern Asia and eastern north America. *J. Arnold Arbor.* **66**: 73–94.

—— & Manchester, S. R. (2001). The use of geological and paleontological evidence in evaluating plant phytogeographic hypothyses in the Northern Hemisphere Tertiary. *Int. J. Pl. Sci.* **162**(6): 3–17.

Tripepi, R. R. & Mitchell, C. A. (1984). Metabolic response of river birch and European birch roots to hypoxia. *Pl. Physiol.* **76**: 31–35.

Tsuda, Y. & Ide, Y. (2005). Wide-range analysis of genetic structure of *Betula maximowicziana*, a long lived pioneer tree species and noble hardwood in the cool temperate zone of Japan. *Molec. Ecol.* **14**(13): 3929–3941.

Tsuyuzaki, S. & Goto, M. (2001). Persistence of seed bank under thick volcanic deposits twenty years after eruptions of Mount Usu, Japan, Hokkaido Island, Japan. *Amer. J. Bot.* **88**(10): 1813–1817.

Tudge, C. (1989). The rise and fall of *Homo sapiens sapiens*. *Philos. Trans.*, *Ser. B* **435**: 479–488.

—— (1990). Good breeding doesn't always show. *New Sci.* (17 Feb.) issue 1704: 67–68.

—— (2004). It's a meat market. *New Sci.* (13 Mar.) issue 2438: 19.

Tuley, G. (1973). A taxonomic history of the British birch tree. *Trans. Roy. Soc. Edinburgh* **42**: 31–41.

Tutin, T. G., Heywood, V. H., Burges, N. A., Valentine, D. H., Walters, S. M., Webb, D. A., Ball, P. W. & Chater, A. O. (1964). *Flora Europaea.* Cambridge University Press.

Tzedakis, P. C. & Bennett, K. D. (1995). Interglacial vegetation succession: a view from southern Europe. *Quatern. Sci. Rev.* **14**: 967–982.

Uphof, J. C. TH. (1959). *Dictionary of economic plants.* H. R. Engelmann (J. Cramer), Weinheim.

Vaarama, A. & Valanne, T. (1973). On the taxonomy, biology and origin of *Betula tortuosa* Ledeb. *Rep. Kevo Subarctic Res. Sta.* **10**: 70–84.

Valkama, E., Salminen, J-P., Koricheva, J. & Pihlaja, K. (2003). Comparative analysis of leaf trichome structure and composition of epicuticular flavonoids in Finnish birch species. *Ann. Bot.* **91**: 643–655.

Van Der Kelen, G. (1993). Autecology and silviculture of indigenous birches in Quebec. In: D. Hunt, *Proceedings of the IDS Betula Symposium*, pp. 41–50. International Dendrological Society.

Vandewoestijne, S., Baguette, M., Brakefield, P. M. & Saccheri, I. J. (2004). Phylogeography of *Aglais urticae* (Lepidoptera) based on DNA sequences of the mitochondrial COI gene and control region. *Molec. Phylogenet. Evol.* **31**: 630–646.

van Gelderen, D. M. (1994). Maple species and infraspecific taxa. In: D. M. van Gelderen, P. C. de Jong & H. J. Oterdoom, *Maples of the world.* Timber Press, Portland, Oregon.

Väre, H. (2001). Mountain birch taxonomy and floristics of mountain birch woodlands. In: F. E. Wielgolaski (ed.), *Nordic Mountain Birch Ecosystems*, Ch. 4. Man and the Biosphere Series, **27**. UNESCO, Paris.

Vassiljev, V. N. (1941). The stone-birch (*Betula ermanii* Cham. s.l.) (Ecology and coenology). *J. Bot. de L'USSR* **26** (2–3): 207–291.

Vilá, C., Sundqvist, A-K., Flagstad, Ø., Seddon, J., Bjornerfeldt, S., Kojola, I., Casulli, A., Sand, H., Wabakken, P. & Ellegren, H. (2003). Rescue of a severely bottlenecked wolf (*Canis lupus*) population by a single immigrant. *Proc. Roy. Soc.* **270** (no. 1510): 91–97.

Vinogradova, Yu. K., Kuklina, A. G. Pimenov, M. G. Sytin, A. K. Kamelin, R. V. & Yurtsev, B. A. (2005). Alexey Konstantinovich Skvortsov to the 85th anniversary of his birth. *Bot. Zhurn.* **90**(1): 125–137. www.salicicola.com/servlet/SaxonServlet?source=/translations/skvbiography.xml&style=/XSL/Translations04.xsl&clea

Vogel, S. (2009). Leaves in the lowest and highest winds: temperature, force and shape. *New Phytol.* **183**: 13–26.

Vvedensky, A. I. (1972). *Conspectus Florae Asiae Mediae.* Fan, Tashkent.

Walter, R. & Epperson, B. K. (2001). Geographic pattern of genetic variation in *Pinus resinosa*: area of greatest diversity is not the origin of postglacial populations. *Molec. Ecol.* **10**: 103–111.

Walter, K. & Gillett, H. J. (eds) (1997). *IUCN Red List of Threatened Plants.* Compiled by the World Conservation Monitoring Centre, IUCN — The World Conservation Union, Gland, Switzerland & Cambridge, UK.

Walters, S. M. (1975). *Betula.* In: C. A. Stace, *Hybridization and the Flora of the British Isles.* Academic Press, London.

Wang, C-W. (1961). *The Forests of China.* Maria Moors Cabot Foundation, Publication No. 5. Harvard University, Cambridge, Massachusetts.

Wang, N., McAllister, H. A., Bartlett, P. & Buggs, R. (2016). Molecular phylogeny and genome sizes of the genus *Betula* L. (Betulaceae). *Ann. Bot.* in press.

Wang, T., Hagqvist, R. & Tigerstedt, M. A. (1999). Inbreeding depression in three generations of selfed families of silver birch (*Betula pendula*) *Canad. J. Forest Res.* **29**(6): 662–668.

Wearstler, K. A. & Barnes, B. V. (1977). Genetic diversity of yellow birch seedlings in Michigan. *Canad. J. Bot.* **55**(22): 2778–2788.

Weaver, R. E. (1978). The ornamental birches. *Arnoldia* **38**(4): 117–131.

Weber, W. A. (2003). The middle Asian element in the southern Rocky Mountain flora of the western United States: a critical biogeographical review. *J. Biogeogr.* **30**: 649–685.

Wei, X-X. & Wang, X-Q. (2004). Recolonisation and radiation in *Larix* (Pinaceae): evidence from nuclear ribosomal DNA paralogues. *Molec. Ecol.* **13**: 3115–3123.

Wen, J. (2001). Evolution of Eastern Asian-Eastern North American biogeographic disjunctions: a few additional issues. *Int. J. Pl. Sci.* **162**(6): 117–122.

Wendle, J. E. (2000). Genome evolution in polyploids. *Pl. Molec. Biol.* **42**: 225–249.

Wheeler, Q. D. (2004). Taxonomic triage and the poverty of phylogeny. *Philos. Trans., Ser. B* **359**: 571–583.

Whitlock, C. & Dawson, M. R. (1990). Pollen and vertebrates of the Early Neogene Haughton Formation, Devon Island, Arctic Canada. *Arctic* **43**(4): 324–330.

Wielgolaski, F. E. & Sonesson, M. (2001). Nordic birch ecosystems — a conceptual overview. In: F. E. Wielgolaski (ed.), *Nordic Mountain Birch Ecosystems*, Ch. 31. Man and the Biosphere Series, **27**. UNESCO, Paris.

Wiens, J. J. & Donoghue, M. J. (2004). Historical biogeography, ecology and species richness. *Trends Ecol. Evol.* **19**(12): 639–644.

Wilkomirski, B., Dubielecka, B. & Mazan, D. (1998). Significance of flavonoid levels in a phenetic taxonomy of the genus *Betula*. *Plant Biosystems* **132**(3): 233–238.

Williams, D. A. (1981). *Chemotaxonomy and cytology of birches in Scotland*. PhD Thesis. University of Aberdeen.

Williams, J. H. & Arnold, M. L. (2001). Sources of genetic structure in the woody perennial *Betula occidentalis*. *Int. J. Pl. Sci.* **162**: 1097–1109.

Williams, J. W., Jackson, S. T. & Kutzbach, J. E. (2007) Projected distribution of novel and disappearing climates by 2100 AD. *Proc. Natl. Acad. Sci. U.S.A.* **104**: 5738–5742.

Willis, K. J., Bennett, K. D. & Walker, D. (eds) (2004). The Evolutionary Legacy of the Ice Ages. *Philos. Trans., Ser. B* (Thematic issue) **359**.

—— & van Andel, T. H. (2004). Trees or no trees? The environments of central and eastern Europe during the Last Glaciation. *Quatern. Sci. Rev.* **23**: 2369–2387.

—— & Niklas, K. J. (2004). The role of Quaternary environmental change in plant macroevolution: the exception or the rule? *Philos. Trans., Ser. B* **359**: 159–172.

——, Rudner, E. & Sümergi, P. (2000). The Full-Glacial forests of central and southeastern Europe. *Quatern. Res.* **53**: 203–213.

Wilson, E. O. (2004). Taxonomy as a fundamental discipline. *Philos. Trans., Ser. B* **359**: 739.

Witcher, I. N. & Wen, J. (2001). Phylogeny and biogeography of *Corylus* (Betulaceae): inferences from ITS sequences. *Syst. Bot.* **26**(2): 283–298.

Wolfe, J. A. (1969). Neogene floristics and vegetational history of the pacific northwest. *Madroño* **20**(3): 83–110.

—— (1973). Fossil forms of the Amentiferae. *Brittonia* 25: 324–355.

—— (1975). Some aspects of plant geography of the Northern Hemisphere during the Late Crataceus and Tertiary. *Ann. Missouri Bot. Gard.* **62**(2): 264–279.

Woodworth, R. H. (1929). Cytological studies in the Betulaceae I. *Betula*. *Bot. Gaz.* **87**(3): 331–363.

—— (1930). Cytological studies in the Betulaceae IV. *Betula*. *Bot. Gaz.* **90**: 108–115.

—— (1931). Polyploidy in the Betulaceae. *J. Arnold Arbor.* **12**: 206–217.

Wu Cheng-yih (1983). *Betula* in *Flora Xizanica* I: 480–485. Science Press, Beijing.

Xiang, Q-P., Xiang, Q-Y., Liston, A. & Zhang, X-C. (2003). Phylogenetic relationships in *Abies* (Pinaceae): evidence from PCR_RFLP of the nuclear ribosomal DNA internal transcribed spacer region. *Bot. J. Linn. Soc.* **145**: 425–435.

Xiang, Q.-Y., Soltis, D. E. & Soltis, P. S. (1998). The eastern Asian and eastern and western North American floristic disjunction: congruent phylogenetic patterns in seven different genera. *Molec. Phylogenet. Evol.* **10**(2): 178–190.

—— & —— (2001). Dispersal-vicariance analysis of intercontinental disjuncts: historical biogeographical implications for Angiosperms in the Northern Hemisphere. *Int. J. Pl. Sci.* **162**(6): 29–39.

Xu, J-X., Ferguson, D. K., Li, C-S. & Wang, Y. F. (2008). Late Miocene vegetation of the Lühe region in Yunnan, southwestern China. *Rev. Paleobot. Palynol.* **148**: 36–59.

Yasuda, S. P., Vogel, P., Tsuchiya, K., Han, S-H., Lin, L-K. & Suzuki, H. (2005). Phylogenetic patterning of mtDNA in the widely distributed harvest mouse (*Micromys minutus*) suggests dramatic cycles of range contraction and expansion during the mid- to late Pleistocene. *Canad. J. Zool.* **83**(11): 1411–1420.

Yamamoto, S. I. (2000). Forest gap dynamics and tree regeneration. *J. Forest. Res.* **5**: 223–229.

——, Nishimura, N. & Matsui, K. (1995). Natural Disturbance and Tree Species Coexistence in an Old-Growth Beech — Dwarf Bamboo Forest, Southwestern Japan. *J. Veg. Sci.* **6**(6): 875–886.

Yaroshenko, A. U., Potapov, P. V. & Turubanova, S. A. (2001). *Intact forest landscapes of northern European Russia.* Greenpeace Russia & Global Forest Watch.

Young, I. M., Montagu, K., Conroy, J. & Bengough, A. G. (1997). Mechanical impedance of root growth directly reduces leaf elongation rates of cereals. *New Phytol.* **135**: 613–619.

Zang, Z., Fan. L., Yang, J., Hao, X. & Gu, Z. (2006). Alkaloid polymorphism and ITS sequence variation in the *Spiraea japonica* complex (Rosaceae) in China: traces of the biological effects of the Himalaya-Tibet Plateau uplift. *Amer. J. Bot.* **93**: 762–769.

Zastawniak, E. & Walther, H. (1998). Betulaceae from Sosnica near Wroclaw (Poland) — a revision of Goeppert's original materials and a study of more recent collections. *Acta Paleobot.* **38**(1): 87–145.

Zeng, J., Li, J-H. & Chen, Z-D. (2008). A new species of *Betula* section *Betulaster* (Betulaceae) from China. *Bot. J. Linn. Soc.* **156**: 523–528.

Zhang, Y-W., Ryder, O. A. & Zhang, Y-P. (2001). Genetic divergence of orangutan subspecies (*Pongo pygmaeus*). *J. Molec. Evol.* **52**: 516–526.

Zhang Y., Zheng Z-H. & Zhang Z-X. (2007). Community structure and regeneration types of *Betula dahurica* forest in Badaling forest center of Beijing. *Forest. Stud. China* **9**(2): 152–156.

Zhilin, S. G. (1989). History of the development of the temperate forest flora in Kazakhstan, U.S.S.R. from the Oligocene to the Early Miocene. *Bot. Rev.* (*Lancaster*) **55**(4): 205–330.

Zink, R. M., Drovetski, S. V. & Rohwer, S. (2002). Phylogeographic patterns in the great spotted woodpecker *Dendrocopos major* across Eurasia. *J. Avian Biol.* **33**: 175–178.

WEB SITES

http://www.efloras.org/object_page.aspx?object_id=2191&flora_id=2 accessed 17/01/2007

http://www.suddenoakdeath.org/pdf/P.ramorum.hosts.June.2003.pdf. Hosts of *Phytophthora ramorum*.

INDEX OF SCIENTIFIC NAMES

Names and authorities (standardised according to IPNI, www.ipni.org) of all species mentioned in the text, including fossil species.

Accepted *Betula* names are in **bold,** all synonyms and other species mentioned in the text are in *italics*. The probable/supposed/suggested nearest relatives of fossil species are given after a colon in square brackets [ ].

Page numbers for illustrations are in **bold**.

INDEX OF COMMON NAMES

INDEX OF CULTIVAR NAMES AND HYBRIDS